Body Sensations

Ferenc Köteles

Body Sensations

The Conscious Aspects of Interoception

Springer

Ferenc Köteles
Faculty of Education and Psychology
Eötvös Loránd University
Budapest, Hungary

ISBN 978-3-030-63200-7 ISBN 978-3-030-63201-4 (eBook)
https://doi.org/10.1007/978-3-030-63201-4

This Springer imprint is published by the registered company Springer Nature Switzerland AG
The registered company address is: Gewerbestrasse 11, 6330 Cham, Switzerland

To Reni, Gergő, and Bazsi

Foreword

How do we know how we feel? To what degree do our internal sensations guide our thoughts, feelings, and behaviors? What is that slight pain I am feeling below my ribcage? The nature of interoception is one of the most vexing puzzles in psychology and medicine. Unlike our perceptions of light or sound or touch, which others can experience, our internal sensations are only directly available to us. We can try to describe the feelings to others but words often fail us.

Every generation or so, a book comes along which provides a new window into the ways people perceive and respond to their bodies' symptoms and sensations. *Body Sensations: The Conscious Aspects of Interoception* by Ferenc Köteles will be the authoritative voice for the next generation of researchers and practitioners. In this impressive volume, Köteles dissects each of the organ and muscular systems that provide perceptible signals to our brains. In addition to laying out the physiology of the neural pathways, he expertly guides the reader into the roles that learning, early experience, and various perceptual biases and distortions can play in shaping what we feel—or think we feel.

In previous generations, great works on interoception were published by some of the top scholars of the time. Wilhelm Wundt and his students in Leipzig in the late 1800s defined the field by trying to map the relationships between objectively measured stimuli (such as the brightness of light, the loudness of tones, pressure on the skin) to people's sensory experiences of brightness, loudness, and pain. In well-controlled laboratory environments, their group suggested that interoception was straightforward—the magnitude of stimuli was predictably related to the magnitude of the sensation.

Over the next 100 years, Wundt's optimistic view gradually diminished as researchers studied interoception among people and other organisms in real-world settings. William James (1890) argued that the conscious awareness of internal cues was the result of behavior and not the other way around. Pavlov and Luria found that an organism's sensory experiences were shaped by multiple pathways in the brain and body, including stimuli that had been conditioned from past experiences. By the mid-1900s, the stark awareness that sensory experiences were influenced by ongoing shifts in people's situations ushered in a new cognitive revolution (e.g., Gibson 1966; Schachter and Singer 1962).

It was not until the late 1960s that there was a general acknowledgment that interoception was a challenge for all areas of psychology and medicine. The most influential force behind this idea was György Ádám. In his pioneering 1967 book, *Interoception and Behavior*, Ádám argued that sensory cues from our internal organs influenced almost all of our actions, including eating, balance, and attention. Although many of these signals were not consciously noticed, many perhaps most could be perceived if the person paid close attention. Over almost 50 years, he published a series of books on interoception that brought together the theoretical rigor of pioneers in physiology, perception, social psychology, evolution, and neurology. He was also a creative methodologist and experimental scientist on his own who devoted himself to understanding the biology and psychology of interoception.

Ádám, however, was far more than a theoretically driven scientist. He was trained as a physician and was someone deeply interested in people. Unlike most interoception researchers, he observed the complexities that patients have in explaining their symptoms to physicians. At the same time, he appreciated the difficulties of physicians trying to translate their patients' symptoms into frameworks that make sense from a medical perspective. His own social skills and natural warmth encouraged patients and physicians to tell him their stories so that he could better understand the clinical challenges of interoception.

Through a lucky accident, I was able to spend some time with Professor Ádám on a visit to Budapest in 1996. His 1980 book, *Perception, Consciousness, and Memory*, had a big impact on me while I was writing my first book on physical symptoms. When we met, I felt we had been friends for years. He had a twinkle in his eye that reflected the absolute joy he had in talking about interoception and the state of the world. I could see why those who knew him always said he was a remarkable teacher, mentor, and friend.

I bring this up because one of Ádám's students was Ferenc Köteles. In reading this book, you will see many of the same traits and intellectual interests as his mentor. With each chapter, Professor Köteles reminds us of the long and varied history of the field of interoception. This is the most thorough book of the topic available covering some of the most recent breakthroughs in science. At the same time, you will appreciate many of the practical applications that may help you as a scientist, practitioner, or even as a regular human being who is awake in the middle of the night trying to figure out what that pain just below your ribcage might mean.

The University of Texas at Austin,
Austin, TX, USA
September 21, 2020

James W. Pennebaker

References

Gibson JJ (1966) The senses considered as perceptual systems. Houghton Mifflin, Oxford, England

James W (1890) The Principles of Psychology. Macmillan, London

Schachter S, Singer JE (1962) Cognitive, social, and physiological determinants of emotional state. Psychol Rev 69:379–399

Preface

This work presents a long journey. As we will see, many important aspects of body sensations, the major topic of this book, cannot satisfactorily be explained without having a closer look at our evolutionary heritage. In evolutionary terms, the entire story started about 500–600 million years ago (mya), increased in complexity approximately 220 mya, and might have reached its most sophisticated form only circa 50,000 years ago, perhaps even more recently. That is quite a perspective.

My personal journey was also long although not that long. Still, I quite often have the impression that at least several hundreds of years would be necessary to read all the relevant books and scientific papers written on this topic. I could only devote approximately two decades to various aspects of this research area so far; my personal research interest includes the placebo and nocebo phenomenon, body awareness, symptom reporting, interoception, and the effects of complementary and mind–body interventions. It appears to be a mixed and rather over-sized bag at first sight, but all these areas have at least one thing in common: they are characterized by the complicated relationship between subjective experience and actual physiological changes. I have a master's degree in both biology and psychology, and this complex background—combined with the influence of two significant mentors—strongly determines my scientific and everyday approach to pretty much everything. On top of this, the philosophical aspects of human consciousness always fascinated me. My older mentor, Professor György Ádám, was a physician and an early adept in interoception. He originally received training in the Pavlovian (classical conditioning) approach to interoception in the former Soviet Union; we still have his doctoral thesis written in Cyrillic script in our lab (Fig. 1). Later, he completed his knowledge with instrumental conditioning in Germany and the United States. His experimental findings and philosophical thoughts were summarized in three excellent monographs in English. Although he was primarily involved in the details of renal and gastrointestinal interoception, conditioning, and signal detection, he always attempted to put his findings in a broader context and tried to deal not only with the behavioral but also the psychological and even philosophical (i.e., consciousness-related) aspects of interoception. My younger mentor, Professor György Bárdos (himself mentored by György Ádám), is a biologist who started his

Fig. 1 György Ádám and the title page of his doctoral (Candidate of Sciences/PhD/) dissertation written in 1955 in the Pavlov Institute of the Academy of Sciences of the Soviet Union. (Courtesy of the Ádám György Psychophysiology Lab)

carrier with animal experimentation at the lab of György Ádám and worked with B.T. Engel, one of the early adepts of biofeedback research in the United States. Later in his carrier, he primarily focused on human-specific and pathological aspects of interoception. His knowledge was summarized in two monographs in Hungarian. The conversations and cooperation with these two excellent scholars who were not only experienced professionally but also very supporting personally had a decisive impact on my research interest and carrier. I am particularly indebted to them for their encouraging attitude and personal warmth.

I prefer scientific books to papers. Papers are focused and concise, obviously representing the most effective option for scientific discussion and progress in the strict sense. Books, however, can tell an entire story and thus provide the reader with a more comprehensive understanding of the topic at hand. Book authors have enough room to express their subjective evaluation and hunches (which often rely on decades of rigorous work), which renders books more interesting, enjoyable, and edifying to me. I particularly like old school monographs; it turns out again and again that seemingly novel ideas were already described, or at least suspected, decades ago, and we just forgot them for a while. There also appears to be a temporal fluctuation; a research area often gains momentum with each generation of researchers. Why? Perhaps because every joke is new for a newborn; or because the old questions can be asked and answered in new ways due to technological and theoretical advances. Still, I am convinced that the excellent monographs of György Ádám, Seymour Fisher, or James Pennebaker, dealing with various aspects of the topic, are worthy of studying or even teaching today (accompanied by a critical evaluation based on our recent knowledge, of course).

This monograph aims to summarize the recent scientific knowledge on conscious aspects of interoception. Actually, I prefer the term body sensations or body feelings (a more precise definition comes in Chap. 2). I believe that these conscious experiences are not suspicious artifacts or biased self-reports that should be eliminated from scientific research or replaced with processes that can be assessed more accurately and objectively. They are heavily influenced by top-down factors for sure; however, my conviction is that those factors themselves are worthy of investigation. For example, we will see later in the book that the discrepancy between actual and felt body events is a phenomenon with multiple consequences, both positive and negative. This deserves more research attention, and in fact should be taught in the elementary school to help people to make good health-related decisions later. I was always surprised by the fact that such a large discrepancy can be demonstrated in practically all areas of body-related research. In my own studies, it appeared in almost all areas from heartbeat detection ability to placebo and nocebo reactions and mind–body interventions. These findings will be discussed in a bit more detail in the book—the option of being subjective is one of the benefits and rewards of writing a monograph.

Unfortunately, this work will not be without mistakes and errors. As it ranges from philosophy and various areas of psychology to neuroanatomy and medicine, perhaps nobody is able (or has the capacity) to understand each and every nuance of this broad area perfectly. Moreover, beyond personal limitations, the room is simply never enough for the detailed discussion of each and every detail. In return, I will do my best to draw a coherent picture on the topic and put it in the broadest possible perspective.

The majority of the chapters of this work were written in Glenbrook Beach, New Zealand, and Kemence, Hungary. Long walks in these areas, characterized by the overwhelming presence of nature, were an incredible help to think about the details of certain findings, as well as to see the entire material from a distance that is necessary to identify the really relevant points and conclusions.

I would like to thank many colleagues for their contribution to my understanding of this complex topic. Beyond my two mentors, I am indebted to Renáta Szemerszky, Eszter Ferentzi, Michael Witthöft, Áron Horváth, and Benedek Tihanyi T. for the conversations and research cooperation that substantially shaped my approach to interoception. Also, I am grateful to Zsuzsanna Dömötör, Barbara Csala, Luca Vig, Zsuzsanna Szabolcs, Péter Babulka, Szilvia Boros, and other former or recent members of the *Ádám György Psychophysiology Lab* for their help in designing and running the experiments. Finally, I thank Zsolt Demetrovics, the Dean of the Faculty of Education and Psychology, Eötvös Loránd University, for providing me with the opportunity of a 1-year-long sabbatical to prepare this work.

Budapest, Hungary
September 28, 2020

Ferenc Köteles

Contents

Chapter 1
Introduction

1.1 The Importance of Body Sensations

One of the most commonly experienced fact with respect to body sensations as well as with emotional feelings is that we do not have the proper words to describe them, even to ourselves (Ádám 1998; Leder 2018). On top of this, we often cannot even realize them or clearly differentiate between affective feelings and body sensations. This impairment in interpreting and consequently handling these internal experiences can sometimes lead to detrimental consequences.

The major traditions of medicine, psychiatry, and psychology faced this issue many decades (or rather centuries) ago and attempted to understand and solve it through a variety of approaches. You will find a more systematic summary of the history of research in the next section. Now, I just try to illustrate my point with several examples. A physician (particularly a general practitioner or an internist) faces subjective symptoms or complaints for which there is no biomedical explanation on a daily basis, in approximately one-third to two-third of the cases (Kroenke and Mangelsdorff 1989; Garralda 2011). This is a surprisingly large proportion (and there are even higher estimates); the proper handling and explanation of these symptoms is a practical question of substantial importance. A psychoanalyst tries to utilize the ambiguity of internal feelings by explaining them symbolically; as we will see, the body can serve as an ideal canvas for such projections. Unfortunately, this can easily lead to mistaken attributions and even implicit (sometimes explicit) blaming of the client for his or her complaints. Certain complementary healers and healing systems are also prone to do this. Psychiatrists and clinical psychologists need to rely on self-reported symptoms and affective feelings to a great extent during the diagnostic procedure and in the determination of appropriate therapeutic intervention. Perhaps the most desperate debates in the history of the scientific research on emotions have been tightly connected to affective feelings and body sensations. Are they the central features of emotions or just negligible side effects? Are body feelings specific to emotions or not? Do they play a causal role in behavior or not?

F. Köteles, *Body Sensations*, https://doi.org/10.1007/978-3-030-63201-4_1

Furthermore, even light body sensations are interpreted as indicators of serious diseases for some people, which leads to further amplification of these sensations, more suffering and the overuse of health-care services. A more recent approach (aka body-oriented therapy), however, assumes that those vague feelings are messages from ourselves and should be observed with an open and neutral attitude. Finally, certain body feelings can also be explained as manifestations of some kind of energy (called qi or prana) in the recently popular Eastern meditation tradition, which energizes and even determines the functioning of the physical part of the body.

Just to mention a few significant persons, Sigmund Freud, perhaps the most influential figure in psychology and psychiatry in the first half of the twentieth century, based his entire development theory on body feelings, the pleasure originating in certain areas of the body (Freud 1922, 1936). Another early giant of psychotherapy, Carl Rogers, emphasized the importance of internal feelings and affective evaluation, considering these the output of the so-called organismic valuing process (Rogers 1959). If those judgments are not accepted as his or her own by the person, the discrepancy may lead to impaired functioning, even to psychopathology. More recently, Antonio Damasio, the renowned neuropsychologist, proposed a model called somatic marker hypothesis (Damasio 1994, 1996), which is also based on patterns of body feelings and plays a central role in decision-making (all these approaches will be discussed in more detail in various chapters of this book). Obviously, body sensations play a decisive, still often overlooked, role in human psychological functioning.

To continue with more common experiences, our preferences and basic attitudes are heavily determined by our so-called gut feelings. To illustrate this, I often ask students about their favorite food, hobby, or romantic partner, also raising the question why they like them. They are often in trouble when trying to answer such a simple question, and are relieved when I tell them that the proper answer is "just because." The process itself is not conscious; what is conscious is its output, a feeling. In a similar manner, if we are asked about our current condition, we briefly focus inside, monitor our internal body and affective feelings, and try to summarize them and translate them into words. Our actual well-being is primarily determined by such body and affective feelings. Many authors also proposed that the ultimate source of creativity and artistic performance is the plethora of body-related feelings (Fisher 1970; Ádám 1998). Finally, body sensations, including perceived symptoms, influence our entire lifestyle; in the long run, this can lead to serious (positive or negative) consequences with respect to our physical health, too (Prince et al. 2007).

Despite their wide presence and considerable impact on many areas of our functioning, scientific research on body sensations is still in a relatively early phase. Findings and theories from various areas of research are often difficult to integrate because of terminological, methodological, and conceptual differences. In the following section, I will try summarize the major traditions that to a great extent determine the recent approach to the topic.

1.2 A Brief History of the Scientific Research on Interoception

Sensu stricto, interoception encompasses the processing of information that is picked up by sensory receptors innervating the internal organs (traditionally: organs belonging to the cardiovascular, respiratory, gastrointestinal, and urogenital systems) and transmitted by ascending pathways of the autonomic nervous system (Ádám 1998). This information primarily refers to the internal state of the body, thus it mainly serves physiological (i.e., homeostatic) purposes. Conscious and communicable perception of visceral events is rarely part of this process; under everyday conditions and within the normal range of changes, we have no direct access to the majority of visceral events, such as blood pressure or the contractions of our stomach. In some cases, however, we can feel the actual condition of certain internal organs and systems; the classical examples are breathlessness during exercise, the satiation of hunger and thirst, and the fullness of the bladder and the rectum (Paintal 1986). These sensations are usually accompanied by a negative affective state which makes them able to influence behavior (Ádám 1998). We receive even more (and more unpleasant) information in pathological cases such as angina pectoris or irritable bowel syndrome: this is called visceral pain; however, it is still considerably less defined and localized than cutaneous pain (Procacci et al. 1986; Cervero and Laird 1999; Cervero 2009).

The systematic scientific research on interoception started with the work of Charles Sherrington more than 100 years ago. Before Sherrington's work, the dominant view in neuroanatomy, shared by prominent scientists of the field, for example, Santiago Ramón y Cajal or John Newport Langley, was that the autonomic nervous system was of exclusively efferent character (Ádám 1980). In other words, no afferent functions and consequently any kind of receptors were assumed. Sherrington, in contrast, was interested in various aspects of interoception (the term itself was not in use that time though); most importantly for our topic, he described the existence of various interoceptive reflexes and proposed three major categories of sensory receptors, i.e., exteroceptors, proprioceptors, and interoceptors (Sherrington 1906). The latter category referred to receptors localized in the internal organs, i.e., the viscera. In other words, interoceptors were (and still are) synonyms for visceroceptors for the majority of physicians and psychophysiologists. Interoceptors are primarily involved in the maintenance of homeostasis; consequently, their characteristics (e.g., specificity, sensitivity, adaptation) represent an important determinant of homeostatic regulation (Dworkin 2007). According to Sherrington, however, interoceptive stimuli are not accompanied by subjective sensations (Ádám 1967).

Three facts are worthy of mentioning in order to illustrate the weak points of the Sherringtonian categorization. First, the gastrointestinal tract, more precisely, the lumen of the stomach and the intestines, belongs more to the external world than to the internal from a physiological point of view as its content is not yet absorbed. Thus, chemoreceptors sensitive to certain ingredients of food and mechanoreceptors assessing the amount of food through the tension of certain parts of the intestines in fact measure characteristics of external factors. The status of the respiratory system is very similar; the air we breathe in, including everything it contains from oxygen to microorganisms, actually represents an external factor to our body. Second, as already mentioned by Chernigovskiy (1967), it is sometimes difficult to identify the anatomical boundary between exteroceptors and interoceptors. For example, sensations from the upper part of the esophagus reach consciousness more readily and belong rather to the exteroceptive senses than those from the lower part. In line with this observation, distension of the upper and lower part of the esophagus was differently perceived and processed in a brain imaging study (Aziz et al. 2000). Third, the brain often integrates information originating in different receptors (e.g., visceroceptors and mechanoreceptors) to track the changes of internal organs. For example, such multimodal integration is described for the perception of heartbeat and breathlessness (this will be discussed in more detail in Chaps. 3 and 4). In such cases, disentangling visceroception and exteroception is a demanding task; from the viewpoint of body sensations, it is often meaningless.

Decades before the publication of Sherrington's work, Ivan Mikhaylovich Sechenov, a distinguished Russian physiologist, mentioned "obscure" or "faint" sensations originating from the thoracic and abdominal cavities as examples of visceral sensations in his classic work (Sechenov 1965), originally published in Russian in 1863. Interestingly enough, these sensations were originally considered "unsconscious" (Ádám 1967) because they could not be verbalized. This approach to the unconscious (i.e., the lack of words that can describe the experience) was in accordance with the view of Watson, a prominent figure of behaviorism (Watson 1925). It gradually turned out that the information picked up by interoceptors and transmitted by afferent vegetative nerves are used for many and quite sophisticated purposes beyond vegetative reflexes and physiological homeostatic regulation (Hölzl et al. 1996); this was also called the extrahomeostatic aspect of interoception (Ádám 1998). Briefly, visceroceptive information reaches the cortex and can influence behavior; it can even be consciously perceived and verbalized in some cases.

Ivan Petrovich Pavlov, the famous Russian physiologist, was convinced that visceral functions are partly controlled by learning and that visceroception can impact behavior. In his own experiments, now taught in the elementary school, visceral changes automatically evoked by exteroceptive stimuli (e.g., salivation evoked by various tastes; this is a so-called unconditioned reflex) became associated with originally neutral stimuli such as tones or lights after several (often many, see Fig. 1.1) repetitions in dogs (Pavlov 1927). In 1926, his student Bykov was able to demonstrate that such an association can also develop for an interoceptive process. An increased diuresis following the rectal injection of 100 ml warm water was measured in a dog called Norka; this is a normal physiological reaction. After a number of repetitions, however, Norka already showed an increase in diuresis when the preparation for the injection started, i.e., when it was put in the experimental room. In summary, the experimental situation became associated with diuresis, and an interoceptive conditioned reflex developed (Bykov 1957). It was also shown with the use of sophisticated experimental techniques that interoceptive stimulation can serve as conditioned stimulus too (Fig. 1.1); for example, rhythmic distension of the

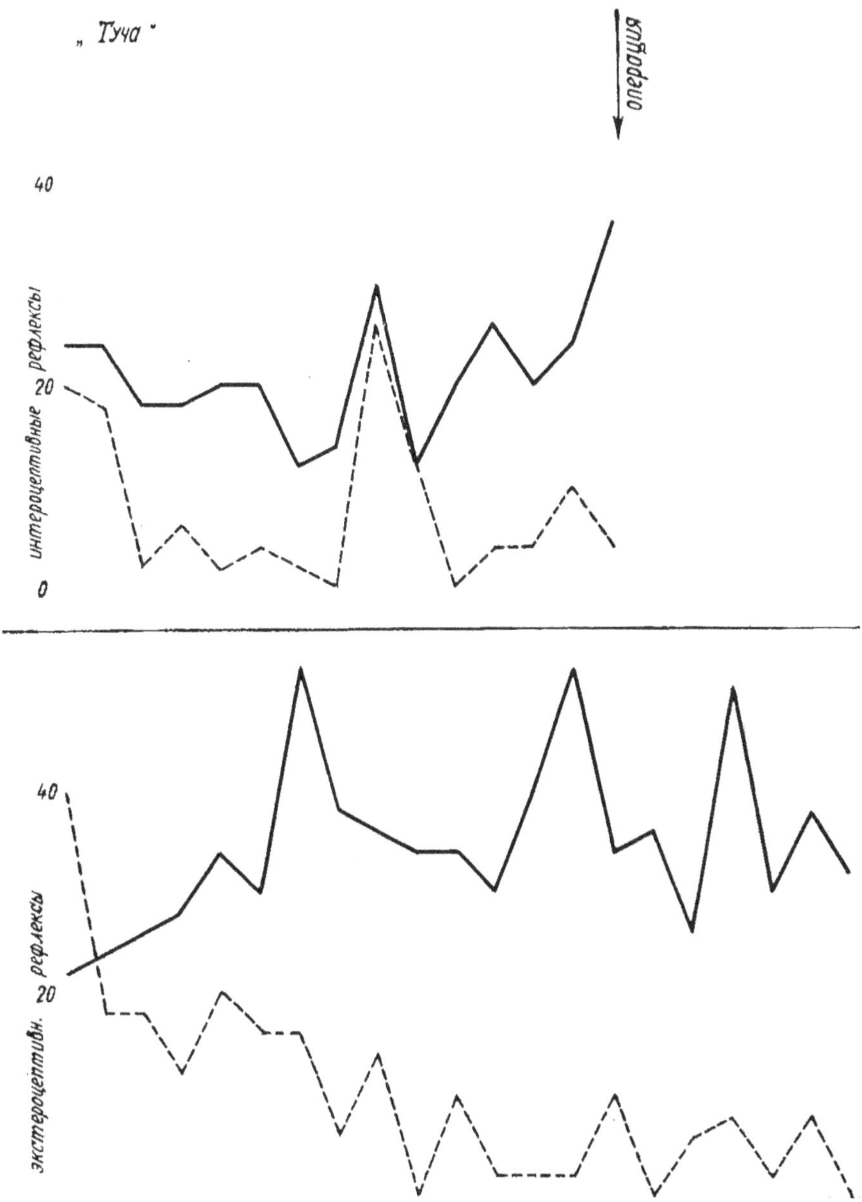

Fig. 1.1 Development of conditioned exteroceptive (sound; lower part) and interoceptive (stimulation of the renal pelvis; upper part) reflexes via classical conditioning (Ádám 1955)

renal pelvis or the carotid sinus can evoke salivation after the usual pairing procedure (Ádám 1967). In the subsequent decades, a wide variety of interoceptive stimuli, for example, pressure, scratching, or irrigation of the mucosa of various internal organs with cold and warm water, were applied in hundreds of experiments in the Soviet Union, both in animals and humans (Razran 1961; Ádám 1967). The evoked changes were measured by classic physiological (e.g., assessing the amount of saliva or dilation of the blood vessels) and electrophysiological (e.g., electroencephalography) methods. Interoceptive conditioned reflexes needed more repetitions to develop than their exteroceptive counterparts; however, once established, they proved to be more stable (i.e., less sensitive to extinction) than exteroceptive reflexes (Bykov 1957; Razran 1961; Ádám 1967). This is an important feature from both a theoretical or a more practical (e.g., pathology-related) point of view. Human participants were often not aware of the visceral stimulation that elicited the conditioned behavioral (e.g., motor or visceral) response; in other words, the reason of their own behavior remained unknown to them. Although the Soviet researchers were skeptical concerning many aspects of the original Freudian theory (e.g., symbolism), they quite early realized that conditioned responses can be interpreted as an unconscious phenomenon; the possible pathological consequences of these findings were also discussed (Airapetyantz and Bykov 1945; Razran 1961; Ádám 1998).

In summary, the discovery of the complex interactions between the viscera and the brain gradually changed the original physiological view that assumed that regulation of the vegetative functions and maintenance of homeostasis are basically independent from the higher functions of the brain. Researchers of the so-called corticovisceral psychophysiology, based on the rules of classical conditioning, gradually became interested in the role that visceroception plays in the regulation of behavior (Razran 1961; Ádám 1998). Unfortunately, the visceral conditioning approach was not without limitations, both methodologically and ideologically; these issues were gradually realized by the scientific community, and, along with the massive political changes (i.e., the end of Stalinism), finally led to the disappearance of the Bykovian school in the 1950s (Ádám 1998).

The other major mechanism of elementary learning, operant conditioning, was also applied to the visceral domain quite early. In fact, some experimental designs used in the Soviet laboratories belonged rather to the operant category than the classical; still, systematic experimentation of the phenomenon started in the United States. In the laboratories of Neal E. Miller, Bernard T. Engel, and others, it was demonstrated that, contrary to the theory of Skinner, autonomic responses can be conditioned instrumentally. For example, the acceleration and deceleration of heart rate were conditioned in animals and humans in an operant way (Shearn 1962; Engel and Hansen 1966; Engel and Chism 1967; Engel and Gottlieb 1970). Blood pressure and other characteristics involved in pathology proved also to be modifiable by operant techniques (Bárdos et al. 1989), thus researchers started to explore and develop possible therapeutic applications in the 1970s. This was called *biofeedback*; in its heyday, it appeared to be a promising paradigm in the treatment of various diseases (e.g., hypertension) and the amelioration of possibly pre-pathological conditions such as the physiological stress reaction. Biofeedback does not make individuals able to consciously sense or monitor the parameter intended to control; even the involvement of interoceptive processes is an open question (Gannon 1977; Cameron 2002). During the acquisition phase, patients learn to *somehow* change an internal parameter relying on visual, auditory, or other exteroceptive feedback. Like riding the bicycle, once research participants acquired the ability, they just used it without being able to explain how. It turned out later that biofeedback techniques that directly target physiological parameters are less effective than originally expected for a variety of reasons, from methodological caveats to physiological limitations and adherence issues (Dworkin and Miller 1986; Cameron 2002; Linden and Moseley 2006).

Similarly to the Bykovian tradition in the 1950s, research on biofeedback lost its momentum to the 1980s (Ádám 1998; Cameron 2002). Again, this was partly due to methodological issues, most importantly, the limited replicability of the early findings (Dworkin and Miller 1986; Cameron 2002). Recently, biofeedback is mainly used for the treatment of a variety of functional and mental disorders, pain, and for stress management (Yu et al. 2018). For the subsequent decades, to the beginning of the twenty-first century, the psychophysiological study of interoception became a less popular field of research.

It is worth mentioning that many methods widely used in today's interoception research, e.g., heartbeat detection paradigms, were originally developed and refined within the framework of biofeedback. The background of this was a seemingly logical assumption. Based on the ideomotor theory of William James, it was proposed by Jasper Brener (1974) that the ability to detect an autonomic response is necessary (and sufficient) for its voluntary control after a period of operant training. As we will see in Chap. 4, this idea was not supported by empirical data (Carroll 1977; Whitehead et al. 1977; Whitehead and Drescher 1980); however, it directed researchers' attention to methodological and measurement issues (Vaitl 1996). It was generally assumed that interoception follows the fundamental laws of psychophysics (Pennebaker 1982; Ádám 1998), originally developed for exteroceptive modalities, and the methodology of signal detection theory was applied to measure detection and discrimination thresholds of interoceptive stimuli, both spontaneous (e.g., heartbeats or gastric contractions) or induced (e.g., bowel distension) (Brener and Jones 1974; Whitehead et al. 1976; Whitehead and Drescher 1980; Hölzl et al. 1996; Ádám et al. 1999). These studies substantially extended the scientific knowledge on interoceptive processes.

In summary, the most important building stones of the classical or psychophysiological approach to interoception (visceroception) are (1) the (more or less exclusive) reliance on information originating in visceroceptors, which (2) can evoke conscious sensations, including near-threshold boundary conditions, that are measurable via special procedures only.

Concerning the second point, some remarks are necessary to be made. At first sight, these seem to be minute nuances of terminology; however, they are of crucial importance for our topic and their elaboration took decades for the researchers. The term sensation may refer to a process; i.e., the processing of interoceptive information by the central nervous system in our context. As it encompasses both low-level (e.g., reflex) and higher-level (including conscious) processing, and the boundary between the two is quite fuzzy, interoception in the broadest sense can be considered as the summary of all such processes (Cameron 2002). It is worth noting here that the term interoception, adopted from Russian in the 1940s, decades after the publication of Sherrington's work, was originally used exactly in this sense (Freeman and Sharp 1941; Airapetyantz and Bykov 1945), and several recent authors also prefer this approach (Kleckner et al. 2017). However, sensation also means feeling; visceral sensations or feelings are private experiences that are less defined than exteroceptive percepts with respect to modality, intensity, and localization. An analogy with the distinction between protopathic and epicritic sensation as described in the somatosensory system (Rivers and Head 1908; Holmes 1941) can be recognized (Ádám 1998; Bárdos 2003). Protopathic (or protocritic) refers to evolutionarily more ancient, vague, poorly localized feelings of pain and temperature, whereas the term epicritic sensibility was proposed for precisely localized and well-discriminable tactile and proprioceptive sensations carried by phylogenetically more recent pathways (Head et al. 1920). Perhaps these special features of interoception, which clearly differ from external perception, led some authors (Hölzl et al. 1996; Ádám 1998) introduce another term, visceral perception. In their view, visceral perception was considered a boundary category between accurately localized and verbally reportable epicritic sensations and vague protopathic feelings, belonging a bit more to the latter. With respect to consciousness, visceral sensations again represent a boundary condition (Ádám 1998). Certain sensations, i.e., unspecific changes in activation and mood as well as involuntary behavior, are out of conscious awareness under everyday conditions; however, they can be sensed by participants in experiments if they are asked to do so (Hölzl et al. 1996). Vague diffuse feelings from the viscera are rarely recognized and reported. However, with the use of more sophisticated paradigms instead of verbal reports, such as signal detection theory or the approximation method by Békésy and Blough, it turned out that a better than random discrimination performance can be achieved even when participants rate the intensity of the feeling zero; this can also be considered preconscious perception (Hölzl et al. 1996). From an electrophysiological point of view, although visceroceptive stimulation can reach the cortex, directly or indirectly, e.g., via activation by the brainstem reticular system, there is a considerable gap between the intensity that causes desynchronization in the EEG and that can be sensed subjectively and reported (Ádám 1967, 1998). In other words, the detection threshold for visceral stimulation is high, and further increases in intensity quickly lead to a different experience, typically pain (these aspects and findings will be discussed in Chap. 4).

From a philosophical point of view, internal sensations are close to qualia (Lewis 1929), i.e., the subjective components of sensory percepts. Qualia represent a primary and inherent component of any sensory experience. They are proposed to have four characteristics: they cannot be directly communicated, are intrinsic, private, and directly apprehensible in consciousness (Dennett 1988). For example, we can reach an agreement with others about the meaning of the word "green" by showing objects characterized by this color to each other; however, the direct internal experience of "greenness" cannot be shared. In the case of internal sensations, however, there are no such external anchors that could help us in communication. Although he did not use the term, Ádám realized the qualia-like features of visceral sensations very well. Further, he proposed that visceral sensations rely "on the inborn experience of the species" (Ádám 1998, p. 13) thus are less influenced by factors usually involved in perception, such as own experiences and memory. This claim will be reconsidered in various chapters of this book.

Another tradition focusing on body feelings or sensations followed a completely different path. The starting point here was that these feelings (1) refer to the actual state of the body and (2) they are subjective, i.e., accessible to the individual only; consequently, they can exclusively be assessed via self-report. Based on this latter feature, we can call this tradition *phenomenological*. In fact, the Aristotelian concept of common sensations (coenaesthesia, Gemeingefühl), often mentioned as the first occurrence of an interoception-like term (Procacci et al. 1986; Ádám 1998), belongs more to this tradition than to the psychophysiology. Other terms, such as coenaesthesis and somesthesis, were used in the same sense and also referred to the perceived state of the body (Ceunen et al. 2016). Similarly to common sensations, the obscure or faint sensations mentioned and attributed to the internal organs by Sechenov in 1863 may also belong to this category; interestingly, he also mentioned an "obscure muscular sense" beyond visceral sensations (Cameron 2002).

An important (and necessary) feature of this approach is that it considers the body unitary, as it is normally perceived as an integrated whole (Leder 2018). Consequently, no clear-cut differentiation among visceral, proprioceptive, and skin-related sensations is possible and made from this perspective. The integration of proprioceptors (and vestibular receptors) and visceroceptors under the umbrella of interoceptors might be surprising, but it is not a novel idea at all; John F. Fulton, the famous American neurophysiologist and Vladimir Nikolaevich Chernigovskiy, the Russian author of an excellent and still recently often cited monograph, followed this path (Fulton 1938; Chernigovskiy 1967). This concept was also preferred in several authors later (Vaitl 1996; Cameron 2001; Khalsa and Lapidus 2016). Recently, the term bodily or somatovisceral afference was proposed that encompasses both visceral (including the olfactory and gustatory modalities) and somatic (proprioceptive and tactile) input (Berntson et al. 2018). As we will see in Chap. 3, recent advances in neuroanatomy clearly show that pain (regardless of its origin), temperature, and some skin-related and muscle-related sensations (affective touch and muscle fatigue, respectively) belong rather to visceroception (interoception) than exteroception or proprioception (Craig 2002, 2003, 2008).

Prominent authors in this field were William James, Wilhelm Wundt, Edward B. Tichtener, and, decades later, Seymour Fisher. James emphasized the role of body feelings in the subjective experience of emotions (1884), whereas Wundt (1903) and Tichtener (1926) focused on the general features of subjective feelings and underlined the prominence of affective quality, i.e., the judgment on their pleasantness or unpleasantness (which is also an Aristotelian idea). With the appearance of behaviorism, the subjective dimension was excluded from scientific research for a while. For example, Walter Cannon, the prominent physiologist defined emotions on an exclusively behavioral (i.e., externally observable) basis (Dror 2014). Later, the role of internal need states in learning and behavior was realized, and the importance of the subjective aspect was reconsidered in the form of drives and motives by Hull (1943), Mowrer (1950), Dollard and Miller (1950), and others (Mason 1961).

Of course, clinicians, who has to deal with individual and necessarily subjective experiences, showed a great interest in emotional feelings and body sensations. Primary principles of the psychoanalytic theory (Freud 1922, 1938), from libido to erogenous zones, are in fact feelings and sensations. Later, the focus was put on various forms of anxiety (Freud 1936), which is another (partly) conscious feeling with important somatic aspects. Sigmund Freud proposed that certain parts of the body (the oral, the anal, and the genital areas) are of special importance during stages of development; later, particularly in stressful periods of life, adults can return or regress to an area and use the satisfaction it offers to bear the situation. He also stated that a substantial proportion of feelings does not enter consciousness (more precisely, cannot be verbalized) because of an active process called repression. Although repression is energy-consuming and biases the perception of reality, it is assumed to have an adaptive goal, i.e., defending the psychological system from disintegration. A related process, called conversion, refers to the translation of dangerous emotional conflicts to somatic symptoms which are less threatening thus manageable to the individual. For psychiatrists and clinical psychologists, belonging to any clinical tradition, the frequent occurrence of body symptoms in multiple organ systems and/or functions is an indicator of pathology, for example, hypochondriasis and somatoform disorder (Barsky 1979; Barsky and Klerman 1983). For physicians, chronic focus on the body and body processes is also a maladaptive characteristic as it can lead to the amplification of symptoms of organic diseases (Barsky 1979).

The first authors who judged normal body feelings worthy of systematic scientific investigation in their own right were Russel E. Mason and Seymour Fisher. They realized that other factors beyond purposeful repression and conversion might be more important in the understanding of the non-verbal nature of feelings. Moreover, they assumed that body feelings can play a substantial role in the functioning of psychologically healthy individuals. They demonstrated empirically that attention focused on various areas of the body shows considerable qualitative and quantitative individual differences, and these differences can influence perception and behavior (for details, see Chap. 5). The entire process is "non-cognitive," i.e., belongs to primary consciousness (see Chap. 2) and often not realized by the individual. They also made considerable effort to translate some of Freud's ideas and phenomenology to scientific terms and investigate them empirically. The monograph of Mason (1961) provides the reader with a very comprehensive summary of the theoretical and empirical knowledge of that time; the books of Fisher (Fisher and Cleveland 1968; Fisher 1970, 1974, 1986) reflect his heroic attempt to understand the phenomenon in its complexity.

Body arousal (perceived physiological activation) and its cognitive evaluation (called labeling) were the building stones of the two-step approach to emotions by Stanley Schachter (Schachter and Singer 1962). Although attribution theorists highlighted the latter, the former was still considered necessary (Cotton 1981; Reisenzein

1983). In the 1980s, the first social-cognitive theories on the development of body-related feelings and symptoms appeared (Leventhal et al. 1979, 1980, 1984; Pennebaker 1982), putting the emphasis on the cognitive processing of somatic sensations. Although the term symptom was sometimes used in a non-pathological sense (Pennebaker 1982), subsequent research focused primarily on the pathological aspects, i.e., the generation and maintenance of subjective somatic symptoms (complaints) and patterns of symptoms (Cioffi 1991; Barsky and Borus 1999).

Concerning the psychophysiological tradition, the first edited monograph on visceral sensation also appeared about that time (Cervero and Morrison 1986). It is striking to see how different these two approaches were. Whereas psychologists, relying on phenomenology, emphasized the ubiquitous nature of somatic sensations and symptoms, physiologists, based on findings obtained by sophisticated experimentation with the stimulation of visceral organs, still considered the phenomenon rare and marginal. It was even stated that "man could go through life without noticing any visceral sensations" (Paintal 1986, p. 3). This apparent lack of awareness of body functions can partly explain the limited interest in visceroception-related research (Cameron 2002). Noteworthy, perhaps this was the first time when the dissociation between objective events and subjective sensations, one of the focal topics of this monograph, was taken seriously. We will see later that it was not only suspected by practicing physicians but had also been demonstrated empirically decades before; still, it did not really enter the consciousness of the scientific community, not to speak of laypeople. Since then, although more than three decades have passed, not much change can be seen; this fact was among the primary motives to write the present monograph.

Continuing our historical overview, emotions and emotional (or "non-cognitive," if you wish) processing became a focal topic of science in the 1990s, based on the work of Robert Zajonc (1980, 1984) and Jaak Pankseep (1982, 1991, 1998), among others. The influential works of Antonio Damasio (1994, 1999) contributed to the re-interpretation of the role of body feelings in healthy functioning. In parallel, the pivotal role of emotional factors (most importantly, negative affectivity) in the experience of somatic symptoms was realized (Watson and Pennebaker 1989; Aronson et al. 2001).

To provide the reader with a more or less complete picture, one more tradition needs to be mentioned briefly in this historical overview. This approach (I will call it *mind-body tradition*) has its origins in the knowledge of ancient Eastern cultures, most importantly the yoga tradition. Followers of this approach emphasize the adaptive aspects and consequences of body focused attention and body feelings, from a better knowledge of our own physiological and psychological needs to the early realization of signs of pathological processes (Bakal 1999; Fogel 2009; Mehling et al. 2009, 2011; Farb et al. 2015). It should be noted at this point that our lacking ability to sense or perceive visceral events and changes was considered adaptive by psychophysiologists; and, the other way around, they regarded the partial regaining of this ability (e.g., by learning) as possibly dangerous (Thomas 1972; Ádám 1980, 1998).

In accordance with the aforementioned developments and the spirit of the current age, interoception is defined in a very inclusive manner recently. Practically everything that is related to the body, from visceral pain to emotional feelings are put under this umbrella regardless of its sensory and perceptual origin (Ceunen et al. 2016). To avoid terminological pitfalls, a clear distinction between visceroception (and visceroceptors) and interoception was proposed (Ceunen et al. 2016). Whereas the former category should refer to the traditional psychophysiological approach to interoception, the latter includes all sources of information about the actual state of the body.

Just to provide the reader with an example, body awareness, a closely related term, is defined as follows: "Body awareness is the perception of bodily states, processes and actions that is presumed to originate from sensory proprioceptive and interoceptive afferents and that an individual has the capacity to be aware of. [...] Body awareness is hypothesized as the product of an interactive and dynamic, emergent process that a) reflects complex afferent, efferent, forward and back-projecting neural activities, b) includes cognitive appraisal and unconscious gating, and c) is shaped by the person's attitudes, beliefs, experience and learning in a social and cultural context" (Mehling et al. 2009, p. 4).

In the last two decades, huge steps in the description of the afferent pathways and brain centers of interoception were made, and the role of conscious sensations was reconsidered from an evolutionary perspective (Craig 2003, 2009, 2011, 2015). Interoception is regarded as a complex and sophisticated interplay between bottom-up sensory and top-down cognitive processes today, where the latter component (called expectations or priors) plays the dominant role. A recently published edited monograph (Tsakiris and De Preester 2018), based on the predictive processing framework (see below), demonstrates this change very well. Focal point of research in the last decade was the role of interoception in various aspects of healthy and pathological functioning (Khalsa et al. 2018).

The most pivotal steps and authors in the research of interoception are summarized in Fig. 1.2.

With the acceptance of an inclusive approach to interoception, however, new issues arise. The concepts and constructs that belong to such a broad category become more and more heterogeneous, which necessitates the introduction of new sub-categories. Recently, three such sub-categories have been proposed; however, the terminology is yet to be unified. The first category roots in the psychophysiological tradition; it refers to the ability to detect body events and changes and/or discriminate between stimuli of different intensity. Logically, it is measured using "objective" behavioral tests. This category is called *interoceptive accuracy* (Ceunen et al. 2013; Garfinkel et al. 2015) or *interoceptive sensitivity* (Garfinkel and Critchley 2013). The second category, belonging primarily to the phenomenological tradition, refers to the perceived ability to sense or detect such changes. It is dubbed *interoceptive awareness* (Ceunen et al. 2013) or *interoceptive sensibility* (Garfinkel and Critchley 2013; Garfinkel et al. 2015) and assessed via self-reports. Finally, *interoceptive awareness* is also used for the metacognitive awareness of objective accuracy (Garfinkel and Critchley 2013). In the present work, the terms interoceptive

Fig. 1.2 Traditions and important figures of interoception research

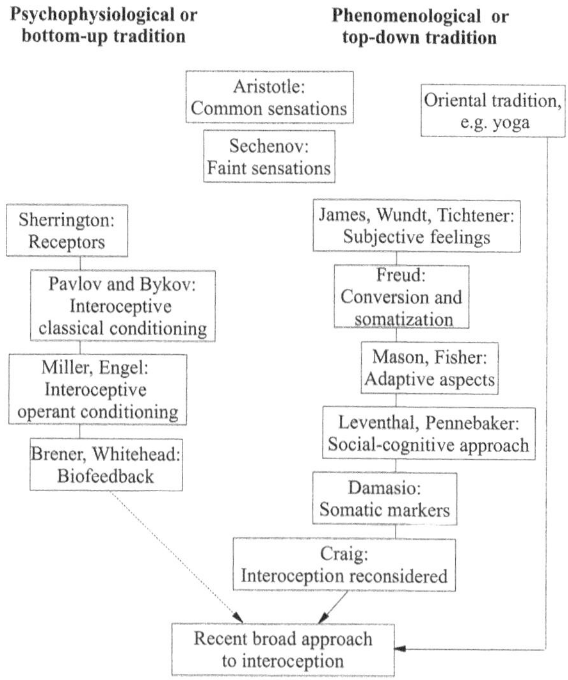

accuracy, interoceptive sensibility, and interoceptive awareness will be used for the three major categories, respectively. We will discuss them in detail in Chaps. 4, 5, and 6, respectively. Extensions to the original tripartite model, such as the actual physiological state of the body (Forkmann et al. 2016), additional preconscious and executive dimensions (Critchley and Garfinkel 2017), evaluative aspects (Farb and Logie 2018; Herbert and Pollatos 2018), and interoceptive learning ability (Canales-Johnson et al. 2015; Yoris et al. 2018), were also proposed. Also, a modified model that differentiates between objectively measured and subjectively assessed aspects of interoception and body focused attention was developed (Murphy et al. 2019a, b). All these inclusions are well justified; this suggests that the original tripartite model is overly simplistic, and a new integrative framework is needed.

My older mentor, György Ádám, would have been shocked by such a broad approach to interoception; my younger mentor, György Bárdos, recognizes and accepts that times change but does not really like this particular development. To emphasize the difference between the classical (i.e., visceroceptor-based or psychophysiological) and the recent inclusive approach to interoception, the term *body sensation* or *body feeling* will be used throughout this book. Although the former appears to be a bit more body-related and specific to me, whereas the latter is more holistic, they will be used interchangeably. In this work, *body sensations or body feelings refer to experiences that are localized or felt in the body (including the skin) from an internal perspective, regardless of their sensory origin.*

1.3 Predictive Processing

In the last two decades, a new paradigm has emerged in neuroscience, called predictive processing or predictive coding (although these two terms are not perfectly identical, we will use them as synonyms throughout the book) (Friston et al. 2006; Friston 2009, 2010). Conceptually, it is based on the observations and ideas of Helmholz (1855), Gregory (1980), and others about the active nature of perception. In this section, I provide the reader with a concise summary of this approach; such an overview is needed for the understanding of many ideas and models presented later. This summary is far from complete; more details can be learned through nice examples from the excellent books of Jacob Hohwy (2014) and Andy Clark (2015).

The basic idea is that the brain is in a very special position: it has to understand and represent external factors (i.e., the causes of its internal changes) exclusively from their effects upon the sensory organs, i.e., sensory patterns (Hohwy 2014). As (theoretically) an infinite number of different patterns of causes can cause the same sensory pattern, the task is basically a reverse-engineering problem. The solution is based on observed statistical regularities; in other words, the brain cannot be completely sure that the original causal pattern was reproduced (and, as a model is always less complex than the modeled phenomenon, the reproduction is never perfect). Thus, the probability of the restoration of the original pattern is never unity, and usually multiple options with different likelihoods are considered.

The best thing that can be done in such a difficult situation is the identification of the characteristics of the assumed causes and their interactions in terms of statistical regularities via developing hypotheses (called generative models) and test them. Once these features are represented with satisfactory accuracy, future sensory patterns can be calculated (predicted) even for novel situations. This is why this paradigm is called predictive processing: instead of passively waiting for incoming information and processing it, the brain actively anticipates sensory input. A huge advantage of this strategy is that only those pieces of the input need to be processed that were not appropriately predicted (in other words, that are surprising). The rest (which is often the lion's share of the total input) can be neglected, which saves the always limited and energy-consuming processing capacity (Clark 2015). As it is discussed in Chap. 2, predictive functioning has further evolutionary advantages; however, it needs a sophisticated (complex, hierarchically built, and expensive) central nervous system.

Now we understand the basic idea, let us see the details. On the abstract level, the predictive processing framework describes perception (and also action, see below) as the product of multilevel hierarchical processing (see Fig. 1.3). At each level, a top-down (descending or backward) signal from the level above, also called prior, describes the system's expectation of the incoming (bottom-up or forward) signal from the level below. As mentioned, the bottom-up signal is not the entire sensory input, just the proportion of it that was not predicted at the previous level; this is called error signal. The comparison of these two signals usually leads to more or less discrepancy or error. To reduce the error, one option is changing the prior (this

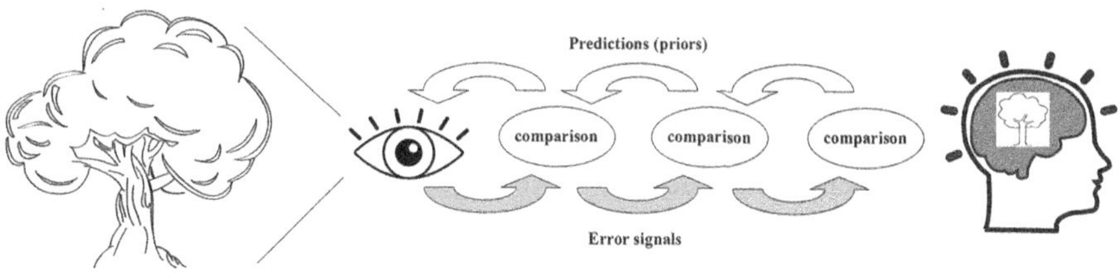

Fig. 1.3 A schematic presentation of the predictive processing approach to the perception of an external object. Ascending and descending information becomes compared at multiple levels of hierarchical processing. The ascending (error) signal is the result of the comparison at the respective level. The descending signal represents the system's predictions (priors) with respect to the ascending signal

will be called posterior or empirical prior), another possibility is changing the input by action, for example, by picking up more information (see below) (Friston 2009; Feldman and Friston 2010). Once the appropriate change is made, the new error code will be sent upward and the new (posterior) prior downward, and the same process takes place at each and every level of the hierarchy until the overall error can be minimized ("explained away"). Again, the brain usually has multiple hypotheses, which means multiple priors with different probabilities. Precision or (un) certainty of the input at hand (technically: the inverse variance of the distribution) is also taken into consideration: signals with higher precision are increased, thus have more impact on the prior(s). In this framework, attention is simply understood as selective sampling of sensory data with (assumed) high precision or optimizing the precision of sensory signals (Feldman and Friston 2010; Clark 2015). It is important to note that the system cannot directly measure the actual precision of the signals, just estimates it, e.g., based on its previous experiences (learning). This feature of the predictive processing approach makes it very (perhaps suspiciously, see later) flexible to explain a wide variety of phenomena.

Levels of the processing hierarchy represent different spatiotemporal patterns: lower-level representations primarily reflect the actual state of affairs (i.e., they are close to actual sensory input), whereas higher levels describe temporally and spatially more and more generalized background factors. Near complete elimination of the errors means that the system is able to represent (predict) the entire situation, i.e., a good representation of the background causes of the input was developed. However, errors usually cannot be simultaneously and completely eliminated at all levels. The best model should find a good balance between explanatory power and kept prediction errors. For example, too perfect matching at low levels leads to an overidentified higher level model with limited generalizability (Hohwy 2014). Moreover, the system—as each and every solution of evolution—is as lazy as possible thus favors quick-and-dirty solutions to save processing time and energy (Clark 2015). This also means that good heuristics will be preferred to perfectly accurate representations. In a novel situation, the brain applies the prior that appears to match the situation best or several priors with different probabilities, and refines its internal model by minimizing the error codes and the estimated reliability of the signal.

Overall, the predictive processing approach assumes the primacy of expectations (priors), which are mainly developed from previous experiences (evolutionary/genetic determination is also allowed, particularly for interoception; Friston 2010; Fotopoulou 2013) in the perception process. From this point of view, perception can also be regarded as a special kind of hallucination or dreaming that is controlled by error codes generated and maintained by sensory input (Hohwy 2014). This does not mean, however, that the brain loses touch with reality. As action is tightly connected to perception at each and every level of the processing hierarchy, a properly functioning brain is able to develop an internal model that matches (for the organism important aspects of) reality very well (Clark 2015). In more detail, this means that multiple action schemata corresponding to alternative hypotheses are developed and maintained in real time at multiple levels of the hierarchy, and the brain actively tests its hypotheses to find the most usable one (Cisek 2007; Cisek and Kalaska 2010). Recently, motor control was also reconceptualized within the predictive processing framework. In this approach, the brain expects (i.e., predicts) an outcome, and the motor system takes care of the rest by finding the pattern of motor commands that minimize the prediction errors through the efferent hierarchy (Clark 2015). This deep involvement of action in the perceptual process thus ensures the continuous adjustment of the model to reality.

This model might be shocking at first sight as it reverses both our common sense model of and the classic biological approach to perception. These classic models assume that sensory input is processed through a hierarchical system in a bottom-up way (this is perhaps best described for the visual modality) thus strongly determines our perception. In the predictive coding approach, however, our prior knowledge and model play the dominant role, and pieces of sensory input can be ignored if they do not fit in the overall picture (at least in well-designed laboratory experiments and interesting sensory illusions). This approach is very appealing as it can handle a number of related phenomena, such as multimodal integration, sensory illusions, and motor control very well (Clark 2015).

The predictive processing framework is quite abstract, thus it can be implemented in many ways. The most likely candidate is Bayesian inference, which continuously updates the probability of hypotheses (assumed outcomes) based on the information available. The flexibility and the apparent explanatory power of the predictive coding approach is surprising which is always a warning signal to me. Biological and psychological phenomena are usually way too complex to be described by one single mechanism. Just like simple computer-based evolutionary models, which easily can support everything depending on the actual pattern of parameters, this framework explains pretty much every aspects of human functioning with apparent ease. To avoid this pitfall, models should be based on empirically obtained or theoretically well-established parameters. Theoretical research in this area so far attempted to translate classical phenomena and constructs, such as attention, salience, and accuracy, into the special computational language of this approach. By doing this, it was demonstrated that this approach is potentially able to explain these phenomena. Simulations were also published; however, empirical data (most importantly, actual values of the crucial parameters) are typically lacking.

Concerning interoception, the most important benefit of the predictive coding approach appears to be its emphasis on top-down processing. To me, this step alone represents an important step forward, as despite of the existence of very convincing and well-established empirical evidence on the effects of beliefs and expectations in the field of placebo and nocebo research (including pain, depression, anxiety, and mood) and the area of medically unexplained symptoms, the idea of dominantly bottom-up processing of interoceptive information still prevails.

Recently, an excellent edited volume on the predictive processing approach to interoception was published (Tsakiris and De Preester 2018). Compared to this work, the present book is rather old-fashioned and almost conservative as it dares to use constructs like homeostasis, emotions, motives, or attention instead of priors, error signals, and estimated precision. In my opinion, those classic constructs are way closer to our everyday experiences from a phenomenological point of view; also, the recent methods of assessment are strongly determined by phenomenology. For example, albeit both refers to salience, research participants can readily answer questions about the direction of their attention or the valence of an actual feeling, whereas asking them about estimated precision would cause trouble, perhaps even embarrassment to them. Although I think that this new paradigm represents a nice step forward, I am convinced that it will take one or two further decades to draw a final conclusion on its practical usability. Therefore, I will apply the conventional terminology throughout the book, and also provide the reader with a brief translation of the most important phenomena into the language of predictive processing in the respective sections.

1.4 The Tenets of the Monograph

In the historical overview above, all issues that I consider significant in relation to body sensations were raised. Some of them were already discussed in some detail, or at least their salience was mentioned. Others were just touched thus their importance remains hidden until a detailed explanation in the respective chapter of the book. In a nutshell, the message of this monograph can be summarized in six tenets (Table 1.1). We will explore and explain these points from multiple directions throughout the book. Right now, I will provide the reader with a brief description of each tenet to gain some insight which makes the understanding of the following chapters more seamless.

The *tenet of realness* relies on the insight that body sensations are not just realistic but real experiences that are localized in a region of one's body or in the entire body. In more formal philosophical terms, they are transparent representations of the body; therefore, at the primary level of consciousness we are not aware of the fact that are the products of our brain. They are so vivid and convincing percepts,

Table 1.1 The six major characteristics (tenets) of body sensations

Tenet	Body sensations…
1. Realness	Are real and very vivid (convincing) experiences
2. Vagueness (ineffability)	Are hard to verbalize and/or to accurately describe
3. Evaluation	Are subject to evaluation
4. Dissociation	Often dissociate from actual (peripheral) body events
5. Malleability	Are particularly sensitive to top-down influences
6. Relevance	Impact human functioning in multiple ways

tightly bound to the self, that it is very hard to overwrite them via abstract knowledge on their origin.

The *tenet of vagueness* consists of two related features. First, body sensations are often weak and vague; second, even if they are marked, such as pain, we have difficulties in putting them into words to others, often even to ourselves (Ruckmick 1936; Mason 1959, 1961; Davitz 1969; Scherer 2004). This is called ineffability by philosophers (Metzinger 2009) and can substantially contribute to the vagueness of the sensations.

The *tenet of evaluation* states that body sensations are usually appraised; most importantly, they can be positive/pleasant or negative/unpleasant. As they belong to the self, their actual state represents an important piece of information.

According to the *tenet of dissociation*, there is often a marked discrepancy between actual (peripheral) body events and states and their perceived counterparts. Sometimes we feel things that do not take place in our body; other times we do not feel things that are actually happening in our body.

The *tenet of malleability* refers to the fact that body sensations can be impacted by information that does not originate in the body, such as knowledge, exteroceptive percepts, or expectations. From a certain point of view, this tenet can be regarded as the consequence of the tenet of vagueness and dissociation. Still, I think that its importance makes it worthy of discussing separately.

Last but not least, the *tenet of relevance* is about the impact of body sensations on human functioning. They influence our decisions, preferences, and behavior in multiple, partly conscious, partly nonconscious ways.

Neither of these six tenets is novel; all of them has been known to (or at least suspected by) philosophers, psychologists, and physicians for centuries. The present monograph is an attempt to understand and explain phenomena related to body sensations in a systematic manner, relying on these axioms.

1.5 Structure of the Book

This monograph, intended to present state-of-the-art knowledge on body sensations (or interoception in the broader sense), is structured as follows. In Chap. 2, we will start with the evolutionary and philosophical aspects of consciousness (the latter will be concise and substantially limited to avoid the labyrinth of related philosophical models and issues). This will be followed by rough neuroanatomy in Chap. 3. This is quite a jump; to maintain the reader's attention, the assumed associations between interoception and the nonconscious generation of the self are also discussed here, again with the necessary inclusion of some philosophy. Chapter 4 presents a summary on our ability to accurately detect and sense visceroceptive and proprioceptive states and changes (i.e., interoceptive accuracy) along with a number of methodological issues. Similarly, Chap. 5 deals with the measurement and implications of interoceptive sensibility, the self-reported dimension of interoception. It also introduces the tingling sensation, a primarily neutral feeling that can be evoked by both bottom-up (sensory stimulation) and top-down (attention) processes. In Chap. 6, the dissociation between objective and subjective body-related events (interoceptive awareness and related phenomena) is presented. Chapter 7 summarizes existing approaches to and models of body events and symptoms, with an outlook to the pathological aspects. Chapter 8 discusses the associations between body sensations and emotions. Subsequently (Chap. 9), negative consequences of the objective-subjective dissociation will be presented; on the one hand, some body changes are not felt, whereas, on the other hand, there are felt changes and body events without any physiological background. Most importantly, the nocebo phenomenon belongs here. Finally, to keep the balance and leave the reader in a good mood, positive aspects of the dissociation (placebo reactions, complementary and alternative medicine, and mind–body techniques) will be discussed in Chap. 10.

To whom is this monograph meant? I expect three distinct groups: (1) laypeople, including students, with a general interest in the topic; (2) health-care professionals, such as physicians, psychiatrists, and clinical psychologists, who face the negative aspects of body sensations and their detrimental consequences on a daily basis; and (3) researchers of the field who may profit from a summary of the available literature and are open to speculations with respect to the interpretation of the often equivocal findings. The amount and detailedness of information presented in the monograph is primarily tailored to the latter group. In order to help readers that are more interested in the overall picture than in open research questions and dilemmas, I will briefly summarize the major points at the end of each chapter. Also, brief take-home messages will be given for each group separately.

Last but not least, I really hope that the reader will enjoy the reading of this work as much as I liked its writing, including the really demanding, sometimes even exhausting, parts of the material.

References

Ádám G (1955) Data about the relationship of the kidney functions and the brain cortex [in Russian]. Candidate of Sciences/PhD/dissertation, Academy of Sciences of the Soviet Union

Ádám G (1967) Interoception and behavior. Akadémiai Kiadó, Budapest

Ádám G (1980) Perception, consciousness, memory. Reflections of a biologist. Akadémia Kiadó, Budapest

Ádám G (1998) Visceral perception: understanding internal cognition. Plenum Press, New York

Ádám G, Fent J, Kajtár M (1999) An attempt for the application of the basic law of psychophysics (Fechner-Stevenes law) in the domain of visceral stimulus in humans. Acta Biol Hung 50:19–25

Airapetyantz E, Bykov KM (1945) Physiological experiments and the psychology of the subconscious. Philos Phenomenol Res 5:577–593

Aronson KR, Barrett LF, Quigley KS (2001) Feeling your body or feeling badly: evidence for the limited validity of the Somatosensory Amplification Scale as an index of somatic sensitivity. J Psychosom Res 51:387–394

Aziz Q, Thompson DG, Ng VW et al (2000) Cortical processing of human somatic and visceral sensation. J Neurosci 20:2657–2663

Bakal D (1999) Minding the body: clinical uses of somatic awareness. Guilford Press, New York

Bárdos G (2003) Pszichovegetatív kölcsönhatások. Viselkedés-élettan 1. Scolar Kft., Budapest

Bárdos G, Talan MI, Engel BT (1989) Learned control of systolic blood pressure raising and lowering in rhesus monkeys. Acta Physiol Hung 73:433–446

Barsky AJ (1979) Patients who amplify bodily sensations. Ann Intern Med 91:63–70. https://doi.org/10.1059/0003-4819-91-1-63

Barsky AJ, Borus JF (1999) Functional somatic syndromes. Ann Intern Med 130:910–921

Barsky AJ, Klerman GL (1983) Overview: hypochondriasis, bodily complaints, and somatic styles. Am J Psychiatry 140:273–283. https://doi.org/10.1176/ajp.140.3.273

Berntson GG, Gianaros PJ, Tsakiris M (2018) Interoception and the autonomic nervous system: bottom-up meets top-down. In: Tsakiris M, De Preester H (eds) The interoceptive mind. From homeostasis to awareness. Oxford University Press, Oxford, England, pp 3–23

Brener J (1974) Factors influencing the specificity of voluntary cardiovascular control. In: Limbic and autonomic nervous systems research. Springer, Boston, pp 335–368

Brener J, Jones JM (1974) Interoceptive discrimination in intact humans: detection of cardiac activity. Physiol Behav 13:763–767

Bykov KM (1957) The cerebral cortex and the internal organs. Chemical Publishing, New York

Cameron OG (2001) Interoception: the inside story - a model for psychosomatic processes. Psychosom Med 63:697–710

Cameron OG (2002) Visceral sensory neuroscience. Interoception. Oxford University Press, New York

Canales-Johnson A, Silva C, Huepe D et al (2015) Auditory feedback differentially modulates behavioral and neural markers of objective and subjective performance when tapping to your heartbeat. Cereb Cortex 25:4490–4503. https://doi.org/10.1093/cercor/bhv076

Carroll D (1977) Cardiac perception and cardiac control. A review. Biofeedback Self Regul 2:349–369

Cervero F (2009) Visceral versus somatic pain: similarities and differences. Dig Dis 27(Suppl 1):3–10. https://doi.org/10.1159/000268115

Cervero F, Laird JM (1999) Visceral pain. Lancet 353:2145–2148. https://doi.org/10.1016/S0140-6736(99)01306-9

Cervero F, Morrison JFB (1986) Visceral sensation. Elsevier, Amsterdam, The Netherlands

Ceunen E, Van Diest I, Vlaeyen JWS (2013) Accuracy and awareness of perception: related, yet distinct (commentary on Herbert et al., 2012). Biol Psychol 92:426–427. https://doi.org/10.1016/j.biopsycho.2012.09.012

Ceunen E, Vlaeyen JWS, Van Diest I (2016) On the origin of interoception. Front Psychol 7:743. https://doi.org/10.3389/fpsyg.2016.00743
Chernigovskiy VN (1967) Interoceptors. American Psychological Association, Washington, DC
Cioffi D (1991) Beyond attentional strategies: cognitive-perceptual model of somatic interpretation. Psychol Bull 109:25–41
Cisek P (2007) Cortical mechanisms of action selection: the affordance competition hypothesis. Philos Trans R Soc Lond Ser B Biol Sci 362:1585–1599. https://doi.org/10.1098/rstb.2007.2054
Cisek P, Kalaska JF (2010) Neural mechanisms for interacting with a world full of action choices. Annu Rev Neurosci 33:269–298. https://doi.org/10.1146/annurev.neuro.051508.135409
Clark A (2015) Surfing uncertainty: prediction, action, and the embodied mind, 1st edn. Oxford University Press, Oxford, England
Cotton JL (1981) A review of research on Schachter's theory of emotion and the misattribution of Arousal. Eur J Soc Psychol 11:365–397. https://doi.org/10.1002/ejsp.2420110403
Craig AD (2002) How do you feel? Interoception: the sense of the physiological condition of the body. Nat Rev Neurosci 3:655–666. https://doi.org/10.1038/nrn894
Craig AD (2003) Interoception: the sense of the physiological condition of the body. Curr Opin Neurobiol 13:500–505. https://doi.org/10.1016/S0959-4388(03)00090-4
Craig AD (2008) Interoception and emotion: a neuroanatomical perspective. In: Lewis M, Haviland-Jones JM, Barrett LF (eds) Handbook of emotions. The Guilford Press, New York, pp 272–288
Craig AD (2009) How do you feel—now? The anterior insula and human awareness. Nat Rev Neurosci 10:59–70. https://doi.org/10.1038/nrn2555
Craig AD (2011) Significance of the insula for the evolution of human awareness of feelings from the body. Ann N Y Acad Sci 1225:72–82. https://doi.org/10.1111/j.1749-6632.2011.05990.x
Craig AD (2015) How do you feel? An interoceptive moment with your neurobiological self. Princeton University Press, Princeton, NJ
Critchley HD, Garfinkel SN (2017) Interoception and emotion. Curr Opin Psychol 17:7–14. https://doi.org/10.1016/j.copsyc.2017.04.020
Damasio A (1994) Descartes's error: emotion, reason, and the human brain. Penguin Books, New York
Damasio A (1996) The somatic marker hypothesis and the possible functions of the prefrontal cortex. Philos Trans R Soc Lond Ser B Biol Sci 351:1413–1420. https://doi.org/10.1098/rstb.1996.0125
Damasio A (1999) The feeling of what happens: body and emotion in the making of consciousness. Harcourt Brace and Co, San Diego, CA
Davitz JR (1969) The language of emotion. Academic, New York
Dennett DC (1988) Quining qualia. In: Marcel AJ, Bisiach E (eds) Consciousness in contemporary science. Oxford University Press, Oxford, England
Dollard J, Miller NE (1950) Personality and psychotherapy. McGraw-Hill, New York
Dror OE (2014) The Cannon–Bard thalamic theory of emotions: a brief genealogy and reappraisal. Emot Rev 6:13–20. https://doi.org/10.1177/1754073913494898
Dworkin BR (2007) Interoception. In: Cacioppo JT, Tassinary LG, Berntson GG (eds) Handbook of psychophysiology, 3rd edn. Cambridge University Press, New York, pp 482–506
Dworkin BR, Miller NE (1986) Failure to replicate visceral learning in the acute curarized rat preparation. Behav Neurosci 100:299–314. https://doi.org/10.1037//0735-7044.100.3.299
Engel BT, Chism RA (1967) Operant conditioning of heart rate speeding. Psychophysiology 3:418–426. https://doi.org/10.1111/j.1469-8986.1967.tb02728.x
Engel BT, Gottlieb SH (1970) Differential operant conditioning of heart rate in the restrained monkey. J Comp Physiol Psychol 73:217–225. https://doi.org/10.1037/h0030217
Engel BT, Hansen SP (1966) Operant conditioning of heart rate slowing. Psychophysiology 3:176–187. https://doi.org/10.1111/j.1469-8986.1966.tb02693.x
Farb NA, Logie K (2018) Interoceptive appraisal and mental health. In: Tsakiris M, De Preester H (eds) The interoceptive mind. From homeostasis to awareness. Oxford University Press, Oxford, England, pp 227–241

Farb NA, Daubenmier JJ, Price CJ et al (2015) Interoception, contemplative practice, and health. Front Psychol 6:1–26. https://doi.org/10.3389/fpsyg.2015.00763
Feldman H, Friston KJ (2010) Attention, uncertainty, and free-energy. Front Hum Neurosci 4:215. https://doi.org/10.3389/fnhum.2010.00215
Fisher S (1970) Body experience in fantasy and behavior. Ardent Media, New York
Fisher S (1974) Body consciousness. J. Aronson, New York
Fisher S (1986) Development and structure of the body image, vol 2. L. Erlbaum Associates, Hillsdale, NJ
Fisher S, Cleveland SE (1968) Body image and personality, 2nd rev edn. Dover Publications, New York
Fogel A (2009) The psychophysiology of self-awareness. Rediscovering the lost art of body sense. W. W. Norton, New York
Forkmann T, Scherer A, Meessen J et al (2016) Making sense of what you sense: disentangling interoceptive awareness, sensibility and accuracy. Int J Psychophysiol 109:71–80. https://doi.org/10.1016/j.ijpsycho.2016.09.019
Fotopoulou A (2013) Beyond the reward principle: consciousness as precision seeking. Neuropsychoanalysis 15:33–38. https://doi.org/10.1080/15294145.2013.10773715
Freeman GL, Sharp LH (1941) Muscular action potentials and the time-error function in lifted weight judgments. J Exp Psychol 29:23
Freud S (1922) Beyond the pleasure principle. Hogarth Press, London
Freud S (1936) The problem of anxiety. Norton, New York
Freud S (1938) The basic writings of Sigmund Freud. Random House, New York
Friston K (2009) The free-energy principle: a rough guide to the brain? Trends Cogn Sci 13:293–301. https://doi.org/10.1016/j.tics.2009.04.005
Friston K (2010) The free-energy principle: a unified brain theory? Nat Rev Neurosci 11:127–138. https://doi.org/10.1038/nrn2787
Friston K, Kilner J, Harrison L (2006) A free energy principle for the brain. J Physiol Paris 100:70–87. https://doi.org/10.1016/j.jphysparis.2006.10.001
Fulton JF (1938) Physiology of the nervous system. Oxford University Press, Oxford, England
Gannon L (1977) The role of interoception in learned visceral control. Biofeedback Self Regul 2:337–347. https://doi.org/10.1007/BF00998621
Garfinkel SN, Critchley HD (2013) Interoception, emotion and brain: new insights link internal physiology to social behaviour. Commentary on: "Anterior insular cortex mediates bodily sensibility and social anxiety" by Terasawa et al. (2012). Soc Cogn Affect Neurosci 8:231–234. https://doi.org/10.1093/scan/nss140
Garfinkel SN, Seth AK, Barrett AB et al (2015) Knowing your own heart: distinguishing interoceptive accuracy from interoceptive awareness. Biol Psychol 104:65–74. https://doi.org/10.1016/j.biopsycho.2014.11.004
Garralda ME (2011) Unexplained physical complaints. Pediatr Clin N Am 58:803–813, ix. https://doi.org/10.1016/j.pcl.2011.06.002
Gregory RL (1980) Perceptions as hypotheses. Philos Trans R Soc Lond Ser B Biol Sci 290:181–197. https://doi.org/10.1098/rstb.1980.0090
Head H, Rivers WHR, Holmes G et al (1920) Studies in neurology. Frowde & Hodder, London
Herbert BM, Pollatos O (2018) The relevance of interoception for eating behavior and eating disorders. In: Tsakiris M, De Preester H (eds) The interoceptive mind. From homeostasis to awareness. Oxford University Press, Oxford, England, pp 165–186
Hohwy J (2014) The predictive mind, 1st edn. Oxford University Press, Oxford, England
Holmes GM (1941) Henry Head, 1861–1940. Obituary Notices of Fellows of the Royal Society 3:665–689. https://doi.org/10.1098/rsbm.1941.0028
Hölzl R, Erasmus LP, Möltner A (1996) Detection, discrimination and sensation of visceral stimuli. Biol Psychol 42:199–214
Hull CL (1943) Principles of behavior. Appleton-Century, New York
James W (1884) What is an emotion? Mind 9:188–205

Khalsa SS, Lapidus RC (2016) Can interoception improve the pragmatic search for biomarkers in psychiatry? Front Psych 7:121. https://doi.org/10.3389/fpsyt.2016.00121

Khalsa SS, Adolphs R, Cameron OG et al (2018) Interoception and mental health: a roadmap. Biol Psychiatry Cogn Neurosci Neuroimaging 3:501–513. https://doi.org/10.1016/j.bpsc.2017.12.004

Kleckner IR, Zhang J, Touroutoglou A et al (2017) Evidence for a large-scale brain system supporting allostasis and interoception in humans. Nat Hum Behav 1:s41562-017-0069-017. https://doi.org/10.1038/s41562-017-0069

Kroenke K, Mangelsdorff AD (1989) Common symptoms in ambulatory care: incidence, evaluation, therapy, and outcome. Am J Med 86:262–266

Leder D (2018) Inside insights: a phenomenology of interoception. In: Tsakiris M, De Preester H (eds) The interoceptive mind. From homeostasis to awareness. Oxford University Press, Oxford, England, pp 307–322

Leventhal H, Brown D, Shacham S, Engquist G (1979) Effects of preparatory information about sensations, threat of pain, and attention on cold pressor distress. J Pers Soc Psychol 37:688–714. https://doi.org/10.1037//0022-3514.37.5.688

Leventhal H, Meyer D, Nerenz D (1980) The common sense representation of illness danger. In: Rachman S (ed) Medical psychology. Pergamon Press, New York, pp 7–30

Leventhal H, Nerenz DR, Steele DS (1984) Illness representations and coping with health threats. In: Baum A, Taylor SE, Singer JE (eds) Handbook of psychology and health. Erlbaum, Hillsdale, NJ, pp 219–252

Lewis CI (1929) Mind and the world order: outline of a theory of knowledge. Charles Scribner's Sons, New York

Linden W, Moseley JV (2006) The efficacy of behavioral treatments for hypertension. Appl Psychophysiol Biofeedback 31:51–63. https://doi.org/10.1007/s10484-006-9004-8

Mason RE (1959) Three studies relating internal sensory experiences and feeling reactions. J Gen Psychol 60:211–228. https://doi.org/10.1080/00221309.1959.9710223

Mason RE (1961) Internal perception and bodily functioning. International Universities Press, New York

Mehling WE, Gopisetty V, Daubenmier JJ et al (2009) Body awareness: construct and self-report measures. PLoS One 4:e5614. https://doi.org/10.1371/journal.pone.0005614

Mehling WE, Wrubel J, Daubenmier JJ et al (2011) Body awareness: a phenomenological inquiry into the common ground of mind-body therapies. Philos Ethics Humanit Med 6:6. https://doi.org/10.1186/1747-5341-6-6

Metzinger T (2009) The ego tunnel. The science of the mind and the myth of the self. Basic Books, New York

Mowrer OH (1950) Learning theory and personality dynamics. Ronald Press, New York

Murphy J, Brewer R, Plans D et al (2019a) Testing the independence of self-reported interoceptive accuracy and attention. Q J Exp Psychol 73:115. https://doi.org/10.1177/1747021819879826

Murphy J, Catmur C, Bird G (2019b) Classifying individual differences in interoception: implications for the measurement of interoceptive awareness. Psychon Bull Rev 26:1467–1471. https://doi.org/10.3758/s13423-019-01632-7

Paintal AS (1986) The visceral sensations—some basic mechanisms. Prog Brain Res 67:3–19. https://doi.org/10.1016/S0079-6123(08)62752-1

Panksepp J (1982) Toward a general psychobiological theory of emotions. Behav Brain Sci 5:407–467. https://doi.org/10.1017/S0140525X00012759

Panksepp J (1991) Affective neurocience: a conceptual framework for the neurobiological study of emotions. In: Strongman K (ed) International reviews of emotion research. Wiley, Chichester, UK, pp 59–99

Panksepp J (1998) Affective neuroscience: the foundations of human and animal emotions, 1st edn. Oxford University Press, New York

Pavlov IP (1927) Conditioned reflexes; an investigation of the physiological activity of the cerebral cortex. Oxford University Press, London

Pennebaker JW (1982) The psychology of physical symptoms. Springer, New York
Prince M, Patel V, Saxena S et al (2007) No health without mental health. Lancet 370:859–877. https://doi.org/10.1016/S0140-6736(07)61238-0
Procacci P, Zoppi M, Maresca M (1986) Clinical approach to visceral sensation. Prog Brain Res 67:21–28. https://doi.org/10.1016/S0079-6123(08)62753-3
Razran G (1961) The observable unconscious and the inferable conscious in current Soviet psychophysiology: interoceptive conditioning, semantic conditioning, and the orienting reflex. Psychol Rev 68:1–147
Reisenzein R (1983) The Schachter theory of emotion: two decades later. Psychol Bull 94:239–264. https://doi.org/10.1037/0033-2909.94.2.239
Rivers WHR, Head H (1908) A human experiment in nerve division. Brain 31:323–450. https://doi.org/10.1093/brain/31.3.323
Rogers C (1959) A theory of therapy, personality, and interpersonal relationships, as developed in the client-centered framework. In: Koch S (ed) Psychology: a study of a science. McGraw-Hill, New York, pp 184–256
Ruckmick CA (1936) The psychology of feeling and emotion. McGraw-Hill, New York
Schachter S, Singer JE (1962) Cognitive, social, and physiological determinants of emotional state. Psychol Rev 69:379–399
Scherer KR (2004) Feelings integrate the central representation of appraisal-driven response organization in emotion. In: Manstead ASR, Frijda N, Fischer A (eds) Feelings and emotions. The Amsterdam Symposium. Cambridge University Press, Cambridge, UK, pp 136–157
Sechenov IM (1965) Reflexes of the brain an attempt to establish the physiological basis of psychological processes. The MIT Press, London
Shearn DW (1962) Operant conditioning of heart rate. Science 137:530–531. https://doi.org/10.1126/science.137.3529.530
Sherrington CS (1906) The integrative action of the nervous system. Yale University Press, New Haven
Thomas L (1972) Autonomy. N Engl J Med 287:90–92
Tichtener EB (1926) A textbook of psychology. Macmillan, New York
Tsakiris M, De Preester H (eds) (2018) The interoceptive mind: from homeostasis to awareness. Oxford University Press, Oxford, England
Vaitl D (1996) Interoception. Biol Psychol 42:1–27
von Helmholtz H (1855) Über das Sehen des Menschen. L. Voss, Leipzig
Watson JB (1925) Behaviorism. People's Institute, New York
Watson D, Pennebaker JW (1989) Health complaints, stress, and distress: exploring the central role of negative affectivity. Psychol Rev 96:234–254
Whitehead WE, Drescher VM (1980) Perception of gastric contractions and self-control of gastric motility. Psychophysiology 17:552–558. https://doi.org/10.1111/j.1469-8986.1980.tb02296.x
Whitehead WE, Drescher VM, Blackwell B (1976) Lack of relationship between Autonomic Perception Questionnaire scores and actual sensitivity for perceiving one's heart beat. Psychophysiology 13:177. https://doi.org/10.1111/j.1469-8986.1984.tb00196.x
Whitehead WE, Drescher VM, Heiman P, Blackwell B (1977) Relation of heart rate control to heartbeat perception. Biofeedback Self Regul 2:371–392. https://doi.org/10.1007/BF00998623
Wundt W (1903) Outlines of psychology. Wilhelm Engelmann, Leipzig
Yoris A, Garcia AM, Salamone P et al (2018) Cardiac interoception in neurological conditions and its relevance for dimensional approaches. In: Tsakiris M, De Preester H (eds) The interoceptive mind. From homeostasis to awareness. Oxford University Press, Oxford, England, pp 187–211
Yu B, Funk M, Hu J et al (2018) Biofeedback for everyday stress management: a systematic review. Front ICT 5:23. https://doi.org/10.3389/fict.2018.00023
Zajonc RB (1980) Feeling and thinking: preferences need no inferences. Am Psychol 35:151–175. https://doi.org/10.1037/0003-066X.35.2.151
Zajonc RB (1984) On the primacy of affect. Am Psychol 39:117–123. https://doi.org/10.1037/0003-066X.39.2.117

Chapter 2
How Did It Start? An Evolutionary Approach to Body Sensations

2.1 The End of the Paradise and Its Consequences

The last (and perhaps only) paradise on Earth existed about 635–540 million years ago (mya) in the shallow waters of the Precambrian ocean (Feinberg and Mallatt 2016). This was called Ediacaran biota (in more poetic terms: "Garden of Ediacara"), consisting mainly of worms and leaf and bag-like animals feeding on the thick, green layer of microbes that covered the bottom of the ocean and the rocks under the water. The entire scenario might have been similar to a small lake that was turned in a green, slimy soup by the sunshine; not exactly resembling our internal picture of the Paradise.

Those very primitive animals were the first bilaterian organisms with two openings and a through-gut; they were barely able to move and had no differentiated sensory organs (Evans et al. 2020). They were not in need of such trickery. These features are quite expensive in terms of energy and other resources; and anyway, they would have been useless in an environment that was characterized by the complete lack of predators. Life meant a safe and relaxed existence in the Ediacaran Garden.

Unfortunately for the inhabitants of that ancient world (but fortunately for us humans), life is in constant change. In the subsequent, relatively short period of time (from approximately 540–520 mya, only a brief second on the timescale of evolution), a huge rise in the number and diversity of species occurred—evolution "switched to hyperdrive" (Feinberg and Mallatt 2016). This is called Cambrian explosion, during which the ancestors and first representatives of the recent animal phyla evolved, from arthropods to the most primitive vertebrates. The Cambrian world was way less friendly than the Ediacaran as the first predators appeared exactly at that time. Gradually, they acquired the ability to swim and actively sought prey; thus, potential game also needed to become able to flight and hide. Swimming opened new ecological niches, which led to the diversification of species. Quick movements and hunting of game need not only sensory organs, muscles, limbs, and

F. Köteles, *Body Sensations*, https://doi.org/10.1007/978-3-030-63201-4_2

claws; a complex nervous system that coordinates the entire activity from sensory input to the adequate motor response is also required. Overall, a biological arms race started, resulting in more and more sophisticated bodies and brains (Smith 1959; Parker 2004; Butler 2008; Plotnick et al. 2010).

Concerning the sensory input, the primary (and often the only) modality of the Precambrian animals might have been the olfactory-gustatory sense (Feinberg and Mallatt 2016). In the more tough subsequent world, the tactile sense (i.e., the ability to sense the vibrations generated by potential predators or prey and transmitted by the water) and vision became obviously more and more important. Vision combined with swimming, even in its elementary forms, is a particularly demanding ability with respect to brain resources. It requires the quick processing of series of images, as well as predicting future states (Llinas 2001). At a given point of time, several decisive evolutionary innovations took place, such as the multiple hierarchical processing of sensory information and the integration of pieces of information coming from different sensory modalities. Multiple (recursive or re-entrant) processing is involved in many neural functions, from motor coordination to perceptual categorization (Edelman and Tononi 2000; Friston 2009; Edelman and Gally 2013). Multisensory or multimodal integration can give the animal an enormous advantage as it can rely on more sources of information which usually means more details, higher accuracy, and reliability (Stein and Stanford 2008; Pasqualotto et al. 2016; Wong 2017). However, integration is not a trivial task, and not without caveats. First, appropriate timing (the adequate temporal synchronization of different pieces of information) is crucial, thus a complex and quick background machinery (i.e., a sophisticated nervous system) is needed. Such a complex brain consumes a lot of energy and nutrients, and perhaps this is why only two clades of animals (certain arthropods and vertebrates) followed this path (Feinberg and Mallatt 2016). Second, integration takes time, at least several tens of milliseconds, which does count in certain situations. In return, it can speed up the final response, for example, due to the improved detection of signals (Stein and Stanford 2008). Third (and this is the most important point for us), sensory information is often noisy and prone to errors. This means that integration necessarily involves comparison and selection of various pieces of input, via multiple (hierarchical) processing. In consequence, the nervous system builds up an internal model (representation) of the external world that best matches the input as a whole (Humphrey 1992; Damasio 1994; Edelman and Tononi 2000; Green and Angelaki 2010; Barrett et al. 2016); for more details, see the summary on the predictive processing approach in Chap. 1. This is a crucially important step. When you step on a piece of lego in the night in the living room and immediately draw up your leg (and also strain your other leg to avoid falling), the entire process runs extremely quickly through the spinal cord. It is a reflex that runs automatically and involuntarily; in fact, the event (more precisely, a part of it) reaches conscious awareness a bit later, when the process already finished. From a conceptual point of view, a simple reflex is also a representation of a frequently occurring event; however, it is rigid (hard-wired) and basically a reaction to the event. In the case of hierarchical processing and multisensory integration, however, an online model of the important aspects of the world is generated and maintained.

Such a representation can be considered the prerequisite or the primordial form of consciousness (or course, reflex responses are still present in higher animals and necessary for normal functioning). The most evident biological adaptive value of consciousness is enabling the organism to response to novel situations in a more versatile way than allowed by simple automatic responses (Baars 1988; Earl 2014). As mentioned, integration takes some time; however, it has another advantage beyond having more information on the world: it opens the gate for predictions. Instead of waiting for events to occur and respond to, the organism can expect future happenings and actively prepare for them. Prospection or "navigating into the future" is a central organizing feature of perception, action, motivation, and affect (Seligman et al. 2013). In return, the organism loses direct touch with reality to some extent—from this point on, the representation of the word becomes the reality. At the perceptual level, the organism is not aware of this replacement; in other words, the representation is transparent (Metzinger 2009). Concerning the topic of this book, this is the origin of the tenet of realness (Chap. 1), i.e., that body sensations are very vivid and convincing experiences. In consequence, we trust our eyes or, remaining closer to our topic, body sensations, and are perfectly convinced that we see or feel the real thing. Or course, logical reasoning based on conceptual knowledge can question the validity of our percepts (this is what we are doing right now), and we can even try to ignore them, but this requires further resources and self-regulation, and usually do not directly change the original percepts. To present the reader with the consequences of this feature through examples, patients with hypochondriasis trust their sensations more than clear diagnostic evidence concerning the lack of somatic disorders and conviction of doctors (Pilowsky 1967; Costa and McCrae 1985). Similarly, patients with hypertension often associate their actual blood pressure with certain symptoms or conditions and use them to assess actual blood pressure. Again, these beliefs, relying on body feelings, are very hard to overwrite even by demonstrating that patients' actual (measured) blood pressures show very weak or no correlation with their felt states (Baumann and Leventhal 1985). The importance of this characteristic (i.e., representation is felt as reality) will be discussed in great detail in various chapters of this book. Moreover, as multimodal integration is not necessarily perfect all the time, and no model is able to completely emulate all aspects of the modeled process, discrepancies and errors occur (Stein and Stanford 2008). Such errors will also be perceived as reality as the representation already replaced direct sensory input. Certain types of errors can also be called sensory illusions, and can be generated at multiple levels of the hierarchical sensory processing, even within a single modality. For example, visual illusions can be the results of such erroneous integration because visual processing is a very complex procedure. Concerning multimodal integration, the famous and intensely studied rubber hand illusion (Botvinick and Cohen 1998) also represents such an integration error. In the classic experimental setting, the rubber hand and the real hand are simultaneously and rhythmically stimulated with a small brush but only the former can be seen by the participant. As a result, an obviously fake hand, or even an imagined hand (Guterstam et al. 2013), can be felt as own if its actual or imagined visual image gets integrated with other, in this case the tactile, modality due to the temporal synchrony between the two. The existence of visceral illusions and hallucinations, perhaps based on similar high-level processes, was also proposed by a number of authors (Ádám 1998; Damasio 2010).

The next crucial step in the evolution of the brain and elementary consciousness took place hundreds of millions of years later, with the appearance of first mammals (appr. 220 mya) and birds (appr. 165 mya). These animals developed homeothermy, i.e., they became able to keep the temperature of their body more or less constant. With respect to the nervous system, this means faster and more complex information processing that allows a more sophisticated representation of the external and internal world (Butler 2008; Gerkema et al. 2013). In return, homeothermy (just like a huge brain, see below) requires much more energy and fairly accurate homeostatic regulation to constantly meet the elevated energetic demand of the organism (Craig 2015). As we will see in Chap. 3, these demands might play a decisive role in the development of the conscious representation of the actual state of the body.

Having an internal model of the world can be considered the prerequisite and perhaps also the most elementary form is consciousness. I know very well that this statement might be surprising, confusing, even disturbing for some readers. The term consciousness is used very differently and inconsistently in various fields, from philosophy and religion to psychology, psychiatry, and medicine. Sometimes, it is considered a synonym for awareness (Griffin 1992); in the medical tradition, however, awareness (i.e., wakefulness) is regarded as a necessary but not sufficient condition of consciousness, and the two terms are clearly distinguished (Ádám 1980; Prigatano and Johnson 2003; Damasio 2010). On top of this, the term conscious awareness is also in use (Cameron 2002). Beyond such elementary terminological issues, we humans automatically identify consciousness with verbalization, symbolization, self-reflection, and similar higher order processes. The Freudian and Pavlovian approaches were quite similar in this respect; they considered the ability to have words for and speak about something, i.e., the existence of secondary associations (Bykov 1957; Ádám 1967), the proof that it reached consciousness. In this vein, behavioral changes demonstrated in animal models were not regarded as signs of conscious processing. In humans, it was found that interoceptive conditioning can evoke changes that run automatically and nonconsciously; to obtain verbal reports, however, more intense stimulation was necessary. Lack of words, also called the ineffability problem (Metzinger 2009), is a general issue with conscious materials and seriously limits both their communicability and our ability to identify internal contents, including body sensations (tenet of vagueness). Still, equating verbal reports with consciousness is a questionable practice (Edelman and Seth 2009). Following this line of thinking, we are prone to think that consciousness is an exclusively human phenomenon, a product of (mostly verbal) communication (Ádám 1980) and necessarily includes self-reflexion (i.e., self-consciousness or self-awareness) (Griffin 1992). Because we humans do possess this high-level consciousness, it is difficult (if not impossible) for us to imagine or emulate the more elementary levels of the phenomenon (Damasio 2010; Panksepp and Biven 2012). But if someone has a dog or a cat, I am sure he or she will be convinced that his or her pet does possess some form of consciousness—it has preferences, moods, anticipates events, and so forth. In the last decades, a great number of animal phenomena have been described that best can be explained via the assumption of consciousness (Griffin 1992; Denton 2005; Feinberg and Mallatt 2016).

To bridge this huge gap between elementary and advanced forms of consciousness, a distinction between primary (also called perceptual) and secondary (dubbed reflective) consciousness was proposed (Lloyd 1989; Farthing 1991). Primary or perceptual consciousness refers to the internal subjective experience which is ongoing and non-verbal (some authors also call this "non-cognitive"), and is based on pictures and sensations. Phenomenal consciousness is another word for this level (Block 1995). The quality of primary consciousness is dominantly determined by the primary sense(s) the animal relies on, e.g., smell in dogs, reflections of ultrasounds in bats, or vision in humans. Donald Griffin lists many examples of animal consciousness in his excellent book (1992) with respect to decision-making, predation-related behavior (e.g., cooperation), or the preparation and use of artifacts. Higher animals are even able to develop expectations (and they show surprise if their expectations are not met, another proof that they run a model of the outer world), use numbers and simple abstract rules. In summary, primary consciousness, although it seems primitive and elementary at first sight, in fact represents a very effective way of problem-solving. Remarkably, these internal calculations take place without the use of language and formal logic, not only in animals but quite often also in us humans.

2.2 The Roots of Subjectivity

What we discussed so far was exteroceptive primary consciousness, i.e., the representation of the external world by *mental images* (image refers here to any sensory modality, not only vision). The most salient characteristic of this representation is isomorphism as the structure of the represented object is preserved (Damasio 2010; Feinberg and Mallatt 2016). It might be surprising, but certain bees and other insects appear to have this form of consciousness too: they can represent difficult spatial structures that are out of vision at the given moment, use this representation to navigate, and they are also able to communicate them to other individuals (Griffin 1992).

Having mental images of the word might be sufficient to build up a mind, but still not enough for a conscious mind; the privacy and subjectivity of the experience is still lacking (Damasio 2010; Damasio and Carvalho 2013; De Preester 2018). To fill this gap, two further representations beyond exteroceptive or perceptual consciousness need to be mentioned; we can call these evaluative (i.e., affective) and body-related (i.e., interoceptive) representations (Feinberg and Mallatt 2016). These representations generate the internal subjective experience that we call *affective feelings* and *body sensations* throughout this book. Concerning the former category, core affect is a feeling that has no object (i.e., is not about something or not reflective, simply a mood state); it can be described along two dimensions, valence and activation (Russell 2003). Thus, affective feelings in their elementary form refer to the basic (good versus bad/pleasant versus unpleasant) overall (i.e., non-isomorphic) evaluation of the actual state of affairs from the viewpoint of the organism (Russell 1980, 2003; Lazarus 1991; Barrett and Russell 1999; Watson et al. 1999). Core affect can be conceptualized as an outcome of the continuous assessment of one's current state; it is determined by internal events, such as hormonal changes, nutritional deficiencies, or the state of the immune system. However, it is important to see that only the output of the assessment, i.e., affect, reaches consciousness. When a sudden change in core affect occurs that is ascribed to an object (this process is automatic), it leads to a secondary, reflective state, called emotion. Emotions have behavioral components, thus at this level basic evaluation is closely associated with (or can be considered the subjective aspect of) two major behavioral categories, i.e., approach (aka appetitive) and withdrawal (avoidance, defensive) (Fowles 1987; Gray 1987; Depue et al. 1994; Lang 1995; Watson et al. 1999; Lang and Davis 2006). The terms positive and negative affect or positive and negative activation basically refer to this construct (Russell 1980; Watson and Tellegen 1985; Watson et al. 1988, 1999; Lang 1995; Barrett and Russell 1999). Whereas the approach system directs organisms toward situations and experiences that yield pleasure and reward, the withdrawal system helps them to avoid aversive stimuli with possibly dangerous consequences (Lang 1995; Watson et al. 1999). The evolutionary advantages of the conscious representation of these processes are obvious. First, as it is stated in motivational priming theory (Lang 1995), they guide behavior (Damasio and Carvalho 2013). Second, they represent the "common currency" for the brain to rank the motivations and determine the quality and order of required behavioral steps (Cabanac 1996). Third, subjective experience mediates the acquisition of new stimulus–response patterns (see below) (Panksepp and Biven 2012). The basic evaluation along the positive–negative dimension is crucially important for our topic because they add a *subjective* dimension to the perceived world (Watson and Clark 1984; Damasio 1994; Feinberg and Mallatt 2016). Affective feelings, the basic two as well as the more differentiated states described below are determined simultaneously by the internal state of the organism and the environment (i.e., external stimuli); put in a different way, they signal the reaction of the organism to the stimulus at hand (Damasio 1994, 1999).

Body sensations can be considered a boundary category between the aforementioned two kinds of representation with respect to isomorphism, the use of mental images, and evaluation (Feinberg and Mallatt 2016). There is an intimate relationship between body feelings and affective feelings. As already described, the subjective component of conscious experience originates in the constant monitoring of internal states (Damasio 2010; Damasio and Carvalho 2013). As affective feelings reflect the actual homeostatic needs of the organism (including social needs in primates), they necessarily rely on body states (Craig 2015). This process results in primordial (already conscious) feelings and the birth of the so-called protoself. The protoself, i.e., the sensation of own internal states, is regarded by Damasio as the most elementary form of subjectivity. The actual or threatening violation of the homeostasis often evokes strong negative feelings to motivate the organism for appropriate behavioral steps (Damasio and Carvalho 2013); from a predictive processing perspective, this is described as a state when sensations increasingly violate the brain's expectation (Joffily and Coricelli 2013). In this case, affective feelings are associated with body sensations; this is the background of the tenet of evaluation (Chap. 1). These homeostatic feelings, which are very close to primary motivations, are also called primal emotions (Liotti et al. 2001).

Which representation of the three appeared first is an open question. Most likely, the complex neural system and the basic mechanisms were developed in response to exteroceptive demands as described by the evolutionary model presented above (Feinberg and Mallatt 2016). However, it can also be argued that the representation and constant monitoring of crucial bodily states, which leads to the development of primary motives, such as hunger or thirst, must have preceded everything else (Denton 2005; Denton et al. 2009; Craig 2015; Corcoran and Hohwy 2018). Assuming simultaneity appears a logical solution of this dilemma. The regulation of internal need states is conducted mostly via automatic mechanisms; consciousness comes into the picture only when behavioral regulation starts to play a substantial role (Ádám 1967, 1998). Behavioral regulation usually means choices and decisions which require motives and intentions. Exteroceptive consciousness without any connection with the internal state of the organism is just a fancy but sterile movie; thus basic evaluation and consideration of the actual needs must have been integrated quite early, and these representations must have coevolved with each other since then. According to the model of Damasio (2010), the perception of external stimuli evokes changes in the state of the organism which is reflected by the protoself. Once the representation of the external stimulus and the changes of the protoself are processed simultaneously and get integrated, a new representation develops which refers to both the stimulus and the organism. This is called core self; it is characterized by self-reference and subjectivity, the most important features of consciousness.

At this point, an incredulous reader may ask why feelings (i.e., conscious representations of internal states) are needed and how do we know (beyond intuition, which is prone to errors) that higher animals do have some kind of consciousness? Emotions represent stereotypic reaction patterns to the most frequently occurring situations in the life of mammals (Panksepp 1998; Damasio 2010; Panksepp and Biven 2012). This definition would have satisfied Walter Cannon and the behaviorists as it does not include subjective states. However, affective neuroscience goes further and with good reasons. It assumes that subjective states, called emotional feelings, are pivotal components of emotions (Panksepp 1998, 2005; Panksepp and Biven 2012). Jaak Panksepp listed and extensively studied seven basic emotional feelings (four positive or pleasant: seeking, play, care, lust; and three negative or unpleasant: anger, panic/grief, fear) claimed to characterize mammals; moreover, he did not exclude the existence of further categories. Primarily, a basic emotion (including the feeling component) is evoked by a typical situation trough an inherited hard-wired connection (e.g., loss of a child or parent generates panic/grief, or sight of a snake generates fear). However, as the ontogenic development of mammals is heavily based on learning, the brain should be able to apply the same typical reaction pattern to new situations and experiences (e.g., the loss of a pet can also generate panic/grief, and the observation of a gun might also evoke fear). These new associations can build up between feelings and features of novel situations, for example, in a conditioned manner. In this process, feelings represent the means trough which well-proven reaction patters can be associated with novel stimuli. Similar to emotional feelings, motives (the boundary between the two is not very well defined) should necessarily reach consciousness in order to impact long-term behavior. A famous example for this is the elephants walking hundreds of kilometers to reach the salt-rich caves of Mount Elgon in Africa. This systematic approaching behavior clearly reflects an *intention* which cannot be explained by simple conditioned reflexes (Denton 2005). Intentions also belong to the domain of consciousness (Denton et al. 2009), and might have a well-defined evolutionary background related to colonization of dry land and facing new challenges to homeostatic regulation (e.g., shortage of water and salt). According to Denton (2005), intentional behavior is based on the so-called primordial emotions, such as thirst, hunger for air and food, pain, sleep, micturition, sexual orgasm, and temperature control. It is worth noting that these body-related feelings are usually accompanied by negative evaluation, i.e., they are unpleasant—unfortunately, this appears to be the best way to directly motivate the organism.

Concerning the second question, i.e., how can we know that animals possess some form of consciousness, we cannot be sure as they cannot speak about it. However, as put by Panksepp and others, we can triangulate between neural (e.g., comparative brain anatomy and physiology), behavioral, and mental analyses, and this assumption has the highest explanatory power (Steiner et al. 2001; Denton et al. 2009; Panksepp and Biven 2012; Damasio and Carvalho 2013). Moreover, the subjective state of the organism (e.g., drug craving) can be used as conditioned stimulus or operant discriminative stimulus in animal models (Bevins and Besheer 2014; Troisi 2015); pigeons were even able to learn to communicate their internal states evoked by a variety of drugs (Lubinski and Thompson 1987). Overall, this indicates that such subjective states do exist in nonhuman animals.

In addition, many animals apparently have an integrated representation of their bodies in the surrounding environment; for example, grizzly bears conceal their entire bodies while watching hunters (Griffin 1992). Moreover, we will see in Chap. 3 that primates, humans included, do have a real-time representation of their homeostatic state in form of body feelings (Craig 2015). These phenomena clearly belong to primary consciousness.

Keeping the major topic of this book in mind, discussion of animal aspects of consciousness might appear overly extensive or even boring to the reader. However, and this is in fact the point I want to make, all these mechanisms are at work in humans, and they heavily impact both higher consciousness (attributions, thoughts, etc.) and behavior. Most importantly, the major topic of this book is very tightly connected to this elementary level of subjective experience.

2.3 Higher Consciousness

What about language, rational thinking, and conscious self-reflection, i.e., secondary or reflective consciousness? As the name indicates, these characteristics add a substantially new quality to the elementary form of consciousness: the contents of the latter are re-represented and handled at a more abstract level or perhaps levels. Beyond experiencing it, we *know* that our experience is subjective, and we are even able to write book chapters on the topic. Moreover, temporal relationships, in form of explicit memories of past experiences and expectations of future states, are also handled at this level. This high-level representation, called autobiographical self by Damasio (2010), mainly characterizes humans. However, its elementary forms can also be found in animals.

To demonstrate how novel these characteristics are compared to primary consciousness, even if we consider its relatively sophisticated form as developed in early mammals 220 mya, let us see the timescale (Fig. 2.1). Logical thinking does not necessarily need language; image-based ways of problem-solving are also possible and well-described in animals (Griffin 1992). Cognitive abilities required for complex analytic thinking might have been available about 6 mya due to the enlargement of prefrontal and temporal cortices (Naskar and Cicurel 2015). Approximately 200 million years versus circa 6 million years, this is a difference of one to two magnitudes. Based on anatomical considerations, paleontologists think that some kind of primitive language might have been existed in some early representatives of the genus *Homo*. The skull of *Homo habilis* (approximately 2 mya) already shows the signs of Broca's area, a brain region which is tightly linked to language production (Tobias 1987). This is a difference of two magnitudes. Neanderthals, living approximately 250,000–25,000 years ago (i.e., 0.25–0.025 mya), might have possessed all abilities (fine control of breathing, position of larynx, etc.) that are necessary for more or less complex speech (MacLarnon and Hewitt 1999; D'Anastasio

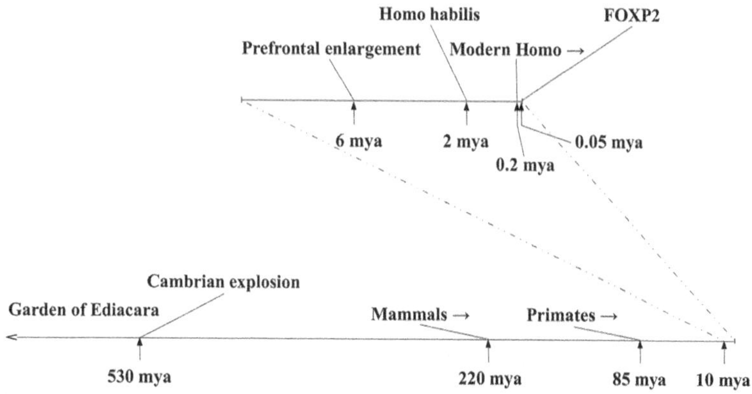

Fig. 2.1 Timescale with the most important evolutionary steps for the development of primary and secondary consciousness

et al. 2013; Papagianni and Morse 2015). Concerning the adaptiveness of language, existing theories emphasize the social aspects; for example, "vocal grooming" is more effective than physical grooming and includes sharing information about others (i.e., gossip) (Dunbar 1998), or it makes coordination of activity of group members more effective (Bickerton 2009). As for *Homo sapiens*, the modern human, a small number of mutations of the famous FOXP2 gene might have opened the gate to modern speech and other advanced human characteristics approximately 50,000 years ago (0.05 mya); this is called the Great Leap Forward theory (Klein 2002). If we consider these sophisticated representation abilities, the temporal relationship to primary consciousness grows to 4–5 magnitudes. The autobiographical or narrative self, armed with language, provides the individual with a higher level of self-reflection, also called meta-consciousness or awareness of awareness (Gallagher 2000; Damasio 2010). Still, it is important to see that the self, as constructed recently at this exclusively human level, is a very novel product of sociocultural evolution even in the timescale of human evolution. For example, Jaynes (1977) argued that linguistic meta-cognition occurred only 3–4 thousand years before, between the birth of the Iliad and the Odyssey.

At first sight, this entire archaeological summary might appear to be self-serving and useless. However, I think it is particularly useful to demonstrate how deeply rooted the processes belonging to primary consciousness are in human thinking. By calculating the magnitudes of differences between the appearance of primary and secondary forms of consciousness, I did not want to suggest that this accurately reflects their relative significance. However, in many domains of human functioning, including health psychology, my primary research area, a large number of issues from risk perception to the nocebo phenomenon and mass psychogenic illness can be explained by such primary processes and primary "thinking." Again, behaviors and beliefs based on primary processes are very difficult to change by providing individuals only with food for secondary consciousness, i.e., information or knowledge regarding their irrationality or even riskiness.

If our feelings play such a crucial role (which I believe is in fact way more influential than we like to confess), are thoughts and language important at all? Seymour Epstein proposed a dual-process theory called cognitive-experiential (self) theory (CEST or later CET) (Epstein et al. 1996; Epstein 2003, 2014). It states that we have two more or less independent information processing systems which run simultaneously and often draw different conclusions regarding the events happening to us or decisions to be made. The evolutionarily older (it is already present in higher animals) intuitive-experiential system, which is quite close to primary consciousness discussed above, mainly relies on associations. Those associations are often based on unimportant or apparent perceptual similarities. The system operates automatically and very quickly, however, its conclusions are often erroneous, particularly in the long run. Unfortunately, those conclusions are quite close to gut feelings, being very strong and often imperative. The cognitive-rational system, a new invention of evolution, is based on language and logical reasoning. It is a slow, deliberate, and effortful process; most importantly, it takes much more aspects and information into consideration, thus it is able to make more adaptive long-term decisions. Which one is the better? In financial decisions and software development, we should perhaps rely exclusively on rationality in an ideal world. In fact, in the lack of cognitive resources and time, experience-based intuition plays a huge role even in these areas of life. Concerning everyday decisions, the two systems should work together and help each other to reach a conclusion within an acceptable timeframe. In other words, both are necessary and useful. I am aware that this theory (and similar dual-process theories, e.g., Daniel Kahneman's (2011) fast and slow thinking systems) was not primarily aimed to explain consciousness. They are developed to understand the basic rules of information processing and decision-making; still, they are very helpful in illustrating the key differences between the two fundamental modes of human mind and pointing out that the ancient mammalian way of information processing plays a substantial role in humans too.

Body feelings (and generally contents of primary consciousness) are difficult to verbalize (this belongs to tenet of vagueness, see Chap. 1)—perhaps this is why our culture, which relies heavily on secondary consciousness and is characterized by issues originating in secondary processing such as schedules and time pressure, is prone to neglect them to a great extent (Leder 2018). According to civilization critics and body-oriented therapists, we are prone to not pay enough attention to the actual state of our body and the messages our body tries to communicate. Right now, the reader might be hungry or sitting in an unpleasant position, but not aware of these signals because this chapter catches his or her full attention (at least I hope so). Perhaps those feelings were realized but intentionally suppressed for a while. The point is that such processes can have harmful consequences in the long run. If we realize postural and muscle-related issues and we are aware of their consequences, we can change our position and avoid chronic low back pain later. Becoming aware of early signs of a migraine attack, the patient might take preventive steps to make it less severe (Bakal 1999). The bottom line is that body sensations often would deserve more attention than actually paid to them. In other cases, however, things are less obvious. As we will see, chronic body focus can lead to constant monitoring for signs and indicators of disease, which results in more symptoms, as in the case of hypochondriasis and idiopathic environmental intolerances. Certain unpleasant body feelings such as disgust can be evoked and used for the purpose of a political campaign (Shook et al. 2017). In summary, focusing to body sensations is a double-edged sword, and their adaptivity is sometimes questionable. Perhaps we have a good (evolutionary) reason to neglect them most of the time and finding a healthy balance between body focus and the suppression of body sensations is a difficult task. This debate will be a recurring topic throughout the monograph.

2.4 Summary

To sum up, we made an attempt to understand and characterize primary consciousness from an evolutionary point of view. It was found that it represents an ancient but very effective way of information processing with special features. Body sensations, as well as affective feelings, were explained as phenomena that clearly belong to this domain. Consequently, we always need to keep their special features in mind. For example, they are very real and convincing experiences (tenet of realness) that are often difficult to verbalize (tenet of vagueness). Moreover, they are tightly associated with evaluative processes (tenet of evaluation). Overall, this might lead to a continuous and complex (partly conscious, partly nonconscious) impact on our everyday functioning (tenet of relevance).

Take-Home Messages

For laypeople: Body sensations, belonging to the ancient part of human consciousness, heavily influence our internal life and behavior. As we cannot easily translate them into words, however, we are often prone to not realize their actual importance.

For clinicians: Questioning the validity and realness of the body sensations of patients relying on logical arguments is rarely successful. As they belong to primary consciousness, they necessarily appear real to them, even in the light of contrary objective evidence. Ignoring or denying them leads to the impression that the clinician lacks empathy thus he or she is not able to understand the patient.

For researchers: The characteristics of and differences between primary and secondary consciousness are rarely considered in interoception research. For example, questionnaires assessing interoceptive sensibility and body awareness are blends of items from the two categories or interoceptive appraisal refers to both direct and conceptual levels of evaluation (Chap. 5).

References

Ádám G (1967) Interoception and behavior. Akadémiai Kiadó, Budapest

Ádám G (1980) Perception, consciousness, memory. Reflections of a biologist. Akadémia Kiadó, Budapest

Ádám G (1998) Visceral perception: understanding internal cognition. Plenum Press, New York

Baars BJ (1988) A cognitive theory of consciousness. Cambridge University Press, Cambridge

Bakal D (1999) Minding the body: clinical uses of somatic awareness. Guilford Press, New York

Barrett LF, Russell JA (1999) The structure of current affect: controversies and emerging consensus. Curr Dir Psychol Sci 8:10–14. https://doi.org/10.1111/1467-8721.00003

Barrett LF, Quigley KS, Hamilton P (2016) An active inference theory of allostasis and interoception in depression. Philos Trans R Soc Lond B Biol Sci 371:20160011. https://doi.org/10.1098/rstb.2016.0011

Baumann LJ, Leventhal H (1985) I can tell when my blood pressure is up, can't I? Health Psychol 4:203–218. https://doi.org/10.1037/0278-6133.4.3.203
Bevins RA, Besheer J (2014) Interoception and learning: import to understanding and treating diseases and psychopathologies. ACS Chem Neurosci 5:624–631. https://doi.org/10.1021/cn5001028
Bickerton D (2009) Adam's tongue: how humans made language, how language made humans, 1st edn. Hill and Wang, New York
Block N (1995) On a confusion about a function of consciousness. Behav Brain Sci 18:227–247. https://doi.org/10.1017/S0140525X00038188
Botvinick M, Cohen J (1998) Rubber hands "feel" touch that eyes see. Nature 391:756. https://doi.org/10.1038/35784
Butler AB (2008) Evolution of brains, cognition, and consciousness. Brain Res Bull 75:442–449. https://doi.org/10.1016/j.brainresbull.2007.10.017
Bykov KM (1957) The cerebral cortex and the internal organs. Chemical Publishing, New York
Cabanac M (1996) On the origin of consciousness, a postulate and its corollary. Neurosci Biobehav Rev 20:33–40
Cameron OG (2002) Visceral sensory neuroscience. Interoception. Oxford University Press, New York
Corcoran AW, Hohwy J (2018) Allostasis, interoception, and the free energy principle: feeling our way forward. In: Tsakiris M, De Preester H (eds) The interoceptive mind. From homeostasis to awareness. Oxford University Press, Oxford, England, pp 272–292
Costa PT, McCrae RR (1985) Hypochondriasis, neuroticism, and aging. When are somatic complaints unfounded? Am Psychol 40:19–28
Craig AD (2015) How do you feel? An interoceptive moment with your neurobiological self. Princeton University Press, Princeton, NJ
D'Anastasio R, Wroe S, Tuniz C et al (2013) Micro-biomechanics of the Kebara 2 hyoid and its implications for speech in Neanderthals. PLoS One 8:e82261. https://doi.org/10.1371/journal.pone.0082261
Damasio A (1994) Descartes's error: emotion, reason, and the human brain. Penguin Books, New York
Damasio A (1999) The feeling of what happens: body and emotion in the making of consciousness. Harcourt Brace and Co, San Diego, CA
Damasio A (2010) Self comes to mind: constructing the conscious brain. Pantheon, New York
Damasio A, Carvalho GB (2013) The nature of feelings: evolutionary and neurobiological origins. Nat Rev Neurosci 14:143–152. https://doi.org/10.1038/nrn3403
De Preester H (2018) Subjectivity as a sentient perspective and the role of interoception. In: Tsakiris M, De Preester H (eds) The interoceptive mind. From homeostasis to awareness. Oxford University Press, Oxford, England, pp 293–306
Denton D (2005) The primordial emotions. Oxford University Press, Oxford, England
Denton DA, McKinley MJ, Farrell M, Egan GF (2009) The role of primordial emotions in the evolutionary origin of consciousness. Conscious Cogn 18:500–514. https://doi.org/10.1016/j.concog.2008.06.009
Depue RA, Luciana M, Arbisi P et al (1994) Dopamine and the structure of personality: relation of agonist-induced dopamine activity to positive emotionality. J Pers Soc Psychol 67:485–498. https://doi.org/10.1037//0022-3514.67.3.485
Dunbar R (1998) Grooming, gossip, and the evolution of language, First Paperback edn. Harvard University Press, Cambridge, MA
Earl B (2014) The biological function of consciousness. Front Psychol 5:697. https://doi.org/10.3389/fpsyg.2014.00697
Edelman GM, Gally JA (2013) Reentry: a key mechanism for integration of brain function. Front Integr Neurosci 7:63. https://doi.org/10.3389/fnint.2013.00063
Edelman DB, Seth AK (2009) Animal consciousness: a synthetic approach. Trends Neurosci 32:476–484. https://doi.org/10.1016/j.tins.2009.05.008

Edelman GM, Tononi G (2000) A universe of consciousness. How matter becomes imagination. Basic Books, New York

Epstein S (2003) Cognitive-experiential self-theory of personality. In: Millon T, Lerner MJ (eds) Comprehensive handbook of psychology, Personality and social psychology, vol 5. Wiley, Hoboken, NJ, pp 159–184

Epstein S (2014) Cognitive-experiential theory. Oxford University Press, Oxford, England

Epstein S, Pacini R, Denes-Raj V, Heier H (1996) Individual differences in intuitive-experiential and analytical-rational thinking styles. J Pers Soc Psychol 71:390–405

Evans SD, Hughes IV, Gehling JG, Droser ML (2020) Discovery of the oldest bilaterian from the Ediacaran of South Australia. Proc Natl Acad Sci U S A 117:7845. https://doi.org/10.1073/pnas.2001045117

Farthing GW (1991) The psychology of consciousness, Facsimile edn. Pearson College Div, Englewood Cliffs, NJ

Feinberg TE, Mallatt JM (2016) The ancient origins of consciousness. How the brain created experience. The MIT Press, Cambridge, MA

Fowles DC (1987) Application of a behavioral theory of motivation to the concepts of anxiety and impulsivity. J Res Pers 21:417–435. https://doi.org/10.1016/0092-6566(87)90030-4

Friston K (2009) The free-energy principle: a rough guide to the brain? Trends Cogn Sci 13:293–301. https://doi.org/10.1016/j.tics.2009.04.005

Gallagher S (2000) Philosophical conceptions of the self: implications for cognitive science. Trends Cogn Sci 4:14–21

Gerkema MP, Davies WIL, Foster RG et al (2013) The nocturnal bottleneck and the evolution of activity patterns in mammals. Proc Biol Sci 280:20130508. https://doi.org/10.1098/rspb.2013.0508

Gray JA (1987) Perspectives on anxiety and impulsivity: a commentary. J Res Pers 21:493–509. https://doi.org/10.1016/0092-6566(87)90036-5

Green AM, Angelaki DE (2010) Multisensory integration: resolving sensory ambiguities to build novel representations. Curr Opin Neurobiol 20:353–360. https://doi.org/10.1016/j.conb.2010.04.009

Griffin GR (1992) Animal minds. Beyond cognition to consciousness. The University of Chicago Press, Chicago

Guterstam A, Gentile G, Ehrsson HH (2013) The invisible hand illusion: multisensory integration leads to the embodiment of a discrete volume of empty space. J Cogn Neurosci 25:1078–1099. https://doi.org/10.1162/jocn_a_00393

Humphrey N (1992) A history of the mind. Evolution and the birth of consciousness. Simon & Schuster, New York

Jaynes J (1977) The origin of consciousness in the breakdown of the bicameral mind. Houghton, Boston, MA

Joffily M, Coricelli G (2013) Emotional valence and the free-energy principle. PLoS Comput Biol 9:e1003094. https://doi.org/10.1371/journal.pcbi.1003094

Kahneman D (2011) Thinking, fast and slow. Farrar, Straus and Giroux, New York

Klein RG (2002) The dawn of human culture, 1st edn. Wiley, New York

Lang PJ (1995) The emotion probe. Studies of motivation and attention. Am Psychol 50:372–385. https://doi.org/10.1037//0003-066x.50.5.372

Lang PJ, Davis M (2006) Emotion, motivation, and the brain: reflex foundations in animal and human research. In: Anders S, Ende G, Junghofer M et al (eds) Progress in Brain Research. Elsevier, Amsterdam, pp 3–29

Lazarus RS (1991) Emotion and adaptation, 1st edn. Oxford University Press, Oxford, England

Leder D (2018) Inside insights: a phenomenology of interoception. In: Tsakiris M, De Preester H (eds) The interoceptive mind. From homeostasis to awareness. Oxford University Press, Oxford, England, pp 307–322

Liotti M, Brannan S, Egan G et al (2001) Brain responses associated with consciousness of breathlessness (air hunger). Proc Natl Acad Sci U S A 98:2035–2040. https://doi.org/10.1073/pnas.98.4.2035

Llinas RR (2001) I of the vortex: from neurons to self by Rodolfo R. Llinas. The MIT Press, Cambridge, MA

Lloyd D (1989) Simple minds. The MIT Press, Cambridge, MA

Lubinski D, Thompson T (1987) An animal model of the interpersonal communication of interoceptive (private) states. J Exp Anal Behav 48:1–15. https://doi.org/10.1901/jeab.1987.48-1

MacLarnon AM, Hewitt GP (1999) The evolution of human speech: the role of enhanced breathing control. Am J Phys Anthropol 109:341–363. https://doi.org/10.1002/(SICI)1096-8644(199907)109:3<341::AID-AJPA5>3.0.CO;2-2

Metzinger T (2009) The ego tunnel. The science of the mind and the myth of the self. Basic Books, New York

Naskar A, Cicurel R (2015) Homo: a brief history of consciousness, 1st edn. CreateSpace Independent Publishing Platform, Scotts Valley, CA

Panksepp J (1998) Affective neuroscience: the foundations of human and animal emotions, 1st edn. Oxford University Press, New York

Panksepp J (2005) Affective consciousness: core emotional feelings in animals and humans. Conscious Cogn 14:30–80. https://doi.org/10.1016/j.concog.2004.10.004

Panksepp J, Biven L (2012) The archaeology of mind. Neuroevolutionary origins of human emotions. W. W. Norton & Company, New York

Papagianni D, Morse MA (2015) Neanderthals rediscovered: how modern science is rewriting their story, Revised and Updated edn. Thames & Hudson, London

Parker A (2004) In the blink of an eye: how vision sparked the big bang of evolution, Reprint edn. Basic Books, New York

Pasqualotto A, Dumitru ML, Myachykov A (2016) Editorial: Multisensory integration: brain, body, and world. Front Psychol 6:2046. https://doi.org/10.3389/fpsyg.2015.02046

Pilowsky I (1967) Dimensions of hypochondriasis. Br J Psychiatry 113:89–93. https://doi.org/10.1192/bjp.113.494.89

Plotnick RE, Dornbos SQ, Chen J (2010) Information landscapes and sensory ecology of the Cambrian radiation. Paleobiology 36:303–317. https://doi.org/10.1666/08062.1

Prigatano GP, Johnson SC (2003) The three vectors of consciousness and their disturbances after brain injury. Neuropsychol Rehabil 13:13–29. https://doi.org/10.1080/09602010244000282

Russell JA (1980) A circumplex model of affect. J Pers Soc Psychol 39:1161–1178. https://doi.org/10.1037/h0077714

Russell JA (2003) Core affect and the psychological construction of emotion. Psychol Rev 110:145–172. https://doi.org/10.1037/0033-295X.110.1.145

Seligman MEP, Railton P, Baumeister RF, Sripada C (2013) Navigating into the future or driven by the past. Perspect Psychol Sci 8:119–141. https://doi.org/10.1177/1745691612474317

Shook NJ, Oosterhoff B, Terrizzi JA Jr, Brady KM (2017) "Dirty politics": the role of disgust sensitivity in voting. Transl Issues Psychol Sci 3:284–297. https://doi.org/10.1037/tps0000111

Smith H (1959) The biology of consciousness. In: Brooks C, Cranefield P (eds) The historical development of physiological thought. The Hafner Publishing Co., New York, pp 109–136

Stein BE, Stanford TR (2008) Multisensory integration: current issues from the perspective of the single neuron. Nat Rev Neurosci 9:255–266. https://doi.org/10.1038/nrn2331

Steiner JE, Glaser D, Hawilo ME, Berridge KC (2001) Comparative expression of hedonic impact: affective reactions to taste by human infants and other primates. Neurosci Biobehav Rev 25:53–74. https://doi.org/10.1016/s0149-7634(00)00051-8

Tobias PV (1987) The brain of Homo habilis: a new level of organization in cerebral evolution. J Hum Evol 16:741–761. https://doi.org/10.1016/0047-2484(87)90022-4

Troisi JR (2015) Sensation within the skin. ACS Chem Neurosci 6:209–210. https://doi.org/10.1021/cn500300a

Watson D, Clark LA (1984) Negative affectivity: the disposition to experience aversive emotional states. Psychol Bull 96:465–490

Watson D, Tellegen A (1985) Toward a consensual structure of mood. Psychol Bull 98:219–235. https://doi.org/10.1037//0033-2909.98.2.219

Watson D, Clark LA, Tellegen A (1988) Development and validation of brief measures of positive and negative affect: the PANAS scales. J Pers Soc Psychol 54:1063–1070. https://doi.org/10.1037/0022-3514.54.6.1063

Watson D, Wiese D, Vaidya J, Tellegen A (1999) The two general activation systems of affect: structural findings, evolutionary considerations, and psychobiological evidence. J Pers Soc Psychol 76:820–838. https://doi.org/10.1037/0022-3514.76.5.820

Wong HY (2017) On proprioception in action: multimodality versus deafferentation: a puzzle about the function of proprioception in action. Mind Lang 32:259–282. https://doi.org/10.1111/mila.12142

Chapter 3
From the Body to the Brain: The Biological Background

3.1 Introduction

The goal of the present chapter is twofold. First, it aims to briefly summarize our current knowledge on the processing of information originating from within the body. This can also be called the bottom-up aspect of interoception. In doing so, two major afferent pathways, i.e., the so-called homeostatic and the somatosensory pathway, will be followed from the periphery (receptors) to the respective cortical centers. In accordance with the topic of this monograph, special attention will be paid to the conscious aspects of processing. Integration within and between these major pathways will also be discussed. Finally, the hypothesized role of body-related information in the development of human self, which takes place at non-conscious levels, will be presented.

Before we start with the first topic, two points need to be clarified. First, the brain uses two major sources of information to build up its representation(s) about the actual state of the body. Beyond sensory receptors that pick up information and transmit it to the central nervous system via afferent nerves, various forms of humoral communication are also at work. Humoral communication uses special molecules (hormones, cytokines, etc.) that are transported via the blood and the cerebrospinal fluid and sensed by the brain. The original Sherringtonian categories (1906) refer only to neural transmission; if asked to choose one category of those, most researcher would regard humoral communication as visceroception. Whereas certain authors explicitly exclude hormones from the umbrella of interoception (Ádám 1998), others think that humoral communication might play a non-negligible role in interoception thus show more interest in these aspects (Damasio 1999, 2010; Cameron 2002, 2009; Critchley and Harrison 2013; Aziz and Ruffle 2018; Colombetti and Harrison 2018). Considering that the recent conceptualization of interoception encompasses emotions and decision-making which might rely more on the overall state of the body (using the terminology of Chap. 2, they are non-isomorphic), inclusion of humoral communication, that is necessarily more

F. Köteles, *Body Sensations*, https://doi.org/10.1007/978-3-030-63201-4_3

"holistic," appears logical to me. Humoral communication was associated with core subjectivity (sentience) exactly because of its non-topographic feature recently (De Preester 2018). Also, hormonal communication plays a crucial role in the stress reaction. Humoral signals (including molecules indicating the actual state of the immune system, such as cytokines) can impact the brain directly, for example, via the so-called classical (circumventricular organs), nonclassical (through the blood–brain barrier), and extraneuronal (microglial) pathways (Cottrell and Ferguson 2004; Fry and Ferguson 2007; Cameron 2009; Critchley and Harrison 2013; Mimee et al. 2013; Aziz and Ruffle 2018). Although humoral signals obviously get integrated with bottom-up neural information, research on the associations between higher aspects of interoception and humoral communication is in its very early phase.

Second, the existence of pure visceroception, i.e., information processing relying exclusively on visceroceptive source(s), can be questioned in many cases. This holds particularly true for near conscious and conscious body sensations. As it was explained in Chap. 2, the brain always tries to integrate all available sources of information, regardless of their origin. For example, the detection of heartbeats was assumed to rely partly on somatosensory (mechanoreceptors in the chest wall and the skin), and perhaps also auditory information beyond visceroception (e.g., changes in blood pressure measured by baroreceptors in the arteries) (Jones et al. 1987; Ádám 1998; Dworkin 2007). Recent empirical evidence shows that information of both visceral and somatosensory origin contributes to the sensation of heartbeats (Khalsa et al. 2009; Salamone et al. 2020); at least four potential sources of heartbeat-evoked potentials (visceroceptive: baroreceptors and receptors in the wall of the heart; somatosensory: mechanoreceptors in the chest wall; vascular-neuro coupling, i.e., changes in microcirculation of the brain) were described (Park and Blanke 2019). These sources obviously cannot be completely separated in a well-functioning organism (Brener and Ring 2016). The same holds true for respiratory resistance (Parshall et al. 2012). Also, receptors in the skin can be sensitive to distension of the colon (Hölzl et al. 1996; Ádám 1998), thus integration of sensations from multiple modalities appears possible here too. From a methodological point of view, this integration poses serious difficulties for visceroception-related research. Researchers interested in the basic steps of interoception tried to exclude somatosensory information via sophisticated designs and the inclusion of control conditions in certain cases (Ádám 1967, 1998; Hölzl et al. 1996; Dworkin 2007). Even if these attempts appear successful (one never can be sure that all factors are known and properly handled), the conclusions drawn from such experiments are limited in scope. In fact, these designs are more appropriate to understand elementary homeostatic mechanisms (e.g., reflex-based regulatory loops) than higher (e.g., behavioral) aspects of interoception. If we are interested in more complex levels of interoception, physiological and anatomical criteria are less usable than functional considerations (Dworkin 2007); in other words, the pathway which the information carries becomes less important than the information itself. We will see that the integration of information already starts at the level of the spinal cord; at higher levels of the system, more or less unitary representation(s) of the body develop. For example, the sensation of our legs or arms during physical exercise is a blend of proprioception (the actual position and movements of joints, tension of muscles, and stretching of tendons), visceroception (the metabolic state and fatigue of the muscles are measured by chemoreceptors), and skin-related information (temperature, stretching, etc.), not to speak of pain. In summary, the nervous system integrates potentially useful pieces of information coming from different sources, regardless of our more or less artificial categories. Integration does not mean that the specificity of information is totally lost; we will see that the ascending pathways and their centers process input in a somatotopic way in both major systems, i.e., spatial relationships within single modalities are preserved. Still, there are functional interconnections, and the consequences of these shortcuts (sometimes secondarily) also contribute to high-level representations of the body, including body sensations.

The two major pathways of the processing of internal information, i.e., the visceral or homeostatic (aka visceral afference) and the somatosensory (dubbed somatic afference), were recently put under the construct of bodily or somatovisceral afference (Berntson et al. 2018); this terminology neatly shows that these two pathways are overlapping and at the same time separate. The most important sources of body-related information are summarized in Table 3.1.

Table 3.1 The most important internal modalities and sources of body-related information and their categorization by the original Sherringtonian and the recent approach. Gustation, touch, and vision are not included, although they also have body-related aspects and become integrated with the interoceptive modalities

Sensory modality	Receptors and their localization	Sherringtonian category	Recent category (primary processing center)
Visceral modalities	Various mechano-, thermo-, and chemoreceptors in the visceral organs (incl. the cardiovascular system)	Interoception	Homeostatic afference (insula)
Temperature	Thermoreceptors in the skin	Exteroception	Homeostatic afference (insula)
Pain	Nociceptors in various tissues	Exteroception	Homeostatic afference (insula/area 3a)
Itch	Histamine-sensitive receptors in the skin	Exteroception	Homeostatic afference (insula)
The metabolic condition of muscles	Ergoreceptors in the striated muscles	Not applicable	Homeostatic afference (insula)
Affective touch	Certain mechanoreceptors in the deep layers of the skin	Exteroception	Homeostatic afference (insula)
Position of joints	Pacinian corpuscles in periarticular tissues, Ruffini end organs in joint capsules	Proprioception	Somatosensory afference (somatosensory cortices)
Tension of striated muscles	Muscle spindles in the muscles, Golgi tendon organs in the tendons	Proprioception	Somatosensory afference (somatosensory cortices)
Vestibular information	Semicircular canals and the otolithic organs in the inner ear	–	Somatosensory afference (?)

3.2 Homeostatic Afferent (Interoceptive) Pathway

In this section, relying primarily on the works and theory of A. D. (Bud) Craig (2002, 2003a, 2008, 2009, 2011, 2015), current knowledge on the central processing of interoceptive information will be briefly summarized. The term interoceptive in this context refers to a broader category than the classic visceroceptive. Beyond input coming from the viscera, pain, itch, thermal information, certain components of muscle-related information, and the so-called affective or sensual touch also belong here. All of these refer to the actual homeostatic state of the body and processed by the same centers, thus Craig applies the term homeostatic afferent pathway to this category. The pathway starts with sensory receptors in the periphery and ends in the dorsal part of the insula in the cortex. An important anatomical fact supporting this inclusive view is that all receptors belonging to this category are connected to sensory neurons via small diameter, slow conducting (so-called Aδ and C) fibers terminating at the most superficial layer(s) of the dorsal horn. In contrast, the somatosensory system is primarily based on large diameter, rapid (Aβ) fibers terminating at deeper layers.

Information is picked up and transformed into electric signals by specific receptors in various tissues in the body. Visceroceptors, localized in the internal organs (traditionally the cardiovascular, respiratory, gastrointestinal, and urogenital systems), are not concentrated in organized structures, they rather form a loose network (Ádám 1998). Generally, their structure is less sophisticated than that of several exteroceptors, and they are often polymodal, i.e., sensitive to more than one modality, such as pain, stretch, and temperature. It is important to note that these characteristics, along with the relatively low number of primary visceral sensory neurons in the dorsal root (Bielefeldt and Gebhart 2007), substantially limit the localizability and specificity of visceral sensations already at this level. A really comprehensive and still recently useful summary on visceroceptors, including morphological and physiological characteristics (e.g., pace of adaptation), was written by Airapetyants (1952). The major categories of visceroceptors are mechanoreceptors, chemoreceptors, and thermoreceptors. Mechanoreceptors are typically sensitive to stretch; they play a pivotal role in the short-term and long-term regulation of blood pressure, respiration, and bowel contractions. Chemoreceptors can be found in certain parts of the cardiovascular system (they are sensitive to the concentration of CO_2 and O_2 in the blood), participate in the regulation of digestion mainly by measuring the concentration of various constituents of food, such as glucose, amino acids, and acids and alkalis (pH), and also in various parts of the respiratory and urogenital systems. Finally, thermoreceptors are primarily localized in the gastrointestinal tract and in various regions of the cardiovascular system. Beyond the periphery, interoceptors can also be found in the brain, most importantly, in the circumventricular organs and in the hypothalamus (Critchley and Harrison 2013). A variety of body sensations evoked by the stimulation of peripheral visceroceptors are described, from dyspnea and weakness of the muscles to satiation of hunger and thirst, heartburn, and various intensities of visceral pain. An early summary of these sensations can be found in the volume edited by Cervero and Morrison (1986). It is important to emphasize that increasing the intensity of visceral stimulation quite quickly results in discomfort and pain; this is a characteristic that remarkably differs from exteroceptive stimulation that usually shows a considerably broader modality-specific range (Ádám 1998).

As mentioned, pain was originally considered an exteroceptive modality by Sherrington. Later, following the first formulation of the gate-control theory (Melzack and Wall 1965), this view was accepted, and pain was explained as a specific pattern of the somatosensory system (i.e., the tactile sense), a consequence of convergence at the level of the spinal cord that is modulated by the brain via descending pathways (Craig 2003b). The term convergence refers to the integration of input from various sources at various levels of the nervous system. Research on visceral pain and new anatomical advances challenge this view; pain is considered a "homeostatic emotion" today, belonging to interoception rather than exteroception (Craig 2003c). Sensory receptors responsive to possible harmful stimuli are often called nociceptors. In fact, this is a rather heuristic (or functional) category as a proportion of nociceptors can also be considered mechanoreceptor, chemoreceptor, or thermoreceptor firing only above a certain (relatively high) threshold, whereas others are sensitive to a wide range of stimulation, from non-painful to painful (Craig 2003b; Robinson and Gebhart 2008; Cameron 2009). Such receptors can be found in many areas of the body, from the skin to the internal organs and the muscles.

Another sense that was classified as exteroceptive by Sherrington but considered interoceptive or homeostatic today is the thermal modality. The thermal sense is intimately related to pain perception as both very cold (below approximately 15 °C) and very cold (above appr. 45 °C) temperatures are accompanied by pain and consequently an unpleasant feeling (Craig 2018). Normal (i.e., innocuous) temperatures are sensed by rapidly adapting free nerve endings of the so-called warm and cold fibers, most sensitive to temperatures of the 30–43 °C and 20–30 °C domain, respectively (Schepers and Ringkamp 2009). Thermoreceptors firing at noxious temperatures can also be considered nociceptors. The integration of thermal information takes place already in the spinal cord which explains the thermal grill illusion (Craig and Bushnell 1994; Craig 2015), originally described by Thunberg (1896). In this illusion, warm and cold bars are arranged in an alternating pattern; when touching the skin, this pattern evokes the feeling of burning hot. Thermosensation results in a wide range of conscious sensations including the painful ice cold and burning hot. Interestingly enough, the pleasantness or unpleasantness of the stimulation in the medium range depends on the current thermoregulatory status of the body (Cabanac et al. 1972): a cup of warm tee can evoke a pleasant feeling in winter and an avoidance reaction in summer. This neatly indicates that evaluation of the actual feeling is the result of a high-level integration process and tightly connected to the actual homeostatic condition of the organism.

The skeletal muscles contain the so-called ergoreceptors, dedicated to measure the mechanical and chemical aspects of muscle work (the latter category is also called metaboreceptors). Metaboreceptors measure the metabolic state of the tissue through H^+/lactate acid and $H_2PO_4^-$ concentration (Craig 2003a; Mense 2008). This information is used in the regulation of microcirculation of the muscles; thus, these receptors, although they are not localized in the viscera, belong more to visceroception than proprioception from a functional point of view (Dworkin 2007).

Finally, the existence of slowly conducting C-fibers originating in certain mechanoreceptors of the skin and terminating on the superficial layers of the dorsal horn of the spinal cord has been well-known decades ago (Bessou et al. 1971; Kumazawa and Perl 1977a, b; Sugiura et al. 1986; Vallbo et al. 1993). This system is activated by slow brushing, such as caress. It was originally supposed that these receptors might contribute to pain perception and light touch. As their activity is simultaneous with that of the other mechanoreceptors of the skin (see below), the understanding of their specific role posed a methodological challenge. The study of a patient with lacking tactile sensation conducted by large myelinated afferents revealed that light stimulation of the hairy but not the glabrous skin evokes a faint sensation accompanied by pleasant affect (Olausson et al. 2002). Cortically, this stimulation is represented in the insula but not in the somatosensory cortex (Olausson et al. 2002; Craig 2015). It is hypothesized that this sensation (called limbic, affective, or sensual touch), primarily occurring in affiliative social interactions, represents a security signal to the organism (Olausson et al. 2002; McGlone et al. 2007; Björnsdotter et al. 2010; Morrison et al. 2010, 2011b). This is in accordance with the well-described social role of grooming in primates (Dunbar 2010). In summary, this information can be considered a homeostatic signal, informing the individual about (social) security thus alleviating anxiety and the stress response (Craig 2015).

Neural signals from the interoceptors described above reach the central nervous system via three major pathways. The first (sympathetic) pathway belongs to the spinal cord (and the trigeminal nerve, which carries analogous information from the head). As mentioned, interoceptors are sensory neurons located in the ganglia of the dorsal root characterized by small diameter, slow conducting (so-called Aδ and C) fibers. Primary sensory neurons relay information to the dorsal horn of the spinal cord. Concerning homeostatic pathways, the most important layer is the so-called lamina I, the most superficial layer. Axons of lamina I afferent neurons ascend in the lateral spinothalamic tract (Craig 2003a). The ventral spinothalamic tract conveys information related to crude touch and movements originating in the deep dorsal horn (lamina V and VII); this pathway will be discussed later, in Sect. 3.4. The second (parasympathetic) pathway of visceroceptive information belongs to cranial nerves, i.e., the facial (VII), the glossopharyngeal (IX), and the vagus (X) nerve, whereas the rarely mentioned sacral dorsal root ganglia (S2–S4) represent the third, also parasympathetic pathway (Cameron 2002, 2009). These three pathways fuse in the brainstem. The nucleus of the solitary tract (NTS) is a sensory center representing the starting point of many homeostatic reflexes and has multiple connections with higher centers involved in autonomic regulation (see below) (Critchley and Harrison 2013). As Damasio (2010) points out, the collection of body-related information is complete at the level of the mid-pons, with the merging of the trigeminal nerve. Parasympathetic nerves, most importantly the vagus, contain more visceral afferent fibers than their sympathetic counterparts. The estimated ratio of parasympathetic to sympathetic afferent neurons is 3:1; this might or might not mean more specificity in the parasympathetic branch than in the sympathetic (Cameron 2002). Moreover, the sympathetic system contains more nociceptive afferents than the parasympathetic.

At this point, Craig made a crucial contribution to our current understanding of body sensations. It turned out that in primates, the most developed branch of mammals including humans, there is a direct, topographically organized input from lamina I spinal cord and analogous trigeminal neurons to two centers of the thalamus (called VMpo and VMb). Evidence regarding a similarly specific representation of the parasympathetic input also exists, however, is less marked (Strigo and (Bud) Craig 2016). In these nuclei, information from both the sympathetic and the parasympathetic branch of the vegetative nervous system is represented in a modality and localization specific (i.e., topographic or somatotopic) way (Craig 2003b; Brooks et al. 2005; Hua et al. 2005; Henderson et al. 2007; Mazzola et al. 2009; Björnsdotter et al. 2009). This means that the brain can identify the origin of the stimuli, at least to some extent, and can use this piece of information to integrate various modalities, e.g., when initiating reflexes. In accordance with this, stimulation within these nuclei evokes localized visceral, cooling, or pain sensations (Hassler and Riechert 1959; Davis et al. 1999; Dostrovsky 2000). In contrast, the thalamus receives indirect information only from the parabrachial nuclei (PB) of the brainstem in non-primate mammals (e.g., rats, and cats); this means that specificity is to a large extent lost (Craig 2003a). Basically the same holds true for the anterior cingulate cortex (aka limbic motor cortex), the major coordinator of homeostatic behavior; in primates, it receives topographic input from lamina I spinal neurons through the thalamus (MDvc), whereas only integrated input from the PB is present in non-primate mammals (Craig 2003a).

Finally, the two thalamic nuclei (VMpo and VMb) project to the posterior region of the insula, a part of the cortex located deeply in the lateral sulcus in primates (Craig 2003a, 2015). According to Craig, this structure can be considered the interoceptive sensory cortex in humans, a homeostatic analogue of the somatosensory cortex.

Lateralization also plays a role here. Whereas activation in the posterior insulae reflects the spatial features of the interoceptive stimuli, further integration steps occur mainly in the right insula. Former EEG studies reported a right hemisphere dominance in tasks involving interoception (more specifically, heartbeat perception) (Davidson et al. 1981; Montgomery and Jones 1984; Hantas et al. 1984; Katkin and Reed 1988; Weisz et al. 1990, 1994; Weisz and Ádám 1993), and the right (non-dominant) anterior insula is considered the primary neural substrate of the meta-representation of the body that is associated with subjective awareness (Craig 2003a, c, 2009; Critchley et al. 2004). However, lateralization with respect to positive/negative affect also described (see below). It is also worth noting that the mid-insula receives information from non-interoceptive modalities, e.g., from the visual, auditory, and somatosensory system, too (Pressnitzer and Hupé 2006; Craig 2009).

Conscious body feelings are generated in the medium part of the right insular cortex, whereas emotion-related aspects of body feelings (i.e., sentience) are located in the anterior part (Craig 2009; Zaki et al. 2012; Terasawa et al. 2013; Adolfi et al. 2017). Diffuse feelings from the body, for example, thirst and hunger, are accompanied by the activation of the medium part of the insula (Craig 2010). Stimulation of the insula evokes well-localized painful and non-painful visceral and somatic sensations (Penfield and Jasper 1954; Penfield and Faulk 1955; Ostrowsky et al. 2002; Mazzola et al. 2011), whereas lesions in this area cause a decrease in pain and temperature sensation (Schmahmann and Leifer 1992; Greenspan et al. 1999). Moreover, the insula becomes activated during a number of conditions with homeostatic relevance such as itch, physical exercise, gastric distension, changes in blood pressure, blood glucose, and osmolarity (i.e., hunger and thirst), and air hunger (Opsommer et al. 2001; Williamson et al. 2001; Henderson et al. 2002, 2007; Craig 2003a; Stephan et al. 2003). Sensual or affective touch also activates this region but not the classic somatosensory areas (Olausson et al. 2002). From a predictive processing point of view, the anterior insula represents a high-level comparator or error module (Paulus and Stein 2006; Gray et al. 2007), whereas the anterior cingulate and the orbitofrontal cortex are assumed to act as prediction units (more details come below) (Seth et al. 2012; Barrett and Simmons 2015).

This concept substantially differs from the classical view, that, based on theoretical considerations and electrophysiological findings, assumed the convergence of somatosensory and viscerosensory information in the major somatosensory areas (S1, S2) of the cortex and mostly ignored the insula (Korn et al. 1966; Ádám 1998). This does not mean, of course, that somatosensory regions do not play a role in interoception (Kern and Shaker 2002; Derbyshire 2003). In the new view, integration of pain-related information occurs secondarily, in the fundus of the central sulcus (area 3a) (Craig 2003a, 2015). Similarly, heartbeat-evoked potentials were identified in the primary somatosensory cortex recently (Kern et al. 2013).

What was the driving force of the development of a direct pathway for homeostatic sensations? According to molecular and taxonomic evidence, primates appeared in the Late Cretaceous period (appr. 85–55 mya) (Springer et al. 2003; Miller et al. 2005; Perelman et al. 2011). They are characterized by large brains relative to body size, in part due to their primary reliance on the visual modality. Detailed processing of visual information (including colors) requires much more brain capacity and energy than other senses. Hominids have disproportionately large brains even within primates; a human brain can use up to 25% of the resting energy budget of the organism (Leonard et al. 2003). As the brain is particularly sensitive to violations of homeostatic balance (for example, with respect to glucose, oxygen, and temperature), detailed information about the current homeostatic state of the body which can initiate specific behavioral steps if necessary might have been clearly adaptive. From this point of view, body feelings such as hunger, thirst, itch, visceral distension, muscle ache, hot and cold, pain, and sensual touch can be considered homeostatic emotions with clear evaluative (pleasant versus unpleasant) and motivational aspects (Craig 2003b, c). This can be the evolutionary origin of the tenet of evaluation (Chap. 1). In contrast, non-primates do not have such distinct body feelings (Craig 2003a); their behavior is organized by less specific mechanisms, i.e., homeostatic integration and arousal localized in the brainstem and the hypothalamus (Norrsell and Craig 1999; Swanson 2000; Craig 2002, 2003a). Beyond its homeostatic significance, the appearance of sophisticated body feelings in the anterior insula might have contributed to the development of human self-awareness (Craig 2009). This region of the insula plays a crucial role in the so-called salience network of the brain (Seeley et al. 2007), which identifies salient stimuli and marks them for additional processing (Menon and Uddin 2010; Uddin et al. 2014). Salient information contributes to higher levels of cognitive processing, such as the regulation of attention, reasoning (including the development of beliefs), explicit motivation, and memory (Balleine and Dickinson 2000; Dolan 2002; Berridge and Robinson 2003; Chun and Turk-Browne 2007; Namkung et al. 2017). Overall, the role of the insula in homeostatic and self-relevant processes, including psychopathology (Namkung et al. 2017), neatly supports the tenet of relevance, i.e., the impact of body sensations on various aspects of human functioning.

Recently, a new model called Embodied Predictive Interoception Coding (EPIC) was proposed, which attempts to put cortical neuroanatomy (based on tracer studies in macaque brain) and the predictive processing framework together (Barrett and Simmons 2015). According to the model, visceromotor regions of the brain (the cingulate, the posterior ventral medial prefrontal and orbitofrontal cortices, and the ventral portions of the anterior insula) send visceromotor predictions (if you wish: high-level commands) to the hypothalamus and the lower homeostatic nuclei of the brainstem (e.g., the periaqueductal gray, PAG) and simultaneously predict changes (in form of efference copies of interoceptive predictions) evoked by those visceromotor commands. Calculation of the discrepancy (error) between expected and actual input takes place at the mid and posterior insula, which propagates the error signal back to the visceromotor regions. Finally, predictions of visceromotor regions are relatively insensitive to prediction errors; in other words, their major way to reduce prediction errors is action, not perception. Basically, this reflects the fact that there are strict biologically determined ("built-in") homeostatic ranges for many internal variables (for more details, see Chap. 8).

3.3 Somatosensory Pathways

Unlike the interoceptive pathway, the classic somatosensory pathway is relatively well-described. It consists of two receptor systems, one for the tactile sensation and another one for proprioception.

The tactile sensation is based on four major types of mechanoreceptors located in the superficial and deep layers of the skin. Rapidly adapting receptors (Meissner corpuscles and Pacinian corpuscles) detect textures and light touch, whereas their slowly adapting counterparts (Ruffini corpuscles and Merkel nerve endings) are mainly sensitive to pressure and stretch (Winkelmann 1986; McGlone et al. 2007; Owens and Lumpkin 2014). There are further receptors around the hair follicles that assess the position changes of the hair. The tactile modality, one of the classic exteroceptive modalities by Sherrington, is characterized by well-defined and well-localized body sensations, which typically refer to external objects. In certain cases (e.g., the sensation of the clothes on different regions of the body), this also yields information on the actual state of the body. Spatial resolution of tactile receptors shows marked differences between body areas, being the highest in the fingers and lips.

The term proprioception was proposed by Sherrington for the perception of joint and body movement as well as for the perception of the position of the body or body segments, in space (Ribeiro and Oliveira 2007). Another term is kinesthesia or kinesthesis, which originally referred to the conscious sensations only (Bastian 1887). Recently, these terms are defined differently by experts of various fields; some consider them synonyms, whereas others regard proprioception as the broadest term that includes kinesthesia (Swanik et al. 2004; Ogard 2011; Ribeiro and Oliveira 2011; Han et al. 2016; Jha et al. 2017). The major functions of the proprioceptive system are protecting joints and muscles against excessive movements and providing a sensory basis for static postures and movements (Knoop et al. 2011). Most importantly to our topic, although automatic processing plays a pivotal role, a proportion of proprioceptive input can reach consciousness in form of body sensations (Ogard 2011; Ribeiro and Oliveira 2011).

Receptors belonging to this sense are located in the tendons (Golgi tendon organs), muscles (muscle spindles), periarticular connective tissues (Pacinian corpuscles), and joint capsules (Ruffini end organs) (Batson 2009; Jha et al. 2017). Free nerve endings detecting the approaching or reaching of potentially dangerous positions (basically nociceptors) and certain mechanoreceptors in the skin also contribute to various aspects of proprioception, for example, the detection of joint position (Lephart et al. 1998; Edin 2001; Ogard 2011). Proprioceptors transform mechanical information (including acceleration and deceleration) into frequency-modulated signals that reach the spinal cord via large diameter, quick Aβ afferent fibers. Articular rapidly adapting mechanoreceptors are responsible to the sensation of joint motion, whereas slowly adapting articular and muscular receptors are sensitive to static positions (Lephart et al. 1998). In the complete absence of changes for a period of time (tens of minutes), even slowly adapting receptors decrease their discharge which leads to the loss of conscious sensation of the locomotor system and its controllability—a condition everybody might be familiar with.

Input from receptors belonging to the somatosensory system enters the dorsal root ganglia and ascends to the brain through the dorsal columns of the spinal cord. As we will see later, this input is already processed at the level of the spinal cord, for example, in the form of reflexes. The neural substrate of this low-level processing is represented by collateral axons that terminate in the deep layers of the dorsal horn. The ascending pathway of the somatosensory system (cuneate and gracile fasciculi; first-order fibers) terminates on neurons of the cuneate and gracile nuclei of the medulla oblongata, respectively. The second-order fibers decussate and terminate in the thalamus (ventral posterolateral nucleus, VPL); this is called the lemniscus medialis pathway. Analogous fibers from the head (via the trigeminal nerve) also terminate near here (ventral posteromedial nucleus, VPM). Input from proprioceptors is processed in an overlying region. Finally, the third-order fibers of the aforementioned nuclei (VPL and VPM) reach the somatosensory regions (S1 area 3b and S2) of the cortex, whereas proprioceptive input is processed in S1 area 3a (Mai and Paxinos 2011; Craig 2015). The entire pathway, from the spinal cord to the cortex, shows somatotopic organization. The most famous representative of this organization is the somatotopic map of cutaneous mechanoreceptors in the area 3b originally described by Wilder Penfield (Penfield and Boldrey 1937); it is often called sensory homunculus. The size of areas representing various areas of the body reflects the density of local innervation; this is why the hands and lips of the homunculus are disproportionately and surreally large.

Information from Golgi tendon organs and muscle spindles also reaches the cerebellum via the spinocerebellar tract, becomes integrated with vestibular information originating in the semicircular canals and the otolithic organs of the inner ear. The former is about rotary movements, whereas the latter is sensitive to linear acceleration. Vestibular information, integrated with proprioceptive and visual information, is the basis of equilibrioception or the sense of balance. This sensation is partly conscious; microstimulation of the cortical part of the system causes the sensation of rotation and translation (Kahane et al. 2003). However, the information is dominantly used in a non-conscious way to control eye movements and maintain posture (Guskiewicz and Perrin 1996). The vestibular system is usually not considered part of the somatosensory system or the other major sensory categories (Lephart et al. 1998).

From the predictive processing perspective, the proprioceptive system and the motor system (input and output) cannot be separated. In this special case, (proprioceptive) predictions are sent from the cortex to the spinal cord, and movements are regarded as the manifestations of predictions (Adams et al. 2013). To make sure that action happens instead of perception (i.e., a change in the prediction), sensory input is simultaneously attenuated (Cisek and Kalaska 2010; Clark 2015). This approach is conceptually equivalent with the classic model of motor control, which uses the term efference copy (or corollary discharge) for the anticipatory element of the regulation (see below).

3.4 Integration

As mentioned in Chap. 2, integration of information from different environmental stimuli is an ancient evolutionary strategy (Stein and Stanford 2008); it reaches its most sophisticated state with the appearance and development of the nervous system. In higher animals, integration is usually discussed with respect to external stimuli and localized in various multisensory areas of the cortex (Ghazanfar and Schroeder 2006; Blanke 2012; Blanke et al. 2015). Concerning interoception in the broader sense, the integration of afferent information both within and between the two major pathways described above occurs at multiple levels of processing. In my view, low-level (i.e., subcortical) integration represents an often overlooked factor in the development of the final representation of the body. Although reflexes localized in the spinal cord and the brainstem represent automatic ("non-conscious") responses, their consequences, such as limb movements, cardiorespiratory changes, or the defense reaction of the abdominal muscles reach consciousness and contribute to the complex representation of the body. Beyond simple reflexes, however, integrated representations of the body are necessary to organize more sophisticated responses. For our purposes, a brief presentation of the most pivotal levels of integration is sufficient. Although we will follow a bottom-up path in doing this, it should be kept in mind that the integration process always involves back and forth communication between lower- and higher-level centers.

The first major structure of the central nervous system where integration occurs is the spinal cord (Critchley and Harrison 2013). For example, the so-called wide-dynamic-range (WDR) cells in lamina V of the dorsal horn receive input from the skin (including temperature), viscera, and muscles. Their receptive fields are modality-ambiguous and musculotopically organized (Schouenborg et al. 1995); they often receive input from a large part of a limb and might play a role in flexor withdrawal mechanisms and sensorimotor integration (Lundberg et al. 1987; Craig 2003b). The axons of lamina V cells terminate in other thalamic nuclei (VP, VPI, and the motor part of the thalamus) as lamina I axons (Craig 2003b). The fact that stimulation of VP evokes localized non-pain-type sensations (e.g., tingling) in healthy humans also indicates some sort of integration.

Somato-visceral integration at the level of the brainstem is also considerable (Sato 1997; Yates and Stocker 1998). Various centers and areas of the brainstem receive multiple inputs from the spinal cord beyond the spinothalamic pathways. Such ascending pathways are the spinoreticular, spinotectal (spinomesencephalic), and the spinosolitaric tracts (Cameron 2002). Nuclei playing a pivotal role in the integration at this level are the NTS, PB, and PAG (Damasio and Carvalho 2013). Gustatory information from cranial nerves (VII, IX, and X), traditionally regarded as exteroceptive but obviously of homeostatic relevance, also becomes integrated at this low level (Craig 2003b, 2009; Avery et al. 2015; Berntson et al. 2018). Homeostatic regulation (including pain regulation) through descending effects and somato-visceral and viscero-visceral reflex responses takes place as the result of complex interactions among these sensory and control centers (Jänig 1996; Hölzl et al. 1999; Cameron 2002; Craig 2003a, 2011, 2018). As described in Chap. 2, some prominent authors (Panksepp 1998a; Panksepp and Northoff 2009; Damasio 2010) hold the view that an elementary form of consciousness, i.e., an integrated and organism-centered (subjective) representation of the environment and the body, including emotion-like responses and feelings, called core self, can build up upon these structures. Damasio also emphasizes the importance of the superior colliculus (Damasio 2010; Damasio and Carvalho 2013) where visual information becomes integrated with other modalities (Wallace et al. 1998). The possibility that an elementary form of multimodal integration and consciousness can develop in such lower brain structures is supported by the behavior of patients with complete destruction of the insula due to herpes simplex encephalitis or that of all telencephalic structures (i.e., cerebral cortex, thalamus, and the basal ganglia) caused by a major stoke in utero (Damasio 2010).

As for higher levels of the brain but still under the cortex, major centers of integration are the thalamus and the amygdala (Brüggemann et al. 1998; Ghazanfar and Schroeder 2006; Stein and Stanford 2008; Pasqualotto et al. 2016). Finally, the somatosensory and homeostatic pathway become integrated in the fundus of the central sulcus (area 3a) and, along with exteroceptive modalities such as vision and audition, in the mid-insula (Craig 2003b, 2015). Whereas the former is mainly involved in viscero-motor (e.g., pain-related) reflexes (Craig 2011), the latter represents the higher-level integration state of body-related information (Craig 2015; Ceunen et al. 2016). Gustatory information also activates this area (Saper 2002; Berntson et al. 2018), whereas the primary sensory cortex of the other chemical sense, olfaction, is located in the anterior insula (Rolls 2015).

As described in Chap. 2, integration of information from different exteroceptive sensory modalities takes place in higher organisms and requires a sophisticated brain. Integration also occurs for interoceptive modalities, which leads to a more complex representation of the body. However, integration, along with predictive processing, also opens the gate for body hallucinations; in consequence, sensed and perceived states of the body can considerably dissociate from actual states (tenet of dissociation) at higher levels of processing, when certain modalities are taken into consideration with higher relative weight and/or top-down factors (priors) play a substantial role.

It is worth noting that the activation of representations of the body are usually accompanied by the activation of control structures at the same level (see the EPIC model above for the cortical part). In the brainstem, sensory structures such as the NTS are tightly connected and often show co-activation with local and higher homeostatic motor centers (the PAG and the hypothalamus, respectively) which guide goal-directed autonomic, neuroendocrine, and behavioral activity (Craig 2003a, 2011). Similarly, insular activation is typically associated with activation of the anterior cingulate cortex, and the somatosensory areas in the cortex are tightly connected to the adjacent motor areas. In other words, input and output are intertwined as the brain maps the actual state of the body (input), initiates changes to maintain homeostasis (output), and receives instant feedback from the changes (input) which might result in further regulatory steps (output). As we will see below, this feature might play an important role in the development of the self.

3.5 Affective-Evaluative Aspects

As mentioned in Chap. 2, positive and negative activation or affect (in body terms: pleasure and pain/displeasure) represent the two major behavioral options, approach and avoidance. Any external cues that are judged important for the organism for whatever reason are automatically categorized along this basic option first (Lazarus 1991; Watson et al. 1999). The position of body sensations, however, is a bit more difficult. On the one hand, at least according to certain theories (see Chap. 8), they are indicators of the evaluation of exteroceptive stimuli. On the other hand, they are also associated with pleasure or displeasure, i.e., subject of evaluation (although this evaluation is often genetically determined) (Damasio 2010; Damasio and Carvalho 2013). I will consider the distal consequences of this unique position later, in the sub-chapter on the birth of the self. Now the two aspects will be presented separately.

Concerning the first aspect and starting with a brief summary of Chap. 2, the primary compass for navigation on the trajectories of internal and external environments for all living organisms is survival (perhaps the only exception to this rule is transmission of genes to the next generation, i.e., reproduction in the broader sense). Survival is possible only if the internal milieu is kept within a manageable and sustainable range; this is called homeostasis (or allostasis). A plethora of automatic mechanisms dedicated to regulate one or another parameter of the internal milieu were evolved by natural selection. Elementary mechanisms operate locally in a more or less blind manner, i.e., without taking into consideration the overall state of the organism. Higher-level regulation, for example, behavioral regulation, is much more complex as it necessarily takes into account more and more aspects of homeostasis. Feelings, such as motives and affect, are subjective experiences that characterize certain aspects of the actual homeostatic state in order to make them comparable (Cabanac 1971) and also enable the organism to associate them with novel cues (learning) (Panksepp and Biven 2012). It necessarily follows from this line of reasoning that the primary or primordial feelings are closely related to the homeostasis of the body (Damasio 2010; Craig 2010; Damasio and Carvalho 2013). The upper part of the brainstem (pons, midbrain) can be activated by emotional stimuli (Damasio et al. 2000); thus, it is reasonable to assume that primordial feelings are generated at the level of the upper brainstem in mammals (Panksepp 1998a; Damasio 2010; Panksepp and Biven 2012). More specifically, the PAG can be considered the first organized integration center of avoidance/approach behavior (Bandler et al. 2000; Craig 2011), as well as perhaps the first center where affective feelings develop (Panksepp 1998b, 2005; Panksepp and Northoff 2009; Damasio and Carvalho 2013). When an affective feeling appears in response to a stimulus, it indicates the automatic evaluation of the stimulus by the organism. This is a systemic, i.e., non-isomorphic, reaction. However, in primates, where certain internal modalities are individually sensed (Craig 2003a, 2015), those sensations and their specific patterns can theoretically reflect a more fine-tuned evaluation as proposed by peripheral theories of emotion (Chap. 8). Paradoxically, "gut feelings" are rather human than animal characteristics.

Let us have a look at the second aspect of body sensations, i.e., their appraisal. Pain, the ultimate body-related emotion, is *per definitionem* unpleasant. The majority of "classic" visceral sensations, for example, the urge to urinate and defecate, are also unpleasant (Ádám 1998; Cameron 2002). Increasing stimulation of the viscera causes discomfort and aversive behavior, then pain (Bárdos 1989). Individual symptoms associated with changes of internal parameters that are often considered non-sensable, such as blood pressure and blood sugar, are also dominantly unpleasant (Pennebaker et al. 1981, 1982; Pennebaker 1982) and so is the over-aroused state of the organism (Marshall and Zimbardo 1979). Muscle ache due to overexertion and work in the anaerobic domain are unpleasant and painful. Temperature sensation beyond the homeostatic boundaries are also accompanied by pain. Obviously, these homeostatic sensations develop and reach consciousness only in emergency situations, when the homeostatic limits are about to reach and behavioral steps need to be taken. However, temperature in the non-noxious range is evaluated relative to the core temperature of the body—the same warmth can be pleasant or unpleasant depending on the homeostatic state (Craig 2015; Strigo and (Bud) Craig 2016).

Sensations with built-in positive evaluation are also known. For example, affective (aka limbic or sensual) touch evokes a very pleasant feeling (Björnsdotter et al. 2010; Morrison et al. 2011a). This connection can be explained by the uttermost importance of belonging to a group and being appreciated by the members of the group for primates. For primates, being alone physically as well as socially (loneliness) for a longer period of time is dangerous and stress-evoking. Jos Brosschot and his colleagues go even further and assume that the default setting of the organism is stress which can be ameliorated by various security signals (Brosschot et al. 2016, 2017, 2018). Whatever the case may be, the social aspect represents a pivotal part of the homeostasis of primates and other animals living in groups thus its representation among the homeostatic signals is logical. Another rarely mentioned example for pleasant body feelings is mild to moderate physical activity. Exercise in the aerobic domain, where the metabolic demands of the muscles are sustainably met, causes an increase in positive affect (Ekkekakis and Petruzzello 1999; Ekkekakis et al. 2011, 2020). However, once the organism reaches the so-called anaerobic threshold, above which the oxygen supply is insufficient for long-term activity and a lactate-based metabolism kicks in, the affective state deteriorates quickly (Ekkekakis and Acevedo 2006; Ekkekakis et al. 2020). The twofold homeostatic background of this regulation is obvious. First, physical activity is associated with adaptive behaviors, such as playing (Panksepp and Biven 2012) and exploration (Raichlen and Alexander 2017), thus it should be motivated and reinforced by a positive feeling state to a (literally) healthy degree, i.e., within the aerobic domain (Cabanac 2006). However, the individual should be demotivated after reaching the possibly dangerous anaerobic range. Finally, sexual satisfaction is also a pleasant body sensation; however, its salience can be best understood from the viewpoint of the survival of the species instead of survival of the individual (Allen and Tsakiris 2018).

The majority of sensations from the proprioceptive system are more neutral. For example, the accuracy of the perception of the actual angle of the elbow joint and the tension of arm flexor muscles not associated with positive and negative affect (Horváth et al. 2019, 2021). Certain movements, however, can be associated with positive or negative affect (Cacioppo et al. 1993; Neumann and Strack 2000). Also, perceived physical abilities and inabilities can trigger secondary positive and negative evaluation (see below).

The existence of cortical lateralization with respect to the actual effective state is well-known. Positive affect or activation is reflected by the relative activation of the left side of the frontal cortex, whereas negative affect or activation is associated with right-side activation (Davidson 1992; Bruder et al. 1997; Tomarken and Keener 1998; Watson et al. 1999). In a similar vein, positive and negative feelings are associated with the anterior part of the left and right insulae, respectively (Craig 2005, 2009). From a physiological point of view, this can reflect parasympathetic (i.e., energy-enriching) and sympathetic (i.e., energy-consuming) activation and emotions (Strigo and (Bud) Craig 2016). However, the two insulae are in continuous communication and interaction with each other; thus, the overall emotional state of the organism is a result of the mutual impact of two opponent systems, just like the regulation of many body functions (Craig 2010; Strigo and (Bud) Craig 2016).

In summary, the association between body sensations and affective evaluation is twofold: they can serve as indicators of evaluation of other (typically external) cues, and they are often subject to evaluation.

3.6 The Origins of the Self

In this section, a number of neurocognitive theories about the self will be presented. It is important to note that the vast majority of the described processes operate at non-conscious levels; in other words, these processes rely on interoceptive information (in the broader sense) but not on conscious body sensations. Although this topic is only marginally related to the major focus of the book, I did not want to completely skip it. Still, we discuss them in less detail. This is why the following brief overview is necessarily and intentionally sketchy, just presenting the reader with the most fundamental ideas and recent interoception-related advances.

We will primarily focus on the so-called knower or pure Ego aspect of the self (self as subject) and neglect the object or known aspect. Concerning the latter, also called material me or material self, we already saw that it is represented at several centers of the brain, most importantly in the right anterior insula (Craig 2010). The former, however, is much more difficult to understand (Christoff et al. 2011); William James (1890) even claimed that it belongs rather to the domain of philosophy and religion than that of science. We do not think so today; however, James' statement clearly indicates that this is a problematic area of research, balancing on the boundaries of philosophy, psychology, and neuroscience (Christoff et al. 2011).

To make a rough distinction, theories about the sense of the self (more precisely, about the minimal requirements of sensation of self as observer and doer, or the first-person subjective experience of ownership) emphasize two or three different aspects, namely agency, the feeling of presence (i.e., the sense of being there), and the sense of ownership (Georgieff and Jeannerod 1998; Gallagher 2000; Seth et al. 2012; Seth 2013). These aspects are in continuous interaction with each other, thus difficult to separate under everyday circumstances. Agency, i.e., the sensation of oneself as the cause of an event, can refer to external as well as internal events. Concerning internal events, the intentionally controllable motor system is perfectly appropriate to distinguish the body from the environment (Christoff et al. 2011). Motor control is based on a so-called forward mechanism; when initiating a movement, the sensory areas are informed and prepared to the change well before it takes place (Wolpert and Flanagan 2001; Wolpert et al. 2003). In other words, the system builds up some sort of (of course, non-conscious) expectation; this is called efference copy or corollary discharge (Sommer and Wurtz 2008). It anticipates the effects of the motor command and compares it to the sensory input (the predictive processing-based equivalent of this model was already described above). If the actual input corresponds to the expected input, it can be concluded that the action was caused by the individual. Based on these considerations, the voluntarily controllable motor system appears to be the best candidate for the understanding of agency (Seth et al. 2012). At a non-conscious level, a new quality can arise from the aforementioned loop that can be called proprioceptive awareness (Gallagher 2003). Proprioceptive awareness is assumed to be non-reflective in its primary form (Gallagher 2005), i.e., the body is rather a subject than an object of perception. It is primarily based on self-induced movements; if we combine it with an ecological framework, it leads to pre-reflective self-consciousness (Gallagher 2005). A similar term is bodily self-consciousness, the pre-reflexive awareness of the body and its functioning relying on sensorimotor integration (Lenggenhager et al. 2007; Aspell et al. 2012). These processes are also at work in higher animals, thus can contribute to a primordial or elementary form of core selfhood (Edelman and Seth 2009).

Simply being able to move something (say, a tool) is clearly not enough to perceive it as belonging to our self (sense of ownership) (de Vignemont 2018a). Sense of ownership can also be the result of multimodal integration of body-related information (Tsakiris 2010). In the case of the rubber hand illusion, an experimental situation where body ownership was replaced from the real hand to a fake limb (Botvinick and Cohen 1998), visual and tactile information gets integrated in the absence of voluntary movements. This exteroception-based model was refined recently with the inclusion of interoceptive modalities (Apps and Tsakiris 2014; Tsakiris 2017; Allen and Tsakiris 2018). For example, the more accurate sensation of heartbeats was associated with less malleability to the rubber hand illusion in one study (Tsakiris et al. 2011), whereas a negative association was described in another (Suzuki et al. 2013). No association was reported in a third study though (Crucianelli et al. 2017). In our own study (Horváth et al. 2020), sense of ownership was also not associated with cardioceptive accuracy. However, those with lower levels of proprioceptive accuracy with respect to the elbow joint (calculated as the variance of individual measurements) reported higher levels of ownership of the rubber hand. This suggests that proprioceptive information might also impact the rubber hand illusion.

As described in Chap. 2, the protoself proposed by Damasio (2010) develops upon elementary or primordial homeostatic sensations. In contrast to external images, the variability of these sensations is limited as homeostatic regulation keeps them between the physiologically acceptable range. In consequence, these sensations are both stable and to some extent variable, resulting in the perception of a more or less stable self (Damasio 1994, 2003, 2010). It is important to note here that temporal stability and at the same time change were considered a fundamental characteristic of self-consciousness by James; he spoke of the stream of consciousness (James 1890). At the level of the core self in the model of Damasio, the protoself reacts to various objects, thus perception of those objects necessarily has a subjective element or self-reference. If those objects belong to the organism itself (i.e., they are felt by the homeostatic or proprioceptive systems), however, a second loop develops: the body reacts to its own changes that in turn causes further changes and so forth. Moreover, we can remember at this point that homeostatic sensory centers are tightly connected to motor centers (see the EPIC model above); thus, some kind of agency (i.e., a close online interaction between input and output) is possible not only in the motor system but also in the autonomic nervous system (Damasio 2010; Christoff et al. 2011; Seth et al. 2012; Craig 2015; Barrett and Simmons 2015). I believe that this is the point from which non-conscious (i.e., not language-based) self-reflection and the primordial sense of ownership can be best explained. Pure sensorimotor integration, in which only the proprioceptive system is involved, is not sufficient because of the lack of (subjective) evaluation. The evaluation of information always takes a homeostatic perspective, the main concern of which are survival and protection of the organism. Thus, reactions to body changes necessarily show a protective direction, i.e., an attempt to avoid harm and reaching the dangerous extremities of the regulated internal states. These principles were also reconsidered recently from the perspective of the predictive processing framework (Allen and Tsakiris 2018).

An essential component of body ownership is the fear of being hurt (de Vignemont 2018a, b). In popular demonstrations of the rubber hand illusion, trying to hit the fake limb with a hammer elicits a withdrawal reaction (it would be interesting to see once what happens if one attempts to hit the actual hand with the hammer). On the phenomenological level, even approaching the boundaries of the safe range of operation can be indicated by negative affect and pain, a homeostatic emotion that is tightly related to body (Craig 2003c). This is in complete accordance with the idea that negative affect has a marked predictive or anticipating quality, i.e., scanning for and awareness of potential threats (Watson et al. 1999). The organism's deeply rooted "intention" to protect itself and avoid harm results in the development of an integrated online (hot) representation of the body with marked affective ("narcissistic") tones (de Vignemont 2018a, b); this is called bodyguard by de Vignemont and assumed to be the origin of body ownership. The original idea of the neuromatrix (Melzack 1989, 1990) as well as its more recent conceptualization (Legrain et al. 2011) represent very similar approaches.

Evaluation is part and parcel of the descriptive functioning of the self too; just like external stimuli, we constantly evaluate our body and its abilities. Body image, as defined by Gallagher (2005), encompasses three components, body percept (perception and representation of the body), body concept (conceptual knowledge about the body), and body affect (evaluation of the body). This "secondary" evaluation can refer to our bodily appearance (Cash and Szymanski 1995; Fredrickson and Roberts 1997; Emanuelsen et al. 2015), physical abilities (Miller et al. 1981; Sonstroem and Morgan 1989; Sonstroem 1997), or the health state of our body (Köteles and Witthöft 2017).

The existence of expectations of future body states or, in the motor system, efference copies, has further consequences. For example, combined with the integrated representation of the body, they enable the brain to simulate the outcomes of various efferent commands instead of actually performing the actions. This simulation process is called *as-if loop* by Antonio Damasio (1994, 2010); compared to the *body loop*, the actual manifestation of the efferent-afferent cycle, it spares time and energy for the organism. Damasio emphasizes (2010) that the result of an as-if loop style processing is less marked than that of the body loop. Still, it explains very well how and why possible but actually non-existing states of the body can be represented and felt in the brain. Damasio also proposes a third way, i.e., the modification of the input, which can lead to "bodily hallucinations" (this appears to be very similar to the tenets of the predictive processing model).

Two new approaches to basic self-related processes were also developed recently. First, the concept of minimal phenomenal selfhood was coined to describe the most elementary form of self-consciousness (Blanke and Metzinger 2009). Based on theoretical considerations and experiments on body illusions, such as the full-body illusion (participants are presented with the real-time image of their own back, and body ownership is transferred to the virtual image due to experimental manipulations such as rhythmic touch), self-identification, self-location, and first-person perspective were proposed as the obligatory components of self-consciousness (Blanke 2012; Blanke et al. 2015). The representation primarily relies on the integration of proprioceptive, vestibular, and visual information.

Second, the crucial importance of rhythmic visceral processes, first of all cardiac activity, was proposed. Lacey and Lacey demonstrated relatively early (1978) that processing of external stimuli partly depends on the timing with respect to the cardiac cycle and assumed that it relies on inhibition of sensorimotor activity by baroreceptors during systole. The moderating effect of cardiac cycle on memory was demonstrated recently (Garfinkel et al. 2013); interestingly, those who were able to sense their heartbeat more accurately were less impacted. Visual processing (Walker and Sandman 1982; Park et al. 2014; Salomon et al. 2016, 2018), as well as timing of saccadic eye movements (i.e., active sampling), was also found to be influenced by rhythmic cardiac activity (Ohl et al. 2016; Kunzendorf et al. 2019; Galvez-Pol et al. 2020). It is worth mentioning that breathing can similarly influence a variety of high-level brain processes (Zelano et al. 2016; Herrero et al. 2018; Tort et al. 2018). A more direct mechanism between rhythmic cardiac activity and subjectivity was also hypothesized, called neural subjective frame (Park and Tallon-Baudry 2014). In this model, multiple representations of cardiac changes yield the basis for the subjective experience; a subsequent version includes rhythmic change from another organ with pacemaker property, the gastrointestinal tract (Babo-Rebelo and Tallon-Baudry 2018; Tallon-Baudry et al. 2018). Although the model is quite novel and somewhat speculative, supporting empirical evidence is accumulating. For example, it turned out that the processing of self-relevant information (including the agentive and descriptive aspects, i.e., I and Me, or the self, respectively) is synchronized with heartbeat-evoked responses in the brain (Babo-Rebelo et al. 2016a, b). In the same vein, participants' recognition of their own face in an ambiguous picture was influenced by rhythmic changes presented in synchrony or asynchrony with their heartbeats (Sel et al. 2017); again, the ability to sense heartbeats was associated with recognition performance. Concerning self-perception, bodily self-consciousness could be manipulated by feedback of heartbeats using various body illusion paradigms (Aspell et al. 2013; Suzuki et al. 2013; Park et al. 2016, 2018; Sel et al. 2017). Similar findings with breathing were reported (Allard et al. 2017). Moreover, evoking the full body illusion caused significant changes in the amplitude of heartbeat-evoked potentials, i.e., the cortical concomitants of cardiac activity (Park et al. 2016, 2018). In conclusion, interoceptive signals might play an important role in the maintenance of the stability of core self-awareness (Allen and Tsakiris 2018). Recent predictive processing-based approaches to the self also emphasize the importance of interoceptive information and use the term interoceptive inference (see Chap. 7 for more details) (Seth 2013; Apps and Tsakiris 2014; Seth and Friston 2016; Allen and Friston 2018; Allen et al. 2019).

Finally, perception of time, a subjective phenomenon intimately associated with consciousness, was also connected with interoception by several authors (for a review, see Wittmann and Meissner 2018). According to the theory of Craig (2010, 2015), the actual states of the body (including the evaluative aspects) are stored as perceptual moments, along with directly preceding and expected future moments. This pattern explains the flow and stability of consciousness described by James (1890). In relaxed states, approximately ten such moments are stored in every second; if salience increases (e.g., in emotionally intense periods), more pictures are needed for the accurate representation of the changes. This leads to perceived slowing down of time.

Overall, although the presented models are fragmented and no final picture can be drawn, this section was intended to demonstrate the crucial role of interoceptive information (using the term in the broader sense) for the knower aspect of the self (self as subject). From a developmental perspective, it is important to note that early experiences, primarily social experiences, might also impact this aspect beyond the knower or object aspect. For example, stressful periods at early life change homeostatic regulation (Fox et al. 2007; Rinaman et al. 2011), and interactions with others impact the integration of interoceptive signals (Fotopoulou and Tsakiris 2017).

3.7 A Final Word on Body Sensations

It is important to note that possessing the ability to sense the actual state of the body in a modality and more or less localization-specific way is not equal with having visceral or other body sensations all the time. According to the evolutionary logic presented in this chapter, the majority of homeostatic demands can be properly handled at lower levels of regulation, like in non-primates. Theoretically, if we provide someone with all factors necessary for homeostasis, from water and food to appropriate temperature, lack of bacteria and emotional upheavals, perhaps it could even be possible to "go through life without noticing any visceral sensations" as stated by Paintal (1986, p. 3). In other words, homeostatic body sensations play a role only if addressing a need state is beyond the capacity of lower-level regulatory mechanisms, thus requires specific behavioral steps. In accordance with this concept, conscious perception thresholds for various visceral modalities are high compared to exteroceptive modalities; moreover, visceral sensations are typically unpleasant or even painful (Ádám 1998; Cameron 2002). If the physiological regulatory system can handle the issue at hand properly, further loading of the already limited attentional capacity would be maladaptive as paying attention to external cues is more important (Mason 1961; Ádám 1967, 1998). If there is a need for behavioral steps, however, marked unpleasantness and painfulness of visceral sensations catch and keep the attention very effectively (Ádám 1998). As presented above, there are exceptions to the rule of unpleasantness, such as affective touch and mild to moderate physical activity; however, the built-in positive evaluation of these modalities can also be very well understood from a homeostatic point of view.

The association between the majority of homeostatic body sensations (most importantly, pain) and negative affect has serious implications with respect to symptom reporting and medically unexplained symptoms. We will see that symptom reports are systematically connected with negative affect, and this connection is bidirectional. As they able to boost and perhaps also to evoke each other, one cannot resist the temptation and speculate that they can easily be confused with each other. This idea, best put by the title of a scientific paper ("Feeling your body or feeling badly"; (Aronson et al. 2001)) will be pursued in more detail later (Chaps. 5 and 7).

3.8 Summary

In this chapter, the two major pathways of neural processing of interoceptive (including visceroceptive-homeostatic and somatosensory) information were described. It was shown that, in accordance with our everyday perception of our body as a unity, integration occurs at many multiple levels of central processing. Although we humans take them for granted, localized interoceptive sensations in fact does not characterize the majority of vertebrates and mammals. Beyond their homeostatic importance, interoceptive stimuli also contribute to the development of both (subject and object) aspects of the human self. Finally, body sensations are involved in emotional processing (see Chap. 8) and are themselves subject to evaluation (Chaps. 5 and 6).

Take-Home Messages

For laypeople: Interoceptive information from within the body is processed and integrated by a number of brain centers through two major pathways. This information is used not only to maintain the internal balance of the body (homeostasis) and perform movements but also to give rise to such a sophisticated phenomenon as the human self.

For clinicians: Body-related information becomes integrated and evaluated in the upper brainstem, thus an elementary form of consciousness can also arise at this relatively low level of processing. The insula represents the major neural substrate of human body sensations, where homeostatic and somatosensory information is integrated with exteroceptive (e.g., visual) modalities. Integration can lead to body illusions in certain cases.

For researchers: One of the hottest points of this area is multisensory (multimodal) integration. Systematic research on body illusions, such as the rubber hand illusion and the full body illusion, can shed more light on details of integration as well as the development and maintenance of human consciousness. I think that the role of proprioception in the rubber hand illusion needs more research attention. Proprioceptive information on the actual position of the hand might not completely be overwritten by visual and tactile cues. As the sense of body ownership is preserved under resting conditions, proprioception might play a substantial role in the maintenance of the self.

References

Ádám G (1967) Interoception and behavior. Akadémiai Kiadó, Budapest

Ádám G (1998) Visceral perception: understanding internal cognition. Plenum Press, New York

Adams RA, Shipp S, Friston KJ (2013) Predictions not commands: active inference in the motor system. Brain Struct Funct 218:611–643. https://doi.org/10.1007/s00429-012-0475-5

Adolfi F, Couto B, Richter F et al (2017) Convergence of interoception, emotion, and social cognition: a twofold fMRI meta-analysis and lesion approach. Cortex 88:124–142. https://doi.org/10.1016/j.cortex.2016.12.019

Airapetyants ES (1952) Higher nervous activity and the receptors of the inner organ. Soviet Academic Publishing House, Leningrad

Allard E, Canzoneri E, Adler D et al (2017) Interferences between breathing, experimental dyspnoea and bodily self-consciousness. Sci Rep 7:9990. https://doi.org/10.1038/s41598-017-11045-y

Allen M, Friston KJ (2018) From cognitivism to autopoiesis: towards a computational framework for the embodied mind. Synthese 195:2459–2482. https://doi.org/10.1007/s11229-016-1288-5

Allen M, Tsakiris M (2018) The body as first prior: interoceptive predictive processing and the primacy of self-models. In: Tsakiris M, De Preester H (eds) The interoceptive mind. From homeostasis to awareness. Oxford University Press, Oxford, England, pp 27–45

Allen M, Levy A, Parr T, Friston KJ (2019) In the body's eye: the computational anatomy of interoceptive inference. bioRxiv 603928. https://doi.org/10.1101/603928

Apps MAJ, Tsakiris M (2014) The free-energy self: a predictive coding account of self-recognition. Neurosci Biobehav Rev 41:85–97. https://doi.org/10.1016/j.neubiorev.2013.01.029
Aronson KR, Barrett LF, Quigley KS (2001) Feeling your body or feeling badly: evidence for the limited validity of the Somatosensory Amplification Scale as an index of somatic sensitivity. J Psychosom Res 51:387–394
Aspell JE, Lenggenhager B, Blanke O (2012) Multisensory perception and bodily self-consciousness: from out-of-body to inside-body experience. In: Murray MM, Wallace MT (eds) The neural bases of multisensory processes. CRC Press/Taylor & Francis, Boca Raton, FL
Aspell JE, Heydrich L, Marillier G et al (2013) Turning body and self inside out: visualized heartbeats alter bodily self-consciousness and tactile perception. Psychol Sci 24:2445–2453. https://doi.org/10.1177/0956797613498395
Avery JA, Kerr KL, Ingeholm JE et al (2015) A common gustatory and interoceptive representation in the human mid-insula. Hum Brain Mapp 36:2996–3006. https://doi.org/10.1002/hbm.22823
Aziz Q, Ruffle JK (2018) The neurobiology of gut feelings. In: Tsakiris M, De Preester H (eds) The interoceptive mind. From homeostasis to awareness. Oxford University Press, Oxford, England, pp 80–101
Babo-Rebelo M, Tallon-Baudry C (2018) Interoceptive signals, brain dynamics, and subjectivity. In: Tsakiris M, De Preester H (eds) The interoceptive mind. From homeostasis to awareness. Oxford University Press, Oxford, England, pp 46–62
Babo-Rebelo M, Richter CG, Tallon-Baudry C (2016a) Neural responses to heartbeats in the default network encode the self in spontaneous thoughts. J Neurosci 36:7829–7840. https://doi.org/10.1523/JNEUROSCI.0262-16.2016
Babo-Rebelo M, Wolpert N, Adam C et al (2016b) Is the cardiac monitoring function related to the self in both the default network and right anterior insula? Philos Trans R Soc Lond B Biol Sci 371:20160004. https://doi.org/10.1098/rstb.2016.0004
Balleine BW, Dickinson A (2000) The effect of lesions of the insular cortex on instrumental conditioning: evidence for a role in incentive memory. J Neurosci 20:8954–8964
Bandler R, Keay KA, Floyd N, Price J (2000) Central circuits mediating patterned autonomic activity during active vs. passive emotional coping. Brain Res Bull 53:95–104. https://doi.org/10.1016/S0361-9230(00)00313-0
Bárdos G (1989) Behavioral consequences of intestinal distention: aversivity and discomfort. Physiol Behav 45:79–85
Barrett LF, Simmons WK (2015) Interoceptive predictions in the brain. Nat Rev Neurosci 16:419–429. https://doi.org/10.1038/nrn3950
Bastian HC (1887) The "muscular sense"; its nature and cortical localisation. Brain 10:1–89. https://doi.org/10.1093/brain/10.1.1
Batson G (2009) Update on proprioception: considerations for dance education. J Dance Med Sci 13:35–41
Berntson GG, Gianaros PJ, Tsakiris M (2018) Interoception and the autonomic nervous system: bottom-up meets top-down. In: Tsakiris M, De Preester H (eds) The interoceptive mind. From homeostasis to awareness. Oxford University Press, Oxford, England, pp 3–23
Berridge KC, Robinson TE (2003) Parsing reward. Trends Neurosci 26:507–513. https://doi.org/10.1016/S0166-2236(03)00233-9
Bessou P, Burgess PR, Perl ER, Taylor CB (1971) Dynamic properties of mechanoreceptors with unmyelinated (C) fibers. J Neurophysiol 34:116–131. https://doi.org/10.1152/jn.1971.34.1.116
Bielefeldt K, Gebhart GF (2007) 3.19 - Visceral afferents A2. In: Kaas JH (ed) Evolution of nervous systems. Academic, Oxford, pp 237–252
Björnsdotter M, Löken L, Olausson H et al (2009) Somatotopic organization of gentle touch processing in the posterior insular cortex. J Neurosci 29:9314–9320. https://doi.org/10.1523/JNEUROSCI.0400-09.2009
Björnsdotter M, Morrison I, Olausson H (2010) Feeling good: on the role of C fiber mediated touch in interoception. Exp Brain Res 207:149–155. https://doi.org/10.1007/s00221-010-2408-y

Blanke O (2012) Multisensory brain mechanisms of bodily self-consciousness. Nat Rev Neurosci 13:556–571. https://doi.org/10.1038/nrn3292

Blanke O, Metzinger T (2009) Full-body illusions and minimal phenomenal selfhood. Trends Cogn Sci 13:7–13. https://doi.org/10.1016/j.tics.2008.10.003

Blanke O, Slater M, Serino A (2015) Behavioral, neural, and computational principles of bodily self-consciousness. Neuron 88:145–166. https://doi.org/10.1016/j.neuron.2015.09.029

Botvinick M, Cohen J (1998) Rubber hands "feel" touch that eyes see. Nature 391:756. https://doi.org/10.1038/35784

Brener J, Ring C (2016) Towards a psychophysics of interoceptive processes: the measurement of heartbeat detection. Philos Trans R Soc Lond B Biol Sci 371:20160015. https://doi.org/10.1098/rstb.2016.0015

Brooks JCW, Zambreanu L, Godinez A et al (2005) Somatotopic organisation of the human insula to painful heat studied with high resolution functional imaging. NeuroImage 27:201–209. https://doi.org/10.1016/j.neuroimage.2005.03.041

Brosschot JF, Verkuil B, Thayer JF (2016) The default response to uncertainty and the importance of perceived safety in anxiety and stress: an evolution-theoretical perspective. J Anxiety Disord 41:22–34. https://doi.org/10.1016/j.janxdis.2016.04.012

Brosschot JF, Verkuil B, Thayer JF (2017) Exposed to events that never happen: generalized unsafety, the default stress response, and prolonged autonomic activity. Neurosci Biobehav Rev 74:287–296. https://doi.org/10.1016/j.neubiorev.2016.07.019

Brosschot JF, Verkuil B, Thayer JF (2018) Generalized unsafety theory of stress: unsafe environments and conditions, and the default stress response. Int J Environ Res Public Health 15:464. https://doi.org/10.3390/ijerph15030464

Bruder GE, Fong R, Tenke CE et al (1997) Regional brain asymmetries in major depression with or without an anxiety disorder: a quantitative electroencephalographic study. Biol Psychiatry 41:939–948. https://doi.org/10.1016/S0006-3223(96)00260-0

Brüggemann J, Shi T, Apkarian AV (1998) Viscerosomatic interactions in the thalamic ventral posterolateral nucleus (VPL) of the squirrel monkey. Brain Res 787:269–276. https://doi.org/10.1016/S0006-8993(97)01512-6

Cabanac M (1971) Physiological role of pleasure. Science 173:1103–1107. https://doi.org/10.1126/science.173.4002.1103

Cabanac M (2006) Exertion and pleasure from an evolutionary perspective. In: Acevedo EO, Ekkekakis P (eds) Psychobiology of physical activity. Human Kinetics, Champaign, IL, pp 79–89

Cabanac M, Massonnet B, Belaiche R (1972) Preferred skin temperature as a function of internal and mean skin temperature. J Appl Physiol 33:699–703. https://doi.org/10.1152/jappl.1972.33.6.699

Cacioppo JT, Priester JR, Berntson GG (1993) Rudimentary determinants of attitudes: II. Arm flexion and extension have differential effects on attitudes. J Pers Soc Psychol 65:5–17. https://doi.org/10.1037/0022-3514.65.1.5

Cameron OG (2002) Visceral sensory neuroscience. Interoception. Oxford University Press, New York

Cameron OG (2009) Visceral brain-body information transfer. NeuroImage 47:787–794. https://doi.org/10.1016/j.neuroimage.2009.05.010

Cash TF, Szymanski ML (1995) The development and validation of the Body-Image Ideals Questionnaire. J Pers Assess 64:466–477. https://doi.org/10.1207/s15327752jpa6403_6

Cervero F, Morrison JFB (1986) Visceral sensation. Elsevier, Amsterdam, The Netherlands

Ceunen E, Vlaeyen JWS, Van Diest I (2016) On the origin of interoception. Front Psychol 7:743. https://doi.org/10.3389/fpsyg.2016.00743

Christoff K, Cosmelli D, Legrand D, Thompson E (2011) Specifying the self for cognitive neuroscience. Trends Cogn Sci 15:104–112. https://doi.org/10.1016/j.tics.2011.01.001

Chun MM, Turk-Browne NB (2007) Interactions between attention and memory. Curr Opin Neurobiol 17:177–184. https://doi.org/10.1016/j.conb.2007.03.005

Cisek P, Kalaska JF (2010) Neural mechanisms for interacting with a world full of action choices. Annu Rev Neurosci 33:269–298. https://doi.org/10.1146/annurev.neuro.051508.135409
Clark A (2015) Surfing uncertainty: prediction, action, and the embodied mind, 1st edn. Oxford University Press, Oxford, England
Colombetti G, Harrison N (2018) From physiology to experience: enriching existing conceptions of "arousal" in affective science. In: Tsakiris M, De Preester H (eds) The interoceptive mind. From homeostasis to awareness. Oxford University Press, Oxford, England, pp 245–258
Cottrell GT, Ferguson AV (2004) Sensory circumventricular organs: central roles in integrated autonomic regulation. Regul Pept 117:11–23. https://doi.org/10.1016/j.regpep.2003.09.004
Craig AD (2002) How do you feel? Interoception: the sense of the physiological condition of the body. Nat Rev Neurosci 3:655–666. https://doi.org/10.1038/nrn894
Craig AD (2003a) Interoception: the sense of the physiological condition of the body. Curr Opin Neurobiol 13:500–505. https://doi.org/10.1016/S0959-4388(03)00090-4
Craig AD (2003b) Pain mechanism: labeled lines versus convergence in central processing. Annu Rev Neurosci 26:1–30. https://doi.org/10.1146/annurev.neuro.26.041002.131022
Craig AD (2003c) A new view of pain as a homeostatic emotion. Trends Neurosci 26:303–307
Craig ADB (2005) Forebrain emotional asymmetry: a neuroanatomical basis? Trends Cogn Sci 9:566–571. https://doi.org/10.1016/j.tics.2005.10.005
Craig AD (2008) Interoception and emotion: a neuroanatomical perspective. In: Lewis M, Haviland-Jones JM, Barrett LF (eds) Handbook of emotions. The Guilford Press, New York, pp 272–288
Craig AD (2009) How do you feel—now? The anterior insula and human awareness. Nat Rev Neurosci 10:59–70. https://doi.org/10.1038/nrn2555
Craig AD (2010) The sentient self. Brain Struct Funct 214:563–577. https://doi.org/10.1007/s00429-010-0248-y
Craig AD (2011) Significance of the insula for the evolution of human awareness of feelings from the body. Ann N Y Acad Sci 1225:72–82. https://doi.org/10.1111/j.1749-6632.2011.05990.x
Craig AD (2015) How do you feel? An interoceptive moment with your neurobiological self. Princeton University Press, Princeton, NJ
Craig AD (2018) Chapter 19 - Central neural substrates involved in temperature discrimination, thermal pain, thermal comfort, and thermoregulatory behavior. In: Romanovsky AA (ed) Handbook of clinical neurology. Elsevier, Amsterdam, pp 317–338
Craig AD, Bushnell MC (1994) The thermal grill illusion: unmasking the burn of cold pain. Science 265:252–255. https://doi.org/10.1126/science.8023144
Critchley HD, Harrison NA (2013) Visceral influences on brain and behavior. Neuron 77:624–638. https://doi.org/10.1016/j.neuron.2013.02.008
Critchley HD, Wiens S, Rotshtein P et al (2004) Neural systems supporting interoceptive awareness. Nat Neurosci 7:189–195. https://doi.org/10.1038/nn1176
Crucianelli L, Krahé C, Jenkinson PM, Fotopoulou A(K) (2017) Interoceptive ingredients of body ownership: affective touch and cardiac awareness in the rubber hand illusion. Cortex. https://doi.org/10.1016/j.cortex.2017.04.018
Damasio A (1994) Descartes's error: emotion, reason, and the human brain. Penguin Books, New York
Damasio A (1999) The feeling of what happens: body and emotion in the making of consciousness. Harcourt Brace and Co, San Diego, CA
Damasio A (2003) Mental self: the person within. Nature 423:227. https://doi.org/10.1038/423227a
Damasio A (2010) Self comes to mind: constructing the conscious brain. Pantheon, New York
Damasio A, Carvalho GB (2013) The nature of feelings: evolutionary and neurobiological origins. Nat Rev Neurosci 14:143–152. https://doi.org/10.1038/nrn3403
Damasio A, Grabowski TJ, Bechara A et al (2000) Subcortical and cortical brain activity during the feeling of self-generated emotions. Nat Neurosci 3:1049–1056. https://doi.org/10.1038/79871
Davidson RJ (1992) Anterior cerebral asymmetry and the nature of emotion. Brain Cogn 20:125–151

Davidson RJ, Horowitz ME, Schwartz GE, Goodman DM (1981) Lateral differences in the latency between finger tapping and the heart beat. Psychophysiology 18:36–41

Davis KD, Lozano RM, Manduch M et al (1999) Thalamic relay site for cold perception in humans. J Neurophysiol 81:1970–1973. https://doi.org/10.1152/jn.1999.81.4.1970

De Preester H (2018) Subjectivity as a sentient perspective and the role of interoception. In: Tsakiris M, De Preester H (eds) The interoceptive mind. From homeostasis to awareness. Oxford University Press, Oxford, England, pp 293–306

de Vignemont F (2018a) Mind the body. An exploration of bodily self-awareness. Oxford University Press, Oxford, England

de Vignemont F (2018b) Was Descartes right after all? An affective background for bodily awareness. In: Tsakiris M, De Preester H (eds) The interoceptive mind. From homeostasis to awareness. Oxford University Press, Oxford, England, pp 259–271

Derbyshire SWG (2003) A systematic review of neuroimaging data during visceral stimulation. Am J Gastroenterol 98:12–20. https://doi.org/10.1111/j.1572-0241.2003.07168.x

Dolan RJ (2002) Emotion, cognition, and behavior. Science 298:1191–1194. https://doi.org/10.1126/science.1076358

Dostrovsky JO (2000) Role of thalamus in pain. In: Progress in brain research. Elsevier, Amsterdam, pp 245–257

Dunbar RIM (2010) The social role of touch in humans and primates: behavioural function and neurobiological mechanisms. Neurosci Biobehav Rev 34:260–268. https://doi.org/10.1016/j.neubiorev.2008.07.001

Dworkin BR (2007) Interoception. In: Cacioppo JT, Tassinary LG, Berntson GG (eds) Handbook of psychophysiology, 3rd edn. Cambridge University Press, New York, pp 482–506

Edelman DB, Seth AK (2009) Animal consciousness: a synthetic approach. Trends Neurosci 32:476–484. https://doi.org/10.1016/j.tins.2009.05.008

Edin BB (2001) Cutaneous afferents provide information about knee joint movements in humans. J Physiol 531:289–297. https://doi.org/10.1111/j.1469-7793.2001.0289j.x

Ekkekakis P, Acevedo EO (2006) Affective responses to acute exercise. In: Acevedo EO, Ekkekakis P (eds) Psychobiology of physical activity. Human Kinetics, Champaign, IL, pp 91–109

Ekkekakis P, Petruzzello SJ (1999) Acute aerobic exercise and affect: current status, problems and prospects regarding dose-response. Sports Med 28:337–374

Ekkekakis P, Parfitt G, Petruzzello SJ (2011) The pleasure and displeasure people feel when they exercise at different intensities: decennial update and progress towards a tripartite rationale for exercise intensity prescription. Sports Med 41:641–671. https://doi.org/10.2165/11590680-000000000-00000

Ekkekakis P, Hartman ME, Ladwig MA (2020) Affective responses to exercise. In: Tenenbaum G, Eklund RC (eds) Handbook of sport psychology, 4th edn. Wiley, Hoboken, NJ, pp 233–253

Emanuelsen L, Drew R, Köteles F (2015) Interoceptive sensitivity, body image dissatisfaction, and body awareness in healthy individuals. Scand J Psychol 56:167–174. https://doi.org/10.1111/sjop.12183

Fotopoulou A, Tsakiris M (2017) Mentalizing homeostasis: the social origins of interoceptive inference. Neuropsychoanalysis 19:3–28. https://doi.org/10.1080/15294145.2017.1294031

Fox NA, Schmidt LA, Henderson HA, Marshall PJ (2007) Developmental psychophysiology: conceptual and methodological issues. In: Handbook of psychophysiology, 3rd edn. Cambridge University Press, New York, pp 453–481

Fredrickson BL, Roberts T-A (1997) Objectification theory. Psychol Women Q 21:173–206. https://doi.org/10.1111/j.1471-6402.1997.tb00108.x

Fry M, Ferguson AV (2007) The sensory circumventricular organs: brain targets for circulating signals controlling ingestive behavior. Physiol Behav 91:413–423. https://doi.org/10.1016/j.physbeh.2007.04.003

Gallagher S (2000) Philosophical conceptions of the self: implications for cognitive science. Trends Cogn Sci 4:14–21

Gallagher S (2003) Bodily self-awareness and object perception. Theoria Et Historia Scientarum 7:53–68

Gallagher S (2005) How the body shapes the mind. Clarendon Press, Oxford

Galvez-Pol A, McConnell R, Kilner JM (2020) Active sampling in visual search is coupled to the cardiac cycle. Cognition 196:104149. https://doi.org/10.1016/j.cognition.2019.104149

Garfinkel SN, Barrett AB, Minati L et al (2013) What the heart forgets: cardiac timing influences memory for words and is modulated by metacognition and interoceptive sensitivity. Psychophysiology 50:505–512. https://doi.org/10.1111/psyp.12039

Georgieff N, Jeannerod M (1998) Beyond consciousness of external reality: a "who" system for consciousness of action and self-consciousness. Conscious Cogn 7:465–477. https://doi.org/10.1006/ccog.1998.0367

Ghazanfar AA, Schroeder CE (2006) Is neocortex essentially multisensory? Trends Cogn Sci 10:278–285. https://doi.org/10.1016/j.tics.2006.04.008

Gray MA, Harrison NA, Wiens S, Critchley HD (2007) Modulation of emotional appraisal by false physiological feedback during fMRI. PLoS One 2:e546. https://doi.org/10.1371/journal.pone.0000546

Greenspan JD, Lee RR, Lenz FA (1999) Pain sensitivity alterations as a function of lesion location in the parasylvian cortex. Pain 81:273–282. https://doi.org/10.1016/s0304-3959(99)00021-4

Guskiewicz KM, Perrin DH (1996) Research and clinical applications of assessing balance. J Sport Rehabil 5:45–63. https://doi.org/10.1123/jsr.5.1.45

Han J, Waddington G, Adams R et al (2016) Assessing proprioception: a critical review of methods. J Sport Health Sci 5:80–90. https://doi.org/10.1016/j.jshs.2014.10.004

Hantas MN, Katkin ES, Reed SD (1984) Cerebral lateralization and heartbeat discrimination. Psychophysiology 21:274–278

Hassler R, Riechert T (1959) Klinische und anatomische Befunde bei stereotaktischen Schmerzoperationen im Thalamus. Archiv für Psychiatrie und Zeitschrift f d ges Neurologie 200:93–122. https://doi.org/10.1007/BF00361358

Henderson LA, Macey PM, Macey KE et al (2002) Brain responses associated with the Valsalva maneuver revealed by functional magnetic resonance imaging. J Neurophysiol 88:3477–3486. https://doi.org/10.1152/jn.00107.2002

Henderson LA, Gandevia SC, Macefield VG (2007) Somatotopic organization of the processing of muscle and cutaneous pain in the left and right insula cortex: a single-trial fMRI study. Pain 128:20–30. https://doi.org/10.1016/j.pain.2006.08.013

Herrero JL, Khuvis S, Yeagle E et al (2018) Breathing above the brain stem: volitional control and attentional modulation in humans. J Neurophysiol 119:145–159. https://doi.org/10.1152/jn.00551.2017

Hölzl R, Erasmus LP, Möltner A (1996) Detection, discrimination and sensation of visceral stimuli. Biol Psychol 42:199–214

Hölzl R, Möltner A, Neidig CW (1999) Somatovisceral interactions in visceral perception: abdominal masking of colonic stimuli. Integr Physiol Behav Sci 34:269–284

Horváth Á, Ferentzi E, Köteles F (2019) Proprioceptive accuracy is not associated with self-reported body awareness, body competence, and affect. Physiol Int 106:347–354. https://doi.org/10.1556/2060.106.2019.33

Horváth Á, Ferentzi E, Bogdány T et al (2020) Proprioception but not cardiac interoception is related to the rubber hand illusion. Cortex 132:361–373. https://doi.org/10.1016/j.cortex.2020.08.026

Horváth Á, Vig L, Ferentzi E, Köteles F (2021) Cardiac and proprioceptive accuracy are not related to body awareness, perceived body competence, and affect. Frontiers in Psychology in Press: https://doi.org/10.3389/fpsyg.2020.575574

Hua LH, Strigo IA, Baxter LC et al (2005) Anteroposterior somatotopy of innocuous cooling activation focus in human dorsal posterior insular cortex. Am J Phys Regul Integr Comp Phys 289:R319–R325. https://doi.org/10.1152/ajpregu.00123.2005

James W (1890) The principles of psychology. Macmillan, London

Jänig W (1996) Neurobiology of visceral afferent neurons: neuroanatomy, functions, organ regulations and sensations. Biol Psychol 42:29–51. https://doi.org/10.1016/0301-0511(95)05145-7

Jha P, Ahamad I, Khurana S et al (2017) Proprioception: an evidence based narrative. Mini review. Res Inves Sports Med. https://doi.org/10.31031/RISM.2017.01.000506

Jones GE, Jones KR, Rouse CH et al (1987) The effect of body position on the perception of cardiac sensations: an experiment and theoretical implications. Psychophysiology 24:300–311. https://doi.org/10.1111/j.1469-8986.1987.tb00300.x

Kahane P, Hoffmann D, Minotti L, Berthoz A (2003) Reappraisal of the human vestibular cortex by cortical electrical stimulation study. Ann Neurol 54:615–624. https://doi.org/10.1002/ana.10726

Katkin ES, Reed SD (1988) Cardiovascular asymmetries and cardiac perception. Int J Neurosci 39:45–52

Kern MK, Shaker R (2002) Cerebral cortical registration of subliminal visceral stimulation. Gastroenterology 122:290–298. https://doi.org/10.1053/gast.2002.30989

Kern M, Aertsen A, Schulze-Bonhage A, Ball T (2013) Heart cycle-related effects on event-related potentials, spectral power changes, and connectivity patterns in the human ECoG. NeuroImage 81:178–190. https://doi.org/10.1016/j.neuroimage.2013.05.042

Khalsa SS, Rudrauf D, Feinstein JS, Tranel D (2009) The pathways of interoceptive awareness. Nat Neurosci 12:1494–1496. https://doi.org/10.1038/nn.2411

Knoop J, Steultjens MPM, van der Leeden M et al (2011) Proprioception in knee osteoarthritis: a narrative review. Osteoarthr Cartil 19:381–388. https://doi.org/10.1016/j.joca.2011.01.003

Korn H, Wendt R, Albe-Fessard D (1966) Somatic projection to the orbital cortex of the cat. Electroencephalogr Clin Neurophysiol 21:209–226. https://doi.org/10.1016/0013-4694(66)90071-x

Köteles F, Witthöft M (2017) Somatosensory amplification – an old construct from a new perspective. J Psychosom Res 101:1–9. https://doi.org/10.1016/j.jpsychores.2017.07.011

Kumazawa T, Perl ER (1977a) Primate cutaneous receptors with unmyelinated (C) fibres and their projection to the substantia gelatinosa. J Physiol Paris 73:287–304

Kumazawa T, Perl ER (1977b) Primate cutaneous sensory units with unmyelinated (C) afferent fibers. J Neurophysiol 40:1325–1338. https://doi.org/10.1152/jn.1977.40.6.1325

Kunzendorf S, Klotzsche F, Akbal M et al (2019) Active information sampling varies across the cardiac cycle. Psychophysiology 56:e13322. https://doi.org/10.1111/psyp.13322

Lacey BC, Lacey JI (1978) Two-way communication between the heart and the brain. Significance of time within the cardiac cycle. Am Psychol 33:99–113

Lazarus RS (1991) Emotion and adaptation, 1st edn. Oxford University Press, Oxford, England

Legrain V, Iannetti GD, Plaghki L, Mouraux A (2011) The pain matrix reloaded: a salience detection system for the body. Prog Neurobiol 93:111–124. https://doi.org/10.1016/j.pneurobio.2010.10.005

Lenggenhager B, Tadi T, Metzinger T, Blanke O (2007) Video ergo sum: manipulating bodily self-consciousness. Science 317:1096–1099. https://doi.org/10.1126/science.1143439

Leonard WR, Robertson ML, Snodgrass JJ, Kuzawa CW (2003) Metabolic correlates of hominid brain evolution. Comp Biochem Physiol A Mol Integr Physiol 136:5–15

Lephart SM, Pincivero DM, Rozzi SL (1998) Proprioception of the ankle and knee. Sports Med 25:149–155

Lundberg A, Malmgren K, Schomburg ED (1987) Reflex pathways from group II muscle afferents. 3. Secondary spindle afferents and the FRA: a new hypothesis. Exp Brain Res 65:294–306. https://doi.org/10.1007/bf00236301

Mai JK, Paxinos G (eds) (2011) The human nervous system, 3rd edn. Academic, Amsterdam

Marshall GD, Zimbardo PG (1979) Affective consequences of inadequately explained physiological arousal. J Pers Soc Psychol 37:970–988. https://doi.org/10.1037/0022-3514.37.6.970

Mason RE (1961) Internal perception and bodily functioning. International Universities Press, New York

Mazzola L, Isnard J, Peyron R et al (2009) Somatotopic organization of pain responses to direct electrical stimulation of the human insular cortex. Pain 146:99–104. https://doi.org/10.1016/j.pain.2009.07.014

Mazzola L, Isnard J, Peyron R, Mauguière F (2011) Stimulation of the human cortex and the experience of pain: Wilder Penfield's observations revisited. Brain 135:631–640. https://doi.org/10.1093/brain/awr265

McGlone F, Vallbo AB, Olausson H et al (2007) Discriminative touch and emotional touch. Can J Exp Psychol 61:173–183

Melzack R (1989) Phantom limbs, the self and the brain (the D. O. Hebb Memorial Lecture). Can Psychol 30:1–16. https://doi.org/10.1037/h0079793

Melzack R (1990) Phantom limbs and the concept of a neuromatrix. Trends Neurosci 13:88–92. https://doi.org/10.1016/0166-2236(90)90179-E

Melzack R, Wall PD (1965) Pain mechanisms: a new theory. Science 150:971–979

Menon V, Uddin LQ (2010) Saliency, switching, attention and control: a network model of insula function. Brain Struct Funct 214:655–667. https://doi.org/10.1007/s00429-010-0262-0

Mense S (2008) Muscle pain: mechanisms and clinical significance. Dtsch Arztebl Int 105:214–219. https://doi.org/10.3238/artzebl.2008.0214

Miller LC, Murphy R, Buss AH (1981) Consciousness of body: private and public. J Pers Soc Psychol 41:397–406. https://doi.org/10.1037/0022-3514.41.2.397

Miller ER, Gunnell GF, Martin RD (2005) Deep time and the search for anthropoid origins. Am J Phys Anthropol Suppl 41:60–95. https://doi.org/10.1002/ajpa.20352

Mimee A, Smith PM, Ferguson AV (2013) Circumventricular organs: targets for integration of circulating fluid and energy balance signals? Physiol Behav 121:96–102. https://doi.org/10.1016/j.physbeh.2013.02.012

Montgomery WA, Jones GE (1984) Laterality, emotionality, and heartbeat perception. Psychophysiology 21:459–465. https://doi.org/10.1111/j.1469-8986.1984.tb00227.x

Morrison I, Löken LS, Olausson H (2010) The skin as a social organ. Exp Brain Res 204:305–314. https://doi.org/10.1007/s00221-009-2007-y

Morrison I, Björnsdotter M, Olausson H (2011a) Vicarious responses to social touch in posterior insular cortex are tuned to pleasant caressing speeds. J Neurosci 31:9554–9562. https://doi.org/10.1523/JNEUROSCI.0397-11.2011

Morrison I, Löken LS, Minde J et al (2011b) Reduced C-afferent fibre density affects perceived pleasantness and empathy for touch. Brain 134:1116–1126. https://doi.org/10.1093/brain/awr011

Namkung H, Kim S-H, Sawa A (2017) The insula: an underestimated brain area in clinical neuroscience, psychiatry, and neurology. Trends Neurosci 40:200–207. https://doi.org/10.1016/j.tins.2017.02.002

Neumann R, Strack F (2000) Approach and avoidance: the influence of proprioceptive and exteroceptive cues on encoding of affective information. J Pers Soc Psychol 79:39–48. https://doi.org/10.1037/0022-3514.79.1.39

Norrsell U, Craig AD (1999) Behavioral thermosensitivity after lesions of thalamic target areas of a thermosensory spinothalamic pathway in the cat. J Neurophysiol 82:611–625. https://doi.org/10.1152/jn.1999.82.2.611

Ogard WK (2011) Proprioception in sports medicine and athletic conditioning. Strength Condition J 33:111–118

Ohl S, Wohltat C, Kliegl R et al (2016) Microsaccades are coupled to heartbeat. J Neurosci 36:1237–1241. https://doi.org/10.1523/JNEUROSCI.2211-15.2016

Olausson H, Lamarre Y, Backlund H et al (2002) Unmyelinated tactile afferents signal touch and project to insular cortex. Nat Neurosci 5:900–904. https://doi.org/10.1038/nn896

Opsommer E, Weiss T, Plaghki L, Miltner WH (2001) Dipole analysis of ultralate (C-fibres) evoked potentials after laser stimulation of tiny cutaneous surface areas in humans. Neurosci Lett 298:41–44. https://doi.org/10.1016/s0304-3940(00)01718-3

Ostrowsky K, Magnin M, Ryvlin P et al (2002) Representation of pain and somatic sensation in the human insula: a study of responses to direct electrical cortical stimulation. Cereb Cortex 12:376–385. https://doi.org/10.1093/cercor/12.4.376

Owens DM, Lumpkin EA (2014) Diversification and specialization of touch receptors in skin. Cold Spring Harb Perspect Med 4:a013656. https://doi.org/10.1101/cshperspect.a013656

Paintal AS (1986) The visceral sensations—some basic mechanisms. Prog Brain Res 67:3–19. https://doi.org/10.1016/S0079-6123(08)62752-1

Panksepp J (1998a) Affective neuroscience: the foundations of human and animal emotions, 1st edn. Oxford University Press, New York

Panksepp J (1998b) The periconscious substrates of consciousness: affective states and the evolutionary origins of the self. J Conscious Stud 5:566–582

Panksepp J (2005) Affective consciousness: core emotional feelings in animals and humans. Conscious Cogn 14:30–80. https://doi.org/10.1016/j.concog.2004.10.004

Panksepp J, Biven L (2012) The archaeology of mind. Neuroevolutionary origins of human emotions. W. W. Norton & Company, New York

Panksepp J, Northoff G (2009) The trans-species core SELF: the emergence of active cultural and neuro-ecological agents through self-related processing within subcortical-cortical midline networks. Conscious Cogn 18:193–215. https://doi.org/10.1016/j.concog.2008.03.002

Park H-D, Blanke O (2019) Heartbeat-evoked cortical responses: underlying mechanisms, functional roles, and methodological considerations. NeuroImage 197:502–511. https://doi.org/10.1016/j.neuroimage.2019.04.081

Park H-D, Tallon-Baudry C (2014) The neural subjective frame: from bodily signals to perceptual consciousness. Philos Trans R Soc Lond Ser B Biol Sci 369:20130208. https://doi.org/10.1098/rstb.2013.0208

Park H-D, Correia S, Ducorps A, Tallon-Baudry C (2014) Spontaneous fluctuations in neural responses to heartbeats predict visual detection. Nat Neurosci 17:612–618. https://doi.org/10.1038/nn.3671

Park H-D, Bernasconi F, Bello-Ruiz J et al (2016) Transient modulations of neural responses to heartbeats covary with bodily self-consciousness. J Neurosci 36:8453–8460. https://doi.org/10.1523/JNEUROSCI.0311-16.2016

Park H-D, Bernasconi F, Salomon R et al (2018) Neural sources and underlying mechanisms of neural responses to heartbeats, and their role in bodily self-consciousness: an intracranial EEG study. Cereb Cortex 28:2351–2364. https://doi.org/10.1093/cercor/bhx136

Parshall MB, Schwartzstein RM, Adams L et al (2012) An official American Thoracic Society statement: update on the mechanisms, assessment, and management of dyspnea. Am J Respir Crit Care Med 185:435–452. https://doi.org/10.1164/rccm.201111-2042ST

Pasqualotto A, Dumitru ML, Myachykov A (2016) Editorial: Multisensory integration: brain, body, and world. Front Psychol 6:2046. https://doi.org/10.3389/fpsyg.2015.02046

Paulus MP, Stein MB (2006) An insular view of anxiety. Biol Psychiatry 60:383–387. https://doi.org/10.1016/j.biopsych.2006.03.042

Penfield W, Boldrey E (1937) Somatic motor and sensory representation in the cerebral cortex of man as studied by electrical stimulation. Brain 60:389–443. https://doi.org/10.1093/brain/60.4.389

Penfield W, Faulk ME (1955) The insula; further observations on its function. Brain 78:445–470. https://doi.org/10.1093/brain/78.4.445

Penfield W, Jasper H (1954) Epilepsy and the functional anatomy of the human brain. AMA Arch NeurPsych 72(5):663–664

Pennebaker JW (1982) The psychology of physical symptoms. Springer, New York

Pennebaker JW, Cox DJ, Gonder-Frederick L et al (1981) Physical symptoms related to blood glucose in insulin-dependent diabetics. Psychosom Med 43:489

Pennebaker JW, Gonder-Frederick L, Stewart H et al (1982) Physical symptoms associated with blood pressure. Psychophysiology 19:201–210. https://doi.org/10.1111/j.1469-8986.1982.tb02547.x

Perelman P, Johnson WE, Roos C et al (2011) A molecular phylogeny of living primates. PLoS Genet 7:e1001342. https://doi.org/10.1371/journal.pgen.1001342

Pressnitzer D, Hupé J-M (2006) Temporal dynamics of auditory and visual bistability reveal common principles of perceptual organization. Curr Biol 16:1351–1357. https://doi.org/10.1016/j.cub.2006.05.054

Raichlen DA, Alexander GE (2017) Adaptive capacity: an evolutionary neuroscience model linking exercise, cognition, and brain health. Trends Neurosci 40:408–421. https://doi.org/10.1016/j.tins.2017.05.001

Ribeiro F, Oliveira J (2007) Aging effects on joint proprioception: the role of physical activity in proprioception preservation. Eur Rev Aging Phys Act 4:71–76. https://doi.org/10.1007/s11556-007-0026-x

Ribeiro F, Oliveira J (2011) Factors influencing proprioception: what do they reveal? In: Klika V (ed) Biomechanics in applications. InTech Open, Rijeka, pp 323–346

Rinaman L, Banihashemi L, Koehnle TJ (2011) Early life experience shapes the functional organization of stress-responsive visceral circuits. Physiol Behav 104:632–640. https://doi.org/10.1016/j.physbeh.2011.04.008

Robinson DR, Gebhart GF (2008) Inside information – the unique features of visceral sensation. Mol Interv 8:242–253. https://doi.org/10.1124/mi.8.5.9

Rolls ET (2015) Taste, olfactory, and food reward value processing in the brain. Prog Neurobiol 127–128:64–90. https://doi.org/10.1016/j.pneurobio.2015.03.002

Salamone PC, Sedeño L, Legaz A et al (2020) Dynamic neurocognitive changes in interoception after heart transplant. Brain Commun. https://doi.org/10.1093/braincomms/fcaa095

Salomon R, Ronchi R, Dönz J et al (2016) The insula mediates access to awareness of visual stimuli presented synchronously to the heartbeat. J Neurosci 36:5115–5127. https://doi.org/10.1523/JNEUROSCI.4262-15.2016

Salomon R, Ronchi R, Dönz J et al (2018) Insula mediates heartbeat related effects on visual consciousness. Cortex 101:87–95. https://doi.org/10.1016/j.cortex.2018.01.005

Saper CB (2002) The central autonomic nervous system: conscious visceral perception and autonomic pattern generation. Annu Rev Neurosci 25:433–469. https://doi.org/10.1146/annurev.neuro.25.032502.111311

Sato A (1997) Neural mechanisms of autonomic responses elicited by somatic sensory stimulation. Neurosci Behav Physiol 27:610–621. https://doi.org/10.1007/bf02463910

Schepers RJ, Ringkamp M (2009) Thermoreceptors and thermosensitive afferents. Neurosci Biobehav Rev 33:205–212. https://doi.org/10.1016/j.neubiorev.2008.07.009

Schmahmann JD, Leifer D (1992) Parietal pseudothalamic pain syndrome. Clinical features and anatomic correlates. Arch Neurol 49:1032–1037. https://doi.org/10.1001/archneur.1992.00530340048017

Schouenborg J, Weng HR, Kalliomäki J, Holmberg H (1995) A survey of spinal dorsal horn neurones encoding the spatial organization of withdrawal reflexes in the rat. Exp Brain Res 106:19–27. https://doi.org/10.1007/bf00241353

Seeley WW, Menon V, Schatzberg AF et al (2007) Dissociable intrinsic connectivity networks for salience processing and executive control. J Neurosci 27:2349–2356. https://doi.org/10.1523/JNEUROSCI.5587-06.2007

Sel A, Azevedo RT, Tsakiris M (2017) Heartfelt self: cardio-visual integration affects self-face recognition and interoceptive cortical processing. Cereb Cortex 27:5144–5155. https://doi.org/10.1093/cercor/bhw296

Seth AK (2013) Interoceptive inference, emotion, and the embodied self. Trends Cogn Sci 17:565–573. https://doi.org/10.1016/j.tics.2013.09.007

Seth AK, Friston KJ (2016) Active interoceptive inference and the emotional brain. Philos Trans R Soc Lond B Biol Sci 371:20160007. https://doi.org/10.1098/rstb.2016.0007

Seth AK, Suzuki K, Critchley HD (2012) An interoceptive predictive coding model of conscious presence. Front Psychol 2:395. https://doi.org/10.3389/fpsyg.2011.00395

Sherrington CS (1906) The integrative action of the nervous system. Yale University Press, New Haven

Sommer MA, Wurtz RH (2008) Brain circuits for the internal monitoring of movements. Annu Rev Neurosci 31:317–338. https://doi.org/10.1146/annurev.neuro.31.060407.125627

Sonstroem RJ (1997) Physical activity and self-esteem. In: Morgan WP (ed) Physical activity and mental health. Taylor and Francis, Bristol, pp 127–144

Sonstroem RJ, Morgan WP (1989) Exercise and self-esteem: rationale and model. Med Sci Sports Exerc 21:329–337

Springer MS, Murphy WJ, Eizirik E, O'Brien SJ (2003) Placental mammal diversification and the Cretaceous–Tertiary boundary. Proc Natl Acad Sci U S A 100:1056–1061. https://doi.org/10.1073/pnas.0334222100

Stein BE, Stanford TR (2008) Multisensory integration: current issues from the perspective of the single neuron. Nat Rev Neurosci 9:255–266. https://doi.org/10.1038/nrn2331

Stephan E, Pardo JV, Faris PL et al (2003) Functional neuroimaging of gastric distention. J Gastrointest Surg 7:740–749

Strigo IA, (Bud) Craig AD (2016) Interoception, homeostatic emotions and sympathovagal balance. Philos Trans R Soc Lond B Biol Sci 371:20160010. https://doi.org/10.1098/rstb.2016.0010

Sugiura Y, Lee CL, Perl ER (1986) Central projections of identified, unmyelinated (C) afferent fibers innervating mammalian skin. Science 234:358–361. https://doi.org/10.1126/science.3764416

Suzuki K, Garfinkel SN, Critchley HD, Seth AK (2013) Multisensory integration across exteroceptive and interoceptive domains modulates self-experience in the rubber-hand illusion. Neuropsychologia 51:2909–2917. https://doi.org/10.1016/j.neuropsychologia.2013.08.014

Swanik C, Lephart S, Rubash H (2004) Proprioception, kinesthesia, and balance after total knee arthroplasty with cruciate-retaining and posterior stabilized prostheses. J Bone Joint Surg Am 86:328–334

Swanson LW (2000) Cerebral hemisphere regulation of motivated behavior. Brain Res 886:113–164. https://doi.org/10.1016/s0006-8993(00)02905-x

Tallon-Baudry C, Campana F, Park H-D, Babo-Rebelo M (2018) The neural monitoring of visceral inputs, rather than attention, accounts for first-person perspective in conscious vision. Cortex 102:139–149. https://doi.org/10.1016/j.cortex.2017.05.019

Terasawa Y, Fukushima H, Umeda S (2013) How does interoceptive awareness interact with the subjective experience of emotion? An fMRI study. Hum Brain Mapp 34:598–612. https://doi.org/10.1002/hbm.21458

Thunberg T (1896) Förnimmelserne vid till samma ställe lokaliserad, samtidigt pågående köld-och värmeretning. Uppsala Läkfören Förh 1:489–495

Tomarken AJ, Keener AD (1998) Frontal brain asymmetry and depression: a self-regulatory perspective. Cognit Emot 12:387–420. https://doi.org/10.1080/026999398379655

Tort ABL, Brankačk J, Draguhn A (2018) Respiration-entrained brain rhythms are global but often overlooked. Trends Neurosci 41:186–197. https://doi.org/10.1016/j.tins.2018.01.007

Tsakiris M (2010) My body in the brain: a neurocognitive model of body-ownership. Neuropsychologia 48:703–712. https://doi.org/10.1016/j.neuropsychologia.2009.09.034

Tsakiris M (2017) The multisensory basis of the self: from body to identity to others [Formula: see text]. Q J Exp Psychol (Hove) 70:597–609. https://doi.org/10.1080/17470218.2016.1181768

Tsakiris M, Tajadura-Jiménez A, Costantini M (2011) Just a heartbeat away from one's body: interoceptive sensitivity predicts malleability of body-representations. Proc Biol Sci 278:2470–2476. https://doi.org/10.1098/rspb.2010.2547

Uddin LQ, Kinnison J, Pessoa L, Anderson ML (2014) Beyond the tripartite cognition-emotion-interoception model of the human insular cortex. J Cogn Neurosci 26:16–27. https://doi.org/10.1162/jocn_a_00462

Vallbo A, Olausson H, Wessberg J, Norrsell U (1993) A system of unmyelinated afferents for innocuous mechanoreception in the human skin. Brain Res 628:301–304. https://doi.org/10.1016/0006-8993(93)90968-s

Walker BB, Sandman CA (1982) Visual evoked potentials change as heart rate and carotid pressure change. Psychophysiology 19:520–527. https://doi.org/10.1111/j.1469-8986.1982.tb02579.x

Wallace MT, Meredith MA, Stein BE (1998) Multisensory integration in the superior colliculus of the alert cat. J Neurophysiol 80:1006–1010. https://doi.org/10.1152/jn.1998.80.2.1006

Watson D, Wiese D, Vaidya J, Tellegen A (1999) The two general activation systems of affect: structural findings, evolutionary considerations, and psychobiological evidence. J Pers Soc Psychol 76:820–838. https://doi.org/10.1037/0022-3514.76.5.820

Weisz J, Ádám G (1993) Hemispheric preference and lateral eye movements evoked by bilateral visual stimuli. Neuropsychologia 31:1299–1306. https://doi.org/10.1016/0028-3932(93)90099-L

Weisz J, Balázs L, Láng E, Adám G (1990) The effect of lateral visual fixation and the direction of eye movements on heartbeat discrimination. Psychophysiology 27:523–527

Weisz J, Balázs L, Adám G (1994) The effect of monocular viewing on heartbeat discrimination. Psychophysiology 31:370–374. https://doi.org/10.1111/j.1469-8986.1994.tb02445.x

Williamson JW, McColl R, Mathews D et al (2001) Hypnotic manipulation of effort sense during dynamic exercise: cardiovascular responses and brain activation. J Appl Physiol 90:1392–1399. https://doi.org/10.1152/jappl.2001.90.4.1392

Winkelmann RK (1986) Sensory Receptors of the skin. In: Yaksh TL (ed) Spinal afferent processing. Springer, Boston, MA, pp 19–57

Wittmann M, Meissner K (2018) The embodiment of time: how interoceptino shapes the perception of time. In: Tsakiris M, De Preester H (eds) The interoceptive mind. From homeostasis to awareness. Oxford University Press, Oxford, England, pp 63–79

Wolpert DM, Flanagan JR (2001) Motor prediction. Curr Biol 11:R729–R732. https://doi.org/10.1016/s0960-9822(01)00432-8

Wolpert DM, Doya K, Kawato M (2003) A unifying computational framework for motor control and social interaction. Philos Trans R Soc Lond B Biol Sci 358:593–602. https://doi.org/10.1098/rstb.2002.1238

Yates BJ, Stocker SD (1998) Integration of somatic and visceral inputs by the brainstem: functional considerations. Exp Brain Res 119:269–275

Zaki J, Davis JI, Ochsner KN (2012) Overlapping activity in anterior insula during interoception and emotional experience. NeuroImage 62:493–499. https://doi.org/10.1016/j.neuroimage.2012.05.012

Zelano C, Jiang H, Zhou G et al (2016) Nasal respiration entrains human limbic oscillations and modulates cognitive function. J Neurosci 36:12448–12467. https://doi.org/10.1523/JNEUROSCI.2586-16.2016

Chapter 4
What Can We Sense? Interoceptive Accuracy

4.1 General Considerations

After reviewing the bottom-up neural processing of body-related information in Chap. 3, in this chapter we will discuss the ability to sense peripheral events, which is called interoceptive sensitivity or accuracy (Ceunen et al. 2013; Garfinkel and Critchley 2013; Garfinkel et al. 2015a). As mentioned at the end of the previous chapter, the fact that interoceptive information reaches higher centers of the brain does not mean that it necessarily reaches conscious awareness and evokes somatic sensations as well. In more technical terms, the main concern of the present chapter is the *detection* and *discrimination* (these two often go hand-in-hand in this area of research) of visceral and other body-related events and changes. Over the decades, considerable efforts were made to develop and refine methods which are appropriate for such elementary sensory measurements; consequently and necessarily, we cannot avoid the detailed discussion of methodological considerations at certain points.

Detection is considered the first step of central processing of information (Pennebaker and Hoover 1984); its output can or cannot reach consciousness. In practical terms, detection is often conceptualized as participants' ability to guess slightly (but statistically already significantly) better than chance about the presence or absence of a stimulus of very weak intensity. Most importantly, participants usually are convinced that they cannot detect the stimulus in question, thus it cannot be verbalized and reported. Verbal reports are typically obtainable at considerable higher stimulus intensities for a number of modalities.

To assess detection ability under the verbalizable level, three basic methods were applied in classic interoception research (Ádám 1998). First, as the involvement of the cerebral cortex was considered mandatory for conscious processing by Pavlov, desynchronization of the resting EEG (see Fig. 4.1) was regarded as the so-called objective threshold in the Bykovian school (Bykov 1957). Second, the Békésy-Blough approximation was used. In this paradigm, originally developed to determine

F. Köteles, *Body Sensations*, https://doi.org/10.1007/978-3-030-63201-4_4

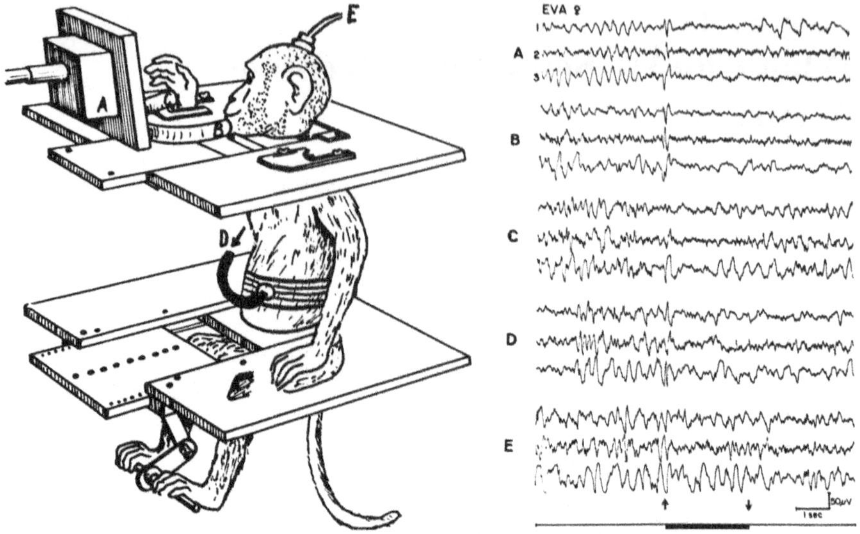

Fig. 4.1 Left: Experimental setting used to measure the electrophysiological and behavioral effects (instrumental interoceptive conditioning) of intestinal stimulation (A, B, C: feeding device with button, D: input of the tube of the balloon, E: cables of implanted EEG electrodes). Right: Registered changes in the EEG evoked by intestinal stimulation in monkey Eva (arrows: start/end of stimulation; 1: fronto-occipital corticalis, 2: lateral-hypothalamic, 3: midbrain formatio reticularis signal). (Source: Ádám 1965)

the auditory threshold, intensity of the stimulus is gradually changed up and down, always based on the feedback of the subject (e.g., verbal report or button press in humans and operantly conditioned behavior in animals). Over the trials, the amplitude of the fluctuations decreases and gradually reaches a final value or range, which is considered the sensory threshold (Békésy 1947; Blough 1958). Third, the methodology of signal detection theory (SDT) was applied. This paradigm describes the actual state of an information processing system as the sum of the signal (information) and noise, which is inherent in all systems (this holds particularly true for visceroception). Detection in this paradigm refers to the ability to discriminate the signal from noise at a given level of intensity. Further, the signal detection approach calculates two basic indices: one (called sensitivity index or *d′*) characterizes the detectability of the signal by the system, whereas the second (dubbed criterion or bias) refers top-down factors systematically influencing participants' decision (Tanner Jr. and Swets 1954; Green and Swets 1966). SDT, widely used in many scientific fields since its development, requires the presentation of hundreds, but at least several tens, of elementary sensory events, thus considerable time and effort from the participants.

In the last several decades, novel paradigms and approaches were developed. Most importantly, the existence of cortical changes evoked by rhythmic contractions of various organs, called viscerosensory-evoked potentials, were revealed and intensely studied (see below). Another promising tool is the startle reaction: afferent information from the cardiovascular, the respiratory and the gastrointestinal system, carried by the vagus nerve, modulates the intensity of the reaction (Richter et al. 2009; Pappens et al. 2013; Schulz et al. 2016, 2017, 2019).

Whereas detection theoretically deals with purely bottom-up information, *perception* refers to the interpretation of this information in the light of past experiences, memories, and external cues (Pennebaker and Hoover 1984). Methods that rely on verbal reports belong more to perception than detection, although the boundary between the two is often fuzzy. In more detail, it is often difficult (if not impossible) to design a pure detection/discrimination paradigm for modalities that easily reach consciousness and/or difficult to manipulate. Moreover, the relationship between signal detection-based measurements and verbal processing is poorly understood; participants in these experiments may use verbal or non-verbal strategies when deciding about the presence or absence of the signal. In other cases, no such "detection" paradigms were developed, thus the available findings were obtained via verbal reports. Therefore, findings obtained by self-report will also be presented in this chapter in certain cases. As we will see at the end of this chapter, the use of perception-type measurements should be preferred for certain modalities.

Also, the majority of body sensations are subject to automatic evaluation (Damasio and Carvalho 2013), thus they have a marked affective-evaluative aspect (e.g., the intrinsic unpleasantness of pain or the genuine pleasantness of sensual touch) beyond the sensory-discriminative aspect. This was also listed among the basic tenets of the present work in Chap. 1. As this evaluation is part and parcel of the neural processing, and, as mentioned in Chap. 3, the affective aspect often appears to be more salient than the discriminative (e.g., it grabs attention and focuses it to the stimuli), I think that it belongs more to detection than perception. In other words, it happens in the early steps of processing of the signal. Thus, a bit surprisingly, the affective-evaluative aspect of body sensations with respect to individual modalities will also be discussed in this chapter.

A second methodological point is that there are huge differences among internal events from a measurement point of view. First, there are events that are generated internally, such as gastric contractions or heartbeats. Receptor-level adaptation is usually not an issue here because of the constant fluctuations. These events can be manipulated to some extent (but not completely), for example, cardiodynamics can be influenced by drugs, posture, and physical activity (see below). However, participants are often aware of the manipulations, and/or multiple trials are difficult to conduct because of the duration of the effect. Second, external manipulation, such as distension of the stomach or the urinary bladder, is usable for a number of modalities. In these cases, fine-tuned multiple manipulations (typically via a balloon inserted in the appropriate region) are possible, and control conditions can be included to partial out non-desired factors, such as tactile information. Third, in certain proprioceptive measurements, such as the angle of a joint or the exertion of a muscle, participants are aware of the actual state, and they have to reproduce a presented state. In these cases, accuracy of reproduction is assessed rather than the sensory detection threshold, which necessarily involves memory.

Even this quick and dirty categorization demonstrates clearly that there are no universally applicable methods in this area of research; in consequence, results obtained with the use of different methods are often difficult to compare with each other and their relationship is not clear. The history of methods designed to assess cardioceptive accuracy and the ongoing debate on their validity illustrate this point very well. Moreover, multiple modalities can be measured for many organs; for example, the ability to sense spontaneous gastric contractions and sensitivity to gastric distension represent two distinct characteristics that belong to the stomach.

And two final remarks, just to illustrate the difficulties of this area of research. First, the majority of the methods developed so far measures detection/discrimination ability regardless of the pathway(s) carrying the information. With the exception of the measurement of colon distension where sophisticated control conditions were designed to exclude skin-related information and some studies on heartbeat detection, most procedures simply measure the totality or integration of information reaching the brain, which often originates in a variety of receptors located in various organs and tissues. This is not really a problem from the viewpoint of this book; however, it poses serious challenges for researchers interested in basic mechanisms. Second, almost all measurements are conducted under resting conditions, to minimize internal and external noise. Considering the fact that most individuals usually perform very poorly even in such optimal conditions and normal stimulus intensities in many tests, it is an obvious question how generalizable these results are to everyday circumstances when our arousal level is elevated, our external senses are bombarded with competing information, and many internal changes occur simultaneously. We will discuss the consequences of this second point later.

The structure of the present chapter is as follows. After a brief summary of early (mostly conditioning related) results, first we will discuss available findings with respect to the classical visceroceptive modalities, i.e., those that belong to the cardiovascular, respiratory, alimentary, and urogenital systems. This will be followed by modalities whose status was recently reconsidered and which were regarded as homeostatic now, such as temperature, itch, pain, sensual touch, and physical exertion. The chemical senses will also be touched briefly. Finally, results obtained from somatosensory modalities and the vestibular system will be presented. For many modalities, the major driving force of research is the understanding of certain pathological conditions, thus I decided to include results obtained from patients too. This might be helpful for understanding the context of the research, as well as for presenting the reader with extreme conditions with respect to the modality in question.

4.2 Early Conditioning Studies

As mentioned in the introduction, interoceptive stimuli can serve as conditioned stimuli in classical conditioning paradigms. Cook and colleagues demonstrated quite early (1960) in the Dog that stimulation of the jejunum can evoke withdrawal of the leg after systematic pairing with electric shocks. Humoral factors, such as adrenaline, noradrenaline, and acetylcholine, were also able to serve as conditioned stimuli. Following the Bykovian school, Ádám (1967) used the rhythmic stimulation of the renal pelvis and the carotid sinus in dogs to evoke conditioned alimentary behaviors. For example, rhythmic dilatation of the renal pelvis at a pressure of 160 mmHg via a chronically implanted balloon let the animal to open the cover of the feeder following the usual pairing procedure (of course, careful steps were taken to exclude other ways of stimulation, such as the auditory and tactile modalities). In a similar way, conditioned operant behavioral responses were controlled by an interoceptive stimulus, the rhythmic distension of the small intestine, in rhesus monkeys (Slucki et al. 1965, 1969). Stimulation was routinely accompanied by EEG measurement; according to the recordings, it was characterized by desynchronization of the resting alpha pattern (Fig. 4.1). It is important to note, however, that desynchronization could have been caused by nonspecific activation of the cortex via brainstem structures too (Ádám 1967; Cameron 2002). In fact, Pavlovian conditioning was demonstrated in decorticate animals quite early (Culler and Mettler 1934; Bromiley 1948), whereas operant conditioning apparently requires cortical structures, such as the insula (DiCara et al. 1970; Thornton and Van-Toller 1973; Lasiter 1983).

Similar experiments were conducted with the participation of human subjects too. Rhythmic stimulation of the duodenum by a balloon caused EEG desynchronization, indicating that the information reached the cortex (Preisich and Ádám 1966). However, participants sensed and reported sensations, such as pressure or (rarely) pain, in 31% of cases only, whereas nothing was felt in 69%. In another experiment investigating the discrimination of two stimuli (i.e., two balloons, 15–20 cm apart, inserted in the duodenum), none of the participants reported any sensation, albeit the EEG indicated clear-cut discrimination (Ádám et al. 1965; Ádám 1967). It was concluded that the majority of intestinal stimuli do not reach consciousness in healthy adults; this conclusion is supported by more recent findings based on the stimulation of the rectum and the esophagus (Kern and Shaker 2002; Kern et al. 2009; Babaei et al. 2013). In another experiments, the threshold of desynchronization (called objective threshold) in terms of balloon pressure was determined, then the intensity of stimulation was gradually increased until it reached consciousness (subjective threshold, assessed via verbal report). In subsequent steps, stimulation under the subjective threshold was gradually weakened, however, verbal reinforcement was given about the occurrence of stimulation. Using this learning procedure, participants' subjective threshold could be lowered to a considerable extent (albeit it usually did not reach the objective threshold) (Ádám 1967; Ádám et al. 1990). Ádám drew the conclusion that the ability to sense visceral events can be acquired (perhaps re-acquired, see below) by learning. He also stated that the same learning process might take place in early childhood when the urge to defecate and urinate, relying on visceroceptive signals, gradually becomes conscious. Since then, the possibility of lowering the detection threshold of visceral events, such as heartbeats or skin temperature, by learning was demonstrated many times (Brener and Jones 1974; Ross and Brener 1981; Schandry and Weitkunat 1990; Guglielmi and Roberts 1994; Schaefer et al. 2014; García-Cordero et al. 2016). Heart transplanted patients showed higher level of cardiac accuracy than controls which was also explained through the involvement of a learning process (Barsky et al. 1998). In a recent longitudinal study, cardioceptive accuracy decreased after heart transplant and reached the original level 1 year later (Salamone et al. 2020). The authors highlight the role of vagal (visceral) innervation and reinnervation in the ability to sense the heartbeats.

Despite their life-long availability (Fisher and Cleveland 1968; Ádám 1998), normal (non-noxious) visceral events are usually not or only vaguely sensible and reportable for healthy adults (this is the tenet of vagueness). This can be the consequence of the weakness of these signals (set by natural selection over millions of years). Another (complementary) explanation emphasizes the relative lack of words for internal sensations; as these percepts cannot be cross-validated by others, they are less elaborated (i.e., less words are available to describe them) and even the available words show huge individual differences in their meaning (Mason 1961). This does not mean that infants do not sense internal events though. It was demonstrated by Soviet researchers in rabbit embryo quite early that interoceptive stimuli can inhibit fetal movements (Moiseeva 1952) and visceral reflexes develop quicker in newborn puppies than later (Moiseeva 1966). These empirical findings let to the conclusion that interoceptive stimuli are at least as sensible as exteroceptive early in life. Perception of internal cues might play a particularly prominent role at early stages of development in animals with long parenting periods as the infant needs to monitor and express its need states to the parents (Ádám 1967). Later, attention turns to the external world and the detection of internal events weakens simultaneously (Ádám 1998). However, early parental care may impact not only the actual homeostatic state of the infant but also its adult physiological responses (Fox et al. 2007; Rinaman et al. 2011) and interoception (Fotopoulou and Tsakiris 2017; von Mohr and Fotopoulou 2018).

Concerning humans, recent research demonstrated that 5-month-old infants are able to discriminate animated characters moving in synchrony or asynchrony with their own heartbeats (Maister et al. 2017). Attempts were made to develop paradigms to assess cardioceptive accuracy in pre-school children (Schaan et al. 2019), as well as for older children (Eley et al. 2004, 2007; Schmitz et al. 2012; Koch and Pollatos 2014a, b; Georgiou et al. 2015; Schauder et al. 2015; De Witte et al. 2016; Mash et al. 2017; Palser et al. 2018; Pile et al. 2018). These studies show that, similar to adults, there are marked individual differences among children in the accuracy of perception of cardiac activity. Moreover, accuracy is associated with emotion regulation and a variety of psychopathologies, such as eating disorders, anxiety, and autism. Adolescence might represent another particularly sensitive period with respect to the development of interoception-related psychopathologies (Murphy et al. 2017).

Interoceptive conditioning studies implicate that although internal signals under the subjective threshold do not reach consciousness, they can evoke changes in physiology and overt behavior in humans. For example, stimulation of various visceral organs through fistulas or swallowed balloons evoked a number of visceral and skeletal reactions in participants after the usual conditioning procedure (Razran 1961). In these cases, the cause of the reactions and/or the reaction itself remain hidden to the individual, depending on the actual pattern of conditioning. Subcategories within interoceptive conditioning were defined by Soviet researchers with respect to the nature of the unconditioned and conditioned stimulus; for example, interoexteroceptive conditioning refers to the pattern where the conditioned stimulus is interoceptive and the unconditioned stimulus is exteroceptive (Razran 1961; Dworkin 2007). Overall, the continuous establishment and extinction of associations between exteroceptive and interoceptive stimuli and internal and behavioral responses was assumed. Functionally, these responses are preparations for future events that challenge the internal balance of the organism, thus they can be considered representatives of predictive homeostasis (Winters et al. 2002). It was also demonstrated that individuals with good heartbeat detection ability show better learning performance in a non-conscious fear conditioning task than poor detectors (Katkin et al. 2001). In other words, heartbeats can be used as conditioned stimuli that predict dangerous situations; however, individuals are aware of their "gut feelings" only but not of the exact visceral stimuli.

It also turned out that visceral stimulation under the conscious threshold can influence behavior in an intensity-dependent way in certain cases. For example, whereas weak stimulation of the gastrointestinal tract and the uterus, below the subjective threshold in humans, has a calming and stress-reducing effect, higher and sensible intensities evoke pain, aggression, and increase reactivity to stress (Kukorelli et al. 1972; Kukorelli and Juhász 1976, 1977; Ádám 1978, 1983; Bárdos et al. 1980; Juhász et al. 1980). In a similar vein, activation of baroreceptors leads to the reduction of arousal (Bonvallet et al. 1954) although changes in blood pressure are not perceived directly (Furedy et al. 1992; Dworkin et al. 1994).

Overall, these results (in concert with the findings presented below) show that interoceptive (visceroceptive) signals under the perception threshold might play a role in the regulation of behavior (tenet of relevance). However, the extent and overall impact of this factors, as well as its psychobiological mechanisms, is yet to be explored.

4.3 Visceral Modalities 1: Heartbeat

The aforementioned methodological problems can perhaps be best illustrated by the history of methods aimed to measure the ability to detect heartbeat, a rhythmic, internally generated event. Thus, we will discuss these methods in detail. Cardiac sensations are located primarily in the anterior lower left part of the chest, i.e., about the position of the heart. The head, the arms, and the center of the abdomen may also be impacted (Khalsa et al. 2009b).

Brener and Jones (1974) gave participants tactile feedback which was synchronized to actual heartbeats or not; the asynchronous condition was stimulation at a constant pace that was equal to the heart rate. Participants had to decide whether short trains of stimuli were in synchrony with their heartbeats or not. Performance, simply measured as the percentage of correct responses, was close to 50%, i.e., random guessing; this was improved to approximately 75% by learning. Clemens and MacDonald (1976) replaced tactile feedback with visual. It also turned out that some participants were able to manipulate their heart rate via respiratory and muscular maneuvers (Carroll 1977). To address this issue, Whitehead et al. (1976, 1977) modified the procedure: participants had to discriminate between series of lights occurring 128 or 384 ms after heartbeats, and participants' ventilation was controlled by a machine (!). The 128 ms condition was considered immediate feedback as this is the approximate delay between the R-peak of the ECG and the mechanical events (i.e., the actual contraction of the ventricles and the progression of the blood pulse in the arteries) that were assumed to be sensed. In addition, SDT was applied to evaluate performance instead of simple percentages. Individual sensitivity indices ranged from 0.05 to 1.56, with a median of 0.33; this indicates large individual differences and a poor group-level performance. Overall, about 25% of participants was categorized as good perceiver, which could be improved only slightly in modified versions of the task (Whitehead et al. 1977; Brener et al. 1993; Windmann et al. 1999; Brener and Ring 2016). Interestingly, it also turned out that the Brener-Jones and the Whitehead task show a very weak association thus they may measure different abilities (Ross and Brener 1981; Jones et al. 1984). Reliability issues were also reported for both tasks (Clemens 1979; Wildman and Jones 1982), although this was later ascribed to fluctuations in subjects with poor cardiac awareness (Jones et al. 1987). Validity was assessed based on the theoretical consideration that increased cardiac output (i.e., stronger heartbeats) should be sensed more readily (Dale and Anderson 1978). This was demonstrated empirically for the Brener-Jones (Montgomery et al. 1984) and the Whitehead methods (Jones et al. 1987). However, manipulations of stroke volume by tilting subjects did not impact performance in the Brener-Kluvitse method (see below) (Ring et al. 1994).

Later, the original method by Whitehead and colleagues were developed further in several steps. Most importantly, based on the argument that the optimal delay for sensation can be longer than 128 ms and individual differences in this respect may exist, variance in delay was included using different stimulus patterns (Katkin et al. 1981, 1982; Yates et al. 1985; Brener and Kluvitse 1988). Moreover, visual feedback was changed to auditory in some tasks to improve participants' temporal judgments. Overall, it turned out that there are marked individual differences in the optimal delay (i.e., that associated with highest performance), and the range from (100–)200 ms to 300(–400) ms from the R-wave appears to be the most convenient for the majority of individuals. Still, performance assessed by these novel tasks remained poor without training (Brener et al. 1993). The method of constant stimuli (MCS), developed by Brener et al. (1993), appeared to be a reliable, relatively short, and sensitive paradigm, thus its use is recommended today (Schneider et al. 1998; Ring et al. 2015; Ring and Brener 2018). Still, further efforts to improve the applicability of the discrimination paradigm are made recently (Kleckner et al. 2015). From a practical point of view, these paradigms are time-consuming and require considerable mental effort from participants; thus, their applicability is limited.

Simultaneously with the appearance of discrimination methods, a different approach to the measurement of heartbeat detection performance, called heartbeat tracking, emerged. Theoretically, this task requires the sensation of each individual heartbeat instead of making a more or less intuitive decision about series of exteroceptive signals. Brener (1974) developed a method where participants had to press a button after each sensed heartbeat; accuracy of perception was calculated from the distribution of latencies between R-waves of the ECG and button presses. This approach was scarcely used in the 1980s (Ludwick-Rosenthal and Neufeld 1985) but gained popularity recently (Melloni et al. 2013; Sedeño et al. 2014; Couto et al. 2014; Canales-Johnson et al. 2015; Hamano et al. 2015; Yoris et al. 2015; Garcia-Cordero et al. 2016; García-Cordero et al. 2017; Yao et al. 2018; Fittipaldi et al. 2020). Unfortunately, a number of calculation methods are used to evaluate accuracy, and the relationships among these indices are not clear. At about the same time, McFarland (1975) asked participants to press a button immediately after each sensed heartbeat, and simply compared the number of presses to actual heartbeats using a sensitivity index: $1 - |(N_{presses} - N_{heartbeats})/N_{heartbeats}|$. Later, this formula became the most widely used way to calculate performance in tracking tasks; an alternative that better handles overestimation tendency was developed by Hart et al. (2013). In the original formula, a value of 0 refers to complete inability to detect individual's heartbeats, whereas 1 indicates perfect performance. Values in the original experiment ranged between 0.48 and 0.98 with a median of 0.85 (McFarland 1975). At first sight, this appears to reflect quite good performance; however, the task was preceded by a trial with visual feedback of heartbeats, and periods without button presses were removed from the analysis. This obviously led to a systematic bias, i.e., the overestimation of detection ability. Later, Dale and Anderson (1978) and Schandry (1981) replaced button presses with simple counting (aka mental counting or mental tracking task). For example, 41% of the participants showed very good (close to perfect) performance in the original Schandry study; however, estimation was also allowed which again leads to potential overestimation. The internal consistency of the Schandry task is usually very good, whereas its test–retest reliability is in the acceptable range (Mussgay et al. 1999; Herbert et al. 2011; Parkin et al. 2013). Its test–retest correlation over a period of 8 weeks was 0.60 in our own study (Ferentzi et al. 2018b). Similar to the discrimination tasks, validity was demonstrated by manipulation of cardiodynamics through changes of posture and body tilt, and sympathetic activation (exercise) (Schandry et al. 1993; Schandry and Bestler 1995). Further, albeit indirect, support for the validity of the Schandry task was received from EEG studies (Gardner et al. 1990). Heartbeat-related changes in the EEG are called heartbeat-evoked potentials (HEP) (Schandry et al. 1986; Montoya et al. 1993). Higher HEP activity was found to be associated with better performance in the Schandry task in several studies (Pollatos and Schandry 2004; Pollatos et al. 2005; Canales-Johnson et al. 2015), which was moderated by attention (Yuan et al. 2007). Further, internally focused attention was associated with higher HEP amplitudes (García-Cordero et al. 2017; Petzschner et al. 2019). Finally, experimental findings concerning the influence of cardiac perception on malleability of body illusions, self-perception, and memory distortion presented in Chap. 3 generally applied this approach to the measurement of cardiac accuracy. This clearly indicates some sort of validity, in spite of the methodological caveats discussed below.

The two approaches presented above represent the two major categories of methods used today, called discrimination (aka forced-choice) and tracking paradigms. It is important to note that most of the studies found no significant association between performances shown in various versions of the two paradigms (Pennebaker and Hoover 1984; Weisz et al. 1988; Phillips et al. 1999; Schulz et al. 2013; Michal et al. 2014; Forkmann et al. 2016; Ring and Brener 2018). A medium level association was reported in several studies (Schaefer et al. 2012; Hart et al. 2013; Garfinkel et al. 2015a, b), whereas Knoll and Hodapp (1992) found associations for very good and very poor perceivers only. A recent meta-analysis, including 23 studies, found a weak association (pooled effect size was 0.21) only (Hickman et al. 2020). This clearly indicates that discrimination and tracking paradigms measure two different constructs; consequently, they should not be considered as equivalent indicators of cardioceptive accuracy (Forkmann et al. 2016; Hickman et al. 2020).

Over time, a number of factors that influence performance in these tasks was described. Beyond instruction (see below) and experimentation-related characteristics (motivation to cooperate, demand characteristics, conformity, etc.), it was assumed that actual focus of attention might play a role in the accuracy of cardiac perception. Self-awareness theory states that internally focused attention makes the perception of internal states more accurate (see in Chap. 5) (Pryor et al. 1977; Gibbons et al. 1979). With respect to cardiac accuracy, self-directed attention evoked by a mirror positioned in front of the participant improved performance in a discrimination task but not in a tracking task (Weisz et al. 1988). In another study using the tracking paradigm, self-focus elicited by a camera improved performance for healthy controls but decreased it for patients with anorexia nervosa (Pollatos et al. 2016). Other studies applying the tracking paradigm reported improvement by self-focus (triggered by a mirror or the image of own face) only for those with low levels of accuracy (Ainley et al. 2012; Maister and Tsakiris 2014); accuracy of tracking was improved by the image of own face and self-relevant words in one study (Ainley et al. 2013). Breathing exercise, another potential means to focus attention on the body, did not enhance cardioceptive accuracy as assessed with a discrimination task (Khalsa et al. 2008). Overall, cardiac accuracy of individuals with a long history of meditation (i.e., presumably high level of self-awareness and superior ability to direct attention on the body) was not better than that of control persons (Nielsen and Kaszniak 2006; Khalsa et al. 2008; Melloni et al. 2013; Otten et al. 2015). Moreover, a short-term intervention with self-focus components was not able to enhance cardiac accuracy in the majority of studies (Parkin et al. 2013), whereas improvement was found in two studies (Bornemann and Singer 2017; Fischer et al. 2017). In our own unpublished study, participants' cardiac accuracy (tracking task) and gastric perception (water load task) were not improved by a 10-week-long yoga intervention. In summary, findings on the impact of self-focused attention on cardiac accuracy are inconclusive.

Cardiac accuracy is influenced by a number of partly overlapping physiological characteristics and conditions, such as cardiovascular reactivity (Pollatos et al.

2007b; Herbert et al. 2010) and stress (Fairclough and Goodwin 2007), systolic blood pressure (O'Brien et al. 1998), resting heart rate and heart rate variability (Knapp-Kline and Kline 2005), and cardiodynamics (most importantly cardiac stroke volume) (Schandry et al. 1993). Moreover, males often perform better in the laboratory than females (Katkin 1985; Jones 1994, 1995; Ehlers et al. 2000), which was partly explained by females' higher reliance on external stimuli (Pennebaker and Roberts 1992). Males' superior performance can also be explained by cardiovascular differences and fitness, as well as by their lower body fat percentage which dampens mechanical stimuli detected by mechanoreceptors located in the chest wall and deep layers of the skin (Rouse et al. 1988; Jones 1995). As no difference between obese and nonobese individuals was found in a well-designed study (Gardner et al. 1990), and no sex differences were reported in other studies (Ferentzi et al. 2018b; Köteles et al. 2020a), this factor can be questioned (Cameron 2002). The accuracy of cardioception, similar to other senses, declines with age (Khalsa et al. 2009c; Murphy et al. 2018b). As it was mentioned in the previous chapters, these factors are important thus should be controlled for if and when basic mechanisms of detection and sensation are investigated. The reader can find more information on these issues from the excellent reviews of Jones (1994, 1995) and Cameron (2002). Also, both somatosensory and interoceptive channels are involved in the detection of heartbeats (Jones et al. 1987; Khalsa et al. 2009a; Salamone et al. 2020). From the integrated point of view of the present work, however, the extent individuals are able to sense their cardiac activity is of primary interest, regardless of the exact mechanism(s).

Over time, both major paradigms received considerable criticism with respect to their validity. These critiques very well reflect the difficulties of visceroceptive measurements as well as the different viewpoints and biases of the researchers.

The major objection toward the discrimination approach is that it requires the share of attention: participants need to simultaneously process exteroceptive (tactile, visual, or auditory) stimuli and compare them to internal sensations (Domschke et al. 2010; Couto et al. 2015; Petzschner et al. 2019). This is for sure a complex task that requires many additional abilities and cognitive steps beyond the sheer sensation of heartbeats. Other authors do not consider this an important limitation (Ring and Brener 2018). Our own observation was that external feedback accompanied by body focused attention often triggers body sensations that interfere with or even suppress genuine heart-related sensations (these attention-induced sensations were discussed in Chap. 5). In other words, consequences of perception cannot be completely eliminated even from discrimination tasks. Another, at first sight surprising, criticism is that results obtained by these paradigms are too "sterile"; i.e., we humans also take into consideration external information in the perception process under everyday circumstances (Pennebaker 1981; Pennebaker and Hoover 1984). Again, this point can be ignored by those who are interested in visceral detection and conduct measurement under lab conditions. However, it does make pretty much sense when we are primarily interested in associations between the perception of cardiac activity and other phenomena such as emotions and various pathologies.

Interactions between input and response (tapping, counting) may exist in the tracking paradigms too (Petzschner et al. 2019). A further critique is here that participants simply reproduce a rhythm, and this is counted instead of actual heartbeats (Yates et al. 1985; Weisz et al. 1988; Kleckner et al. 2015). In line with this, believed heart rate showed moderate level association with counted heartbeats in several studies (Ring and Brener 1996; Dunn et al. 2010), but it was independent of them in another study (Ainley et al. 2014).

Accuracy of time estimation, a conceptually related factor, was independent of cardiac accuracy in one study (Ehlers and Breuer 1992) and showed weak to moderate associations with the Schandry-score in a number of experiments (Knoll and Hodapp 1992; Dunn et al. 2010; Terasawa et al. 2014; Ainley et al. 2014; Shah et al. 2017). Estimation of resting heart rate, a factor that is conceptually related to time estimation, represents another biasing factor. It was demonstrated that heart rate-related beliefs are positively associated with the Schandry score (i.e., higher estimated accuracy) (Pennebaker and Hoover 1984; Ludwick-Rosenthal and Neufeld 1985; Phillips et al. 1999; Ring et al. 2015). Finally, Schandy scores were predicted by beliefs on heart rate rather than actual heart rate in patients with cardiac pacemakers after manipulating actual heart rate (Windmann et al. 1999).

A further critique is that, as accuracy and criterion cannot be separated, tracking methods are generally more prone to perceptual biases (Pennebaker and Hoover 1984; Witthöft et al. 2020). Additionally, button pressing requires motor abilities and its efferent and afferent component can interfere with the sensation of heartbeats (Pennebaker and Epstein 1983; Pennebaker and Hoover 1984). The mental counting task avoids this caveat; however, it makes approximation of heartbeats (e.g., based on an internal rhythm) easier; it is important to note at this point that estimation was allowed in certain instructions of the Schandry task (Schandry 1981; Pollatos et al. 2007c). Indeed, findings of several studies clearly indicate that instruction does impact performance; if participants are explicitly asked to count sensed heartbeats only, accuracy scores decrease (Ehlers et al. 1995; Ehlers and Breuer 1996; Desmedt et al. 2018). In our lab, we also use a rather strict instruction and often receive questions from reviewers about the unusually low Schandry scores (our mean is typically about 0.5, whereas many studies report means in the 0.7 domain). Overall, as the vast majority of participants does not count each heartbeat, heart rate is considerably underestimated by this method (Zamariola et al. 2018; Corneille et al. 2020); this shows that sensation of individual heartbeats is very difficult if not impossible for most people. This underestimation tendency generally supports the notion that participants actually count they sensed heartbeats and miss some of them (Ehlers and Breuer 1992). Still, they might use their knowledge instead of sensations in ambiguous cases (Brener and Ring 2016).

The debate on the validity of the Schandry task gained new momentum recently (Zamariola et al. 2018; Ainley et al. 2020; Zimprich et al. 2020). A number of authors draw the conclusion that this method is not appropriate for the measurement of cardioceptive accuracy (Ring and Brener 1996; Ring et al. 2015). On the other hand, it can be argued that knowledge of heart rate (partly) based on previous experiences and situational cues can be satisfactory under everyday circumstances, thus it represents a usable piece of information (Pennebaker 1981; Pennebaker and Hoover 1984; Ring and Brener 2018).

For some reason, the weak points of discrimination methods receive less attention. Participants need to decide about the correspondence between a number (usually 5–10) of heartbeats and external stimuli. How they reach a yes-or-no decision, however, is not really explored. Different strategies are conceivable here too. For example, if 2–3 heart beats are clearly sensed and are in synchrony with the external stimuli, certain participants might conclude that the correct answer is yes. Others might apply a more strict criterion and press the yes button only when all heartbeats are perceived to be synchronous with the reference signal. This represents an additional source of bias, not assessed by SDT.

Other approaches to the accuracy of sensation of cardiac activity focus more on heart rate. For example, participants are asked to indicate when a change in their heart rate occurs (Mandler and Kahn 1960; Epstein and Stein 1974), discriminate between short periods with higher and lower heart rate (Ashton et al. 1979), or reach synchrony between their actual heart rate and a visual signal by adjusting the latter (Porges and Raskin 1969; Carroll 1977; Gannon 1977, 1980; Carroll and Whellock 1980; Ehlers et al. 1988; Palmer et al. 2019). Finally, the intraindividual approach manipulates participants heart rate in a number of tasks and estimates the association between actual and estimated heart rates for each individual (Tyrer et al. 1980; Pennebaker et al. 1982). Overall, healthy participants' cardiac awareness proved to be poor in these studies too. In our own study (Köteles and Babulka 2014), participants inhaled very tiny doses of three essential oils (lavender, eucalypt, and rosemary) and rated experienced change in heart rate and blood pressure after each intervention, respectively. Actual and perceived changes were not related to each other; however, perceived changes in alertness (another subjectively assessed characteristics) showed a medium to strong association with the perceived physiological changes. It was concluded that in the absence of perceivable bottom-up information, individuals use other available internal information to infer physiological changes. It is also important to note that accuracy of pulse estimation was not related to accuracy of heartbeat detection in medical outpatients (Barsky et al. 1993), thus these paradigms may measure different abilities.

The overall conclusion that can be drawn from the presented findings is that, regardless of the paradigm and method used, the majority of healthy individuals sense cardiac activity fairly poorly under resting conditions. They are usually motivated to achieve good results in experimental situations, thus they try to utilize additional internal and/or external information to improve their estimation. My personal opinion is that this latter process does not take place under everyday circumstances. In other words, we are not curious or do not care about our cardiac activity most of the time. Important exceptions to this rule are physical activity, certain cardiovascular conditions, panic attacks, and perhaps also emotions (this latter topic will be discussed in Chap. 8).

Physical activity is accompanied by sympathetic activation which influences many cardiac parameters thus can improve detection performance (Schandry et al. 1993). On the other hand, internal noise due to the rhythmic movements, breathing, and increased arousal also increases which impairs detection. According to empirical evidence, healthy participants, as well as cardiac patients in the rehabilitation phase, replicate a previously presented physical load with a marked underestimation tendency (i.e., the replicated load is higher than the presented) (Kollenbaum 1994; Kollenbaum et al. 1994, 1996; Köteles et al. 2020a). Moreover, individuals with better perception performance at rest set lower levels of physical load for themselves (Herbert et al. 2007). This latter result was explained by better fine-tuning of activity level using cardiac cues. In our own study (Köteles et al. 2020a), however, an association between resting cardiac perception (Schandry task) and replication performance was revealed only for very low levels of physical activity, i.e., walking at a quick pace. For higher levels of activation, but still in the aerobic domain (i.e., jogging and running at medium pace), this association disappeared. Although participants were clearly instructed to use their perceived heart rate in the replication tasks, they used a variety of other internal cues such as breathing or general feeling of fatigue to estimate their actual level of activation. This shows that more systemic or general indicators such as fatigue, hunger for breath, or actual mood state may be more useful for the estimation and regulation the actual effort than cardiac activity (Mihevic 1981; Jameson and Ring 2000; Ekkekakis and Acevedo 2006; Ekkekakis 2009). Concerning athletes, distance runners were more accurate in perceiving cardiac changes than non-runners (House 1983). Further, male long-distance runners showed superior performance to controls in a study, but females did not (Jones and Hollandsworth 1981). Unfortunately, these promising results could not be replicated later (Montgomery et al. 1984). Considering the fact that the primary strategy of distance runners is dissociation (e.g., distracting attention from the actual situation) (Morgan and Pollock 1977; Lohse and Sherwood 2011; Razon et al. 2014), regulation of actual effort appears to run in a lower, non-conscious level. Thus, even if elite athletes show superior performance in heartbeat detection tasks, this can be due to their cardiovascular characteristics rather than any learned ability.

Heartbeat is often associated with both negative and positive emotions (Nummenmaa et al. 2018); in other words, its valence is relatively neutral in healthy people (Chap. 8). However, in patients with congenital heart disease and panic disorder, heart-related sensations become associated with negative affect, i.e., anxiety (Karsdorp et al. 2007, 2009; Khalsa and Feinstein 2018). In these patients, assessment of the evaluative aspect beyond detection ability would be of primary interest.

4.4 Visceral Modalities 2: Blood Pressure

Although medical textbooks generally emphasize that blood pressure and its changes in the non-extreme range are not perceivable thus should be measured by devices, many people associate changes in mood and activation level with blood pressure (and often mistakenly adjust their medication accordingly). This deeply rooted conviction is very difficult to change by contrary information (Baumann and Leventhal 1985), particularly for those with hypertension (Meyer et al. 1985). As elevated blood pressure is a feature of stress reaction and leads to cardiovascular pathology in the long run, biofeedback studies with animal models and human subjects targeted its controllability mainly in the 60s and 70s. It turned out that the effectiveness of pure biofeedback interventions is smaller than those combined with relaxation techniques; moreover, the decrease in blood pressure achieved by such combined interventions is clearly negligible from a clinical point of view (Seer 1979; Nakao et al. 2003). Researchers started to think about the potential role of interoception in the process. It turned out that estimation of actual blood pressure and accuracy of discrimination is poor in untrained individuals but can be improved via feedback; moreover, it shows large individual differences (Luborsky et al. 1976; Cinciripini et al. 1979; Vidergar et al. 1983; Greenstadt et al. 1986; Näring and van der Staak 1995). No reliable replication of previously presented blood pressure values under physical exercise was reported in two studies (Kollenbaum et al. 1994, 1996). An experimental approach, i.e., the direct manipulation of carotid baroreceptors, led to basically the same conclusions (Furedy et al. 1992). Further, patients with hypertension showed higher accuracy in the perception of blood pressure than normal controls (Koroboki et al. 2010). Within-subject analyses indicated that individuals tend to associate blood pressure changes with indirect indicators such as mood and various symptoms; however, correlations with actual changes are quite low on average (Baumann and Leventhal 1985; Pennebaker and Watson 1988; Müller et al. 1994; Brondolo et al. 1999). Thus, hypertension is not a completely asymptomatic disease; however, there are large individual differences with respect to the quality, quantity, and reliability of the associated symptoms (Reek et al. 1982; Steptoe and Vögele 1992; Pilgrim et al. 1992). The acute inhibitory effect of baroreceptor stimulation on the central nervous system is well-known (Dworkin et al. 1994; O'Brien et al. 1998); subjective changes in symptoms and mood state reported by individuals can be partly explained on this basis. In chronic cases, however, baroreceptors adapt thus cannot measure absolute values of blood pressure (Dworkin 2007). This can partly explain why studies on the associations between blood pressure and mood states (including disorders) are inconclusive (Donner-Banzhoff et al. 1997; Jorm 2001; Lucas et al. 2004; Lee and Hotopf 2009; Ng et al. 2010). However, as it was presented in the previous section, higher blood pressure improves detection of heartbeats (O'Brien et al. 1998; Koroboki et al. 2010); thus, the latter could theoretically serve as a proxy for blood pressure estimation. Beyond indirect internal indicators, estimation of blood pressure is often based on contextual or environmental cues (Fahrenberg et al. 1995). Feedback based on external and internal ("biosituational") information improved most effectively the accuracy of estimation (Barr et al. 1988); and the other way around, expectations evoked by situational cues can substantially bias it (Köteles and Babulka 2014).

In conclusion, individuals' ability to assess their blood pressure and its changes is generally poor. It relies non-interoceptive and indirect interoceptive cues based on believed associations rather than direct access to blood pressure values (Pennebaker and Epstein 1983). The subjective indicators (symptoms) of blood pressure-related information are uniformly negative (unpleasant).

4.5 Visceral Modalities 3: Respiratory Modalities

The respiratory system has a special position among the visceral systems: as respiratory movements (inhalation and exhalation) are controlled by striated muscles, such as the intercostal muscles and the diaphragm, breathing activity can be voluntarily influenced. This also means that once we pay attention we are perfectly aware of our actual breathing. Similar to cardiac perception, as multimodal integration takes place during processing, it is hard to decide whether this ability belongs to proprioception or visceroception. The rhythm of respiration in turn impacts cardiac activity (this is called respiratory sinus arrhythmia) which makes us able to manipulate our actual heart rate to some extent (this issue was already mentioned above). Further, a respiratory pattern that is different from the automatic pattern and that is voluntarily maintained for a while can lead to systemic changes which impact a number of vegetative functions (Ma et al. 2017; Russo et al. 2017; Saoji et al. 2018; Zaccaro et al. 2018), pain (Jafari et al. 2017), attention (Melnychuk et al. 2018), and consciousness (Rhinewine and Williams 2007; Miller and Nielsen 2015; Saoji et al. 2018). From this point of view, respiration can be considered a gate to other autonomic functions and altered states of consciousness; this fact was realized quite early and utilized by various breathing practices in religious and nonreligious traditions. For a researcher, these characteristics pose a methodological challenge. It can be argued, however, that for naïve individuals (i.e., those without long-term experience with breathing practices), the maintenance of a non-normal breathing pattern over a period of time is too demanding; sooner or later we tend to forget it (particularly if we do not think that it is an important factor) and the autonomic nervous system takes back the control of respiration.

Hunger for breath or breathlessness (primarily caused by increased CO_2 level in the blood) is among those rare visceral modalities that can evoke conscious sensations (Paintal 1986). These sensations are accompanied by the activation of the insula, the cingulum, and other limbic/paralimbic regions of the brain (Banzett et al. 2000; Liotti et al. 2001; Evans et al. 2002). In experimental settings, participants are typically asked to breathe air which contains various levels of CO_2 or rebreathe from a bag; then, the association between objective respiratory parameters (e.g., tidal volume, respiratory rate) and subjective reports of the respective characteristic of breathing are calculated (Van den Bergh et al. 2004). Although this index is often called interoceptive accuracy, in fact it is closer to self-report (perception) than a pure detection. Participants perform quite good in this test, those with medically unexplained symptoms (MUS) showing lower levels of accuracy than individuals without MUS (Chap. 7) (Van den Bergh et al. 2004; Bogaerts et al. 2005, 2008).

Another, although not independent, aspect of respiratory activity is the sensation of actual airway resistance, i.e., the internal resistance of the respiratory system influencing the stream of airflow. Under everyday conditions, this factor is mainly determined by the internal diameter of the airways; high resistance due to the narrowing of the airways at various points of the system, such as in the case of various respiratory diseases, needs more muscular effort to overcome. Information on actual resistance is transmitted partly by the vagus, partly via other mechanisms such as respiratory muscle afferents (Noble et al. 1971; Killian et al. 1980; Gandevia et al. 1981; Zhao et al. 2002; Parshall et al. 2012); thus, one cannot speak of pure visceroception in this area either. Airway resistance can be manipulated indirectly (e.g., by narrowing the airways with the use of drugs) or directly (e.g., by asking participants to breathe through a mouthpiece whose resistance can be modified). In the latter case, a number of trials can be conducted and the ability to discriminate between different loads can be measured (Campbell et al. 1961; Bennett et al. 1962; Wiley and Zechman 1966; Zechman and Burki 1976; Yan and Bates 1999; Garfinkel et al. 2016). Peak expiratory flow rate, i.e., the maximum speed of exhaled air, is also used as an indicator of airway resistance. Systematic experimentation in this field started with asthma patients. It turned out that patients with asthma show quite poor average awareness of respiratory functions with large individual differences (McFadden et al. 1973; Rubinfeld and Pain 1976, 1977; Silverman et al. 1987; Kifle et al. 1997; Davenport et al. 2012; Barnes et al. 2019); however, this can be improved by training (Silverman et al. 1987; Reeder et al. 1990; Schandry et al. 1996; Stout et al. 1997). This also means that symptoms reported by asthma patients such as breathlessness often have no clear-cut biological background (i.e., the actual obstruction of the airways) (Rietveld et al. 1999). For example, expectations and emotions can heavily influence them (Meyer et al. 1990; Put et al. 2004; De Peuter et al. 2004, 2007; Giardino et al. 2010; von Leupoldt et al. 2010b; Chan et al. 2012; Tsai et al. 2013), and low-level categorization processes appear to play a substantial role (Petersen et al. 2014; Van den Bergh et al. 2018; Zacharioudakis et al. 2020). Also, patients with asthma sense actual airway resistance less accurately than non-asthmatic individuals (Dahme et al. 1996). As in other modalities, healthy individuals also show considerable individual differences with respect to the accuracy of perception of airway resistance and peak expiratory flow (Webster and Colrain 2000; Köteles et al. 2018), and males perform better than females (Harver et al. 1993). On the other hand, females experience more unpleasant body sensations, such as chest pressure, faintness, dizziness, and trembling, and show more aversive reactions to higher respiratory loads (Miller and Davenport 2015). Discrimination ability can be improved by yoga and meditation practice (Villien et al. 2005; Daubenmier et al. 2013). Respiratory-related evoked potentials are also used in this field. Whereas the P1 amplitudes correlate with the actual resistive load and are generated mainly in the sensorimotor cortex, the P3 components appear to characterize subjective perception of load originating from the parietal cortex (Knafelc and Davenport 1997, 1999; Webster and Colrain 2000; Davenport et al. 2007; von Leupoldt et al. 2010a, b). Severe respiratory load, used as unconditioned stimulus, was associated with interoceptive (slight respiratory load) and exteroceptive (neutral picture) stimuli in experiments on fear conditioning. Whereas the former conditioned stimulus increased the startle reaction only, the latter was associated with more self-reported fear too (Pappens et al. 2013).

Perhaps the most disturbing symptom of the infections of the upper respiratory tract is nasal congestion. We clearly sense how difficult and effortful breathing through our noses is. Unfortunately, even these very concrete sensations are not particularly accurate. The overall association between the severity of subjective symptomatology and that estimated by a physician is very weak in the case of upper respiratory infections (Barsky et al. 1988). Objective measurements of nasal airflow, such as rhinometry, cannot accurately predict subjective sensation (André et al. 2009; Dipak et al. 2011; Baraniuk 2011). In fact, the most important factor shaping the sensation of nasal fullness is the temperature of the mucosa; cold sensation, e.g., evoked by essential oils such as eucalypt, decreases felt nasal patency (Burrow et al. 1983; Sullivan et al. 2014; Bailey et al. 2017). Expectations can also shape this experience to a great extent (Pennebaker and Skelton 1981; Pennebaker 1982).

Overall, respiratory changes are sensed better than those in other visceral modalities; however, there are considerable individual differences (Steptoe and Vögele 1992; Steptoe and Noll 1997), and various factors (including affect) can distort accuracy here too (von Leupoldt and Dahme 2013; Hayen et al. 2015; Faull et al. 2017). Higher levels of CO_2 in the blood, higher respiratory resistance, and nasal fullness indicate a deviation from homeostasis thus are associated with negative evaluation and evoke avoidance behavior (Klein 1993; von Leupoldt and Dahme 2005; Davenport and Vovk 2009; Alius et al. 2013). Interestingly, sighs, i.e., spontaneous deep breaths, have a positive effect on both affect and physiology via resetting respiratory regulation (Vlemincx et al. 2010, 2013, 2016; Wuyts et al. 2011).

4.6 Visceral Modalities 4: Gastric Modalities

In 1912, Walter Cannon and Anton Washburn published a paper on the assumed physiological background of hunger (Cannon and Washburn 1912). They differentiated appetite and hunger, and claimed that "hunger pangs" are caused by rhythmic contractions of the stomach, i.e., they are of local origin and not a general or systemic sensation. As empirical evidence from animal experiments was limited (Hertz 1911), Washburn decided to carry out self-experimentation. First, he became accustomed to a tube with a rubber balloon attached to its lower end that was swallowed for 2–3 h every day for several weeks. When the presence of the equipment became non-disturbing, Washburn started to register his own perceived hunger pangs and stomach contractions assessed through pressure changes in the balloon. Later, activity of the lower esophagus was also assessed. Based on the simultaneity of the sensations and the contractions, it was stated that hunger pangs, usually localized mainly on or below the sternal region of the chest, are evoked by the activity of these parts of the gastrointestinal tract (the cause of the contractions remained unknown though). Fifty years later, differences in the perception of hunger by obese and non-obese individuals were studied (Griggs and Stunkard 1964; Stunkard and Koch 1964). It turned out that the correspondence between the feeling of hunger and gastric motility is not that clear-cut for everyone as experienced by Washburn. Obese participants were prone to "deny" their gastric sensations (i.e., they did not report feelings associated with gastric activity), but this "nay-saying" tendency could be reversed by training. Systematic experimentation led to the conclusion that there is no close relationship between gastric activity and hunger (Stunkard and Fox 1971). The association between gastric activity (the so-called phase III contractions) and hunger pangs are replicated recently; the rumbling sensation was ascribed to the air moving within the gastric lumen during the contraction (Janssen et al. 2011). However, there are more complex physiological mechanisms behind the sensation of hunger than gastric muscular activity (Dagher 2009). Early learning processes and social and emotional factors can substantially alter the perception of the internal state (Bruch 1961, 1969). Overall, stomach-related sensations (i.e., tension and distension) play a more direct role in the feeling of satiation than hunger (Janssen et al. 2011).

As gastric motility cannot be manipulated by breathing, muscular activity, or imagination as easily as heart rate, it was chosen to prove or disprove Brener's theory on the primary role of interoception in visceral control (see Chap. 1) (Whitehead and Drescher 1980). Just like in the Whitehead heartbeat detection task, participants were asked to decide whether a light was presented in synchrony or asynchrony with a gastric contraction; performance was evaluated with the use of signal detection theory. Similar to cardioceptive accuracy, Brener's theory was not supported by the data. More importantly for us, only half of the subjects was able to detect gastric stimuli above chance; in terms of SDT, mean d' was 0.4 but it ranged from 0 to 1.15. Moreover, many participants relied on auditory signals (sounds of contractions) beyond felt contractions. Again, it is important to emphasize that many modalities can be used to detect visceral changes, and a proportion of them is clearly not visceroceptive. Beyond gastric contractions, this was also shown for nausea, a clearly conscious (and unpleasant) experience (Sanger and Andrews 2006; Farmer et al. 2015). Humoral communication also plays a substantial role in the two-way communication between the brain and the gastrointestinal tract (also called brain-gut axis) (Aziz and Thompson 1998; Van Oudenhove et al. 2007; Aziz and Ruffle 2018).

Another important gastric stimulus beyond rhythmic contraction is (dis)tension. As it was mentioned, it plays a role in the regulation of food intake, and listed among the rare cases when visceral conditions are clearly sensed (Paintal 1986). Thus, not the objective detection threshold but differences in sensitivity to gastric stimulation are usually assessed in this area. Gastric tone can be monitored in an invasive and rather inconvenient way, which is called the barostat method (see the next section for more details) (Azpiroz and Malagelada 1985, 1987; Tack et al. 1998). To assess sensitivity to fullness or distension in a less intrusive manner, the so-called water load paradigm was developed. In various versions of this test, participants are asked to drink water (or a caloric drink in several clinical studies) administered in a strict schedule or until discomfort appears, and rate the characteristics of the accompanying gastric feelings. Similar to heartbeat perception, these tests are not without methodological issues (Mimidis 2007; Andresen 2009; van Dyck 2015). Stimulation is often accompanied by the recording of electrogastrographic activity (EGG). Patients with functional dyspepsia show higher sensitivity to gastric fullness than healthy controls (Koch et al. 2000; Boeckxstaens et al. 2001, 2002; Strid et al. 2001; Jones et al. 2003, 2005; Chen et al. 2005; Van Den Elzen et al. 2007). In a similar vein, bulimia nervosa and bulimic symptoms were associated with earlier satiation (Koch et al. 1998; van Dyck et al. 2016). In conclusion, altered perception of gastric tension appears to play a role in these clinical conditions (Mertz et al. 1998; Herbert and Pollatos 2018; Houte et al. 2019). Studies in healthy participants (both children and adults) uniformly revealed marked individual differences in gastric sensitivity (Sood et al. 2002; Hoffman et al. 2009; Herbert et al. 2012; van Dyck et al. 2016; Ferentzi et al. 2018a).

Evoked potential and startle reaction studies indicate that activity and distension of the esophagus and the stomach reaches the cortex (Frieling et al. 1989; Schulz et al. 2017). A recent study even demonstrated associations between gastric basal rhythm and the amplitude of the cortical alpha rhythm (Richter et al. 2017). These findings, along with HEP-related research already presented, indicate that visceral changes do impact the resting activity of the brain.

Distension of the stomach generates sensations that are accompanied by negative affect in healthy participants (Feinle 1998; Ferentzi et al. 2018a) and more so in patients with dyspepsia, reflux, and gastroparesis (Azpiroz and Malagelada 1987; Bradley et al. 1993; Wright et al. 2005). Similarly to respiratory sensations, the evaluative dimension, i.e., the experienced displeasure, might be as important indicator of sensitivity as perceived fullness or satiety of the stomach.

4.7 Visceral Modalities 5: Intestinal Modalities

As it was presented above, the effects of the stimulation of the bowels have been systematically studied for many decades, both in animals and humans. Concerning animals, isolated segments of the small intestine with preserved blood supply and innervation (dubbed Thiry-Vella loop) were stimulated electrically or mechanically. Depending on the intensity and type of stimulation above the so-called behavioral threshold, rats showed marked behavioral changes (Bárdos and Ádám 1978; Bárdos et al. 1980; Bárdos 1989). Like stomach distension, distension of the guts acts as a satiety signal for thirst and hunger (Bárdos et al. 1981, 2002; Bárdos 1989, 1997, 2001; Gyetvai and Bárdos 1999); this supports the idea that this sensation plays a role in the regulation of food and water intake in humans.

As for human experimentation, the most widely used way is the utilization of the barostat method. This typically means multiple manipulation of the (dis)tension at various (one or more) points of the intestines with the use of inserted balloons (Whitehead and Delvaux 1997; van der Schaar et al. 1999). Contractions (spontaneous movements or responses to stimulation) could also be recorded through the balloon system. This experimental setting also enables the researchers to apply the calculations of signal detection theory and differentiate detection/discrimination threshold and bias; some research groups preferred the use of verbal reports though.

The first modern studies in this area targeted the pathophysiology of irritable bowel syndrome (IBS), a functional gastrointestinal disorder. IBS is characterized by abdominal pain or discomfort combined with altered bowel habits (constipation, diarrhea, or both) without identifiable organic cause (Keszthelyi et al. 2012). Pain threshold was lower, whereas the motility of the rectosigmoid and anal sections of the large intestine in response to stimulation was higher in IBS patients than controls in the majority of these studies (Ritchie 1973; Whitehead et al. 1980; Kullmann and Fielding 1981; Bouin et al. 2002; Chang et al. 2003; Kwan et al. 2005; Posserud et al. 2007). Interestingly, no hyperalgesic response to stimulation of the skin by electric current or ice water was found (Cook et al. 1987; Whitehead et al. 1990). No differences between patients with IBS and controls in sensation threshold were revealed; both groups were equally sensitive to distension by a weak (non-painful) stimulus (Whitehead et al. 1980; Posserud et al. 2007). Lower detection thresholds for distension and stimulation with warm water were reported for IBS patients in one study (Li et al. 2004). Negative affectivity (anxiety and depression) was associated with pain ratings but not with rectal sensory thresholds and impacted neural processing of the stimulation (Elsenbruch et al. 2010a, b). Moreover, the size of the referred area on the skin for non-painful sensations was significantly greater for IBS patients than for healthy controls (Posserud et al. 2007). It was concluded that increased attention to bowel-related sensations and the tendency to interpret sensation as pain (i.e., top-down factors) play a decisive role in IBS (Whitehead and Palsson 1998; Posserud et al. 2007; Keszthelyi et al. 2012).

Basic research on this field was done in colostomy patients whose rectal colon was surgically removed, and an artificial anus was formed on the wall of the abdomen because of malignant tumor. Insertion of the balloon system in the artificial anus and measurement of detection threshold of the sigmoid colon were easier for these patients (Ádám et al. 1990, 1999). In these experiments, careful steps were taken to control for skin-related sensations. Overall, participants showed a quite low detection threshold that could be further lowered by training. It also turned out that skin-related masking stimuli significantly impacted detection performance; in other words, sensation was partly generated and transmitted via the exteroceptive (i.e., somatosensory) pathways (Ádám et al. 1990). Sensation turned into discomfort after the increase of the pressure (Ádám et al. 1999). Similar experiments with healthy participants reinforced these findings; it was also revealed that there are considerable individual differences with respect to the correspondence of the detection (as assessed by SDT) and verbally reported sensation thresholds (Hölzl et al. 1994, 1996).

In healthy individuals, the gradual increase in the intensity of stimulation of the sigmoid colon evokes pressure, then pain sensation, referred to the lower abdomen and lower back region; repeated stimulation increases both the perceived intensity of the stimulus and the referred area (Ness et al. 1990). Mental stress increases, whereas relaxation decreases the perception of pressure but not pain (Ford et al. 1995). Hypocapnic hyperventilation also lowers the perception threshold (Bharucha et al. 1996). Stimulation activates the midbrain, the insula, and the somatosensory cortex in healthy individuals (Rosenberger et al. 2009; Hamaguchi et al. 2013), whereas other centers of the pain network (prefrontal cortex, anterior cingulum) are also involved in patients with IBS (Yuan et al. 2003).

Perception of barostat-induced distension in the small intestines is also possible. In this field of research detection thresholds were usually determined by verbal reports. At small intensities, stimulation typically produces the feeling of fullness or abdominal pressure located in the epigastrium and paraumbilical region. Simultaneous stimulation at a nearby location decreases, whereas previous stimulation at the same location increases perception threshold (Serra et al. 1995); stimulating a larger bowel segment intensifies the sensation (Serra et al. 1998). Reflex changes were elicited by stimulation under the perception threshold (Rouillon et al. 1991a). Dissection of the vagus as well as attention focused on the stimulation decreased the sensory threshold (Accarino et al. 1997; Holtmann et al. 1998). As in the case of the large intestines, increasing the pressure turns the sensation into discomfort and pain (Rouillon et al. 1991b).

In summary, sensations originating from the intestines are vague; however, they quickly turn into discomfort and pain with the increase in the intensity of the stimulus.

4.8 Visceral Modalities 6: Urinary System

Although stimulation of the renal pelvis triggered behavioral responses in early conditioning studies (see above), normal activity of the kidneys is not perceivable. Concerning the lower urinary tract, the urge to void the urinary bladder also belongs to the rare conscious sensations indicating the actual state of the viscera. Bladder-related sensations and their behavioral consequences were also investigated in the Bykovian school in animals and humans, and an early tracking paradigm was also elaborated (Razran 1961). Over time, a number of measures were developed to assess these sensations (Medina Lucena and Tincello 2019). Although the use of forced choice assessment accompanied via manipulation of the actual filling of the bladder could theoretically be possible, usually visual analog scales or similar measures (i.e., verbal reports) are used. Some studies assessed perceived fullness (i.e., the sensory aspect of the sensation) (Nagle et al. 2017), whereas participants were asked to rate the intensity of the urge to void the bladder in others (Craggs 2005; Dompeyre et al. 2007; Lowenstein et al. 2008). Individuals without urinary problems can quite accurately sense the actual fullness of the bladder and of course also the urge to void, whereas patients with overactive bladder and other urinary syndromes show poorer acuity (Ness et al. 1998; Yamaguchi et al. 2007; De Wachter et al. 2011; Nagle et al. 2017; Uzun et al. 2019). Sensations evoked by stimulation are referred to suprapubic region; as for the intestines, increases in stimulation quickly lead to discomfort and pain (Ness et al. 1998; Naimi et al. 2019).

Information about the actual state of the urinary bladder mainly originates from mechanoreceptors of the wall of the bladder (Iggo 1955); it is transmitted by the pelvic nerve and the spinothalamic tract. Other receptors and pathways are also involved (Gillespie et al. 2009; De Wachter et al. 2011). Interestingly, patients with a complete division of the spinal cord were also able to vaguely sense the urge through signs of general sympathetic activation and the sensation of abdominal fullness, possibly transmitted by the coeliac branch of the vagus (Nathan 1956). In accordance with this idea, repeated stimulation led to cardiovascular activation (Ness et al. 1998); moreover, intensity and discomfort correlated well with sympathetic electrodermal responses and contractions of the pelvic floor (Reitz et al. 2003; Schmid et al. 2004; Schulz et al. 2019). The sensation becomes temporarily weaker after the voluntary contraction of the pelvic muscles. Subjective sensation, however, is not closely related to objective measures such as electrical sensory thresholds in healthy individuals and patients with disorders of the lower urinary tract (Wyndaele 1992; Creighton et al. 1994; De Wachter and Wyndaele 2001; De Laet et al. 2005). Overall, these findings suggest the involvement of multiple mechanisms, possibly including somatosensory pathways (interestingly, certain authors call bladder-related sensations proprioceptive). As in other cases, the brain appears to use multiple internal information about the state of the bladder, and external (environmental) stimuli also impact it (Brading 2005; Harvey et al. 2012; Eastham and Gillespie 2013).

The stimulation of the bladder evokes cortical potentials (Badr et al. 1982, 1984); neuroimaging studies demonstrated the activation of the insula and the anterior cingulate cortex (Griffiths 2007; Ketai et al. 2016). Central sensitization mechanisms play a role in functional syndromes such as bladder overactivity (Griffiths et al. 2007; Yamaguchi et al. 2007; Griffiths and Tadic 2008).

In summary, slight or more marked discomfort, sometimes also pain, are associated with the actual fullness of the urinary bladder. The affective aspect of the sensation is needed to catch attention and to suspend the actual activity until the bladder in voided.

4.9 Visceral Modalities 7: Sexual Excitement

Sexual excitement (and orgasm) belongs to the definitely conscious and pleasant body sensations. Thus, the actual state (most importantly the excited state) of the genital organs appears to be quite accurately sensible. As we will see, however, this only partly holds true. Considering the nature and the temporal dynamics of the sexual response, repeated stimulation and/or manipulation of the intensity or the actual state of the external genital organs to obtain data for the calculation of detection/discrimination threshold are not feasible. Thus, verbal reports and sometimes physiological measurements are used to characterize the sensations. Overall, the correspondence between objective and subjective measures of sexual arousal (also called sexual concordance) appears weaker in women than in men (Woodard and Diamond 2009; Suschinsky and Lalumière 2012). According to the results of a meta-analysis, the average correlation is 0.66 for males and as low as 0.26 for females (Chivers et al. 2010). Excitation state was measured using a variety of devices, typically measuring the volume or circumference of the penis and the changes of blood flow in the vagina (vaginal pulse amplitude) (Zuckerman 1971; Geer et al. 1974; Hoon et al. 1976; Kuban et al. 1999). In women, the association between actual physiological excitation and self-rated sexual arousal typically ranged from nonsignificant to medium in early studies (Geer et al. 1974; Wincze et al. 1976; Morokoff and Heiman 1980; Palace and Gorzalka 1990; Laan et al. 1994; Meston and Gorzalka 1995; Wouda et al. 1998); a strong association was reported in one study with the participation of eight individuals only (Henson et al. 1979). Later, the equivocal results were partly explained by large individual differences in response (Rellini et al. 2005). Recently, somewhat better correspondence was shown with the use of novel physiological (Suschinsky et al. 2015; Bouchard et al. 2017) and statistical (Handy and Meston 2016) methods.

Beyond methodological issues in the measurement of both objective indicators and subjective assessment of sexual excitation, the difference between sexes can be explained by the multiple meanings of sexual stimuli for women, as well as their higher dependence on situational and social cues relevant to safety, commitment, and intimacy (Laan et al. 2007). This also means that pleasure associated with these sensations is of complex origin, determined only partly by the input from the genital organs.

4.10 Visceral Modalities 8: Other Modalities

In this section, physiological conditions that do not fit in the categories above will be briefly discussed. First of all, two basic states belong here: thirst and hunger. These conditions, although they do refer to the actual state of the body and the insula plays a substantial role in their processing (Craig 2002, 2015; Uher et al. 2006), are traditionally considered motives or primordial emotions rather than interoceptive sensations (Denton 2005; Denton et al. 2009; Young et al. 2019). This reflects the fact that the regulation of drinking and eating behavior, including the perception of thirst and hunger, is a very complex process determined partly by genetic factors (de Castro 1993a, b, c; Cooling and Blundell 1998; de Castro and Lilenfeld 2005) and partly by personal experience such as parental influences (Bruch 1961, 1969; Birch et al. 2003; Harshaw 2008; Brannigan et al. 2015). Concerning both hunger and thirst, the brain is informed about the actual state of the body partly via humoral processes, including the direct measurement of certain substances of food of the blood by the circumventricular organs (see Chap. 3) (Cameron 2009; Mimee et al. 2013). Both conditions are accompanied by a negative affective state to motivate the organism to reduce them. Hunger is regulated in a very complex way, as it energizes the uptake of a number of different molecules with different roles in the human body. Thus, it is probably impossible to find one single physiological parameter behind it that could be measured in an interoceptive experiment. For thirst, the actual hydration status of the body as assessed from the blood could theoretically play this role. Interestingly, the most important body sensation associated with thirst is dry mouth; this feeling, along with perceived fullness of the stomach (Bárdos et al. 1981; Bárdos 1997), partly determines the cessation of drinking before plasma dilution becomes significant (Rolls et al. 1980; Labbe et al. 2009). Moreover, there are clearly non-homeostatic instances of fluid intake, for example, drinking during eating (Engell 1988; Brannigan et al. 2015); this also shows that the two processes are not independent from each other (McKiernan et al. 2008). Generally, there are large individual differences in the perceived sensitivity to signals of hunger and thirst (Francis and Stevenson 2011; Brannigan et al. 2015); however, to my knowledge no experiments to measure detection or discrimination thresholds in a systematic manner have been conducted to date.

Blood glucose level, a variable loosely connected to both hunger and thirst, is regulated quite tightly by a hormonal feedback loop in the human body (Rodríguez-Rodríguez et al. 2018). Serious deviations from the normal range, i.e., extremely low and high levels of blood glucose, occur rarely in healthy individuals; they usually indicate issues with regulation, such as diabetes mellitus. Hypoglycemia is characterized by trembling, sweating, confusion, speech problems, uncoordination, and hunger, whereas typical symptoms of hyperglycemia are thirst, dry mouth, alertness, tingling sensation in the extremities, frequent urination, and stomach cramps (Gonder-Frederick and Cox 1991). Moreover, a number of nonspecific symptoms and sensations, such as blurred vision, fatigue, weakness, dyspnea, and nausea, are experienced in both cases. In summary, a serious deviation from homeostasis is associated with negative affect, somatic symptoms, and cognitive impairment for hypoglycemia (Taylor and Rachman 1988; Matthew et al. 1997). Many diabetic patients have strong beliefs about their ability to accurately sense the actual levels of glucose in the blood, and they often try to control blood glucose through behavioral means (Gonder-Frederick et al. 1986; O'Connell et al. 1990). Unfortunately, these body sensations are not appropriate (in technical terms: both their validity and reliability are way too low) for this purpose and a proportion of the steps patients take is also mistaken (O'Connell et al. 1984). Just like blood pressure, body sensations and the associated beliefs very firmly resist contrary information (Gonder-Frederick and Cox 1991). Moreover, even reliable indicators (symptoms) of high and low blood glucose are idiosynchratic, i.e., they vary from individual to individual, thus a group-level analysis is usually not able to demonstrate their existence (Pennebaker et al. 1981; Pennebaker 1982; Cox et al. 1985; Schandry et al. 1996). Perhaps the only common characteristic of these symptoms is unpleasantness, indicating that something is wrong in the homeostatic regulation (Wiebe et al. 1994). The proportion of right and clinically significant estimates of actual blood glucose (based on the so-called error grid analysis) was 46% under hospitalized conditions and 60% at home in a study with diabetic patients; accuracy was improved by environmental cues in the latter case (Gonder-Frederick and Cox 1991). This level of accuracy is clearly not satisfying from a clinical point of view, as failures in the detection of extreme conditions may lead to serious consequences. Based on promising initial findings, training interventions were designed and carried out to improve individuals' accuracy of estimation of actual glucose levels (Cox et al. 1991, 1994, 1995, 2001). Both internal and external cues were used in these interventions. Although increased awareness of blood glucose proved to be beneficial, its accuracy remained insufficient to replace external monitoring via medical instruments. In conclusion, as the actual level of glucose in the blood cannot be sensed accurately enough, it should be monitored via devices in diabetes mellitus (Kovatchev et al. 2015; Rodríguez-Rodríguez et al. 2018). In healthy individuals (i.e., those without diabetes mellitus), changes of actual blood glucose levels are way less marked thus practically impossible to sense.

Thirst, hunger and low blood glucose can impact a variety of human behaviors, from altruism to decision-making (Briers et al. 2006; Dewall et al. 2008; Aarøe and Petersen 2013; Shabat-Simon et al. 2018). Although null findings were also reported (Häusser et al. 2019), these results might represent neat examples of the tenet of relevance, i.e., the (not necessarily conscious) impact of interoception on human functioning.

Finally, a number of attempts were made to assess detection and perception of overall (sympathetic) physiological activation or arousal. As this is a systemic feature and its research is tightly connected to emotion and stress research, it will be discussed in detail in Chap. 8. Here, only the methodological considerations and the conclusions will be presented.

A considerable difficulty in this area is posed by the fact that arousal is a poorly defined construct (Andreassi 2007; Berntson and Cacioppo 2007). Whereas it is often handled as a unitary characteristic in psychological research, distinctions between physiological, cortical, and behavioral arousal were made quite early (Lacey 1967). Activation of the formatio reticularis in the brainstem, an often mentioned source of (cortical) arousal, is not unspecific either; the activity of various neurotransmitter systems, such as the noradrenergic, dopaminergic, serotonergic, and cholinergic system, has well separable effects (Aston-Jones et al. 1996; Robbins et al. 1998; Berntson and Cacioppo 2007). On top of this, physiological measures of arousal, such as electrodermal response and heart rate, often dissociate under various conditions (this is called directional fractionation) (Lacey 1959). Thus, there are serious issues even with the selection of the appropriate objective measure of arousal. Concerning the possible physiological indicators, heart rate and blood pressure are barely sensible; electrodermal activity, including spontaneous changes which can reflect both emotional and cognitive changes (Crider 2008), is not an exception (Pennebaker 1982; Pennebaker et al. 1982; Steptoe and Vögele 1992; Steptoe and Noll 1997; Andor et al. 2008; Lau et al. 2015; Krautwurst et al. 2016; Kleckner et al. 2017). Overall, perceived or experienced arousal, assessed by self-report, is often just loosely related to the actual physiological arousal (Colombetti and Harrison 2018). Higher levels of sympathetic activation are accompanied by a number of symptoms, such as irritability, anxiety, and issues with concentration (Cameron et al. 2000). Finally, it is important to note that the aroused state without appropriate explanation (attribution) is perceived as unpleasant (Marshall and Zimbardo 1979; Maslach 1979).

4.11 Further Homeostatic Modalities 1: Thermosensation

It is well-known that the skin-related senses (i.e., the tactile, thermal, painful, and pruritic modalities) have a dual character; they have a discriminative (most pronounced for the tactile sense) and an affective (most important in the case of pain) aspect (McGlone and Spence 2010). As discussed in Chap. 2, affect always refers to the adaptive salience of the stimulus from the viewpoint of the organism, and it has a motivating aspect (Craig 2009; McGlone and Spence 2010). The direction of the evaluation can be more or less stabile (sensual touch is pleasant; pain and itch are mainly unpleasant) or, as for thermosensation, may depend on the actual state of the organism.

Thermosensation was classified as exteroceptive by Sherrington as it informs us about the temperature of touched objects (and indirectly also the quality of them; for example, metals are identified based on their felt coolness). The sensitivity of human haptic thermosensation is very high, with respect to both resolution (about 0.1 °C) and localization (Craig 2018). However, thermosensation is not absolute. External temperatures are automatically compared to the neutral temperature of the skin (33 °C); temperatures above this value are felt warm and hot, whereas lower temperatures are sensed cool and cold within the non-painful range; more extreme temperatures evoke pain (Craig 2018). Evaluation of the felt temperature also depends on the actual state of the organism; the same cold sensation can be pleasant under warm conditions and unpleasant in cold (Cabanac et al. 1972). This clearly indicates that thermosensation has a built-in motivational aspect; beyond exteroception, it also plays a role in the regulation of the core temperature of the body. As this regulation often requires behavioral steps (e.g., seeking shade or sunlight), the actual state of the organism is consciously accessible and subject to evaluation. Substantial deviation from the normal range evokes negative affect; if we find a place or activity that shifts our temperature to the desirable direction, it will be evaluated as pleasant (Cabanac et al. 1972). These evaluative aspects make conscious thermosensation quite tricky; in neuroanatomical terms, although the actual temperature of the stimulus is represented fairly accurately in the middle/posterior dorsal insula in a somatotopic way, subjective ratings correlate stronger with activation of the right anterior insula (Craig et al. 2000; Hua et al. 2005).

Despite our high thermal discrimination ability, the conscious sensation of the actual temperature of the skin is malleable to top-down influences. In the experiment of Pennebaker (1982), tiny spontaneous changes (increases and decreases) in temperature were sensed better if they were in accordance with participants' (manipulated) expectations. Participants in the control group having no specific expectations did not really sense these fluctuations. The average correlation between finger pulse volume (actual temperature of the skin depends on its blood supply) and the feeling of warm hands was only moderate in another study and showed huge individual differences (Pennebaker et al. 1982).

Taken together, these findings indicate that the pure sensation of changes of skin temperature is not very accurate in the physiological range. These slight and local changes appear to be unimportant under everyday circumstances. If the sensed temperature is of homeostatic relevance, for example, the entire skin surface is felt hot or a factor in the environment can influence our temperature, a marked evaluation emerges beyond the sole sensation. Valence and intensity of this evaluation provides the organism with more important information about the necessity of behavioral steps than sheer temperature values.

In several groundbreaking experiments, the implicit impact of holding hot and cold things (e.g., cup of coffee) on social warmth was reported (Williams and Bargh 2008; Ijzerman and Semin 2009; Bargh and Shalev 2012), and the role of the insula of the process was assumed (Kang et al. 2011). Unfortunately, these appealing findings could not (or just partly) be replicated later in larger samples (Lynott et al. 2014; Schilder et al. 2014; Donnellan et al. 2015; Chabris et al. 2019), thus the status of the idea that physical and social warmth can impact each other is somewhat uncertain. If it receives convincing empirical support, it can be another example of the tenet of relevance.

4.12 Further Homeostatic Modalities 2: Itch

Although it is generated in the superficial layers of skin, the itching sensation primarily belongs to the homeostatic modalities (Craig 2003a; McGlone et al. 2007). It can be evoked by mechanical, thermal, chemical, and electrical stimulation (von Frei 1922; Rothman 1922; Shelley and Arthur 1957; Blythe et al. 2019). The relationship between itch and pain is complicated: they rarely coexist in the same region of the skin, and mild (e.g., scratching) or more severe pain can eliminate the itching sensation (McMahon and Koltzenburg 1992; Ikoma et al. 2003). The exact neurophysiological background of this interaction is yet to be explored (Akiyama and Carstens 2013; Green and Dong 2016; Lee et al. 2016).

Itch is primarily generated by histamine-sensitive (and mechanically insensitive) receptors and transmitted by very slow conducting C-fibers of a specific subpopulation of lamina I neurons of the spinothalamic tract (Handwerker et al. 1991; Andrew and Craig 2001). Non-histaminic fibers sensitive to itch are polymodal, i.e., they also respond to mechanic and thermal stimulation (Davidson et al. 2012; Dhand and Aminoff 2014). The central processing of the sensation involves various parts of the thalamus (including the VPL, see Chap. 3), the primary somatosensory cortex, and the insula; activity of the latter correlates well with the felt intensity of the sensation (Drzezga et al. 2001; Craig 2002; Davidson et al. 2012; Lee et al. 2016).

Itch, like pain, is an unpleasant sensation with a well-defined motivating aspect and behavioral consequences (Craig 2003a; Twycross et al. 2003; Yosipovitch et al. 2003; Green and Dong 2016). Thus, the sensation, usually evoked by histamine or cowhage in the laboratory and assessed with a visual analog scale, the so-called verbal magnitude estimation technique or a dedicated questionnaire, corresponds well with the intensity of stimulation; it also shows considerable individual and gender differences (Simone et al. 1987, 1991; Darsow et al. 1996; Drzezga et al. 2001; Johanek et al. 2007; Kosteletzky et al. 2009; LaMotte et al. 2009; Sikand et al. 2009; Hartmann et al. 2015). Sensation-related differences between histamine and cowhage-induced itch were also reported, which might indicate (partly) separate processing of the stimuli (Johanek et al. 2008; Namer et al. 2008; Kosteletzky et al. 2009; Ringkamp et al. 2011; Ringkamp and Meyer 2014). Still, the itching sensation, just like pain, is sensitive to top-down influences such as expectations evoked by verbal suggestions and/or conditioning (van Laarhoven et al. 2011; Bartels et al. 2014, 2017; Napadow et al. 2015; Meeuwis et al. 2019; Blythe et al. 2019) or observation of others scratching themselves (contagious itch) (Ogden and Zoukas 2009; Holle et al. 2012). Psychogenic itch, i.e., pathological itching sensation, caused and maintained dominantly by top-down factors also exists (Misery et al. 2018). Personality characteristics, mainly those under the umbrella of negative affect, such as neuroticism and anxiety, were associated with the intensity and/or unpleasantness of the sensation in healthy participants as well as in patients with chronic itch (Ogden and Zoukas 2009; Bartels et al. 2014, 2015). This is in accordance with the idea that the conscious, affective-motivational aspect of itching is more prominent that the sensory-discriminative.

Overall, the sensory-discriminative and affective-evaluative aspect of the itching sensation can barely be differentiated. It strongly and effectively motivates us to scratch the impacted area of the skin; everyone knows how hard is to resist the temptation. The temptation itself, however, is motivated primarily by the evaluative aspect, i.e., the unpleasantness of the sensation.

4.13 Further Homeostatic Modalities 3: Pain

Pain, defined as "an unpleasant sensory and emotional experience associated with actual or potential tissue damage, or described in terms of such damage" (IASP 1994), has a very special status among body sensations. Its perhaps most important feature is unpleasantness, i.e., the negative affective quality (Cameron 2002). This evaluative aspect is so prominent that pain was also called a homeostatic emotion (Craig 2003b; Panerai 2011). Nociception (i.e., activation of the nociceptors) evokes well-known and observable local and global defense reactions, such as visceral and motor reflexes and behavioral avoidance (Sherrington 1906; Schouenborg 2002; Rhudy et al. 2005). However, pain is a complex conscious experience in humans and higher vertebrates (National Research Council 2009; Vierck et al. 2013; Feinberg and Mallatt 2016), whose relation with nociception is far from isomorphic (Melzack and Wall 1965; Moseley 2007; Wiech 2016; Lee et al. 2020). The subjective aspect of clinical pain is usually described along the dimensions of intensity, discomfort, localization, and quality, the latter encompassing both sensory and affective aspects (Melzack 1975, 1987). In experimental situations, usually intensity and discomfort are assessed (Drewes et al. 2003). These two dimensions of pain are closely associated; their exact relationship depends on the characteristics of the stimulation, e.g., its duration and the availability of assurances (Price 2000). Pain experience is determined by an interaction of bottom-up and top-down factors, the latter descending from cortical and limbic structures through brainstem centers (most importantly, the PAG) to the spinal cord (Brooks and Tracey 2005; Eippert et al. 2009; Tinnermann et al. 2017). Central structures involved in pain processing in the broader sense (primary and secondary somatosensory, insular, anterior cingulate, and prefrontal cortices and the thalamus; see Chap. 3) are often referred as neuromatrix or pain matrix (Melzack 1990; Apkarian et al. 2005; Tracey and Mantyh 2007; Legrain et al. 2011; Bingel et al. 2012; Aziz and Ruffle 2018); it has a sensory-discriminative (aka lateral) and an affective-evaluative (aka medial) component (Rainville et al. 1997; Auvray et al. 2010; Morton et al. 2016). However, these two components are not completely separable from each other (Wiech 2016). The most recent approach to the experience of pain is based on the predictive processing framework and as per usual emphasizes the importance of top-down elements (Anchisi and Zanon 2015; Wiech 2016).

There are substantial differences in the experience of pain with respect to its origin. Whereas localization of the so-called visceral pain (and second pain, the late component of cutaneous pain) is poor with a marked evaluative-emotional aspect (this is also called protopathic, see Chap. 1), and it is often referred to somatic structures, the first component of somatic or cutaneous pain, as well as joint and muscle pain, is well-localized (Cameron 2002; Robinson and Gebhart 2008; Cervero 2009). Moreover, the former is more unpleasant and more often evokes anxiety (Ness and Gebhart 1990; Strigo et al. 2002; Dunckley et al. 2005a, b).

Pain not only reaches consciousness, it powerfully grabs attention to facilitate escape behavior (Eccleston and Crombez 1999). Distraction of attention to another target usually (although not always) reduces both the intensity and unpleasantness of the painful stimulus (Miron et al. 1989; Bantick et al. 2002; Villemure and Bushnell 2002; Dunckley et al. 2005a). Another factor that influences the pain experience is affect, although it is not completely independent from attention (Villemure and Bushnell 2002). Whereas positive mood states reduce pain perception under experimental conditions, negatives states increase it (Zelman et al. 1991; Weisenberg et al. 1998; de Wied and Verbaten 2001; Meagher et al. 2001; Villemure and Bushnell 2002; Villemure et al. 2003; Rainville et al. 2005; Rhudy et al. 2006, 2008; Wiech and Tracey 2009). The presence of anxious and depressive feelings is associated with increased pain perception in healthy individuals as well as in patients with mood and anxiety disorders (Nelson and Novy 1997; Carter et al. 2002; Carroll et al. 2004; Poole et al. 2009; Woo 2010; McIver et al. 2018). Moreover, anxiety or fear of pain (McCracken and Dhingra 2002; Albaret et al. 2004; Lundberg et al. 2011; Knaster et al. 2012; Zale et al. 2013) as well as pain catastrophization tendency (Sullivan et al. 1995, 2001; Huber et al. 2010) heavily contributes to the experience. These findings support the notion that pain experience overlaps with emotion to a great extent (Donaldson et al. 2003; Petrovic et al. 2005; Zweyer et al. 2006; Benedetti 2008; Wager et al. 2011).

As pain is a conscious experience, its most widely used measures are based on self-report. The sensory-discriminative aspect (pain intensity) is typically assessed using numeric, visual analog, or verbal descriptor scales; pain unpleasantness is measured via similar scales and/or via behavioral measures such as tolerance (i.e., the intensity of stimulation that evokes escape reaction) (Gracely 1999; Drewes et al. 2003; Staahl and Drewes 2004). Studies that are not interested in different aspects of pain perception often use one single scale only; this shows that these two aspects are overlapping to a great extent. Moreover, as conscious experience and verbal reports are malleable to top-down influences (Gracely 1999), more objective physiological measures such as electrodermal activity, pupil dilation, heart rate, blood pressure, or facial expressions are also applied beyond self-report (Donaldson et al. 2003; Rhudy et al. 2008; Prkachin 2009; Breimhorst et al. 2011; Loggia et al. 2011; Geuter et al. 2014). Overall, empirical evidence shows that classic physiological measures and activation of pain processing centers of the brain usually show good correspondence with self-reports (Lang and Davis 2006; Aslaksen et al. 2007; Manchikanti et al. 2011; Bjørkedal and Flaten 2011; Geuter et al. 2014).

Pain can be evoked in many ways. Typically, mechanical (e.g., pressure, distension), thermal (heat or cold), electrical, or chemical stimulation is applied to various organs (skin, muscles, viscera) (Staahl and Drewes 2004; Arendt-Nielsen and Yarnitsky 2009). Whereas there is a moderate to strong association between pain threshold (i.e., the minimum intensity of stimulation that is felt painful) and tolerance within the same modality, cross-modality associations are typically weak or nonsignificant (Harris and Rollman 1983; Lautenbacher and Rollman 1993; Janal et al. 1994; Fillingim et al. 2004; Hastie et al. 2005; Bhalang et al. 2005; Krishnan et al. 2012; Cruz-Almeida et al. 2013). Stronger associations were reported in several studies (Clark and Bindra 1956; Haslam 1967; Brown et al. 1973). Females often show higher sensitivity to experimental pain than males; however, this partly depends on modality and the outcome measures (Fillingim et al. 1999, 2009; Rollman et al. 2004; Hastie et al. 2005; Mogil 2012; Racine et al. 2012; Boerner et al. 2014). The extent of individual differences in pain sensitivity is substantial; it might be determined by biological (Fukudo et al. 2009; Fukudo and Kanazawa 2011; Schaub et al. 2013; Kilpatrick et al. 2015) and psychological factors (Melzack 1973; Benedetti 2011; Ravn et al. 2012).

In summary, pain represents a body sensation with very high salience (Lee et al. 2020); this is clearly indicated by the negative affect and the behavioral reactions automatically evoked by acute pain.

4.14 Further Homeostatic Modalities 4: Affective or Limbic Touch

The existence of mechanoreceptors with unmyelinated C fibers is well-known in mammals, such as cats (Zotterman 1939) and primates (Bessou et al. 1971; Kumazawa and Perl 1977). They were identified in the human skin in the late 80s (Johansson et al. 1988; Nordin 1990). It also turned out that a proportion of these receptors are sensitive to very weak stimulation (it can be as weak as 0.06–0.23 g in the skin of the face), and they respond particularly well to gentle stroke characterized by a velocity of 1–10 cm/s (Nordin 1990; McGlone et al. 2007; Löken et al. 2009; Morrison et al. 2011). They are localized in areas of the body covered by hairy skin and lack in glabrous skin (Vallbo et al. 1993; McGlone et al. 2007). Their stimulation (processed by the insula, not the somatosensory cortices) evokes positive sensations, affiliative responses, and reduction of pain (Francis et al. 1999;

Olausson et al. 2002; Nummenmaa et al. 2016; Krahé et al. 2016, 2018; von Mohr et al. 2018). Based on these characteristics, the phenomenon is dubbed affective, sensual, or limbic touch by different authors.

On the basis of the aforementioned effects, the facilitating role of sensual touch in social relationships was assumed (Björnsdotter et al. 2010; Morrison et al. 2010). This is particularly adaptive for primates, including humans, as they live in groups. For them (both for infants and adults), affective touch can be considered a signal of social security, an important aspect of homeostasis (Craig 2015). Rated pleasantness of the stimulation is closely associated with the neural discharge rate (Löken et al. 2009) as well as afferent density (Morrison et al. 2011; Kass-Iliyya et al. 2017). This and the comparatively low sensory threshold indicate that, similar to its negative counterpart, pain, sensual touch appears to be a dominantly conscious experience by design. As the effective component appears more important here than the sheer perceptual sensation (which component does grab your attention when you are gently caressed by your romantic partner?), empirical studies routinely assess intensity of pleasantness by self-report (McGlone et al. 2007; Löken et al. 2009; Morrison et al. 2011); sometimes more objective measures, such as facial electromyography, are also used (Pawling et al. 2017). So far, research has focused primarily on the general characteristics of this modality rather than the existence and correlates of individual differences.

4.15 Further Homeostatic Modalities 5: Muscle Fatigue and Perceived Exertion

Muscle work requires energy and oxygen; the availability of the latter determines the metabolic way that provides the muscles with energy-reach molecules (Powers and Howley 2014). For slight to mediate loads (in the so-called aerobic domain), oxygen transported by the cardiovascular system is sufficient for the complete oxidation of glucose, even in the long run. In the case of intense muscle work (particularly when many large muscle groups are working simultaneously), the oxygen supply is insufficient, thus the system gradually changes to anaerobic metabolism. The boundary of the two is called anaerobic threshold. The by-product of this process, lactate, starts to accumulate in the muscles and pushes the internal milieu to acidic direction. Simultaneously, glucose reserves of the muscle gradually deplete (Al-Mulla et al. 2011; Powers and Howley 2014). As a consequence, performance of the muscle drops, which is manifested in an increase of the perceived effort necessary to exert a desired force and an eventual inability to produce this force (Enoka and Stuart 1992). Metaboreceptors (aka ergoreceptors) in the muscles primarily monitor acidity; this information is needed for the regulation of local and systemic (cardiorespiratory) changes (Wilson et al. 2002; Craig 2003a). At the conscious level, it significantly contributes to the sensation of muscle fatigue (Al-Mulla et al. 2011). However, other peripheral and central processes, such as the depletion of the neurotransmitter supply in the neuromuscular junction and the so-called central fatigue, are also involved in the sensation (Barry and Enoka 2007; Laurent and Green 2009; Briki and Majed 2019). In addition, burning pain evokes sooner or later, due to metabolic changes, spasms, and micro-injuries (Light and Perl 2003; Mense 2008). Characteristics of fatigue also depend on the fitness and the muscle composition of the individual, which is partly genetically determined (Powers and Howley 2014). Considering these issues, the fact that no empirical data on interoceptive accuracy related to muscle fatigue is available in this field is not surprising.

Muscle work is characterized by a variety of local and systemic changes; thus, the actual level of perceived exertion necessarily integrates information from various sources (Mihevic 1981; Borg 1982; St Clair Gibson and Noakes 2004). As the biological changes evoked by physical exertion are closely tied to each other, ratings of the perceived exertion scale (RPE) (Borg 1970), the most widely used measurement tool, show strong associations with heart rate and blood lactate level (Ekblom and Goldbarg 1971; Borg 1982). The RPE, unlike most scales used in psychological assessment, is an interval scale, i.e., differences between ratings are directly comparable to each other. Further, perceived exertion ratings partly depend on the kind of exercise (Ekblom and Goldbarg 1971; Mihevic 1981), external cues (Pennebaker and Lightner 1980; Briki and Majed 2019), and direction of attention (Razon et al. 2010, 2014; Lohse and Sherwood 2011). To the knowledge of the author of this book, no systematic experimentation on discrimination thresholds in this area was conducted.

Perceived fatigue of the muscles is associated with a negative affective state, i.e., it motivates us to cease exercise or at least decrease its intensity. On the other hand, exercise in the aerobic domain, when oxygen is available thus no lactate is produced, generates positive affect. Once the anaerobic threshold is passed, the affective state turns in negative as described above (Ekkekakis and Petruzzello 1999; Ekkekakis 2003, 2009; Ekkekakis and Acevedo 2006; Ekkekakis et al. 2008, 2011, 2020). The physiological background of this association is not clear; humoral (such as increased release of monoamines and endorphins) and homeostatic-interoceptive (increase of core temperature and afferent impulses) mechanisms are assumed to play a role (Paluska and Schwenk 2000; Cabanac 2006; Meeusen 2006). Phenomenological studies show that mild to moderate physical activity is accompanied by both peripheral (pleasure in the lower limbs in the case of jogging) and more central (pleasure in the chest, referring to the activity of cardiorespiratory system) positive sensations (Mihevic 1981). Similar to perceived exertion, affective state can be the result of integration of peripheral and central inputs; all these characteristics are well explainable from a homeostatic-evolutionary perspective (for more details, see Chap. 3) (Ekkekakis 2003; Cabanac 2006). Of course, exteroceptive information and cognitive factors (e.g., self-efficacy, self-determined intensity) may also play a role (Pennebaker and Lightner 1980; Ekkekakis and Acevedo 2006; Ekkekakis et al. 2011). Most importantly to our topic, the relationship between the actual load and the intensity of the experienced affect shows considerable individual differences, particularly around the threshold (Ekkekakis et al. 2005, 2006, 2011, 2020; Lind et al. 2005; Herbert et al. 2007; Ekkekakis 2009; Köteles et al. 2020b).

Physical activity in the aerobic domain, similar to sensual touch (and sexual satisfaction), belongs to the rare cases when interoceptive information appears to evoke a positive affective state. As this relationship appears biologically determined, both physical activity and sensual touch can be used to improve the affective state of individuals with high levels of negative affect (in more detail, see Chap. 7). In my view, this important practical aspect alone is sufficient to pay more attention to the sensory background of these sensations, as well as to their relationship.

4.16 Further Homeostatic Modalities 6: Smell and Taste

Contrary to the Sherringtonian categorization, the olfactory and gustatory senses can also be regarded as visceroceptive or homeostatic, based on their anatomical and functional properties (i.e., they rely on contact receptors located within the body) as well as their connection with the insular cortex (see Chap. 3) (King et al. 1999; Craig 2002; Avery et al. 2015; Rolls 2015; Berntson et al. 2018). The chemical sense might represent the most ancient sense (Chap. 2) and has a marked evaluative component; it impacts behavior, attention, learning, and mood (Jellinek 1997; Soudry et al. 2011; Maffei et al. 2012; Miranda 2012; Frank et al. 2013; Avery et al. 2015). This evaluation might have originally referred to the quality of the food. Later in the course of evolution, it was generalized to other stimuli and factors; in humans, marked social and pathological aspects are also present, for example, with respect to disgust (Smith et al. 2011; Bianchi and Carter 2012; Weck et al. 2014; Shook et al. 2017) or hypersensitivity to odors (Nordin et al. 2004, 2013, 2014).

The sensory-discriminative aspect of the olfactory sense can be measured by Sniffin' Sticks (Kobal et al. 1996; Hummel et al. 1997; Albrecht et al. 2008; Koeppel et al. 2020); analog methods are available for gustation (Bartoshuk 1978; Webb et al. 2015). Concerning the evaluative aspect, the majority of odors and tastes can be meaningfully characterized along the pleasantness–unpleasantness dimension (Köteles and Babulka 2014). The substantial overlap between emotional and olfactory processing is widely recognized and accepted (Weber and Heuberger 2008; Negoias et al. 2010; Soudry et al. 2011; Krusemark et al. 2013; Takahashi et al. 2015). In our own studies (Ferentzi et al. 2017, 2018a), the association between perceived intensity and unpleasantness of the bitter taste was very strong, which clearly indicates the importance of the affective-evaluative aspect. As in the case of emotions, valence and arousal can be assessed independently (Pollatos et al. 2007a, d).

Smells and tastes are on the boundary of exteroception and interoception. They inform us about external objects and their salience; however, substances from those objects directly interact with our own molecules and evoke marked emotional feelings, sometimes accompanied by body reactions (e.g., nausea and facial expressions).

4.17 Somatosensory Modalities 1: Muscle Tension and Exertion

Similar to many other modalities, systematic research on the association between actual and felt muscle tension partly was a by-product of biofeedback studies. Biofeedback-assisted relaxation was assumed to be more effective and considerably shorter than previous methods, such as progressive relaxation and autogenic training (Kinsman et al. 1975). As muscles belong to the motor system that can be controlled voluntarily, the general assumption was that the actual relaxation/tension state of the muscles can be sensed. Just like heart rate, this ability was thought to be necessary for voluntary control in early biofeedback studies (Stilson et al. 1980). For example, it was argued that patients with migraine first need to realize the tense state of their muscles to relax them (Sime and DeGood 1977). Progressive relaxation training also claims that awareness of the actual state of the muscles improves during training (Jacobson 1938). As this assumption appeared obvious, many relaxation studies did not assess the change using objective measurements; they simply relied on self-reported states (Hillenberg and Collins 1982).

However, it turned out that the actual relaxation/tension state of the muscles (usually the frontalis and the forearm extensor muscles were monitored), as assessed by electromyography, shows no or just weak association with the cognitive relaxation state (Lader and Mathews 1971; Alexander et al. 1975; Shedivy and Kleinman 1977; Sime and DeGood 1977; Le Boeuf 1980). These findings represented indirect evidence only as perceived state of the muscles was not assessed. To address the issue in a more direct way, Kinsman and his colleagues (1975) subjected participants to various forms of feedback-guided relaxation training consisting of short trials and asked them to evaluate whether the muscles of their forehead became more relaxed compared to the preceding trial or not. Ratio of the correct guesses was not particularly high (about 0.6) but already above chance level at baseline and was further improved by biofeedback as opposed to verbal feedback. It was also realized, however, that these findings might have been the results of a guessing strategy as participants usually expected an increase in relaxation over time. This assumption was supported by the data of a second study; it was concluded that "changes in muscle tension alone are not at all well perceived" (Staudenmayer and Kinsman 1976, p. 198). Findings of another study somewhat tempered this strict conclusion (Sime and DeGood 1977); still, participants' awareness of their actual muscle tension was fairly low. A subsequent study reported a near zero correlation ($r = -0.12$) between changes assessed by electromyography and a Likert-type self-report scale (Taylor and Lee 1991). In conclusion, despite of our strong conviction based on perceived body changes, our ability to sense the actual tension state of our resting muscles is very limited, to say the least.

Another line of research was interested in the association between actual and felt force or exertion. It is important to note here that this area of research overlaps to a great extent with the study of perceived exertion discussed above. Still, as findings were often obtained and discussed in the context of relaxation, their presentation appears more appropriate here. Participants were typically asked to replicate previously presented forces on a hand dynamometer. They performed quite well in these tasks; however, it turned out that the association (in accordance with the laws of psychophysics) is nonlinear, and there are considerable individual differences in performance (Stevens and Mack 1959; Eisler 1962). Subsequent experimentation revealed that awareness of effort is more involved in muscle control than awareness of the actual state (Stilson et al. 1980). This is in complete accordance with the feed-forward control mechanism of the motor system described in Chap. 3 (Wolpert et al. 2003; Sommer and Wurtz 2008).

In a similar vein, patients with fibromyalgia were asked to discriminate different intensities of self-produced isometric contractions of the trapezius muscle. Mean discrimination ability, calculated as correlation between electromyography and self-report, was about 0.58 with a standard deviation of 0.29 which indicates substantial individual differences. Following biofeedback training, discrimination ability improved with a simultaneous drop of variance (0.78 ± 0.11) (Sarnoch et al. 1997). Using the same paradigm, both patients with somatoform disorder and healthy controls showed fairly good correspondence between objective and felt measures of muscle tension, with the superiority of the former group in another study (Scholz et al. 2001). Again, the extent of individual differences was substantial.

A third line of research focused on hypothesized differences between those with and without schizophrenia with respect to proprioception, i.e., the ability to discriminate between weights (Rosenbaum et al. 1959). Schizophrenics' performance was poorer than controls for lighter weights only (Rosenbaum et al. 1965; Ritzler and Rosenbaum 1974). Percentage of correct decisions for healthy controls was higher (about 0.8–0.9) than in relaxation experiments, reflecting a quite good performance. Unfortunately, measures of individual differences are not reported in these studies. As often in science, subsequent studies could not replicate the inferior performance of schizophrenics in weight discrimination tasks (Ritzler 1977; Leventhal et al. 1982); however, biological relatives of persons with schizophrenia performed worse than controls in a more recent study (Chang and Lenzenweger 2005). Standard deviations of the mean performance index in these studies again indicate the existence of substantial individual differences.

As mentioned, what these studies measured was a blend of proprioceptive input and motor output at best (in fact, dominance of the latter is very likely). Because of the functional characteristic of the motor system, these components are difficult to separate in tasks that involve muscle work. The overall conclusion is that we are fairly precisely aware of our muscular effort; for example, d' was 1.42 ± 0.306 for healthy controls in the weight discrimination task (Chang and Lenzenweger 2005). However, awareness of muscle tension under passive conditions (i.e., when the muscles are not working) is surprisingly poor.

Concerning the evaluative aspects of muscle tension, stress and negative affective states might make certain muscles more tense than usual; however, we sense only gross changes. Interestingly enough, certain movements can cause changes in evaluative processing; whereas flexion of the arms and moving toward the screen of the computer was associated with a positive, arm extension, and moving away was related to a negative bias (Cacioppo et al. 1993; Neumann and Strack 2000). Thus, certain muscles and movements might be associated with the two major behavioral systems, avoidance and approach.

4.18 Somatosensory Modalities 2: Joint Position and Movement

Perception of the angles of the joints relies on multiple systems; beyond proprioceptors in the joints, mechanoreceptors detecting the tension of the affected muscles and mechanoreceptors located in the skin above the joints are also involved (Lephart et al. 1998; Edin 2001; Ribeiro and Oliveira 2011; Proske and Gandevia 2012). The typical testing paradigms in this field (dubbed joint reproduction tests), are (1) presenting a joint position, resetting the joint to the initial position, and asking the participant to reproduce the presented position (ipsilateral condition in the case of joints of the limbs), (2) presenting a position with one limb and asking the participant to reproduce it with the other limb (contralateral condition), or (3) the combination of the former two (Goble 2010). Usually the knee, the ankle, and the elbow joints are tested; paradigm 1 is also used to measure the reproduction acuity of the position of the head in the horizontal (Revel et al. 1994; Loudon et al. 1997) or the sagittal (Cheng et al. 2010) plane. Necessary limitations of the testing that can impact measured performance are the utilization of memory (case 1) and the involvement of communication between the two hemispheres and possible differences between the accuracy of the right and left joint (case 2). Moreover, if the participant is asked to move his or her limb, instead of letting it move by the measurement device, the motor command may also contribute to the performance (Gandevia et al. 2006; Han et al. 2016). Other paradigms are also used, for example, reproduction can be passive or only the threshold of detection of passive movement is assessed (Grob et al. 2002; Lund et al. 2008; Han et al. 2016; Jha et al. 2017).

In these tests, healthy participant's average detection threshold is usually as low as 1°–2° (Grob et al. 2002; Lund et al. 2008), whereas acuity of the reproduction is in the 2 –4° domain. The performance declines with age, which can contribute to the increased risk of falls in the elderly (Skinner et al. 1984; Kaplan et al. 1985; Hurley et al. 1998; Ribeiro and Oliveira 2007; Goble et al. 2009a, 2011, 2012). This deterioration can be partly due to cognitive factors posed by the task itself, such as memory and cognitive load (Boisgontier et al. 2012). Osteoarthritis, chronic pain, and injuries can also impair proprioception of the joints (Jerosch and Prymka 1996; Marks 1996; Sharma and Pai 1997; Lund et al. 2008; Knoop et al. 2011; Stanton et al. 2016), whereas physical activity helps maintain proprioceptive abilities (Ribeiro and Oliveira 2007; Goble et al. 2009a). Athletes usually show better proprioceptive acuity than controls (Han et al. 2014, 2015), and it can be further improved by special training (Pánics et al. 2008; Ogard 2011). Higher proprioceptive accuracy is associated with less sport-related injury (Payne et al. 1997; Cameron et al. 2003). Interestingly, people are more accurate with their non-dominant hand than the dominant with respect to the elbow joint; this was ascribed to the dominant hand's relatively higher reliance on visual information in the control of fine motor movements, such as writing (Goble et al. 2006, 2009b; Goble and Brown 2007, 2008, 2009). A comparatively higher dependence on visual feedback was also reported in fibromyalgia patients (Bardal et al. 2016).

Other paradigms target the dynamic component of joint-related proprioception, i.e., the ability to sense passive movements and detect motion (Kokmen et al. 1978; Lephart et al. 1998; Goble and Brown 2009; Han et al. 2013; Boisgontier and Nougier 2013). The presence of individual differences is demonstrated here too. Interesting is the fact that various aspects of proprioceptive accuracy or performance with respect to different joints often show only weak associations with each other (Kramer et al. 1997; Pincivero et al. 2001; Grob et al. 2002; Boerboom et al. 2008; Knoop et al. 2011). This might be partly due to measurement-related issues and measurement error, and partly due to differences at various levels of processing (Han et al. 2013, 2016). Even different aspects of proprioceptive abilities as assessed with different tests with respect to the same joint appear independent from each other (Yang et al. 2020).

Concerning the evaluative aspects of proprioceptive processing, little research has been conducted to date. In our own studies, proprioceptive acuity with respect to the elbow joint appeared to be independent from the actual and chronic affective condition in healthy individuals (Horváth et al. 2019, 2021). Also, everyday experience suggests that affective changes, even if they exist, are less important here than the sensory-discriminative aspect.

4.19 Somatosensory Modalities 3: Vestibular Information

Vestibular information, although it activates distinct regions of the posterior insula (zu Eulenburg et al. 2013), becomes integrated with visual and proprioceptive input in early (subcortical) steps of processing (Balaban and Thayer 2001; Balaban 2002; Ferrè et al. 2013). Thus, the measurement of sensation of pure vestibular input is a demanding task. It requires the passive movement of the body in a fixed position with blinded eyes to exclude the majority of information from other modalities (Fitzpatrick and McCloskey 1994; Lakie and Loram 2006; MacNeilage et al. 2010; Palla and Lenggenhager 2014). On top of this, amplitude, direction, and velocity of movement need to be controlled. Alternatively, the vestibular system can be manipulated by posture of the head (otoliths in the inner ear cannot distinguish between gravity and acceleration) (Schwabe and Blanke 2008), electric current (Fitzpatrick et al. 1994; Utz et al. 2010), or caloric stimulation by inserting warm or cold water/air in the external auditory canal (Been et al. 2007; Zapala et al. 2008). The latter methods, as well as the stimulation of the vestibular cortex (Kahane et al. 2003), evoke motion-related body sensations, such as the feeling of rotation and translation. These interventions are useful in the diagnostic assessment of the vestibular function, in the study of integration of body-related information (Ferrè et al. 2013; Coelho and Balaban 2015; Schönherr and May 2016), and out-of-body experiences (Schwabe and Blanke 2008); caloric stimulation can evoke phantom limb experiences too (Le Chapelain et al. 2001; André et al. 2001). However, these paradigms cannot really be used the establish ecologically meaningful detection (or perception) thresholds. Findings from experiments that passively move the body revealed that accuracy of perception of movement based on vestibular information only is lower than that relying on visual or proprioceptive modalities (Fitzpatrick and McCloskey 1994; Peterka 2002; Lakie and Loram 2006; MacNeilage et al. 2010). Similarly, albeit the maintenance of balance relies on proprioceptive, visual, and vestibular information, the contribution of the vestibular system appears less marked than that of the former two (Guskiewicz and Perrin 1996; Riemann et al. 2002).

A proportion of vestibular information readily reaches consciousness, and it can sometimes have unpleasant effects such as vertigo, motion sickness, nausea, even anxiety (Balaban and Jacob 2001; Balaban 2002; Klosterhalfen et al. 2005). However, these negative effects are not directly tied to the vestibular modality; they are usually evoked secondarily by discrepancies during integration or other aspects of processing.

4.20 Somatosensory Modalities 4: Tactile Information

The tactile modality primarily refers to discriminative information about external objects. However, this modality relies on direct contact between the object and the skin, thus the actual state of the organism directly impacts the process. Moreover, tactile information is often body-related; for example, sensations from clothes or from the contact with the floor or the surface of the seat contributes to the feeling of posture. Thus, tactile information can be used to assess characteristics of somatosensory processing in various conditions.

A signal detection-based tactile task (dubbed somatosensory signal detection task) was developed to assess characteristics of processing of near threshold tactile stimulation (vibration) (Lloyd et al. 2008). Detection threshold is individually measured, and stimulation is adjusted in this paradigm; thus, the focus was not on differences in detection ability but on the proneness to feel something in the absence of stimulation (false alarms). It turned out that a simultaneous visual signal close to the location of tactile stimulation, body-focused attention, and the sight of own hand increases the perception of nonexisting tactile stimulation; this can be considered a somatosensory illusion (Mirams et al. 2010, 2012, 2017; McKenzie et al. 2010). The occurrence of such somatosensory illusions was associated with higher levels of self-reported somatoform dissociations and medically unexplained symptoms (Brown et al. 2007, 2010; Katzer et al. 2011, 2012). In these conditions, body-related stimuli generally are not evaluated as neutral; they are considered symptoms indicating a possible threat to health, thus they are associated with negative affect. Moreover, the predictive-precautionary characteristic of negative affect (see Chap. 2) renders the criterion of perception less strict which results in more false alarms. In accordance with this idea, mindfulness meditation, which attempts to apply a more neutral or non-judgmental approach to external and internal events, lowered the frequency of such illusions (Mirams et al. 2013). More importantly for our topic, however, individuals without pathology show less negative bias. Thus, as in other somatosensory modalities, the sensory-discriminative aspect of the sensation appears much more salient than the evaluative in healthy individuals.

Finally, as it was discussed in Chap. 2, the tactile modality often becomes integrated with another exteroceptive modality, vision. As the rubber hand illusion demonstrates, this integration is so strong that it can even overwrite proprioception (at least under resting conditions), which leads to mistaken localization of the hand (called proprioceptive drift) and felt ownership toward the fake hand and disownership toward the real hand (Tsakiris 2010, 2017). Also, patients with abnormal skin sensations (infestation) were more prone to the rubber hand illusion (Eccles et al. 2015). These and other examples (e.g., the full body illusion also described in Chap. 3) neatly indicate the contribution of the visual modality to the integrated representation of the own body.

4.21 The Question of Generalizability

After reviewing the most important interoceptive modalities from the viewpoint of detection and/or discrimination ability, now we will focus on the relationship between various modalities. A logical assumption is that those people who detect, say, heartbeats above the average will show superior performance with respect to other modalities too. This association is implicitly or explicitly assumed in studies which use one single interoceptive channel, most often heartbeat tracking ability, as a generalized indicator of interoceptive accuracy (Tsakiris et al. 2011; Ainley and Tsakiris 2013; Herbert and Pollatos 2014; Emanuelsen et al. 2015; Ainley et al. 2016; Pollatos et al. 2016; Murphy et al. 2017). Historically, this assumption may rely on constructs that assumed generalized individual differences or strategies in information processing, such as augmenting-reducing (Petrie 1967, 1978; von Knorring et al. 1978), sensitization-repression (Byrne 1961, 1964; Byrne et al. 1968), or the arousal-based models of extraversion–intraversion and hypochondriasis (Eysenck 1967; Hanback and Revelle 1978). Another possibility is that the integrated way we perceive our bodies intuitively leads to the belief that detection of its changes must also be a general ability.

Compared to the overall number of studies on interoceptive accuracy, only a very small proportion assessed and compared interoceptive accuracy across multiple modalities; moreover, they typically included two or three modalities only. In the study of Whitehead and Drescher (1980), accuracy of detection of heartbeats showed a medium-level association with that of stomach contractions (forced choice tests were used for both cases). Similar results were reported recently in a pilot study (Garfinkel et al. 2017). This finding was also supported by additional studies

(Herbert et al. 2012; van Dyck et al. 2016) that demonstrated associations between cardiac and gastric interoception, using the Schandry task and the water load paradigm, respectively. This association can be the consequence of neighboring and partly overlapping insular representations of the two organs (Avery et al. 2015). Results from other studies are less promising from the viewpoint of generalized ability. Respiratory resistance-related accuracy was moderately associated with cardioceptive accuracy as assessed by the Schandry task (Steptoe and Noll 1997), but not with cardiac detection ability (forced choice task) (Harver et al. 1993; Garfinkel et al. 2016, 2017). There were no significant associations between sexual concordance, the accuracy of detection of heart rate and respiration rate (Suschinsky and Lalumière 2012, 2014). The acuity of detecting changes in skin conductance was moderately associated with respiratory resistance accuracy but not with cardioceptive accuracy (Steptoe and Noll 1997; Krautwurst et al. 2014, 2016). Interoceptive accuracy with respect to heart rate was not associated with that of finger temperature, skin conductance, and respiration rate; finger temperature-related accuracy was not associated with respiration rate and showed a weak to moderate negative association with skin conductance (sweaty hands) (Pennebaker 1982). Cardioceptive accuracy showed a weak to moderate association with various aspects of smelling ability (Koeppel et al. 2020). Finally, sensitivity to pain and cardioceptive accuracy were unrelated in some studies (Werner et al. 2009; Ferentzi et al. 2020), whereas a positive association was reported in others (Pollatos et al. 2012; Weiss et al. 2014).

Even measures within the same modality or organ system, such as accuracy of sensation of respiratory volume and frequency (Bogaerts et al. 2005) or accuracy of sensation of heart rate, heartbeats, and arrhythmias (Barsky et al. 1993), were independent of each other. Under physical loading, accuracy of reproduction of heart rate was independent of that of blood pressure (Kollenbaum et al. 1996); moreover, heartbeat perception ability at rest was associated with accuracy of reproduction for very slight physical load only (Köteles et al. 2020a). Similarly, no associations between various measurements of proprioceptive accuracy of the same joint (Grob et al. 2002; Jong et al. 2005; Elangovan et al. 2014; Li et al. 2016; Niespodziński et al. 2018; Yang et al. 2020) and different joints using the same method were found (Waddington and Adams 1999; Goble 2010; Han et al. 2013). Finally, the same conclusion was drawn with respect to pain (see above).

Our own studies including more modalities showed no associations either. No significant correlations between heartbeat perception (Schandry task), gastric sensitivity (water load test), balancing ability, sensitivity to pain, and proprioceptive accuracy were revealed, and exploratory factor analysis indicated the clear separation of modalities (Ferentzi et al. 2017, 2018a).

Overall, these results suggest that the accuracy of detection of interoceptive stimuli is not generalizable across modalities. The only exception might be the association between heartbeat and gastric-related accuracy; however, even this connection appears to be way too weak to speak of a generalized ability. At first sight, these results are in opposition with the idea of Craig (see Chap. 3 for more details) regarding the existence (and evolutionary adaptivity) of a generalized body sensation (Craig 2002, 2015). In my opinion, there is no conflict here whatsoever. Interoceptive accuracy refers to the near conscious detection of internal stimuli for some modalities, and conscious detection for others, depending to the applied method and the characteristics of the modality. Generalized body sensation, however, is a conscious representation of the actual overall state of the body. The contribution of bottom-up information from various modalities to this representation is not necessarily equal or even (Ferentzi et al. 2018a). Further, a conscious representation is always influenced by top-down processes (Chap. 7) and already includes the evaluation of the state. This latter point was clearly shown by the aforementioned findings of Craig, when activation of the middle/posterior insula corresponded well with thermal stimulation; however, subjective evaluations were associated more with right anterior activation (Craig et al. 2000).

Moreover, the aforementioned studies focus on the sensory-discriminative aspect of interoception. As a matter of fact, the sheer strength of a stimulus does not necessarily reflect its actual salience to the organism. The latter is indicated more precisely by the affective-evaluative aspect. Thus, generalization along the affective dimension, reflecting some sort of sensitivity to threats to the integrity of the body, appears more likely (we can follow this line in Chap. 5). This idea is supported by the associations between dyspnea and pain (Gracely et al. 2007; von Leupoldt and Dahme 2007; Schön et al. 2008b; von Leupoldt et al. 2009), which is also related to negative affect (Schön et al. 2008a). Results demonstrating an association between sensitivity to pain and the bitter taste (indicating the presence of dangerous substances in the food) also in accordance with this notion (Erden et al. 2007; Perkins et al. 2016). In our own studies (Ferentzi et al. 2017, 2018a), however, these associations could not be replicated.

4.22 The Pathological Aspects of Interoceptive Accuracy

Concerning pathologies, the DSM-5, the most recent version of the diagnostic and statistical manual of the American Psychiatric Association (2013), mentions the involvement of interoception in anxiety, mood, eating, and somatic symptom disorders (Khalsa and Feinstein 2018). However, whereas cardioceptive accuracy shows a positive association with anxiety-related pathological conditions (Domschke et al. 2010), it is decreased for mood (Barrett et al. 2016) and eating (Pollatos et al. 2008; Herbert and Pollatos 2014, 2018) disorders, and not or only marginally impacted in the case of somatic symptom disorder (Barsky et al. 1995; Witthöft et al. 2020). As we will see in Chap. 8, atypical interoceptive accuracy can also characterize other conditions, such as alexithymia (Murphy et al. 2017, 2018a) and autism (DuBois et al. 2016; Mul et al. 2018). It is important to realize, however, that direction of causality is not determined in these conditions (Schulz and Vögele 2015). Thus, changes in accuracy might causally contribute to the condition in certain cases or can be a consequence of the disorder in others, such as in neurological conditions (García-Cordero et al. 2016; Yoris et al. 2018); alternatively, both can reflect an underlying third variable. Concerning these points, it seems unlikely to me that all these conditions could be explained from the viewpoint of interoceptive accuracy in a satisfactory manner or that atypical interoception would represent the common susceptibility factor ("p-factor") to psychopathology as proposed recently by some authors (Brewer et al. 2016; Murphy et al. 2017).

4.23 Summary

In this chapter, empirical findings concerning our "objective" ability to detect interoceptive stimuli or discriminate between stimuli of different intensity were summarized. Of the basic tenets this work relies on, the vagueness (i.e., limited detectability) of actual body events and the tendency to evaluate perceived information were elaborated in more detail.

In my view, the most important conclusion that can be drawn from these findings is that interoceptive accuracy cannot be conceptualized and operationalized in a uniform way for all modalities (see Table 4.1). At least three categories can be identified. First, evolution rendered some of the discussed modalities to be able to have a direct impact on conscious experience (i.e., a comparatively weak stimulation can evoke conscious body sensations for healthy persons). Consequently, interoceptive

Table 4.1 Detectability and/or discriminability of the most important internal modalities

Modality	Sensory threshold	Affective (evaluative) aspect
Visceral modalities	Generally high (i.e., detectability is low) with several exceptions such as tension of the rectum and the urinary bladder, fullness of the stomach, and breathlessness	Generally absent; definitely negative for the exceptions
Temperature	Low for external objects, undetermined for internal changes	Present and depends on the actual thermal state of the organism
Itch	Low with marked individual differences	Marked negative
Pain	Low with marked individual differences	Marked negative
Affective touch	Low with marked individual differences	Marked positive
Physical exertion	Low with marked individual differences	Marked; direction and intensity depend on the load
Tension of striated muscles	High for inactive states, low for active work with marked individual differences	Less marked
Position of joints	Low with marked individual differences in discrimination ability	Less marked
Vestibular information	Low	Less marked
Tactile information	Low with marked individual differences	Less marked

accuracy necessarily overlaps with interoceptive sensibility (see Chap. 5) for these modalities. For example, distension of the stomach, the rectum, and the urinary bladder, the sensation of temperature, itch, and pain, the actual level of physical activity, and sensual touch evoke well-defined conscious sensations. In these cases, the measurement of pure detection or discrimination thresholds might be (at least theoretically) possible for certain modalities and very challenging, even impossible for others. However, the practical importance of pure discrimination is questionable in these cases as these sensations are rarely neutral (or only for very weak stimuli): they are associated with pleasure or displeasure as primary motivational forces. For sensations so closely tied to evaluation, both functionally and phenomenologically, the affective-evaluative aspect appears to be at least as salient as the discriminative (Herbert et al. 2010, 2011; Herbert and Pollatos 2012, 2018; Leder 2018). Also, some sort of generalizability along the affective dimension appears more plausible to me; preliminary evidence for this was found in our study where somatosensory amplification (basically an amalgam of interoceptive sensibility and negative effect, see Chap. 5) was associated with the evaluative but not the sensory aspect of bitterness (Ferentzi et al. 2017).

Second, modalities belonging to motor functions in the broader sense, such as the sense of the angle of joints, the exertion of individual muscles, vestibular sensation, and skin-related sensations referring to posture, also readily reach consciousness. In contrast to the first category, however, even if they have evaluative aspects, those are less marked and very well distinguishable from the discriminative dimension. Thus, accuracy refers to the accuracy of discrimination between different states in these cases. Theoretically, forced choice paradigms can be used here but not without problems. For example, considering that errors in perception can happen in both directions (i.e., overestimation or underestimation of the presented joint angle or muscle tension are equally possible), understanding the bias component poses conceptual challenges in this area. Moreover, participants need to compare two states, the actual one and one stored in the memory or represented by another body part (e.g., the contralateral joint); this introduces considerable additional error. Further, indicators of accuracy can refer to average performance (e.g., the mean difference between presented and a reproduced states) or the consistency of reproduction (technically the variance or the standard deviation of replications) in these cases.

Third, as mentioned by Paintal (1986), the majority of classic visceroceptive modalities in the healthy domain of functioning do not impact conscious functioning in a direct way, i.e., stimuli from these modalities do not reach conscious awareness. Apparently no evolutionary pressure existed in this direction as regulatory issues can be reliably solved at lower levels of the hierarchy. I sometimes have the impression that the ability to sense these changes often might be a simple anatomical side effect. For example, sensation of heartbeats in the chest or at different points of the extremities happens only because rhythmic mechanical changes occur in the proximity of the tactile receptors of the skin. Would evolution have been able to design a better mechanism without changes in pressure (for example, a pump which operates with rotation), we could barely sense it. In a similar vein, we can hear and sometimes sense borborygmi as they generate noise and stimulate the mechanoreceptors of the skin. This is not to say, of course, that the brain cannot or will not use these pieces of information for various purposes or that overall information on these functions cannot have indirect effects on conscious functioning, for example, via changes of arousal (Ádám 1998). Similarly, subliminal information on rhythmic changes can impact information processing and contribute to the feeling of the self and bodily self-awareness as mentioned in Chap. 3.

The meaningfulness of discrimination and detection thresholds as determined by forced choice paradigms and signal detection theory in this area of research seems quite limited to me. For example, the sensory threshold of a certain modality (i.e., the intensity of the stimulus that can be sensed slightly better than chance) contains no information about the exact relationship between above threshold intensities and sensations. For rhythmic internal signals, such as heartbeat or gastric contractions, for which the elementary events can be detected in a yes-or-no manner, measurements are typically made under resting conditions, and results are generalized to activated states. Attempts were made to demonstrate the validity of the basic laws of phychophysics in the interoceptive domain (Hölzl et al. 1996; Ádám et al. 1999); however, these studies are interested in common characteristics of information processing rather than individual differences. On top of this, current detection/discrimination paradigms assume a tight temporal connection between stimulation and sensation, which is not always the case (Hölzl et al. 1996; Khalsa et al. 2009b). Moreover, at the level of conscious processing and functioning, absolute values of a certain internal variable are usually less important than their relationship to the overall condition of the body. In summary, a sensory detection threshold (the "accuracy" of interoception) assessed under sterile laboratory conditions is meaningless without taking into consideration the actual state of the individual and the environment.

I know very well that these are unorthodox thoughts considering the widespread use and claimed methodological superiority of these methods. However, it is important to emphasize once again that the majority of participants perform very poorly in these visceroceptive tasks even under well-controlled laboratory circumstances. These already vague sensations might then completely disappear in everyday life when we are bombarded with external stimulation. Even if we focus on our body, stimuli from the somatosensory and the motor system will dominate as their perceptual thresholds are way lower and the sheer amount of stimuli they generate is way higher.

Nature invented a much more effective means to catch and direct attention on stimuli that are of relevance for conscious functioning; this is affect. The same objective change can evoke more intense sensations in certain individuals than in others; still, if the sensation is not associated (inherently or secondarily) with evaluation, it will remain unimportant and will easily be overlooked or ignored. The evaluative dimension might even impact participants' and patients' detection/discrimination performance for many modalities, particularly those with marked affective aspect.

Concerning this point, the available evidence is far from complete as the evaluative aspect is usually not assessed in a systematic way. For example, the affective-evaluative dimension of perception is typically not measured directly in studies investigating cardioceptive accuracy. The positive association between trait anxiety and cardioceptive accuracy (Pollatos et al. 2007b, e, 2009; Domschke et al. 2010) might reflect a learned association but yields no direct information regarding the direct evaluation of heartbeats. In fact, not every patient with anxiety disorder and individual with high level of trait anxiety shows better cardioceptive accuracy, and not everyone with high cardioceptive accuracy is characterized by anxiety. Further, state anxiety is usually not associated with better cardioceptive accuracy (Pollatos et al. 2007b; Garfinkel et al. 2016) and only marked perceivable elevations of heart rate (evoked by isoproterenol, a non-selective beta adrenergic agonist) lead to anxiety in healthy individuals (Cameron and Minoshima 2002; Khalsa et al. 2009b). However, evidence on the importance of the evaluation of felt heartbeats is available for patients with panic attack (Khalsa and Feinstein 2018).

In the case of symptom reporting tendency and somatoform disorders, a negative association with cardioceptive accuracy was found in several studies (Mailloux and Brener 2002; Pollatos et al. 2011; Schaefer et al. 2012; Weiss et al. 2014) and no association in others (Barsky et al. 1995; Aronson et al. 2001; Ferentzi et al. 2017; Witthöft et al. 2020). Although symptom reporting is connected with negative affect, this body of evidence cannot be used to draw conclusion about the connection between cardioception and affective evaluation. It is important to note that symptom reporting refers to the experience of multiple symptoms belonging to multiple organ systems; for these symptoms, the presence of specific pathophysiology can be excluded. Under such vague circumstances, negative affect might act as a biasing factor, which drives the development of illness-related cognitive schemata that leads to the perception of multiple symptoms. Concerning modalities with well-defined body sensations, such as pain, itch, distension of the stomach, the urinary bladder, and the rectum (and perhaps also the sensation of heartbeats for certain individuals), however, negative affect can make perception more accurate (Smith et al. 1995; Mora et al. 2007). These options are discussed in much more detail in Chap. 7.

Finally, as interoceptive accuracy cannot be generalized across senses, always the modality that best matches the research question at hand should be used (Ferentzi et al. 2018a). The position of heartbeat detection, the most widely used and misused modality, is particularly challenging here. First, it is often considered an appropriate indicator of self-identity or attentiveness to the body from an internal perspective in the field of self-objectification, alexithymia, eating disorders, and body illusions without convincing theoretical support (Tsakiris et al. 2011; Ainley et al. 2013; Herbert and Pollatos 2014; Emanuelsen et al. 2015; Pollatos et al. 2016; Murphy et al. 2017). Second, there are two major paradigms that measure clearly different abilities, and the choice between the two is often not theory-based. It also happens that both methods are applied and always the result that supports the respective hypothesis is reported (Kandasamy et al. 2016). This is obviously bad practice; more research would be needed to understand better the differences between these two paradigms.

Take-Home Messages

For laypeople: Our ability to accurately sense the actual condition and changes of our internal organs is very limited. Exceptions to this rule are the distension of the urinary bladder, the rectum, and the stomach, and hunger for air. In these cases, displeasure makes the sensation stand out more.

For clinicians: There is no such thing as a general interoceptive ability. Assumed interoceptive deficits in certain pathological conditions should always be investigated with respect to the interoceptive modality involved in the pathology. For example, the water load test, assessing sensitivity to stomach distension, is a more appropriate method than heartbeat perception tests to study possible interoceptive deficits behind eating disorders.

For researchers: Perhaps the most prominent issue in this area of research is the plethora and limited comparability of the available methods. A consensus of experts would be desirable on the paradigms that should be applied in future studies. If conscious experience and functioning is involved in the studied condition(s), methods based on self-report appear to be more appropriate than forced choice tests.

References

Aarøe L, Petersen MB (2013) Hunger games: fluctuations in blood glucose levels influence support for social welfare. Psychol Sci 24:2550–2556. https://doi.org/10.1177/0956797613495244

Accarino AM, Azpiroz F, Malagelada JR (1997) Attention and distraction: effects on gut perception. Gastroenterology 113:415–422. https://doi.org/10.1053/gast.1997.v113.pm9247458

Ádám G (1965) Interoception and behavior [in Hungarian]. Doctor of Sciences (DSc) dissertation, Hungarian Academy of Sciences

Ádám G (1967) Interoception and behavior. Akadémiai Kiadó, Budapest

Ádám G (1978) Visceroception, awareness, and behavior. In: Schwartz GE, Shapiro D (eds) Consciousness and self-regulation: advances in research and theory, vol 2. Springer, Boston, MA, pp 199–213

Ádám G (1983) Intestinal afferent influence on behavior. In: Hölzl R, Whitehead WE (eds) Psychophysiology of the gastrointestinal tract. Experimental and clinical applications. Plenum Press, New York, pp 351–360

Ádám G (1998) Visceral perception: understanding internal cognition. Plenum Press, New York

Ádám G, Preisich P, Kukorelli T, Kelemen V (1965) Changes in human cerebral electrical activity in response to mechanical stimulation of the duodenum. Electroencephalogr Clin Neurophysiol 18:409–411. https://doi.org/10.1016/0013-4694(65)90060-X

Ádám G, Balázs L, Vidos T, Keszler P (1990) Detection of colon distension in colonostomy patients. Psychophysiology 27:451–456. https://doi.org/10.1111/j.1469-8986.1990.tb02342.x

Ádám G, Fent J, Kajtár M (1999) An attempt for the application of the basic law of psychophysics (Fechner-Stevenes law) in the domain of visceral stimulus in humans. Acta Biol Hung 50:19–25

Ainley V, Tsakiris M (2013) Body conscious? Interoceptive awareness, measured by heartbeat perception, is negatively correlated with self-objectification. PLoS One 8:e55568. https://doi.org/10.1371/journal.pone.0055568

Ainley V, Tajadura-Jiménez A, Fotopoulou A, Tsakiris M (2012) Looking into myself: the effect of self-focused attention on interoceptive sensitivity. Psychophysiology 49:1504–1508. https://doi.org/10.1111/j.1469-8986.2012.01468.x

Ainley V, Maister L, Brokfeld J et al (2013) More of myself: manipulating interoceptive awareness by heightened attention to bodily and narrative aspects of the self. Conscious Cogn 22:1231–1238. https://doi.org/10.1016/j.concog.2013.08.004

Ainley V, Brass M, Tsakiris M (2014) Heartfelt imitation: high interoceptive awareness is linked to greater automatic imitation. Neuropsychologia 60C:21–28. https://doi.org/10.1016/j.neuropsychologia.2014.05.010

Ainley V, Apps MAJ, Fotopoulou A, Tsakiris M (2016) "Bodily precision": a predictive coding account of individual differences in interoceptive accuracy. Philos Trans R Soc Lond B Biol Sci 371:20160003. https://doi.org/10.1098/rstb.2016.0003

Ainley V, Tsakiris M, Pollatos O et al (2020) Comment on "Zamariola et al. (2018), Interoceptive Accuracy Scores are Problematic: Evidence from Simple Bivariate Correlations"—The empirical data base, the conceptual reasoning and the analysis behind this statement are misconceived and do not support the authors' conclusions. Biol Psychol 152:107870. https://doi.org/10.1016/j.biopsycho.2020.107870

Akiyama T, Carstens E (2013) Neural processing of itch. Neuroscience 250:697–714. https://doi.org/10.1016/j.neuroscience.2013.07.035

Albaret M-C, Muñoz Sastre MT, Cottencin A, Mullet E (2004) The Fear of Pain questionnaire: factor structure in samples of young, middle-aged and elderly European people. Eur J Pain 8:273–281. https://doi.org/10.1016/j.ejpain.2003.09.005

Albrecht J, Anzinger A, Kopietz R et al (2008) Test–retest reliability of the olfactory detection threshold test of the Sniffin' Sticks. Chem Senses 33:461–467. https://doi.org/10.1093/chemse/bjn013

Alexander AB, French CA, Goodman NJ (1975) A comparison of auditory and visual feedback in biofeedback assisted muscular relaxation training. Psychophysiology 12:119–123. https://doi.org/10.1111/j.1469-8986.1975.tb01259.x

Alius MG, Pané-Farré CA, Von Leupoldt A, Hamm AO (2013) Induction of dyspnea evokes increased anxiety and maladaptive breathing in individuals with high anxiety sensitivity and suffocation fear. Psychophysiology 50:488–497. https://doi.org/10.1111/psyp.12028

Al-Mulla MR, Sepulveda F, Colley M (2011) A review of non-invasive techniques to detect and predict localised muscle fatigue. Sensors (Basel) 11:3545–3594. https://doi.org/10.3390/s110403545

Anchisi D, Zanon M (2015) A Bayesian perspective on sensory and cognitive integration in pain perception and placebo analgesia. PLoS One 10:e0117270. https://doi.org/10.1371/journal.pone.0117270

Andor T, Gerlach AL, Rist F (2008) Superior perception of phasic physiological arousal and the detrimental consequences of the conviction to be aroused on worrying and metacognitions in GAD. J Abnorm Psychol 117:193–205. https://doi.org/10.1037/0021-843X.117.1.193

André JM, Martinet N, Paysant J et al (2001) Temporary phantom limbs evoked by vestibular caloric stimulation in amputees. Neuropsychiatry Neuropsychol Behav Neurol 14:190–196

André RF, Vuyk HD, Ahmed A et al (2009) Correlation between subjective and objective evaluation of the nasal airway. A systematic review of the highest level of evidence. Clin Otolaryngol 34:518–525. https://doi.org/10.1111/j.1749-4486.2009.02042.x

Andreassi JL (2007) Psychophysiology: human behavior & physiological response, 5th edn. Lawrence Erlbaum Associates Publishers, Mahwah, NJ

Andresen V (2009) Visceral sensitivity testing. Best Pract Res Clin Gastroenterol 23:313–324. https://doi.org/10.1016/j.bpg.2009.04.007

Andrew D, Craig AD (2001) Spinothalamic lamina I neurons selectively sensitive to histamine: a central neural pathway for itch. Nat Neurosci 4:72–77. https://doi.org/10.1038/82924

APA (2013) Diagnostic and statistical manual of mental disorders, 5th edn. American Psychiatric Association, Arlington, VA

Apkarian AV, Bushnell MC, Treede R-D, Zubieta J-K (2005) Human brain mechanisms of pain perception and regulation in health and disease. Eur J Pain 9:463–484. https://doi.org/10.1016/j.ejpain.2004.11.001
Arendt-Nielsen L, Yarnitsky D (2009) Experimental and clinical applications of quantitative sensory testing applied to skin, muscles and viscera. J Pain 10:556–572. https://doi.org/10.1016/j.jpain.2009.02.002
Aronson KR, Barrett LF, Quigley KS (2001) Feeling your body or feeling badly: evidence for the limited validity of the Somatosensory Amplification Scale as an index of somatic sensitivity. J Psychosom Res 51:387–394
Ashton R, White KD, Hodgson G (1979) Sensitivity to heart rate: a psychophysical study. Psychophysiology 16:463–466. https://doi.org/10.1111/j.1469-8986.1979.tb01505.x
Aslaksen PM, Myrbakk IN, Høifødt RS, Flaten MA (2007) The effect of experimenter gender on autonomic and subjective responses to pain stimuli. Pain 129:260–268. https://doi.org/10.1016/j.pain.2006.10.011
Aston-Jones G, Rajkowski J, Kubiak P et al (1996) Role of the locus coeruleus in emotional activation. Prog Brain Res 107:379–402
Auvray M, Myin E, Spence C (2010) The sensory-discriminative and affective-motivational aspects of pain. Neurosci Biobehav Rev 34:214–223. https://doi.org/10.1016/j.neubiorev.2008.07.008
Avery JA, Kerr KL, Ingeholm JE et al (2015) A common gustatory and interoceptive representation in the human mid-insula. Hum Brain Mapp 36:2996–3006. https://doi.org/10.1002/hbm.22823
Aziz Q, Ruffle JK (2018) The neurobiology of gut feelings. In: Tsakiris M, De Preester H (eds) The interoceptive mind. From homeostasis to awareness. Oxford University Press, Oxford, England, pp. 80–101
Aziz Q, Thompson DG (1998) Brain-gut axis in health and disease. Gastroenterology 114:559–578. https://doi.org/10.1016/S0016-5085(98)70540-2
Azpiroz F, Malagelada JR (1985) Physiological variations in canine gastric tone measured by an electronic barostat. Am J Physiol-Gastrointest Liver Physiol 248:G229–G237. https://doi.org/10.1152/ajpgi.1985.248.2.G229
Azpiroz F, Malagelada JR (1987) Gastric tone measured by an electronic barostat in health and postsurgical gastroparesis. Gastroenterology 92:934–943
Babaei A, Siwiec RM, Kern M et al (2013) Intrinsic functional connectivity of the brain swallowing network during subliminal esophageal acid stimulation. Neurogastroenterol Motil 25:992–e779. https://doi.org/10.1111/nmo.12238
Badr G, Carlsson CA, Fall M et al (1982) Cortical evoked potentials following stimulation of the urinary bladder in man. Electroencephalogr Clin Neurophysiol 54:494–498. https://doi.org/10.1016/0013-4694(82)90034-7
Badr GG, Fall M, Carlsson CA et al (1984) Cortical evoked potentials obtained after stimulation of the lower urinary tract. J Urol 131:306–309. https://doi.org/10.1016/s0022-5347(17)50358-4
Bailey RS, Casey KP, Pawar SS, Garcia GJM (2017) Correlation of nasal mucosal temperature with subjective nasal patency in healthy individuals. JAMA Facial Plast Surg 19:46–52. https://doi.org/10.1001/jamafacial.2016.1445
Balaban CD (2002) Neural substrates linking balance control and anxiety. Physiol Behav 77:469–475. https://doi.org/10.1016/S0031-9384(02)00935-6
Balaban CD, Jacob RG (2001) Background and history of the interface between anxiety and vertigo. J Anxiety Disord 15:27–51. https://doi.org/10.1016/S0887-6185(00)00041-4
Balaban CD, Thayer JF (2001) Neurological bases for balance-anxiety links. J Anxiety Disord 15:53–79
Bantick SJ, Wise RG, Ploghaus A et al (2002) Imaging how attention modulates pain in humans using functional MRI. Brain 125:310–319. https://doi.org/10.1093/brain/awf022
Banzett RB, Mulnier HE, Murphy K et al (2000) Breathlessness in humans activates insular cortex. Neuroreport 11:2117–2120
Baraniuk JN (2011) Subjective nasal fullness and objective congestion. Proc Am Thorac Soc 8:62–69. https://doi.org/10.1513/pats.201006-042RN

Bardal EM, Roeleveld K, Ihlen E, Mork PJ (2016) Micro movements of the upper limb in fibromyalgia: the relation to proprioceptive accuracy and visual feedback. J Electromyogr Kinesiol 26:1–7. https://doi.org/10.1016/j.jelekin.2015.12.006
Bárdos G (1989) Behavioral consequences of intestinal distention: aversivity and discomfort. Physiol Behav 45:79–85
Bárdos G (1997) Fluid intake and behavioral changes in rats associated with the distension of the small and large intestine. Behav Neurosci 111:834–844
Bárdos G (2001) Conditioned taste aversion to gut distension in rats. Physiol Behav 74:407–413. https://doi.org/10.1016/S0031-9384(01)00483-8
Bárdos G, Ádám G (1978) Visceroceptive control of operant behavior in rats. Physiol Behav 20:369–375. https://doi.org/10.1016/0031-9384(78)90315-3
Bárdos G, Nagy J, Ádám G (1980) Thresholds of behavioral reactions evoked by intestinal and skin stimuli in rats. Physiol Behav 24:661–665. https://doi.org/10.1016/0031-9384(80)90393-5
Bárdos G, Laszy J, Ádám G (1981) Satiation evoked by distension of the small intestine in water deprived rats. Physiol Behav 26:7–12. https://doi.org/10.1016/0031-9384(81)90071-8
Bárdos G, Gyetvai B, Móricz K et al (2002) Discomfort: not pain but still unpleasant feelings from the gut. Acta Biol Hung 53:423–434. https://doi.org/10.1556/ABiol.53.2002.4.4
Bargh JA, Shalev I (2012) The substitutability of physical and social warmth in daily life. Emotion 12:154–162. https://doi.org/10.1037/a0023527
Barnes PJ, Szefler SJ, Reddel HK, Chipps BE (2019) Symptoms and perception of airway obstruction in asthmatic patients: clinical implications for use of reliever medications. J Allergy Clin Immunol 144:1180–1186. https://doi.org/10.1016/j.jaci.2019.06.040
Barr M, Pennebaker JW, Watson D (1988) Improving blood pressure estimation through internal and environmental feedback. Psychosom Med 50:37–45. https://doi.org/10.1097/00006842-198801000-00005
Barrett LF, Quigley KS, Hamilton P (2016) An active inference theory of allostasis and interoception in depression. Philos Trans R Soc Lond B Biol Sci 371:20160011. https://doi.org/10.1098/rstb.2016.0011
Barry BK, Enoka RM (2007) The neurobiology of muscle fatigue: 15 years later. Integr Comp Biol 47:465–473. https://doi.org/10.1093/icb/icm047
Barsky AJ, Goodson JD, Lane RS, Cleary PD (1988) The amplification of somatic symptoms. Psychosom Med 50:510–519
Barsky AJ, Cleary PD, Brener J, Ruskin JN (1993) The perception of cardiac activity in medical outpatients. Cardiology 83:304–315
Barsky AJ, Brener J, Coeytaux RR, Cleary PD (1995) Accurate awareness of heartbeat in hypochondriacal and non-hypochondriacal patients. J Psychosom Res 39:489–497
Barsky AJ, Ahern DK, Brener J et al (1998) Palpitations and cardiac awareness after heart transplantation. Psychosom Med 60:557–562. https://doi.org/10.1097/00006842-199809000-00007
Bartels DJP, van Laarhoven AIM, Haverkamp EA et al (2014) Role of conditioning and verbal suggestion in placebo and nocebo effects on itch. PLoS One 9:e91727. https://doi.org/10.1371/journal.pone.0091727
Bartels DJP, van Laarhoven AIM, van de Kerkhof PCM, Evers AWM (2015) Placebo and nocebo effects on itch: effects, mechanisms, and predictors. Eur J Pain. https://doi.org/10.1002/ejp.750
Bartels DJP, van Laarhoven AIM, Stroo M et al (2017) Minimizing nocebo effects by conditioning with verbal suggestion: a randomized clinical trial in healthy humans. PLoS One 12:e0182959. https://doi.org/10.1371/journal.pone.0182959
Bartoshuk LM (1978) The psychophysics of taste. Am J Clin Nutr 31:1068–1077. https://doi.org/10.1093/ajcn/31.6.1068
Baumann LJ, Leventhal H (1985) I can tell when my blood pressure is up, can't I? Health Psychol 4:203–218. https://doi.org/10.1037/0278-6133.4.3.203
Been G, Ngo TT, Miller SM, Fitzgerald PB (2007) The use of tDCS and CVS as methods of non-invasive brain stimulation. Brain Res Rev 56:346–361. https://doi.org/10.1016/j.brainresrev.2007.08.001

Békésy G (1947) A new audiometer. Acta Otolaryngol 35:411–422. https://doi.org/10.3109/00016484709123756

Benedetti F (2008) Mechanisms of placebo and placebo-related effects across diseases and treatments. Annu Rev Pharmacol Toxicol 48:33–60. https://doi.org/10.1146/annurev.pharmtox.48.113006.094711

Benedetti F (2011) The patient's brain. The neuroscience behind the doctor-patient relationship. Oxford University Press, New York

Bennett ED, Jayson MI, Rubenstein D, Campbell EJ (1962) The ability of man to detect added non-elastic loads to breathing. Clin Sci 23:155–162

Berntson GG, Cacioppo JT (2007) Integrative physiology: homeostasis, allostasis, and the orchestration of systemic physiology. In: Handbook of psychophysiology, 3rd edn. Cambridge University Press, New York, NY, pp 433–452

Berntson GG, Gianaros PJ, Tsakiris M (2018) Interoception and the autonomic nervous system: bottom-up meets top-down. In: Tsakiris M, De Preester H (eds) The interoceptive mind. From homeostasis to awareness. Oxford University Press, Oxford, England, pp. 3–23

Bessou P, Burgess PR, Perl ER, Taylor CB (1971) Dynamic properties of mechanoreceptors with unmyelinated (C) fibers. J Neurophysiol 34:116–131. https://doi.org/10.1152/jn.1971.34.1.116

Bhalang K, Sigurdsson A, Slade GD, Maixner W (2005) Associations among four modalities of experimental pain in women. J Pain 6:604–611. https://doi.org/10.1016/j.jpain.2005.04.006

Bharucha AE, Camilleri M, Ford MJ et al (1996) Hyperventilation alters colonic motor and sensory function: effects and mechanisms in humans. Gastroenterology 111:368–377. https://doi.org/10.1053/gast.1996.v111.pm8690201

Bianchi KN, Carter MM (2012) An experimental analysis of disgust sensitivity and fear of contagion in Spider and Blood Injection Injury Phobia. J Anxiety Disord 26:753–761. https://doi.org/10.1016/j.janxdis.2012.06.004

Bingel U, Tracey I, Wiech K (2012) Neuroimaging as a tool to investigate how cognitive factors influence analgesic drug outcomes. Neurosci Lett 520:149–155. https://doi.org/10.1016/j.neulet.2012.04.043

Birch LL, Fisher JO, Davison KK (2003) Learning to overeat: maternal use of restrictive feeding practices promotes girls' eating in the absence of hunger. Am J Clin Nutr 78:215–220. https://doi.org/10.1093/ajcn/78.2.215

Bjørkedal E, Flaten MA (2011) Interaction between expectancies and drug effects: an experimental investigation of placebo analgesia with caffeine as an active placebo. Psychopharmacology 215:537–548. https://doi.org/10.1007/s00213-011-2233-4

Björnsdotter M, Morrison I, Olausson H (2010) Feeling good: on the role of C fiber mediated touch in interoception. Exp Brain Res 207:149–155. https://doi.org/10.1007/s00221-010-2408-y

Blough DS (1958) A method for obtaining psychophysical thresholds from the pigeon. J Exp Anal Behav 1:31–43. https://doi.org/10.1901/jeab.1958.1-31

Blythe JS, Peerdeman KJ, Veldhuijzen DS et al (2019) Placebo and nocebo effects on itch: a review of experimental methods. Itch 4:e27. https://doi.org/10.1097/itx.0000000000000027

Boeckxstaens GE, Hirsch DP, Van Den Elzen BDJ et al (2001) Impaired drinking capacity in patients with functional dyspepsia: relationship with proximal stomach function. Gastroenterology 121:1054–1063. https://doi.org/10.1053/gast.2001.28656

Boeckxstaens GE, Hirsch DP, Kuiken SD et al (2002) The proximal stomach and postprandial symptoms in functional dyspeptics. Am J Gastroenterol 97:40–48. https://doi.org/10.1111/j.1572-0241.2002.05421.x

Boerboom AL, Huizinga MR, Kaan WA et al (2008) Validation of a method to measure the proprioception of the knee. Gait Posture 28:610–614. https://doi.org/10.1016/j.gaitpost.2008.04.007

Boerner KE, Birnie KA, Caes L et al (2014) Sex differences in experimental pain among healthy children: a systematic review and meta-analysis. Pain 155:983–993. https://doi.org/10.1016/j.pain.2014.01.031

Bogaerts K, Notebaert K, Van Diest I et al (2005) Accuracy of respiratory symptom perception in different affective contexts. J Psychosom Res 58:537–543. https://doi.org/10.1016/j.jpsychores.2004.12.005

Bogaerts K, Millen A, Li W et al (2008) High symptom reporters are less interoceptively accurate in a symptom-related context. J Psychosom Res 65:417–424. https://doi.org/10.1016/j.jpsychores.2008.03.019

Boisgontier MP, Nougier V (2013) Ageing of internal models: from a continuous to an intermittent proprioceptive control of movement. Age (Dordr) 35:1339–1355. https://doi.org/10.1007/s11357-012-9436-4

Boisgontier MP, Olivier I, Chenu O, Nougier V (2012) Presbypropria: the effects of physiological ageing on proprioceptive control. Age (Dordr) 34:1179–1194. https://doi.org/10.1007/s11357-011-9300-y

Bonvallet M, Dell P, Hiebel G (1954) Tonus sympathique et activité électrique corticale. Electroencephalogr Clin Neurophysiol 6:119–144. https://doi.org/10.1016/0013-4694(54)90011-5

Borg GA (1970) Perceived exertion as an indicator of somatic stress. Scand J Rehabil Med 2:92–98

Borg GA (1982) Psychophysical bases of perceived exertion. Med Sci Sports Exerc 14:377–381

Bornemann B, Singer T (2017) Taking time to feel our body: steady increases in heartbeat perception accuracy and decreases in alexithymia over 9 months of contemplative mental training. Psychophysiology 54:469–482. https://doi.org/10.1111/psyp.12790

Bouchard KN, Chivers ML, Pukall CF (2017) Effects of genital response measurement device and stimulus characteristics on sexual concordance in women. J Sex Res 54:1197–1208. https://doi.org/10.1080/00224499.2016.1265641

Bouin M, Plourde V, Boivin M et al (2002) Rectal distention testing in patients with irritable bowel syndrome: sensitivity, specificity, and predictive values of pain sensory thresholds. Gastroenterology 122:1771–1777. https://doi.org/10.1053/gast.2002.33601

Brading AF (2005) The role of urinary urgency and its measurement in the overactive bladder symptom syndrome: current concepts and future prospects. BJU Int 96:441–442. https://doi.org/10.1111/j.1464-410X.2005.05755_6.x

Bradley LA, Richter JE, Pulliam TJ et al (1993) The relationship between stress and symptoms of gastroesophageal reflux: the influence of psychological factors. Am J Gastroenterol 88:11–19

Brannigan M, Stevenson RJ, Francis H (2015) Thirst interoception and its relationship to a Western-style diet. Physiol Behav 139:423–429. https://doi.org/10.1016/j.physbeh.2014.11.050

Breimhorst M, Sandrock S, Fechir M et al (2011) Do intensity ratings and skin conductance responses reliably discriminate between different stimulus intensities in experimentally induced pain? J Pain 12:61–70. https://doi.org/10.1016/j.jpain.2010.04.012

Brener J (1974) A general model of voluntary control applied to the phenomena of learned cardiovascular change. In: Obrist PA, Black AH, Brener J, DiCara LV (eds) Cardiovascular psychophysiology: current issues in response mechanisms, biofeedback and methodology. Aldine Transaction, New Brunswick, NJ, pp 365–391

Brener J, Jones JM (1974) Interoceptive discrimination in intact humans: detection of cardiac activity. Physiol Behav 13:763–767

Brener J, Kluvitse C (1988) Heartbeat detection: judgments of the simultaneity of external stimuli and heartbeats. Psychophysiology 25:554–561. https://doi.org/10.1111/j.1469-8986.1988.tb01891.x

Brener J, Ring C (2016) Towards a psychophysics of interoceptive processes: the measurement of heartbeat detection. Philos Trans R Soc Lond B Biol Sci 371:20160015. https://doi.org/10.1098/rstb.2016.0015

Brener J, Liu X, Ring C (1993) A method of constant stimuli for examining heartbeat detection: comparison with the Brener-Kluvitse and Whitehead methods. Psychophysiology 30:657–665

Brewer R, Cook R, Bird G (2016) Alexithymia: a general deficit of interoception. R Soc Open Sci 3:150664. https://doi.org/10.1098/rsos.150664

Briers B, Pandelaere M, Dewitte S, Warlop L (2006) Hungry for money: the desire for caloric resources increases the desire for financial resources and vice versa. Psychol Sci 17:939–943. https://doi.org/10.1111/j.1467-9280.2006.01808.x

Briki W, Majed L (2019) Adaptive effects of seeing green environment on psychophysiological parameters when walking or running. Front Psychol 10:252. https://doi.org/10.3389/fpsyg.2019.00252

Bromiley RB (1948) Conditioned responses in a dog after removal of neocortex. J Comp Physiol Psychol 41:102–110. https://doi.org/10.1037/h0060930

Brondolo E, Rosen RC, Kostis JB, Schwartz JE (1999) Relationship of physical symptoms and mood to perceived and actual blood pressure in hypertensive men: a repeated-measures design. Psychosom Med 61:311–318. https://doi.org/10.1097/00006842-199905000-00010

Brooks J, Tracey I (2005) From nociception to pain perception: imaging the spinal and supraspinal pathways. J Anat 207:19–33. https://doi.org/10.1111/j.1469-7580.2005.00428.x

Brown RA, Fader K, Barber TX (1973) Responsiveness to pain: stimulus-specificity versus generality. Psychol Rec 23:1–7. https://doi.org/10.1007/BF03394130

Brown RJ, Poliakoff E, Kirkman MA (2007) Somatoform dissociation and somatosensory amplification are differentially associated with attention to the tactile modality following exposure to body-related stimuli. J Psychosom Res 62:159–165. https://doi.org/10.1016/j.jpsychores.2006.08.008

Brown RJ, Brunt N, Poliakoff E, Lloyd DM (2010) Illusory touch and tactile perception in somatoform dissociators. J Psychosom Res 69:241–248. https://doi.org/10.1016/j.jpsychores.2009.11.010

Bruch H (1961) Conceptual confusion in eating disorders. J Nerv Ment Dis 133:46–54. https://doi.org/10.1097/00005053-196107000-00005

Bruch H (1969) Hunger and instinct. J Nerv Ment Dis 149:91–114. https://doi.org/10.1097/00005053-196908000-00002

Burrow A, Eccles R, Jones AS (1983) The effects of camphor, eucalyptus and menthol vapour on nasal resistance to airflow and nasal sensation. Acta Otolaryngol 96:157–161. https://doi.org/10.3109/00016488309132886

Bykov KM (1957) The cerebral cortex and the internal organs. Chemical Publishing, New York

Byrne D (1961) The repression-sensitization scale: rationale, reliability and validity. J Pers 29:334–349

Byrne D (1964) Repression-sensitization as a dimension of personality. Prog Exp Pers Res 72:169–220

Byrne D, Steinberg MA, Schwartz MS (1968) Relationship between repression-sensitization and physical illness. J Abnorm Psychol 73:154–155

Cabanac M (2006) Exertion and pleasure from an evolutionary perspective. In: Acevedo EO, Ekkekakis P (eds) Psychobiology of physical activity. Human Kinetics, Champaign, IL, pp 79–89

Cabanac M, Massonnet B, Belaiche R (1972) Preferred skin temperature as a function of internal and mean skin temperature. J Appl Physiol 33:699–703. https://doi.org/10.1152/jappl.1972.33.6.699

Cacioppo JT, Priester JR, Berntson GG (1993) Rudimentary determinants of attitudes: II. Arm flexion and extension have differential effects on attitudes. J Pers Soc Psychol 65:5–17. https://doi.org/10.1037/0022-3514.65.1.5

Cameron OG (2002) Visceral sensory neuroscience. Interoception. Oxford University Press, New York

Cameron OG (2009) Visceral brain-body information transfer. NeuroImage 47:787–794. https://doi.org/10.1016/j.neuroimage.2009.05.010

Cameron OG, Minoshima S (2002) Regional brain activation due to pharmacologically induced adrenergic interoceptive stimulation in humans. Psychosom Med 64:851–861

Cameron OG, Zubieta JK, Grunhaus L, Minoshima S (2000) Effects of yohimbine on cerebral blood flow, symptoms, and physiological functions in humans. Psychosom Med 62:549–559. https://doi.org/10.1097/00006842-200007000-00014

Cameron M, Adams R, Maher C (2003) Motor control and strength as predictors of hamstring injury in elite players of Australian football. Phys Ther Sport 4:159–166. https://doi.org/10.1016/S1466-853X(03)00053-1

Campbell EJ, Freedman S, Smith PS, Taylor ME (1961) The ability of man to detect added elastic loads to breathing. Clin Sci 20:223–231

Canales-Johnson A, Silva C, Huepe D et al (2015) Auditory feedback differentially modulates behavioral and neural markers of objective and subjective performance when tapping to your heartbeat. Cereb Cortex 25:4490–4503. https://doi.org/10.1093/cercor/bhv076

Cannon WB, Washburn AL (1912) An explanation of hunger. Am J Physiol Legacy Content 29:441–454. https://doi.org/10.1152/ajplegacy.1912.29.5.441

Carroll D (1977) Cardiac perception and cardiac control. A review. Biofeedback Self Regul 2:349–369

Carroll D, Whellock J (1980) Heart rate perception and the voluntary control of heart rate. Biol Psychol 11:169–180

Carroll LJ, Cassidy JD, Côté P (2004) Depression as a risk factor for onset of an episode of troublesome neck and low back pain. Pain 107:134–139. https://doi.org/10.1016/j.pain.2003.10.009

Carter LE, McNeil DW, Vowles KE et al (2002) Effects of emotion on pain reports, tolerance and physiology. Pain Res Manag 7:21–30

Cervero F (2009) Visceral versus somatic pain: similarities and differences. Dig Dis 27(Suppl 1):3–10. https://doi.org/10.1159/000268115

Ceunen E, Van Diest I, Vlaeyen JWS (2013) Accuracy and awareness of perception: related, yet distinct (commentary on Herbert et al., 2012). Biol Psychol 92:426–427. https://doi.org/10.1016/j.biopsycho.2012.09.012

Chabris CF, Heck PR, Mandart J et al (2019) No evidence that experiencing physical warmth promotes interpersonal warmth: two failures to replicate. Social Psychol (Gottingen) 50:127–132. https://doi.org/10.1027/1864-9335/a000361

Chan P-YS, von Leupoldt A, Bradley MM et al (2012) The effect of anxiety on respiratory sensory gating measured by respiratory-related evoked potentials. Biol Psychol 91:185–189. https://doi.org/10.1016/j.biopsycho.2012.07.001

Chang BP, Lenzenweger MF (2005) Somatosensory processing and schizophrenia liability: proprioception, exteroceptive sensitivity, and graphesthesia performance in the biological relatives of schizophrenia patients. J Abnorm Psychol 114:85–95. https://doi.org/10.1037/0021-843X.114.1.85

Chang L, Berman S, Mayer EA et al (2003) Brain responses to visceral and somatic stimuli in patients with irritable bowel syndrome with and without fibromyalgia. Am J Gastroenterol 98:1354–1361. https://doi.org/10.1111/j.1572-0241.2003.07478.x

Chen CL, Lin HH, Chen MC, Huang LC (2005) Dyspeptic symptoms and water load test in patients with functional dyspepsia and reflux disease. Scand J Gastroenterol 40:28–32

Cheng C-H, Wang J-L, Lin J-J et al (2010) Position accuracy and electromyographic responses during head reposition in young adults with chronic neck pain. J Electromyogr Kinesiol 20:1014–1020. https://doi.org/10.1016/j.jelekin.2009.11.002

Chivers ML, Seto MC, Lalumière ML et al (2010) Agreement of self-reported and genital measures of sexual arousal in men and women: a meta-analysis. Arch Sex Behav 39:5–56. https://doi.org/10.1007/s10508-009-9556-9

Cinciripini PM, Epstein LH, Martin JE (1979) The effects of feedback on blood pressure discrimination. J Appl Behav Anal 12:345–353. https://doi.org/10.1901/jaba.1979.12-345

Clark JW, Bindra D (1956) Individual differences in pain thresholds. Can J Psychol 10:69–76. https://doi.org/10.1037/h0083660

Clemens WJ (1979) Assessment, learning, and retention of heart beat discrimination. Psychophysiology 16:333–338. https://doi.org/10.1111/j.1469-8986.1979.tb01472.x

Clemens WJ, MacDonald DF (1976) Relationship between heart beat discrimination and heart rate control. Psychophysiology 13:176

Coelho CM, Balaban CD (2015) Visuo-vestibular contributions to anxiety and fear. Neurosci Biobehav Rev 48:148–159. https://doi.org/10.1016/j.neubiorev.2014.10.023

Colombetti G, Harrison N (2018) From physiology to experience: enriching existing conceptions of "arousal" in affective science. In: Tsakiris M, De Preester H (eds) The interoceptive mind. From homeostasis to awareness. Oxford University Press, Oxford, England, pp. 245–258

Cook L, Davidson A, Davis DJ, Kelleher RG (1960) Epinephrine, norepinephrine, and acetylcholine as conditioned stimuli for avoidance behavior. Science 131:990–991. https://doi.org/10.1126/science.131.3405.990

Cook IJ, van Eeden A, Collins SM (1987) Patients with irritable bowel syndrome have greater pain tolerance than normal subjects. Gastroenterology 93:727–733. https://doi.org/10.1016/0016-5085(87)90434-3

Cooling J, Blundell J (1998) Are high-fat and low-fat consumers distinct phenotypes? Differences in the subjective and behavioural response to energy and nutrient challenges. Eur J Clin Nutr 52:193–201. https://doi.org/10.1038/sj.ejcn.1600539

Corneille O, Desmedt O, Zamariola G et al (2020) A heartfelt response to Zimprich et al. (2020), and Ainley et al. (2020)'s commentaries: acknowledging issues with the HCT would benefit interoception research. Biol Psychol 152:107869. https://doi.org/10.1016/j.biopsycho.2020.107869

Couto B, Salles A, Sedeño L et al (2014) The man who feels two hearts: the different pathways of interoception. Soc Cogn Affect Neurosci 9:1253–1260. https://doi.org/10.1093/scan/nst108

Couto B, Adolfi F, Sedeño L et al (2015) Disentangling interoception: insights from focal strokes affecting the perception of external and internal milieus. Front Psychol 6:503. https://doi.org/10.3389/fpsyg.2015.00503

Cox DJ, Clarke WL, Gonder-Frederick L et al (1985) Accuracy of perceiving blood glucose in IDDM. Diabetes Care 8:529–536. https://doi.org/10.2337/diacare.8.6.529

Cox DJ, Gonder-Frederick L, Julian D et al (1991) Intensive versus standard blood glucose awareness training (BGAT) with insulin-dependent diabetes: mechanisms and ancillary effects. Psychosom Med 53:453–462. https://doi.org/10.1097/00006842-199107000-00010

Cox DJ, Gonder-Frederick L, Julian DM, Clarke W (1994) Long-term follow-up evaluation of blood glucose awareness training. Diabetes Care 17:1–5. https://doi.org/10.2337/diacare.17.1.1

Cox D, Gonder-Frederick L, Polonsky W et al (1995) A multicenter evaluation of blood glucose awareness training-II. Diabetes Care 18:523–528. https://doi.org/10.2337/diacare.18.4.523

Cox DJ, Gonder-Frederick L, Polonsky W et al (2001) Blood glucose awareness training (BGAT-2): long-term benefits. Diabetes Care 24:637–642. https://doi.org/10.2337/diacare.24.4.637

Craggs MD (2005) Objective measurement of bladder sensation: use of a new patient-activated device and response to neuromodulation. BJU Int 96(Suppl 1):29–36. https://doi.org/10.1111/j.1464-410X.2005.05649.x

Craig AD (2002) How do you feel? Interoception: the sense of the physiological condition of the body. Nat Rev Neurosci 3:655–666. https://doi.org/10.1038/nrn894

Craig AD (2003a) Interoception: the sense of the physiological condition of the body. Curr Opin Neurobiol 13:500–505. https://doi.org/10.1016/S0959-4388(03)00090-4

Craig AD (2003b) A new view of pain as a homeostatic emotion. Trends Neurosci 26:303–307

Craig AD (2009) How do you feel—now? The anterior insula and human awareness. Nat Rev Neurosci 10:59–70. https://doi.org/10.1038/nrn2555

Craig AD (2015) How do you feel? An interoceptive moment with your neurobiological self. Princeton University Press, Princeton, NJ

Craig AD (2018) Chapter 19 - Central neural substrates involved in temperature discrimination, thermal pain, thermal comfort, and thermoregulatory behavior. In: Romanovsky AA (ed) Handbook of clinical neurology. Elsevier, Amsterdam, pp 317–338

Craig AD, Chen K, Bandy D, Reiman EM (2000) Thermosensory activation of insular cortex. Nat Neurosci 3:184–190. https://doi.org/10.1038/72131

Creighton SM, Plevnik S, Stanton SL (1994) Urethral sensitivity in the aetiology of sensory urgency. Br J Urol 73:190–195. https://doi.org/10.1111/j.1464-410X.1994.tb07491.x
Crider A (2008) Personality and electrodermal response lability: an interpretation. Appl Psychophysiol Biofeedback 33:141–148. https://doi.org/10.1007/s10484-008-9057-y
Cruz-Almeida Y, Riley JL, Fillingim RB (2013) Experimental pain phenotype profiles in a racially and ethnically diverse sample of healthy adults. Pain Med 14:1708–1718. https://doi.org/10.1111/pme.12203
Culler E, Mettler FA (1934) Conditioned behavior in a decorticate dog. J Comp Psychol 18:291–303. https://doi.org/10.1037/h0072306
Dagher A (2009) The neurobiology of appetite: hunger as addiction. Int J Obes 33(Suppl 2):S30–S33. https://doi.org/10.1038/ijo.2009.69
Dahme B, Richter R, Mass R (1996) Interoception of respiratory resistance in asthmatic patients. Biol Psychol 42:215–229. https://doi.org/10.1016/0301-0511(95)05156-2
Dale A, Anderson D (1978) Information variables in voluntary control and classical conditioning of heart rate: field dependence and heart-rate perception. Percept Mot Skills 47:79–85. https://doi.org/10.2466/pms.1978.47.1.79
Damasio A, Carvalho GB (2013) The nature of feelings: evolutionary and neurobiological origins. Nat Rev Neurosci 14:143–152. https://doi.org/10.1038/nrn3403
Darsow U, Ring J, Scharein E, Bromm B (1996) Correlations between histamine-induced wheal, flare and itch. Arch Dermatol Res 288:436–441. https://doi.org/10.1007/bf02505231
Daubenmier JJ, Sze J, Kerr CE et al (2013) Follow your breath: respiratory interoceptive accuracy in experienced meditators. Psychophysiology 50:777–789. https://doi.org/10.1111/psyp.12057
Davenport PW, Vovk A (2009) Cortical and subcortical central neural pathways in respiratory sensations. Respir Physiol Neurobiol 167:72–86. https://doi.org/10.1016/j.resp.2008.10.001
Davenport PW, Chan P-YS, Zhang W, Chou Y-L (2007) Detection threshold for inspiratory resistive loads and respiratory-related evoked potentials. J Appl Physiol 102:276–285. https://doi.org/10.1152/japplphysiol.01436.2005
Davenport KL, Huang CH, Davenport MP, Davenport PW (2012) Relationship between respiratory load perception and perception of nonrespiratory sensory modalities in subjects with life-threatening asthma. Pulm Med 2012:310672. https://doi.org/10.1155/2012/310672
Davidson S, Zhang X, Khasabov SG et al (2012) Pruriceptive spinothalamic tract neurons: physiological properties and projection targets in the primate. J Neurophysiol 108:1711–1723. https://doi.org/10.1152/jn.00206.2012
de Castro JM (1993a) A twin study of genetic and environmental influences on the intake of fluids and beverages. Physiol Behav 54:677–687. https://doi.org/10.1016/0031-9384(93)90076-r
de Castro JM (1993b) Genetic influences on daily intake and meal patterns of humans. Physiol Behav 53:777–782. https://doi.org/10.1016/0031-9384(93)90188-l
de Castro JM (1993c) Independence of genetic influences on body size, daily intake, and meal patterns of humans. Physiol Behav 54:633–639. https://doi.org/10.1016/0031-9384(93)90070-V
de Castro JM, Lilenfeld LRR (2005) Influence of heredity on dietary restraint, disinhibition, and perceived hunger in humans. Nutrition 21:446–455. https://doi.org/10.1016/j.nut.2004.07.010
de Jong A, Kilbreath SL, Refshauge KM, Adams R (2005) Performance in different proprioceptive tests does not correlate in ankles with recurrent sprain. Arch Phys Med Rehabil 86:2101–2105. https://doi.org/10.1016/j.apmr.2005.05.015
De Laet K, De Wachter S, Wyndaele J-J (2005) Current perception thresholds in the lower urinary tract: sine- and square-wave currents studied in young healthy volunteers. Neurourol Urodyn 24:261–266. https://doi.org/10.1002/nau.20095
De Peuter S, Van Diest I, Lemaigre V et al (2004) Dyspnea: the role of psychological processes. Clin Psychol Rev 24:557–581. https://doi.org/10.1016/j.cpr.2004.05.001
De Peuter S, Put C, Lemaigre V et al (2007) Context-evoked overperception in asthma. Psychol Health 22:737–748. https://doi.org/10.1080/14768320601151702

De Wachter S, Wyndaele JJ (2001) Can the sensory threshold toward electrical stimulation be used to quantify the subjective perception of bladder filling? A study in young healthy volunteers. Urology 57:655–8; discussion 658–9. https://doi.org/10.1016/s0090-4295(01)00914-1

De Wachter SG, Heeringa R, van Koeveringe GA, Gillespie JI (2011) On the nature of bladder sensation: the concept of sensory modulation. Neurourol Urodyn 30:1220–1226. https://doi.org/10.1002/nau.21038

de Wied M, Verbaten MN (2001) Affective pictures processing, attention, and pain tolerance. Pain 90:163–172. https://doi.org/10.1016/S0304-3959(00)00400-0

De Witte NAJ, Sütterlin S, Braet C, Mueller SC (2016) Getting to the heart of emotion regulation in youth: the role of interoceptive sensitivity, heart rate variability, and parental psychopathology. PLoS One 11:e0164615. https://doi.org/10.1371/journal.pone.0164615

Denton D (2005) The primordial emotions. Oxford University Press, Oxford, England

Denton DA, McKinley MJ, Farrell M, Egan GF (2009) The role of primordial emotions in the evolutionary origin of consciousness. Conscious Cogn 18:500–514. https://doi.org/10.1016/j.concog.2008.06.009

Desmedt O, Luminet O, Corneille O (2018) The heartbeat counting task largely involves non-interoceptive processes: evidence from both the original and an adapted counting task. Biol Psychol 138:185–188. https://doi.org/10.1016/j.biopsycho.2018.09.004

Dewall CN, Baumeister RF, Gailliot MT, Maner JK (2008) Depletion makes the heart grow less helpful: helping as a function of self-regulatory energy and genetic relatedness. Personal Soc Psychol Bull 34:1653–1662. https://doi.org/10.1177/0146167208323981

Dhand A, Aminoff MJ (2014) The neurology of itch. Brain 137:313–322. https://doi.org/10.1093/brain/awt158

DiCara LV, Braun JJ, Pappas BA (1970) Classical conditioning and instrumental learning of cardiac and gastrointestinal responses following removal of neocortex in the rat. J Comp Physiol Psychol 73:208–216. https://doi.org/10.1037/h0030200

Dipak S, Prepageran N, Haslinda S et al (2011) The role of sensation in subjective and objective evaluation of nasal patency. Indian J Otolaryngol Head Neck Surg 63:10–14. https://doi.org/10.1007/s12070-010-0067-8

Dompeyre P, Fritel X, Bader G et al (2007) Bladder sensitivity testing using a visual analogue scale: comparative cystometric study on women. Neurourol Urodyn 26:350–355. https://doi.org/10.1002/nau.20301

Domschke K, Stevens S, Pfleiderer B, Gerlach AL (2010) Interoceptive sensitivity in anxiety and anxiety disorders: an overview and integration of neurobiological findings. Clin Psychol Rev 30:1–11. https://doi.org/10.1016/j.cpr.2009.08.008

Donaldson GW, Chapman CR, Nakamura Y et al (2003) Pain and the defense response: structural equation modeling reveals a coordinated psychophysiological response to increasing painful stimulation. Pain 102:97–108. https://doi.org/10.1016/s0304-3959(02)00351-2

Donnellan MB, Lucas RE, Cesario J (2015) On the association between loneliness and bathing habits: nine replications of Bargh and Shalev (2012) Study 1. Emotion 15:109–119. https://doi.org/10.1037/a0036079

Donner-Banzhoff N, Chan Y, Szalai JP, Hilditch JR (1997) Low blood pressure associated with low mood: a red herring? J Clin Epidemiol 50:1175–1181. https://doi.org/10.1016/S0895-4356(97)00157-1

Drewes AM, Gregersen H, Arendt-Nielsen L (2003) Experimental pain in gastroenterology: a reappraisal of human studies. Scand J Gastroenterol 38:1115–1130. https://doi.org/10.1080/00365520310004399

Drzezga A, Darsow U, Treede R-D et al (2001) Central activation by histamine-induced itch: analogies to pain processing: a correlational analysis of O-15 H2O positron emission tomography studies. Pain 92:295–305. https://doi.org/10.1016/S0304-3959(01)00271-8

DuBois D, Ameis SH, Lai M-C et al (2016) Interoception in autism spectrum disorder: a review. Int J Dev Neurosci 52:104–111. https://doi.org/10.1016/j.ijdevneu.2016.05.001

Dunckley P, Wise RG, Aziz Q et al (2005a) Cortical processing of visceral and somatic stimulation: differentiating pain intensity from unpleasantness. Neuroscience 133:533–542. https://doi.org/10.1016/j.neuroscience.2005.02.041
Dunckley P, Wise RG, Fairhurst M et al (2005b) A comparison of visceral and somatic pain processing in the human brainstem using functional magnetic resonance imaging. J Neurosci 25:7333–7341. https://doi.org/10.1523/JNEUROSCI.1100-05.2005
Dunn BD, Stefanovitch I, Evans D et al (2010) Can you feel the beat? Interoceptive awareness is an interactive function of anxiety- and depression-specific symptom dimensions. Behav Res Ther 48:1133–1138. https://doi.org/10.1016/j.brat.2010.07.006
Dworkin BR (2007) Interoception. In: Cacioppo JT, Tassinary LG, Berntson GG (eds) Handbook of psychophysiology, 3rd edn. Cambridge University Press, New York, pp 482–506
Dworkin BR, Elbert T, Rau H et al (1994) Central effects of baroreceptor activation in humans: attenuation of skeletal reflexes and pain perception. Proc Natl Acad Sci U S A 91:6329–6333
Eastham JE, Gillespie JI (2013) The concept of peripheral modulation of bladder sensation. Organogenesis 9:224–233. https://doi.org/10.4161/org.25895
Eccles JA, Garfinkel SN, Harrison NA et al (2015) Sensations of skin infestation linked to abnormal frontolimbic brain reactivity and differences in self-representation. Neuropsychologia 77:90–96. https://doi.org/10.1016/j.neuropsychologia.2015.08.006
Eccleston C, Crombez G (1999) Pain demands attention: a cognitive-affective model of the interruptive function of pain. Psychol Bull 125:356–366. https://doi.org/10.1037/0033-2909.125.3.356
Edin BB (2001) Cutaneous afferents provide information about knee joint movements in humans. J Physiol 531:289–297. https://doi.org/10.1111/j.1469-7793.2001.0289j.x
Ehlers A, Breuer P (1992) Increased cardiac awareness in panic disorder. J Abnorm Psychol 101:371–382
Ehlers A, Breuer P (1996) How good are patients with panic disorder at perceiving their heartbeats? Biol Psychol 42:165–182
Ehlers A, Margraf J, Roth WT et al (1988) Anxiety induced by false heart rate feedback in patients with panic disorder. Behav Res Ther 26:1–11. https://doi.org/10.1016/0005-7967(88)90028-9
Ehlers A, Breuer P, Dohn D, Fiegenbaum W (1995) Heartbeat perception and panic disorder: possible explanations for discrepant findings. Behav Res Ther 33:69–76
Ehlers A, Mayou RA, Sprigings DC, Birkhead J (2000) Psychological and perceptual factors associated with arrhythmias and benign palpitations. Psychosom Med 62:693–702
Eippert F, Finsterbusch J, Bingel U, Büchel C (2009) Direct evidence for spinal cord involvement in placebo analgesia. Science 326:404. https://doi.org/10.1126/science.1180142
Eisler H (1962) Subjective scale of force for a large muscle group. J Exp Psychol 64:253–257. https://doi.org/10.1037/h0048323
Ekblom B, Goldbarg AN (1971) The influence of physical training and other factors on the subjective rating of perceived exertion. Acta Physiol Scand 83:399–406. https://doi.org/10.1111/j.1748-1716.1971.tb05093.x
Ekkekakis P (2003) Pleasure and displeasure from the body: perspectives from exercise. Cogn Emot 17:213–239. https://doi.org/10.1080/02699930302292
Ekkekakis P (2009) Let them roam free? Physiological and psychological evidence for the potential of self-selected exercise intensity in public health. Sports Med 39:857–888. https://doi.org/10.2165/11315210-000000000-00000
Ekkekakis P, Acevedo EO (2006) Affective responses to acute exercise. In: Acevedo EO, Ekkekakis P (eds) Psychobiology of physical activity. Human Kinetics, Champaign, IL, pp 91–109
Ekkekakis P, Petruzzello SJ (1999) Acute aerobic exercise and affect: current status, problems and prospects regarding dose-response. Sports Med 28:337–374
Ekkekakis P, Hall EE, Petruzzello SJ (2005) Some like it vigorous: measuring individual differences in the preference for and tolerance of exercise intensity. J Sport Exerc Psychol 27:350–374. https://doi.org/10.1123/jsep.27.3.350

Ekkekakis P, Lind E, Joens-Matre RR (2006) Can self-reported preference for exercise intensity predict physiologically defined self-selected exercise intensity? Res Q Exerc Sport 77:81–90. https://doi.org/10.1080/02701367.2006.10599334

Ekkekakis P, Hall EE, Petruzzello SJ (2008) The relationship between exercise intensity and affective responses demystified: to crack the 40-year-old nut, replace the 40-year-old nutcracker! Ann Behav Med 35:136–149. https://doi.org/10.1007/s12160-008-9025-z

Ekkekakis P, Parfitt G, Petruzzello SJ (2011) The pleasure and displeasure people feel when they exercise at different intensities: decennial update and progress towards a tripartite rationale for exercise intensity prescription. Sports Med 41:641–671. https://doi.org/10.2165/11590680-000000000-00000

Ekkekakis P, Hartman ME, Ladwig MA (2020) Affective responses to exercise. In: Tenenbaum G, Eklund RC (eds) Handbook of sport psychology, 4th edn. Wiley, Hoboken, NJ, pp 233–253

Elangovan N, Herrmann A, Konczak J (2014) Assessing proprioceptive function: evaluating joint position matching methods against psychophysical thresholds. Phys Ther 94:553–561. https://doi.org/10.2522/ptj.20130103

Eley TC, Stirling L, Ehlers A et al (2004) Heart-beat perception, panic/somatic symptoms and anxiety sensitivity in children. Behav Res Ther 42:439–448. https://doi.org/10.1016/S0005-7967(03)00152-9

Eley TC, Gregory AM, Clark DM, Ehlers A (2007) Feeling anxious: a twin study of panic/somatic ratings, anxiety sensitivity and heartbeat perception in children. J Child Psychol Psychiatry 48:1184–1191. https://doi.org/10.1111/j.1469-7610.2007.01838.x

Elsenbruch S, Rosenberger C, Bingel U et al (2010a) Patients with irritable bowel syndrome have altered emotional modulation of neural responses to visceral stimuli. Gastroenterology 139:1310–1319. https://doi.org/10.1053/j.gastro.2010.06.054

Elsenbruch S, Rosenberger C, Enck P et al (2010b) Affective disturbances modulate the neural processing of visceral pain stimuli in irritable bowel syndrome: an fMRI study. Gut 59:489–495. https://doi.org/10.1136/gut.2008.175000

Emanuelsen L, Drew R, Köteles F (2015) Interoceptive sensitivity, body image dissatisfaction, and body awareness in healthy individuals. Scand J Psychol 56:167–174. https://doi.org/10.1111/sjop.12183

Engell D (1988) Interdependency of food and water intake in humans. Appetite 10:133–141. https://doi.org/10.1016/0195-6663(88)90064-5

Enoka RM, Stuart DG (1992) Neurobiology of muscle fatigue. J Appl Physiol 72:1631–1648. https://doi.org/10.1152/jappl.1992.72.5.1631

Epstein LH, Stein DB (1974) Feedback-influenced heart rate discrimination. J Abnorm Psychol 83:585–588. https://doi.org/10.1037/h0037065

Erden V, Basaranoglu G, Korkut Y et al (2007) Relation between bitter taste sensitivity and incidence or intensity of propofol injection pain. Eur J Anaesthesiol 24:516–520. https://doi.org/10.1017/S0265021506001396

Evans KC, Banzett RB, Adams L et al (2002) BOLD fMRI identifies limbic, paralimbic, and cerebellar activation during air hunger. J Neurophysiol 88:1500–1511

Eysenck HJ (1967) The biological basis of personality. Charles C. Thomas, Springfield

Fahrenberg J, Franck M, Baas U, Jost E (1995) Awareness of blood pressure: interoception or contextual judgement? J Psychosom Res 39:11–18. https://doi.org/10.1016/0022-3999(94)00070-L

Fairclough SH, Goodwin L (2007) The effect of psychological stress and relaxation on interoceptive accuracy: implications for symptom perception. J Psychosom Res 62:289–295. https://doi.org/10.1016/j.jpsychores.2006.10.017

Farmer AD, Ban VF, Coen SJ et al (2015) Visually induced nausea causes characteristic changes in cerebral, autonomic and endocrine function in humans. J Physiol Lond 593:1183–1196. https://doi.org/10.1113/jphysiol.2014.284240

Faull OK, Hayen A, Pattinson KTS (2017) Breathlessness and the body: neuroimaging clues for the inferential leap. Cortex 95:211–221. https://doi.org/10.1016/j.cortex.2017.07.019

Feinberg TE, Mallatt JM (2016) The ancient origins of consciousness. How the brain created experience. The MIT Press, Cambridge, MA

Feinle C (1998) Role of intestinal chemoreception in the induction of gastrointestinal sensations. DTW Dtsch Tierarztl Wochenschr 105:441–444

Ferentzi E, Köteles F, Csala B et al (2017) What makes sense in our body? Personality and sensory correlates of body awareness and somatosensory amplification. Personal Individ Differ 104:75–81. https://doi.org/10.1016/j.paid.2016.07.034

Ferentzi E, Bogdány T, Szabolcs Z et al (2018a) Multichannel investigation of interoception: sensitivity is not a generalizable feature. Front Hum Neurosci 12:223. https://doi.org/10.3389/fnhum.2018.00223

Ferentzi E, Drew R, Tihanyi BT, Köteles F (2018b) Interoceptive accuracy and body awareness – temporal and longitudinal associations in a non-clinical sample. Physiol Behav 184:100–107. https://doi.org/10.1016/j.physbeh.2017.11.015

Ferentzi E, Geiger M, Mai-Lippold S, et al (2020) Interaction between sex and cardiac interoceptive accuracy in measures of induced pain. Manuscript submitted for publication

Ferrè ER, Bottini G, Iannetti GD, Haggard P (2013) The balance of feelings: vestibular modulation of bodily sensations. Cortex 49:748–758. https://doi.org/10.1016/j.cortex.2012.01.012

Fillingim RB, Edwards RR, Powell T (1999) The relationship of sex and clinical pain to experimental pain responses. Pain 83:419–425. https://doi.org/10.1016/S0304-3959(99)00128-1

Fillingim RB, Ness TJ, Glover TL et al (2004) Experimental pain models reveal no sex differences in pentazocine analgesia in humans. Anesthesiology 100:1263–1270. https://doi.org/10.1097/00000542-200405000-00031

Fillingim RB, King CD, Ribeiro-Dasilva MC et al (2009) Sex, gender, and pain: a review of recent clinical and experimental findings. J Pain 10:447–485. https://doi.org/10.1016/j.jpain.2008.12.001

Fischer D, Messner M, Pollatos O (2017) Improvement of interoceptive processes after an 8-week body scan intervention. Front Hum Neurosci 11:452. https://doi.org/10.3389/fnhum.2017.00452

Fisher S, Cleveland SE (1968) Body image and personality, 2nd rev. edn. Dover Publications, New York

Fittipaldi S, Abrevaya S, de la Fuente A et al (2020) A multidimensional and multi-feature framework for cardiac interoception. NeuroImage 212:116677. https://doi.org/10.1016/j.neuroimage.2020.116677

Fitzpatrick R, McCloskey DI (1994) Proprioceptive, visual and vestibular thresholds for the perception of sway during standing in humans. J Physiol 478:173–186

Fitzpatrick R, Burke D, Gandevia SC (1994) Task-dependent reflex responses and movement illusions evoked by galvanic vestibular stimulation in standing humans. J Physiol Lond 478(Pt 2):363–372. https://doi.org/10.1113/jphysiol.1994.sp020257

Ford MJ, Camilleri M, Zinsmeister AR, Hanson RB (1995) Psychosensory modulation of colonic sensation in the human transverse and sigmoid colon. Gastroenterology 109:1772–1780. https://doi.org/10.1016/0016-5085(95)90743-2

Forkmann T, Scherer A, Meessen J et al (2016) Making sense of what you sense: disentangling interoceptive awareness, sensibility and accuracy. Int J Psychophysiol 109:71–80. https://doi.org/10.1016/j.ijpsycho.2016.09.019

Fotopoulou A, Tsakiris M (2017) Mentalizing homeostasis: the social origins of interoceptive inference. Neuropsychoanalysis 19:3–28. https://doi.org/10.1080/15294145.2017.1294031

Fox NA, Schmidt LA, Henderson HA, Marshall PJ (2007) Developmental psychophysiology: conceptual and methodological issues. In: Handbook of psychophysiology, 3rd edn. Cambridge University Press, New York, pp 453–481

Francis HM, Stevenson RJ (2011) Higher reported saturated fat and refined sugar intake is associated with reduced hippocampal-dependent memory and sensitivity to interoceptive signals. Behav Neurosci 125:943–955. https://doi.org/10.1037/a0025998

Francis S, Rolls ET, Bowtell R et al (1999) The representation of pleasant touch in the brain and its relationship with taste and olfactory areas. Neuroreport 10:453–459. https://doi.org/10.1097/00001756-199902250-00003

Frank S, Kullmann S, Veit R (2013) Food related processes in the insular cortex. Front Hum Neurosci 7:499. https://doi.org/10.3389/fnhum.2013.00499

Frieling T, Enck P, Wienbeck M (1989) Cerebral responses evoked by electrical stimulation of the esophagus in normal subjects. Gastroenterology 97:475–478

Fukudo S, Kanazawa M (2011) Gene, environment, and brain-gut interactions in irritable bowel syndrome. J Gastroenterol Hepatol 26(Suppl 3):110–115. https://doi.org/10.1111/j.1440-1746.2011.06631.x

Fukudo S, Kanazawa M, Mizuno T et al (2009) Impact of serotonin transporter gene polymorphism on brain activation by colorectal distention. NeuroImage 47:946–951. https://doi.org/10.1016/j.neuroimage.2009.04.083

Furedy JJ, Rau H, Roberts LE (1992) Physiological and psychological differentiation of bidirectional baroreceptor carotid manipulation in humans. Physiol Behav 52:953–958. https://doi.org/10.1016/0031-9384(92)90376-D

Gandevia SC, Killian KJ, Campbell EJ (1981) The effect of respiratory muscle fatigue on respiratory sensations. Clin Sci 60:463–466. https://doi.org/10.1042/cs0600463

Gandevia SC, Smith JL, Crawford M et al (2006) Motor commands contribute to human position sense. J Physiol 571:703–710. https://doi.org/10.1113/jphysiol.2005.103093

Gannon L (1977) The role of interoception in learned visceral control. Biofeedback Self Regul 2:337–347. https://doi.org/10.1007/BF00998621

Gannon L (1980) Cardiac perception and voluntary control of heart rate. Physiol Psychol 8:509–514

Garcia-Cordero I, Sedeño L, Fittipaldi S, Ibanez A (2016) The neuroanatomical organization of interoceptive dimensions: multimodal evidence from neurodegeneration and stroke. Int J Psychophysiol 108:128–129. https://doi.org/10.1016/j.ijpsycho.2016.07.379

García-Cordero I, Sedeño L, de la Fuente L et al (2016) Feeling, learning from and being aware of inner states: interoceptive dimensions in neurodegeneration and stroke. Philos Trans R Soc Lond B Biol Sci 371:20160006. https://doi.org/10.1098/rstb.2016.0006

García-Cordero I, Esteves S, Mikulan EP et al (2017) Attention, in and out: scalp-level and intracranial EEG correlates of interoception and exteroception. Front Neurosci 11:411. https://doi.org/10.3389/fnins.2017.00411

Gardner RM, Morrell JA, Watson DN, Sandoval SL (1990) Cardiac self-perception in obese and normal persons. Percept Mot Skills 70:1179–1186

Garfinkel SN, Critchley HD (2013) Interoception, emotion and brain: new insights link internal physiology to social behaviour. Commentary on: "Anterior insular cortex mediates bodily sensibility and social anxiety" by Terasawa et al. (2012). Soc Cogn Affect Neurosci 8:231–234. https://doi.org/10.1093/scan/nss140

Garfinkel SN, Seth AK, Barrett AB et al (2015a) Knowing your own heart: distinguishing interoceptive accuracy from interoceptive awareness. Biol Psychol 104:65–74. https://doi.org/10.1016/j.biopsycho.2014.11.004

Garfinkel SN, Tiley C, O'Keeffe S et al (2015b) Discrepancies between dimensions of interoception in autism: implications for emotion and anxiety. Biol Psychol 114:117–126. https://doi.org/10.1016/j.biopsycho.2015.12.003

Garfinkel SN, Manassei MF, Hamilton-Fletcher G et al (2016) Interoceptive dimensions across cardiac and respiratory axes. Philos Trans R Soc Lond B Biol Sci 371:20160014. https://doi.org/10.1098/rstb.2016.0014

Garfinkel SN, Manassei MF, Engels M et al (2017) An investigation of interoceptive processes across the senses. Biol Psychol 129:371–372. https://doi.org/10.1016/j.biopsycho.2017.08.010

Geer JH, Morokoff P, Greenwood P (1974) Sexual arousal in women: the development of a measurement device for vaginal blood volume. Arch Sex Behav 3:559–564. https://doi.org/10.1007/bf01541137

Georgiou E, Matthias E, Kobel S et al (2015) Interaction of physical activity and interoception in children. Front Psychol 6:502. https://doi.org/10.3389/fpsyg.2015.00502

Geuter S, Gamer M, Onat S, Büchel C (2014) Parametric trial-by-trial prediction of pain by easily available physiological measures. Pain 155:994–1001. https://doi.org/10.1016/j.pain.2014.02.005

Giardino ND, Curtis JL, Abelson JL et al (2010) The impact of panic disorder on interoception and dyspnea reports in chronic obstructive pulmonary disease. Biol Psychol 84:142–146. https://doi.org/10.1016/j.biopsycho.2010.02.007

Gibbons FX, Carver CS, Scheier MF, Hormuth SE (1979) Self-focused attention and the placebo effect: fooling some of the people some of the time. J Exp Soc Psychol 15:263–274. https://doi.org/10.1016/0022-1031(79)90037-4

Gillespie JI, van Koeveringe GA, de Wachter SG, de Vente J (2009) On the origins of the sensory output from the bladder: the concept of afferent noise. BJU Int 103:1324–1333. https://doi.org/10.1111/j.1464-410X.2009.08377.x

Goble DJ (2010) Proprioceptive acuity assessment via joint position matching: from basic science to general practice. Phys Ther 90:1176–1184. https://doi.org/10.2522/ptj.20090399

Goble DJ, Brown SH (2007) Task-dependent asymmetries in the utilization of proprioceptive feedback for goal-directed movement. Exp Brain Res 180:693–704. https://doi.org/10.1007/s00221-007-0890-7

Goble DJ, Brown SH (2008) Upper limb asymmetries in the matching of proprioceptive versus visual targets. J Neurophysiol 99:3063–3074. https://doi.org/10.1152/jn.90259.2008

Goble DJ, Brown SH (2009) Dynamic proprioceptive target matching behavior in the upper limb: effects of speed, task difficulty and arm/hemisphere asymmetries. Behav Brain Res 200:7–14. https://doi.org/10.1016/j.bbr.2008.11.034

Goble DJ, Lewis CA, Brown SH (2006) Upper limb asymmetries in the utilization of proprioceptive feedback. Exp Brain Res 168:307–311. https://doi.org/10.1007/s00221-005-0280-y

Goble DJ, Coxon JP, Wenderoth N et al (2009a) Proprioceptive sensibility in the elderly: degeneration, functional consequences and plastic-adaptive processes. Neurosci Biobehav Rev 33:271–278. https://doi.org/10.1016/j.neubiorev.2008.08.012

Goble DJ, Noble BC, Brown SH (2009b) Proprioceptive target matching asymmetries in left-handed individuals. Exp Brain Res 197:403–408. https://doi.org/10.1007/s00221-009-1922-2

Goble DJ, Mousigian MA, Brown SH (2011) Compromised encoding of proprioceptively determined joint angles in older adults: the role of working memory and attentional load. Exp Brain Res 216:35–40. https://doi.org/10.1007/s00221-011-2904-8

Goble DJ, Coxon JP, Impe AV et al (2012) The neural basis of central proprioceptive processing in older versus younger adults: an important sensory role for right putamen. Hum Brain Mapp 33:895–908. https://doi.org/10.1002/hbm.21257

Gonder-Frederick LA, Cox DJ (1991) Symptom perception, symptom beliefs, and blood glucose discrimination in the self-treatment of insulin-dependent diabetes. In: Skelton JA, Croyle RT (eds) Mental representation in health and illness. Springer, New York, pp 220–246

Gonder-Frederick LA, Cox DJ, Bobbitt SA, Pennebaker JW (1986) Blood glucose symptom beliefs of diabetic patients: accuracy and implications. Health Psychol 5:327–341. https://doi.org/10.1037//0278-6133.5.4.327

Gracely RH (1999) Pain measurement. Acta Anaesthesiol Scand 43:897–908. https://doi.org/10.1034/j.1399-6576.1999.430907.x

Gracely RH, Undem BJ, Banzett RB (2007) Cough, pain and dyspnoea: similarities and differences. Pulm Pharmacol Ther 20:433–437. https://doi.org/10.1016/j.pupt.2006.12.005

Green D, Dong X (2016) The cell biology of acute itch. J Cell Biol 213:155–161. https://doi.org/10.1083/jcb.201603042

Green DM, Swets JW (1966) Signal detection theory and psychophysics. Wiley, New York

Greenstadt L, Shapiro D, Whitehead R (1986) Blood pressure discrimination. Psychophysiology 23:500–509. https://doi.org/10.1111/j.1469-8986.1986.tb00665.x

Griffiths D (2007) Imaging bladder sensations. Neurourol Urodyn 26:899–903. https://doi.org/10.1002/nau.20488

Griffiths D, Tadic SD (2008) Bladder control, urgency, and urge incontinence: evidence from functional brain imaging. Neurourol Urodyn 27:466–474. https://doi.org/10.1002/nau.20549

Griffiths D, Tadic SD, Schaefer W, Resnick NM (2007) Cerebral control of the bladder in normal and urge-incontinent women. NeuroImage 37:1–7. https://doi.org/10.1016/j.neuroimage.2007.04.061

Griggs RC, Stunkard A (1964) The interpretation of gastric motility: II. Sensitivity and bias in the perception of gastric motility. Arch Gen Psychiatry 11:82–89. https://doi.org/10.1001/archpsyc.1964.01720250084011

Grob KR, Kuster MS, Higgins SA et al (2002) Lack of correlation between different measurements of proprioception in the knee. J Bone Joint Surg Br 84-B:614–618. https://doi.org/10.1302/0301-620X.84B4.0840614

Guglielmi RS, Roberts AH (1994) Volitional vasomotor lability and vasomotor control. Biol Psychol 39:29–44. https://doi.org/10.1016/0301-0511(94)90055-8

Guskiewicz KM, Perrin DH (1996) Research and clinical applications of assessing balance. J Sport Rehabil 5:45–63. https://doi.org/10.1123/jsr.5.1.45

Gyetvai B, Bárdos G (1999) Modulation of taste reactivity by intestinal distension in rats. Physiol Behav 66:529–535. https://doi.org/10.1016/S0031-9384(98)00319-9

Hamaguchi T, Kano M, Kanazawa M et al (2013) Effects of preceding stimulation on brain activation in response to colonic distention in humans. Psychosom Med 75:453–462. https://doi.org/10.1097/PSY.0b013e3182926682

Hamano S, Ochi E, Tsuchiya Y et al (2015) Relationship between performance test and body composition/physical strength characteristic in sprint canoe and kayak paddlers. Open Access J Sports Med 6:191–199. https://doi.org/10.2147/OAJSM.S82295

Han J, Anson J, Waddington G, Adams R (2013) Proprioceptive performance of bilateral upper and lower limb joints: side-general and site-specific effects. Exp Brain Res 226:313–323. https://doi.org/10.1007/s00221-013-3437-0

Han J, Anson J, Waddington G, Adams R (2014) Sport attainment and proprioception. Int J Sports Sci Coach 9:159–170. https://doi.org/10.1260/1747-9541.9.1.159

Han J, Waddington G, Anson J, Adams R (2015) Level of competitive success achieved by elite athletes and multi-joint proprioceptive ability. J Sci Med Sport 18:77–81. https://doi.org/10.1016/j.jsams.2013.11.013

Han J, Waddington G, Adams R et al (2016) Assessing proprioception: a critical review of methods. J Sport Health Sci 5:80–90. https://doi.org/10.1016/j.jshs.2014.10.004

Hanback JW, Revelle W (1978) Arousal and perceptual sensitivity in hypochondriacs. J Abnorm Psychol 87:523–530

Handwerker HO, Forster C, Kirchhoff C (1991) Discharge patterns of human C-fibers induced by itching and burning stimuli. J Neurophysiol 66:307–315. https://doi.org/10.1152/jn.1991.66.1.307

Handy AB, Meston CM (2016) Interoceptive awareness moderates the relationship between perceived and physiological genital arousal in women. J Sex Med 13:1907–1914. https://doi.org/10.1016/j.jsxm.2016.09.018

Harris G, Rollman GB (1983) The validity of experimental pain measures. Pain 17:369–376. https://doi.org/10.1016/0304-3959(83)90168-9

Harshaw C (2008) Alimentary epigenetics: a developmental psychobiological systems view of the perception of hunger, thirst and satiety. Dev Rev 28:541–569. https://doi.org/10.1016/j.dr.2008.08.001

Hart N, McGowan J, Minati L, Critchley HD (2013) Emotional regulation and bodily sensation: interoceptive awareness is intact in borderline personality disorder. J Personal Disord 27:506–518. https://doi.org/10.1521/pedi_2012_26_049

Hartmann EM, Handwerker HO, Forster C (2015) Gender differences in itch and pain-related sensations provoked by histamine, cowhage and capsaicin. Acta Derm Venereol 95:25–30. https://doi.org/10.2340/00015555-1894

Harver A, Katkin ES, Bloch E (1993) Signal-detection outcomes on heartbeat and respiratory resistance detection tasks in male and female subjects. Psychophysiology 30:223–230

Harvey J, Finney S, Stewart L, Gillespie J (2012) The relationship between cognition and sensation in determining when and where to void: the concept of cognitive voiding. BJU Int 110:1756–1761. https://doi.org/10.1111/j.1464-410X.2012.11078.x

Haslam DR (1967) Heat pain and pressure pain. Psychon Sci 9:567–568. https://doi.org/10.3758/BF03327892

Hastie BA, Riley JL, Robinson ME et al (2005) Cluster analysis of multiple experimental pain modalities. Pain 116:227–237. https://doi.org/10.1016/j.pain.2005.04.016

Häusser JA, Stahlecker C, Mojzisch A et al (2019) Acute hunger does not always undermine prosociality. Nat Commun 10:4733. https://doi.org/10.1038/s41467-019-12579-7

Hayen A, Herigstad M, Wiech K, Pattinson KTS (2015) Subjective evaluation of experimental dyspnoea—effects of isocapnia and repeated exposure. Respir Physiol Neurobiol 208:21–28. https://doi.org/10.1016/j.resp.2014.12.019

Henson DE, Rubin HB, Henson C (1979) Analysis of the consistency of objective measures of sexual arousal in women. J Appl Behav Anal 12:701–711. https://doi.org/10.1901/jaba.1979.12-701

Herbert BM, Pollatos O (2012) The body in the mind: on the relationship between interoception and embodiment. Top Cogn Sci 4:692–704. https://doi.org/10.1111/j.1756-8765.2012.01189.x

Herbert BM, Pollatos O (2014) Attenuated interoceptive sensitivity in overweight and obese individuals. Eat Behav 15:445–448. https://doi.org/10.1016/j.eatbeh.2014.06.002

Herbert BM, Pollatos O (2018) The relevance of interoception for eating behavior and eating disorders. In: Tsakiris M, De Preester H (eds) The interoceptive mind. From homeostasis to awareness. Oxford University Press, Oxford, England, pp 165–186

Herbert BM, Ulbrich P, Schandry R (2007) Interoceptive sensitivity and physical effort: implications for the self-control of physical load in everyday life. Psychophysiology 44:194–202. https://doi.org/10.1111/j.1469-8986.2007.00493.x

Herbert BM, Pollatos O, Flor H et al (2010) Cardiac awareness and autonomic cardiac reactivity during emotional picture viewing and mental stress. Psychophysiology 47:342–354. https://doi.org/10.1111/j.1469-8986.2009.00931.x

Herbert BM, Herbert C, Pollatos O (2011) On the relationship between interoceptive awareness and alexithymia: is interoceptive awareness related to emotional awareness? J Pers 79:1149–1175. https://doi.org/10.1111/j.1467-6494.2011.00717.x

Herbert BM, Muth ER, Pollatos O, Herbert C (2012) Interoception across modalities: on the relationship between cardiac awareness and the sensitivity for gastric functions. PLoS One 7:e36646. https://doi.org/10.1371/journal.pone.0036646

Hertz AF (1911) The sensibility of the alimentary canal. Oxford University Press, London

Hickman L, Seyedsalehi A, Cook JL et al (2020) The relationship between heartbeat counting and heartbeat discrimination: a meta-analysis. Biol Psychol. https://doi.org/10.1016/j.biopsycho.2020.107949

Hillenberg JB, Collins FL (1982) A procedural analysis and review of relaxation training research. Behav Res Ther 20:251–260. https://doi.org/10.1016/0005-7967(82)90143-7

Hoffman I, Vos R, Tack J (2009) Normal values for the satiety drinking test in healthy children between 5 and 15 years. Neurogastroenterol Motil 21:517–5e6. https://doi.org/10.1111/j.1365-2982.2007.01057.x

Holle H, Warne K, Seth AK et al (2012) Neural basis of contagious itch and why some people are more prone to it. Proc Natl Acad Sci U S A 109:19816–19821. https://doi.org/10.1073/pnas.1216160109

Holtmann G, Goebell H, Jockenhoevel F, Talley N (1998) Altered vagal and intestinal mechanosensory function in chronic unexplained dyspepsia. Gut 42:501–506

Hölzl R, Möltner A, Erasmus L-P et al (1994) Detection, discrimination, and sensation of visceral stimuli. University of Mannheim, Mannheim

Hölzl R, Erasmus LP, Möltner A (1996) Detection, discrimination and sensation of visceral stimuli. Biol Psychol 42:199–214

Hoon PW, Wincze JP, Hoon EF (1976) Physiological assessment of sexual arousal in women. Psychophysiology 13:196–204. https://doi.org/10.1111/j.1469-8986.1976.tb00097.x

Horváth Á, Ferentzi E, Köteles F (2019) Proprioceptive accuracy is not associated with self-reported body awareness, body competence, and affect. Physiol Int 106:347–354. https://doi.org/10.1556/2060.106.2019.33

Horváth Á, Vig L, Ferentzi E, Köteles F (2021) Cardiac and proprioceptive accuracy are not related to body awareness, perceived body competence, and affect. Frontiers in Psychology in Press: https://doi.org/10.3389/fpsyg.2020.575574

House JD (1983) Perception of voluntary control of cardiac function in distance runners and non-runners. Percept Mot Skills 57:25–26. https://doi.org/10.2466/pms.1983.57.1.25

Houte KVD, Carbone F, Tack J (2019) Postprandial distress syndrome: stratification and management. Expert Rev Gastroenterol Hepatol 13:37–46. https://doi.org/10.1080/17474124.2019.1543586

Hua LH, Strigo IA, Baxter LC et al (2005) Anteroposterior somatotopy of innocuous cooling activation focus in human dorsal posterior insular cortex. Am J Phys Regul Integr Comp Phys 289:R319–R325. https://doi.org/10.1152/ajpregu.00123.2005

Huber C, Kunz M, Artelt C, Lautenbacher S (2010) Attentional and emotional mechanisms of pain processing and their related factors: a structural equations approach. Pain Res Manag 15:229–237

Hummel T, Sekinger B, Wolf SR et al (1997) "Sniffin" sticks': olfactory performance assessed by the combined testing of odor identification, odor discrimination and olfactory threshold. Chem Senses 22:39–52. https://doi.org/10.1093/chemse/22.1.39

Hurley MV, Rees J, Newham DJ (1998) Quadriceps function, proprioceptive acuity and functional performance in healthy young, middle-aged and elderly subjects. Age Ageing 27:55–62

IASP (1994) Classification of chronic pain "Part III: Pain terms, a current list with definitions and notes on usage". IASP Press, Seattle

Iggo A (1955) Tension receptors in the stomach and the urinary bladder. J Physiol 128:593–607

Ijzerman H, Semin GR (2009) The thermometer of social relations: mapping social proximity on temperature. Psychol Sci 20:1214–1220. https://doi.org/10.1111/j.1467-9280.2009.02434.x

Ikoma A, Rukwied R, Ständer S et al (2003) Neurophysiology of pruritus: interaction of itch and pain. Arch Dermatol 139:1475–1478. https://doi.org/10.1001/archderm.139.11.1475

Jacobson E (1938) Progressive relaxation, 2nd edn. University of Chicago Press, Oxford, England

Jafari H, Courtois I, Van den Bergh O et al (2017) Pain and respiration: a systematic review. Pain 158:995–1006. https://doi.org/10.1097/j.pain.0000000000000865

Jameson C, Ring C (2000) Contributions of local and central sensations to the perception of exertion during cycling: effects of work rate and cadence. J Sports Sci 18:291–298. https://doi.org/10.1080/026404100365027

Janal MN, Glusman M, Kuhl JP, Clark WC (1994) On the absence of correlation between responses to noxious heat, cold, electrical and ischemic stimulation. Pain 58:403–411. https://doi.org/10.1016/0304-3959(94)90135-x

Janssen P, Vanden Berghe P, Verschueren S et al (2011) Review article: The role of gastric motility in the control of food intake. Aliment Pharmacol Ther 33:880–894. https://doi.org/10.1111/j.1365-2036.2011.04609.x

Jellinek JS (1997) Psychodynamic odour effects and their mechanisms. Cosmetics & Toiletries 112:51–61

Jerosch J, Prymka M (1996) Proprioception and joint stability. Knee Surg Sports Traumatol Arthrosc 4:171–179

Jha P, Ahamad I, Khurana S et al (2017) Proprioception: an evidence based narrative review. Res Inves Sports Med. https://doi.org/10.31031/RISM.2017.01.000506

Johanek LM, Meyer RA, Hartke T et al (2007) Psychophysical and physiological evidence for parallel afferent pathways mediating the sensation of itch. J Neurosci 27:7490–7497. https://doi.org/10.1523/JNEUROSCI.1249-07.2007

Johanek LM, Meyer RA, Friedman RM et al (2008) A role for polymodal C-fiber afferents in nonhistaminergic itch. J Neurosci 28:7659–7669. https://doi.org/10.1523/JNEUROSCI.1760-08.2008

Johansson RS, Trulsson M, Olsson KÅ, Westberg K-G (1988) Mechanoreceptor activity from the human face and oral mucosa. Exp Brain Res 72:204–208. https://doi.org/10.1007/BF00248518

Jones GE (1994) Perception of visceral sensations: a review of recent findings, methodologies, and future directions. In: Jennings JR, Ackles PK, Coles MGH (eds) Advances in psychophysiology: a research annual, vol 5. Jessica Kingsley Publishers, London, pp 55–191

Jones GE (1995) Constitutional and physiological factors in heartbeat perception. In: Vaitl D, Schandry R (eds) From the heart to the brain: the psychophysiology of circulation-heart interaction. Peter Lang, Frankfurt, pp 173–192

Jones GE, Hollandsworth JG (1981) Heart rate discrimination before and after exercise-induced augmented cardiac activity. Psychophysiology 18:252–257

Jones GE, O'Leary RT, Pipkin BL (1984) Comparison of the Brener-Jones and Whitehead procedures for assessing cardiac awareness. Psychophysiology 21:143–148. https://doi.org/10.1111/j.1469-8986.1984.tb00196.x

Jones GE, Jones KR, Rouse CH et al (1987) The effect of body position on the perception of cardiac sensations: an experiment and theoretical implications. Psychophysiology 24:300–311. https://doi.org/10.1111/j.1469-8986.1987.tb00300.x

Jones MP, Hoffman S, Shah D et al (2003) The water load test: observations from healthy controls and patients with functional dyspepsia. Am J Physiol Gastrointest Liver Physiol 284:G896–G904. https://doi.org/10.1152/ajpgi.00361.2002

Jones MP, Roth LM, Crowell MD (2005) Symptom reporting by functional dyspeptics during the water load test. Am J Gastroenterol 100:1334–1339. https://doi.org/10.1111/j.1572-0241.2005.40802.x

Jorm AF (2001) Association of hypotension with positive and negative affect and depressive symptoms in the elderly. Br J Psychiatry 178:553–555. https://doi.org/10.1192/bjp.178.6.553

Juhász G, Kukorelli T, Détári L (1980) Relationships between aggressive behavior, visceral afferentation and sleep in cats. Proc Int Union Physiol Sciences 14:497

Kahane P, Hoffmann D, Minotti L, Berthoz A (2003) Reappraisal of the human vestibular cortex by cortical electrical stimulation study. Ann Neurol 54:615–624. https://doi.org/10.1002/ana.10726

Kandasamy N, Garfinkel SN, Page L et al (2016) Interoceptive ability predicts survival on a London trading floor. Sci Rep 6:32986. https://doi.org/10.1038/srep32986

Kang Y, Williams LE, Clark MS et al (2011) Physical temperature effects on trust behavior: the role of insula. Soc Cogn Affect Neurosci 6:507–515. https://doi.org/10.1093/scan/nsq077

Kaplan FS, Nixon JE, Reitz M et al (1985) Age-related changes in proprioception and sensation of joint position. Acta Orthop Scand 56:72–74

Karsdorp PA, Kindt M, Everaerd W, Mulder BJM (2007) Preattentive processing of heart cues and the perception of heart symptoms in congenital heart disease. Behav Res Ther 45:1893–1902. https://doi.org/10.1016/j.brat.2007.02.010

Karsdorp PA, Kindt M, Rietveld S et al (2009) False heart rate feedback and the perception of heart symptoms in patients with congenital heart disease and anxiety. Int J Behav Med 16:81–88. https://doi.org/10.1007/s12529-008-9001-9

Kass-Iliyya L, Leung M, Marshall A et al (2017) The perception of affective touch in Parkinson's disease and its relation to small fibre neuropathy. Eur J Neurosci 45:232–237. https://doi.org/10.1111/ejn.13481

Katkin ES (1985) Blood, sweat, and tears: individual differences in autonomic self-perception. Psychophysiology 22:125–137. https://doi.org/10.1111/j.1469-8986.1985.tb01573.x

Katkin ES, Blascovich J, Goldband S (1981) Empirical assessment of visceral self-perception: individual and sex differences in the acquisition of heartbeat discrimination. J Pers Soc Psychol 40:1095–1101. https://doi.org/10.1037/0022-3514.40.6.1095

Katkin ES, Morell MA, Goldband S et al (1982) Individual differences in heartbeat discrimination. Psychophysiology 19:160–166. https://doi.org/10.1111/j.1469-8986.1982.tb02538.x

Katkin ES, Wiens S, Ohman A (2001) Nonconscious fear conditioning, visceral perception, and the development of gut feelings. Psychol Sci 12:366–370

Katzer A, Oberfeld D, Hiller W, Witthöft M (2011) Tactile perceptual processes and their relationship to medically unexplained symptoms and health anxiety. J Psychosom Res 71:335–341. https://doi.org/10.1016/j.jpsychores.2011.03.009

Katzer A, Oberfeld D, Hiller W et al (2012) Tactile perceptual processes and their relationship to somatoform disorders. J Abnorm Psychol 121:530–543. https://doi.org/10.1037/a0026536

Kern MK, Shaker R (2002) Cerebral cortical registration of subliminal visceral stimulation. Gastroenterology 122:290–298. https://doi.org/10.1053/gast.2002.30989

Kern M, Chai K, Lawal A, Shaker R (2009) Effect of esophageal acid exposure on the cortical swallowing network in healthy human subjects. Am J Physiol Gastrointest Liver Physiol 297:G152–G158. https://doi.org/10.1152/ajpgi.00062.2009

Keszthelyi D, Troost FJ, Masclee AA (2012) Irritable bowel syndrome: methods, mechanisms, and pathophysiology. Methods to assess visceral hypersensitivity in irritable bowel syndrome. Am J Physiol Gastrointest Liver Physiol 303:G141–G154. https://doi.org/10.1152/ajpgi.00060.2012

Ketai LH, Komesu YM, Dodd AB et al (2016) Urgency urinary incontinence and the interoceptive network: a functional magnetic resonance imaging study. Am J Obstet Gynecol 215:449.e1–449.e17. https://doi.org/10.1016/j.ajog.2016.04.056

Khalsa SS, Feinstein JS (2018) The somatic error hypothesis of anxiety. In: Tsakiris M, De Preester H (eds) The interoceptive mind. From homeostasis to awareness. Oxford University Press, Oxford, England, pp 144–164

Khalsa SS, Rudrauf D, Damasio AR et al (2008) Interoceptive awareness in experienced meditators. Psychophysiology 45:671–677. https://doi.org/10.1111/j.1469-8986.2008.00666.x

Khalsa SS, Rudrauf D, Feinstein JS, Tranel D (2009a) The pathways of interoceptive awareness. Nat Neurosci 12:1494–1496. https://doi.org/10.1038/nn.2411

Khalsa SS, Rudrauf D, Sandesara C et al (2009b) Bolus isoproterenol infusions provide a reliable method for assessing interoceptive awareness. Int J Psychophysiol 72:34–45. https://doi.org/10.1016/j.ijpsycho.2008.08.010

Khalsa SS, Rudrauf D, Tranel D (2009c) Interoceptive awareness declines with age. Psychophysiology 46:1130–1136. https://doi.org/10.1111/j.1469-8986.2009.00859.x

Kifle Y, Seng V, Davenport PW (1997) Magnitude estimation of inspiratory resistive loads in children with life-threatening asthma. Am J Respir Crit Care Med 156:1530–1535. https://doi.org/10.1164/ajrccm.156.5.9703011

Killian KJ, Mahutte CK, Campbell EJ (1980) Resistive load detection during passive ventilation. Clin Sci 59:493–495. https://doi.org/10.1042/cs0590493

Kilpatrick LA, Mayer EA, Labus JS et al (2015) Serotonin transporter gene polymorphism modulates activity and connectivity within an emotional arousal network of healthy men during an aversive visceral stimulus. PLoS One 10:e0123183. https://doi.org/10.1371/journal.pone.0123183

King AB, Menon RS, Hachinski V, Cechetto DF (1999) Human forebrain activation by visceral stimuli. J Comp Neurol 413:572–582

Kinsman RA, O'Banion K, Robinson S, Staudenmayer H (1975) Continuous biofeedback and discrete posttrial verbal feedback in frontalis muscle relaxation training. Psychophysiology 12:30–35. https://doi.org/10.1111/j.1469-8986.1975.tb03055.x

Kleckner IR, Wormwood JB, Simmons WK et al (2015) Methodological recommendations for a heartbeat detection-based measure of interoceptive sensitivity. Psychophysiology 52:1432–1440. https://doi.org/10.1111/psyp.12503

Kleckner IR, Zhang J, Touroutoglou A et al (2017) Evidence for a large-scale brain system supporting allostasis and interoception in humans. Nat Hum Behav 1:s41562-017-0069-017. https://doi.org/10.1038/s41562-017-0069

Klein DF (1993) False suffocation alarms, spontaneous panics, and related conditions. An integrative hypothesis. Arch Gen Psychiatry 50:306–317. https://doi.org/10.1001/archpsyc.1993.01820160076009

Klosterhalfen S, Kellermann S, Stockhorst U et al (2005) Latent inhibition of rotation chair-induced nausea in healthy male and female volunteers. Psychosom Med 67:335–340. https://doi.org/10.1097/01.psy.0000156930.00201.e0

Knafelc ME, Davenport PW (1997) Relationship between resistive loads and P1 peak of respiratory-related evoked potential. J Appl Physiol 83:918–926. https://doi.org/10.1152/jappl.1997.83.3.918

Knafelc ME, Davenport PW (1999) Relationship between magnitude estimation of resistive loads, inspiratory pressures, and the RREP P(1) peak. J Appl Physiol 87:516–522. https://doi.org/10.1152/jappl.1999.87.2.516

Knapp-Kline K, Kline JP (2005) Heart rate, heart rate variability, and heartbeat detection with the method of constant stimuli: slow and steady wins the race. Biol Psychol 69:387–396. https://doi.org/10.1016/j.biopsycho.2004.09.002

Knaster P, Estlander A-M, Karlsson H et al (2012) Temperament traits and chronic pain: the association of harm avoidance and pain-related anxiety. PLoS One 7:e45672. https://doi.org/10.1371/journal.pone.0045672

Knoll JF, Hodapp V (1992) A comparison between two methods for assessing heartbeat perception. Psychophysiology 29:218–222

Knoop J, Steultjens MPM, van der Leeden M et al (2011) Proprioception in knee osteoarthritis: a narrative review. Osteoarthr Cartil 19:381–388. https://doi.org/10.1016/j.joca.2011.01.003

Kobal G, Hummel T, Sekinger B et al (1996) "Sniffin' sticks": screening of olfactory performance. Rhinology 34:222–226

Koch A, Pollatos O (2014a) Cardiac sensitivity in children: sex differences and its relationship to parameters of emotional processing. Psychophysiology 51:932–941. https://doi.org/10.1111/psyp.12233

Koch A, Pollatos O (2014b) Interoceptive sensitivity, body weight and eating behavior in children: a prospective study. Front Psychol 5:1003. https://doi.org/10.3389/fpsyg.2014.01003

Koch KL, Bingaman S, Tan L, Stern RM (1998) Visceral perceptions and gastric myoelectrical activity in healthy women and in patients with bulimia nervosa. Neurogastroenterol Motil 10:3–10. https://doi.org/10.1046/j.1365-2982.1998.00080.x

Koch KL, Hong SP, Xu L (2000) Reproducibility of gastric myoelectrical activity and the water load test in patients with dysmotility-like dyspepsia symptoms and in control subjects. J Clin Gastroenterol 31:125–129. https://doi.org/10.1097/00004836-200009000-00007

Koeppel CJ, Ruser P, Kitzler H et al (2020) Interoceptive accuracy and its impact on neuronal responses to olfactory stimulation in the insular cortex. Hum Brain Mapp 41:2898–2908. https://doi.org/10.1002/hbm.24985

Kokmen E, Bossemeyer RW, Williams WJ (1978) Quantitative evaluation of joint motion sensation in an aging population. J Gerontol 33:62–67. https://doi.org/10.1093/geronj/33.1.62

Kollenbaum V-E (1994) A clinical method for the assessment of interoception of cardiovascular strain in CHD patients. J Psychophysiol 8:121–130

Kollenbaum V-E, Dahme B, Kirchner G et al (1994) Assessment of current cardiovascular stress in healthy, young people. Z Exp Angew Psychol 41:78–97

Kollenbaum V-E, Dahme B, Kirchner G (1996) 'Interoception' of heart rate, blood pressure, and myocardial metabolism during ergometric work load in healthy young subjects. Biol Psychol 42:183–197. https://doi.org/10.1016/0301-0511(95)05154-6

Koroboki E, Zakopoulos N, Manios E et al (2010) Interoceptive awareness in essential hypertension. Int J Psychophysiol 78:158–162. https://doi.org/10.1016/j.ijpsycho.2010.07.003

Kosteletzky F, Namer B, Forster C, Handwerker HO (2009) Impact of scratching on itch and sympathetic reflexes induced by cowhage (Mucuna pruriens) and histamine. Acta Derm Venereol 89:271–277. https://doi.org/10.2340/00015555-0624

Köteles F, Babulka P (2014) Role of expectations and pleasantness of essential oils in their acute effects. Acta Physiol Hung 101:329–340. https://doi.org/10.1556/APhysiol.101.2014.3.8

Köteles F, Babulka P, Szemerszky R et al (2018) Inhaled peppermint, rosemary and eucalyptus essential oils do not change spirometry in healthy individuals. Physiol Behav 194:319–323. https://doi.org/10.1016/j.physbeh.2018.06.022

Köteles F, Éliás I, Szabolcs Z et al (2020a) Accuracy of reproduction of physical training load is not associated with resting heartbeat perception in healthy individuals. Biol Psychol 150:107831. https://doi.org/10.1016/j.biopsycho.2019.107831

Köteles F, Teufel B, Körmendi J et al (2020b) Cardioceptive accuracy is associated with arousal but not with valence and perceived exertion under physical load. Psychophysiology 57:e13620. https://doi.org/10.1111/psyp.13620

Kovatchev BP, Patek SD, Ortiz EA, Breton MD (2015) Assessing sensor accuracy for non-adjunct use of continuous glucose monitoring. Diabetes Technol Ther 17:177–186. https://doi.org/10.1089/dia.2014.0272

Krahé C, Drabek MM, Paloyelis Y, Fotopoulou A (2016) Affective touch and attachment style modulate pain: a laser-evoked potentials study. Philos Trans R Soc Lond B Biol Sci 371:20160009. https://doi.org/10.1098/rstb.2016.0009

Krahé C, von Mohr M, Gentsch A et al (2018) Sensitivity to CT-optimal, affective touch depends on adult attachment style. Sci Rep 8:14544. https://doi.org/10.1038/s41598-018-32865-6

Kramer J, Handfield T, Kiefer G et al (1997) Comparisons of weight-bearing and non-weight-bearing tests of knee proprioception performed by patients with patello-femoral pain syndrome and asymptomatic individuals. Clin J Sport Med 7:113–118. https://doi.org/10.1097/00042752-199704000-00007

Krautwurst S, Gerlach AL, Gomille L et al (2014) Health anxiety—an indicator of higher interoceptive sensitivity? J Behav Ther Exp Psychiatry 45:303–309. https://doi.org/10.1016/j.jbtep.2014.02.001

Krautwurst S, Gerlach AL, Witthöft M (2016) Interoception in pathological health anxiety. J Abnorm Psychol 125:1179–1184. https://doi.org/10.1037/abn0000210

Krishnan S, Salter A, Sullivan T et al (2012) Comparison of pain models to detect opioid-induced hyperalgesia. J Pain Res 5:99–106. https://doi.org/10.2147/JPR.S27738

Krusemark EA, Novak LR, Gitelman DR, Li W (2013) When the sense of smell meets emotion: anxiety-state-dependent olfactory processing and neural circuitry adaptation. J Neurosci 33:15324–15332. https://doi.org/10.1523/JNEUROSCI.1835-13.2013

Kuban M, Barbaree HE, Blanchard R (1999) A comparison of volume and circumference phallometry: response magnitude and method agreement. Arch Sex Behav 28:345–359. https://doi.org/10.1023/A:1018700813140

Kukorelli T, Juhász G (1976) Electroencephalographic synchronization induced by stimulation of small intestine and splanchnic nerve in cats. Electroencephalogr Clin Neurophysiol 41:491–500

Kukorelli T, Juhász G (1977) Sleep induced by intestinal stimulation in cats. Physiol Behav 19:355–358. https://doi.org/10.1016/0031-9384(77)90278-5

Kukorelli T, Adám G, Gimes R, Tóth F (1972) Uteral stimulation and vigilance level in humans. Acta Physiol Acad Sci Hung 42:403–410

Kullmann G, Fielding JF (1981) Rectal distensibility in the irritable bowel syndrome. Ir Med J 74:140–142

Kumazawa T, Perl ER (1977) Primate cutaneous sensory units with unmyelinated (C) afferent fibers. J Neurophysiol 40:1325–1338. https://doi.org/10.1152/jn.1977.40.6.1325

Kwan CL, Diamant NE, Mikula K, Davis KD (2005) Characteristics of rectal perception are altered in irritable bowel syndrome. Pain 113:160–171. https://doi.org/10.1016/j.pain.2004.10.005

Laan E, Everaerd W, van Bellen G, Hanewald G (1994) Women's sexual and emotional responses to male- and female-produced erotica. Arch Sex Behav 23:153–169. https://doi.org/10.1007/bf01542096

Laan E, Janssen E, Janssen E (2007) How do men and women feel? Determinants of subjective experience of sexual arousal. In: The psychophysiology of sex. Indiana University Press, Bloomington, IN, pp 278–290

Labbe D, Almiron-Roig E, Hudry J et al (2009) Sensory basis of refreshing perception: role of psychophysiological factors and food experience. Physiol Behav 98:1–9. https://doi.org/10.1016/j.physbeh.2009.04.007

Lacey JI (1959) Psychophysiological approaches to the evaluation of psychotherapeutic process and outcome. In: Research in psychotherapy. American Psychological Association, Washington, DC, pp 160–208

Lacey JI (1967) Somatic response patterning and stress: some revisions of activation theory. In: Appley MH, Thumbull R (eds) Psychological stress: issues in research. Appletin-Century-Crofts, New York, pp 14–42

Lader MH, Mathews AM (1971) Electromyographic studies of tension. J Psychosom Res 15:479–486. https://doi.org/10.1016/0022-3999(71)90030-4

Lakie M, Loram ID (2006) Manually controlled human balancing using visual, vestibular and proprioceptive senses involves a common, low frequency neural process. J Physiol 577:403–416. https://doi.org/10.1113/jphysiol.2006.116772

LaMotte RH, Shimada SG, Green BG, Zelterman D (2009) Pruritic and nociceptive sensations and dysesthesias from a spicule of cowhage. J Neurophysiol 101:1430–1443. https://doi.org/10.1152/jn.91268.2008

Lang PJ, Davis M (2006) Emotion, motivation, and the brain: reflex foundations in animal and human research. In: Anders S, Ende G, Junghofer M, et al. (eds) Progress in brain research. Elsevier, Amsterdam pp. 3–29

Lasiter PS (1983) Gastrointestinal reactivity in rats lacking anterior insular neocortex. Behav Neural Biol 39:149–154. https://doi.org/10.1016/s0163-1047(83)90790-2

Lau P, Miesen M, Wunderlich R et al (2015) The relevance of interoception in chronic tinnitus: analyzing interoceptive sensibility and accuracy. Biomed Res Int. https://doi.org/10.1155/2015/487372

Laurent CM, Green JM (2009) Multiple models can concurrently explain fatigue during human performance. Int J Exerc Sci 2:280–293

Lautenbacher S, Rollman GB (1993) Sex differences in responsiveness to painful and non-painful stimuli are dependent upon the stimulation method. Pain 53:255–264. https://doi.org/10.1016/0304-3959(93)90221-a

Le Boeuf A (1980) Effects of frontalis biofeedback on subjective ratings of relaxation. Percept Mot Skills 50:99–103. https://doi.org/10.2466/pms.1980.50.1.99

Le Chapelain L, Beis JM, Paysant J, André JM (2001) Vestibular caloric stimulation evokes phantom limb illusions in patients with paraplegia. Spinal Cord 39:85–87. https://doi.org/10.1038/sj.sc.3101093

Leder D (2018) Inside insights: a phenomenology of interoception. In: Tsakiris M, De Preester H (eds) The interoceptive mind. From homeostasis to awareness. Oxford University Press, Oxford, England, pp 307–322

Lee WE, Hotopf M (2009) No prospective association between hypotension and idiopathic fatigue. J Psychosom Res 67:153–158. https://doi.org/10.1016/j.jpsychores.2009.01.008

Lee JS, Han JS, Lee K et al (2016) The peripheral and central mechanisms underlying itch. BMB Rep 49:474–487. https://doi.org/10.5483/bmbrep.2016.49.9.108

Lee I-S, Necka EA, Atlas LY (2020) Distinguishing pain from nociception, salience, and arousal: how autonomic nervous system activity can improve neuroimaging tests of specificity. NeuroImage 204:116254. https://doi.org/10.1016/j.neuroimage.2019.116254

Legrain V, Iannetti GD, Plaghki L, Mouraux A (2011) The pain matrix reloaded: a salience detection system for the body. Prog Neurobiol 93:111–124. https://doi.org/10.1016/j.pneurobio.2010.10.005

Lephart SM, Pincivero DM, Rozzi SL (1998) Proprioception of the ankle and knee. Sports Med 25:149–155

Leventhal DB, Schuck JR, Clemons JT, Cox M (1982) Proprioception in schizophrenia. J Nerv Ment Dis 170:21–26

Li Y, Wang Y, Zuo X et al (2004) Visceral perception thresholds after rectal thermal and pressure stimuli in irritable bowel syndrome patients. J Gastroenterol Hepatol 19:187–191. https://doi.org/10.1111/j.1440-1746.2004.03225.x

Li L, Ji Z-Q, Li Y-X, Liu W-T (2016) Correlation study of knee joint proprioception test results using common test methods. J Phys Ther Sci 28:478–482. https://doi.org/10.1589/jpts.28.478

Light AR, Perl ER (2003) Unmyelinated afferent fibers are not only for pain anymore. J Comp Neurol 461:137–139. https://doi.org/10.1002/cne.10691

Lind E, Joens-Matre RR, Ekkekakis P (2005) What intensity of physical activity do previously sedentary middle-aged women select? Evidence of a coherent pattern from physiological, perceptual, and affective markers. Prev Med 40:407–419. https://doi.org/10.1016/j.ypmed.2004.07.006

Liotti M, Brannan S, Egan G et al (2001) Brain responses associated with consciousness of breathlessness (air hunger). Proc Natl Acad Sci U S A 98:2035–2040. https://doi.org/10.1073/pnas.98.4.2035

Lloyd DM, Mason L, Brown RJ, Poliakoff E (2008) Development of a paradigm for measuring somatic disturbance in clinical populations with medically unexplained symptoms. J Psychosom Res 64:21–24. https://doi.org/10.1016/j.jpsychores.2007.06.004

Loggia ML, Juneau M, Bushnell MC (2011) Autonomic responses to heat pain: heart rate, skin conductance, and their relation to verbal ratings and stimulus intensity. Pain 152:592–598. https://doi.org/10.1016/j.pain.2010.11.032

Lohse KR, Sherwood DE (2011) Defining the focus of attention: effects of attention on perceived exertion and fatigue. Front Psychol 2:332. https://doi.org/10.3389/fpsyg.2011.00332

Löken LS, Wessberg J, Morrison I et al (2009) Coding of pleasant touch by unmyelinated afferents in humans. Nat Neurosci 12:547–548. https://doi.org/10.1038/nn.2312

Loudon JK, Ruhl M, Field E (1997) Ability to reproduce head position after whiplash injury. Spine 22:865–868. https://doi.org/10.1097/00007632-199704150-00008

Lowenstein L, FitzGerald MP, Kenton K et al (2008) Validation of a real-time urodynamic measure of urinary sensation. Am J Obstet Gynecol 198:661.e1–4; discussion 661.e4–5. https://doi.org/10.1016/j.ajog.2008.02.023

Luborsky L, Brady JP, McClintock M et al (1976) Estimating one's own systolic blood pressure: effects of feedback training. Psychosom Med 38:426. https://doi.org/10.1097/00006842-197611000-0

Lucas KE, Rowe PC, Coresh J et al (2004) Prospective association between hypotension and idiopathic chronic fatigue. J Hypertens 22:691–695. https://doi.org/10.1097/00004872-200404000-00010

Ludwick-Rosenthal R, Neufeld RW (1985) Heart beat interoception: a study of individual differences. Int J Psychophysiol 3:57–65

Lund H, Juul-Kristensen B, Hansen K et al (2008) Movement detection impaired in patients with knee osteoarthritis compared to healthy controls: a cross-sectional case-control study. J Musculoskelet Neuronal Interact 8:391–400

Lundberg M, Grimby-Ekman A, Verbunt J, Simmonds MJ (2011) Pain-related fear: a critical review of the related measures. Pain Res Treat 2011:494196. https://doi.org/10.1155/2011/494196

Lynott D, Corker KS, Wortman J et al (2014) Replication of "Experiencing physical warmth promotes interpersonal warmth" by Williams and Bargh (2008). Soc Psychol 45:216–222. https://doi.org/10.1027/1864-9335/a000187

Ma X, Yue Z-Q, Gong Z-Q et al (2017) The effect of diaphragmatic breathing on attention, negative affect and stress in healthy adults. Front Psychol 8:874. https://doi.org/10.3389/fpsyg.2017.00874

MacNeilage PR, Banks MS, DeAngelis GC, Angelaki DE (2010) Vestibular heading discrimination and sensitivity to linear acceleration in head and world coordinates. J Neurosci 30:9084–9094. https://doi.org/10.1523/JNEUROSCI.1304-10.2010

Maffei A, Haley M, Fontanini A (2012) Neural processing of gustatory information in insular circuits. Curr Opin Neurobiol 22:709–716. https://doi.org/10.1016/j.conb.2012.04.001

Mailloux J, Brener J (2002) Somatosensory amplification and its relationship to heartbeat detection ability. Psychosom Med 64:353–357

Maister L, Tsakiris M (2014) My face, my heart: cultural differences in integrated bodily self-awareness. Cogn Neurosci 5:10–16. https://doi.org/10.1080/17588928.2013.808613

Maister L, Tang T, Tsakiris M (2017) Neurobehavioral evidence of interoceptive sensitivity in early infancy. eLife 6:e25318. https://doi.org/10.7554/eLife.25318

Manchikanti L, Giordano J, Fellows B, Hirsch JA (2011) Placebo and nocebo in interventional pain management: a friend or a foe—or simply foes? Pain Physician 14:E157–E175

Mandler G, Kahn M (1960) Discrimination of changes in heart rate: two unsuccessful attempts. J Exp Anal Behav 3:21–25. https://doi.org/10.1901/jeab.1960.3-21

Marks R (1996) Further evidence of impaired position sense in knee osteoarthritis. Physiother Res Int 1:127–136

Marshall GD, Zimbardo PG (1979) Affective consequences of inadequately explained physiological arousal. J Pers Soc Psychol 37:970–988. https://doi.org/10.1037/0022-3514.37.6.970

Mash LE, Schauder KB, Cochran C et al (2017) Associations between interoceptive cognition and age in autism spectrum disorder and typical development. J Cogn Educ Psychol 16:23–37. https://doi.org/10.1891/1945-8959.16.1.23

Maslach C (1979) Negative emotional biasing of unexplained arousal. J Pers Soc Psychol 37:953–969. https://doi.org/10.1037/0022-3514.37.6.953

Mason RE (1961) Internal perception and bodily functioning. International Universities Press, New York

Matthew T, Sherwin RS, Murphy J, Kerr D (1997) Importance of cerebral blood flow to the recognition of and physiological responses to hypoglycemia. Diabetes 46:829–833. https://doi.org/10.2337/diabetes.46.5.829

McCracken LM, Dhingra L (2002) A short version of the Pain Anxiety Symptoms Scale (PASS-20): preliminary development and validity. Pain Res Manag 7:45–50. https://doi.org/10.1155/2002/517163

McFadden ER, Kiser R, DeGroot WJ (1973) Acute bronchial asthma. N Engl J Med 288:221–225. https://doi.org/10.1056/NEJM197302012880501

McFarland RA (1975) Heart rate perception and heart rate control. Psychophysiology 12:402–405. https://doi.org/10.1111/j.1469-8986.1975.tb00011.x

McGlone F, Spence C (2010) The cutaneous senses: touch, temperature, pain/itch, and pleasure. Neurosci Biobehav Rev 34:145–147. https://doi.org/10.1016/j.neubiorev.2009.08.008

McGlone F, Vallbo AB, Olausson H et al (2007) Discriminative touch and emotional touch. Can J Exp Psychol 61:173–183

McIver TA, Kornelsen J, Stroman PW (2018) Diversity in the emotional modulation of pain perception: an account of individual variability. Eur J Pain 22:319–332. https://doi.org/10.1002/ejp.1122

McKenzie KJ, Poliakoff E, Brown RJ, Lloyd DM (2010) Now you feel it, now you don't: how robust is the phenomenon of illusory tactile experience? Perception 39:839–850. https://doi.org/10.1068/p6401

McKiernan F, Houchins JA, Mattes RD (2008) Relationships between human thirst, hunger, drinking, and feeding. Physiol Behav 94:700–708. https://doi.org/10.1016/j.physbeh.2008.04.007

McMahon SB, Koltzenburg M (1992) Itching for an explanation. Trends Neurosci 15:497–501. https://doi.org/10.1016/0166-2236(92)90102-E

Meagher MW, Arnau RC, Rhudy JL (2001) Pain and emotion: effects of affective picture modulation. Psychosom Med 63:79–90. https://doi.org/10.1097/00006842-200101000-00010
Medina Lucena H, Tincello DG (2019) Methods of assessing and recording bladder sensation: a review of the literature. Int Urogynecol J 30:3–8. https://doi.org/10.1007/s00192-018-3760-x
Meeusen R (2006) Physical activity and neuotransmitter release. In: Acevedo EO, Ekkekakis P (eds) Psychobiology of physical activity. Human Kinetics, Champaign, IL, pp 129–143
Meeuwis SH, van Middendorp H, van Laarhoven AIM et al (2019) Effects of open- and closed-label nocebo and placebo suggestions on itch and itch expectations. Front Psych 10:436. https://doi.org/10.3389/fpsyt.2019.00436
Melloni M, Sedeño L, Couto B et al (2013) Preliminary evidence about the effects of meditation on interoceptive sensitivity and social cognition. Behav Brain Funct 9:47. https://doi.org/10.1186/1744-9081-9-47
Melnychuk MC, Dockree PM, O'Connell RG et al (2018) Coupling of respiration and attention via the locus coeruleus: effects of meditation and pranayama. Psychophysiology 55:e13091. https://doi.org/10.1111/psyp.13091
Melzack R (1973) The puzzle of pain. Basic Books, New York
Melzack R (1975) The McGill Pain Questionnaire: major properties and scoring methods. Pain 1:277–299. https://doi.org/10.1016/0304-3959(75)90044-5
Melzack R (1987) The short-form McGill Pain Questionnaire. Pain 30:191–197. https://doi.org/10.1016/0304-3959(87)91074-8
Melzack R (1990) Phantom limbs and the concept of a neuromatrix. Trends Neurosci 13:88–92. https://doi.org/10.1016/0166-2236(90)90179-E
Melzack R, Wall PD (1965) Pain mechanisms: a new theory. Science 150:971–979
Mense S (2008) Muscle pain: mechanisms and clinical significance. Dtsch Arztebl Int 105:214–219. https://doi.org/10.3238/artzebl.2008.0214
Mertz H, Fullerton S, Naliboff B, Mayer EA (1998) Symptoms and visceral perception in severe functional and organic dyspepsia. Gut 42:814–822. https://doi.org/10.1136/gut.42.6.814
Meston CM, Gorzalka BB (1995) The effects of sympathetic activation on physiological and subjective sexual arousal in women. Behav Res Ther 33:651–664. https://doi.org/10.1016/0005-7967(95)00006-j
Meyer D, Leventhal H, Gutmann M (1985) Common-sense models of illness: the example of hypertension. Health Psychol 4:115–135. https://doi.org/10.1037//0278-6133.4.2.115
Meyer R, Kröner-Herwig B, Spörkel H (1990) The effect of exercise and induced expectations on visceral perception in asthmatic patients. J Psychosom Res 34:455–460. https://doi.org/10.1016/0022-3999(90)90069-G
Michal M, Reuchlein B, Adler J et al (2014) Striking discrepancy of anomalous body experiences with normal interoceptive accuracy in depersonalization-derealization disorder. PLoS One 9:e89823. https://doi.org/10.1371/journal.pone.0089823
Mihevic PM (1981) Sensory cues for perceived exertion: a review. Med Sci Sports Exerc 13:150–163
Miller S, Davenport PW (2015) Subjective ratings of prolonged inspiratory resistive loaded breathing in males and females. Psychophysiology 52:90–97. https://doi.org/10.1111/psyp.12297
Miller T, Nielsen L (2015) Measure of significance of holotropic breathwork in the development of self-awareness. J Altern Complement Med 21:796–803. https://doi.org/10.1089/acm.2014.0297
Mimee A, Smith PM, Ferguson AV (2013) Circumventricular organs: targets for integration of circulating fluid and energy balance signals? Physiol Behav 121:96–102. https://doi.org/10.1016/j.physbeh.2013.02.012
Mimidis K (2007) Drinking tests in functional dyspepsia: what do they really measure? Neurogastroenterol Motil 19:947–950. https://doi.org/10.1111/j.1365-2982.2007.01034.x
Mirams L, Poliakoff E, Brown RJ, Lloyd DM (2010) Vision of the body increases interference on the somatic signal detection task. Exp Brain Res 202:787–794. https://doi.org/10.1007/s00221-010-2185-7

Mirams L, Poliakoff E, Brown RJ, Lloyd DM (2012) Interoceptive and exteroceptive attention have opposite effects on subsequent somatosensory perceptual decision making. Q J Exp Psychol (Hove) 65:926–938. https://doi.org/10.1080/17470218.2011.636823

Mirams L, Poliakoff E, Brown RJ, Lloyd DM (2013) Brief body-scan meditation practice improves somatosensory perceptual decision making. Conscious Cogn 22:348–359

Mirams L, Poliakoff E, Lloyd DM (2017) Spatial limits of visuotactile interactions in the presence and absence of tactile stimulation. Exp Brain Res 235:2591–2600. https://doi.org/10.1007/s00221-017-4998-0

Miranda MI (2012) Taste and odor recognition memory: the emotional flavor of life. Rev Neurosci 23:481–499. https://doi.org/10.1515/revneuro-2012-0064

Miron D, Duncan GH, Bushnell MC (1989) Effects of attention on the intensity and unpleasantness of thermal pain. Pain 39:345–352. https://doi.org/10.1016/0304-3959(89)90048-1

Misery L, Dutray S, Chastaing M et al (2018) Psychogenic itch. Transl Psychiatry 8:52. https://doi.org/10.1038/s41398-018-0097-7

Mogil JS (2012) Sex differences in pain and pain inhibition: multiple explanations of a controversial phenomenon. Nat Rev Neurosci 13:859–866. https://doi.org/10.1038/nrn3360

Moiseeva NA (1952) Interoceptive reflex in embryogenesis. Dokl Akad Nauk SSSR 37:321

Moiseeva NA (1966) Visceral analysis in the early stages of ontogenesis. Voprosy Sravnitel'Noi Fiziologii Analizatorov 2:128–145

Montgomery WA, Jones GE, Hollandsworth JG (1984) The effects of physical fitness and exercise on cardiac awareness. Biol Psychol 18:11–22. https://doi.org/10.1016/0301-0511(84)90022-X

Montoya P, Schandry R, Müller A (1993) Heartbeat evoked potentials (HEP): topography and influence of cardiac awareness and focus of attention. Electroencephalogr Clin Neurophysiol 88:163–172

Mora PA, Halm E, Leventhal H, Ceric F (2007) Elucidating the relationship between negative affectivity and symptoms: the role of illness-specific affective responses. Ann Behav Med 34:77–86. https://doi.org/10.1007/bf02879923

Morgan WP, Pollock ML (1977) Psychologic characterization of the elite distance runner. Ann N Y Acad Sci 301:382–403. https://doi.org/10.1111/j.1749-6632.1977.tb38215.x

Morokoff PJ, Heiman JR (1980) Effects of erotic stimuli on sexually functional and dysfunctional women: multiple measures before and after sex therapy. Behav Res Ther 18:127–137. https://doi.org/10.1016/0005-7967(80)90107-2

Morrison I, Löken LS, Olausson H (2010) The skin as a social organ. Exp Brain Res 204:305–314. https://doi.org/10.1007/s00221-009-2007-y

Morrison I, Löken LS, Minde J et al (2011) Reduced C-afferent fibre density affects perceived pleasantness and empathy for touch. Brain 134:1116–1126. https://doi.org/10.1093/brain/awr011

Morton DL, Sandhu JS, Jones AK (2016) Brain imaging of pain: state of the art. J Pain Res 9:613–624. https://doi.org/10.2147/JPR.S60433

Moseley GL (2007) Reconceptualising pain according to modern pain science. Phys Ther Rev 12:169–178. https://doi.org/10.1179/108331907X223010

Mul C-L, Stagg SD, Herbelin B, Aspell JE (2018) The feeling of me feeling for you: interoception, alexithymia and empathy in autism. J Autism Dev Disord 48:2953–2967. https://doi.org/10.1007/s10803-018-3564-3

Müller A, Montoya P, Schandry R, Hartl L (1994) Changes in physical symptoms, blood pressure and quality of life over 30 days. Behav Res Ther 32:593–603. https://doi.org/10.1016/0005-7967(94)90013-2

Murphy J, Brewer R, Catmur C, Bird G (2017) Interoception and psychopathology: a developmental neuroscience perspective. Dev Cogn Neurosci 23:45–56. https://doi.org/10.1016/j.dcn.2016.12.006

Murphy J, Catmur C, Bird G (2018a) Alexithymia is associated with a multidomain, multidimensional failure of interoception: evidence from novel tests. J Exp Psychol Gen 147:398–408. https://doi.org/10.1037/xge0000366

Murphy J, Geary H, Millgate E et al (2018b) Direct and indirect effects of age on interoceptive accuracy and awareness across the adult lifespan. Psychon Bull Rev 25:1193–1202. https://doi.org/10.3758/s13423-017-1339-z

Mussgay L, Klinkenberg N, Rüddel H (1999) Heart beat perception in patients with depressive, somatoform, and personality disorders. J Psychophysiol 13:27–36. https://doi.org/10.1027//0269-8803.13.1.27

Nagle AS, Speich JE, De Wachter SG et al (2017) Non-invasive characterization of real-time bladder sensation using accelerated hydration and a novel sensation meter: an initial experience. Neurourol Urodyn 36:1417–1426. https://doi.org/10.1002/nau.23137

Naimi HA, Nagle AS, Vinod NN et al (2019) An innovative, non-invasive sensation meter allows for a more comprehensive understanding of bladder sensation events: a prospective study in participants with normal bladder function. Neurourol Urodyn 38:208–214. https://doi.org/10.1002/nau.23831

Nakao M, Yano E, Nomura S, Kuboki T (2003) Blood pressure-lowering effects of biofeedback treatment in hypertension: a meta-analysis of randomized controlled trials. Hypertens Res 26:37–46. https://doi.org/10.1291/hypres.26.37

Namer B, Carr R, Johanek LM et al (2008) Separate peripheral pathways for pruritus in man. J Neurophysiol 100:2062–2069. https://doi.org/10.1152/jn.90482.2008

Napadow V, Li A, Loggia ML et al (2015) The imagined itch: brain circuitry supporting nocebo-induced itch in atopic dermatitis patients. Allergy. https://doi.org/10.1111/all.12727

Näring GW, van der Staak CP (1995) Perception of heart rate and blood pressure: the role of alexithymia and anxiety. Psychother Psychosom 63:193–200

Nathan PW (1956) Sensations associated with micturition. Br J Urol 28:126–131. https://doi.org/10.1111/j.1464-410x.1956.tb04744.x

National Research Council (2009) Recognition and alleviation of pain in laboratory animals. National Academies Press, Washington, DC

Negoias S, Croy I, Gerber J et al (2010) Reduced olfactory bulb volume and olfactory sensitivity in patients with acute major depression. Neuroscience 169:415–421. https://doi.org/10.1016/j.neuroscience.2010.05.012

Nelson DV, Novy DM (1997) Self-report differentiation of anxiety and depression in chronic pain. J Pers Assess 69:392–407. https://doi.org/10.1207/s15327752jpa6902_10

Ness TJ, Gebhart GF (1990) Visceral pain: a review of experimental studies. Pain 41:167–234

Ness TJ, Metcalf AM, Gebhart GF (1990) A psychophysiological study in humans using phasic colonic distension as a noxious visceral stimulus. Pain 43:377–386. https://doi.org/10.1016/0304-3959(90)90035-C

Ness TJ, Richter HE, Varner RE, Fillingim RB (1998) A psychophysical study of discomfort produced by repeated filling of the urinary bladder. Pain 76:61–69. https://doi.org/10.1016/s0304-3959(98)00023-2

Neumann R, Strack F (2000) Approach and avoidance: the influence of proprioceptive and exteroceptive cues on encoding of affective information. J Pers Soc Psychol 79:39–48. https://doi.org/10.1037/0022-3514.79.1.39

Ng T-P, Feng L, Niti M, Yap K-B (2010) Low blood pressure and depressive symptoms among Chinese older subjects: a population-based study. Am J Med 123:342–349. https://doi.org/10.1016/j.amjmed.2009.09.026

Nielsen L, Kaszniak AW (2006) Awareness of subtle emotional feelings: a comparison of long-term meditators and nonmeditators. Emotion 6:392–405. https://doi.org/10.1037/1528-3542.6.3.392

Niespodziński B, Kochanowicz A, Mieszkowski J et al (2018) Relationship between joint position sense, force sense, and muscle strength and the impact of gymnastic training on proprioception. Biomed Res Int 2018:1–10. https://doi.org/10.1155/2018/5353242

Noble MIM, Frankel HL, Else W, Guz A (1971) The ability of man to detect added resistive loads to breathing. Clin Sci 41:285–287

Nordin M (1990) Low-threshold mechanoreceptive and nociceptive units with unmyelinated (C) fibres in the human supraorbital nerve. J Physiol Lond 426:229–240. https://doi.org/10.1113/jphysiol.1990.sp018135

Nordin S, Millqvist E, Löwhagen O, Bende M (2004) A short Chemical Sensitivity Scale for assessment of airway sensory hyperreactivity. Int Arch Occup Environ Health 77:249–254. https://doi.org/10.1007/s00420-004-0504-7

Nordin S, Claeson A-S, Andersson M et al (2013) Impact of health-risk perception on odor perception and cognitive performance. Chem Percept 6:190–197. https://doi.org/10.1007/s12078-013-9153-0

Nordin S, Neely G, Olsson D, Sandström M (2014) Odor and noise intolerance in persons with self-reported electromagnetic hypersensitivity. Int J Environ Res Public Health 11:8794–8805. https://doi.org/10.3390/ijerph110908794

Nummenmaa L, Tuominen L, Dunbar R et al (2016) Social touch modulates endogenous μ-opioid system activity in humans. NeuroImage 138:242–247

Nummenmaa L, Hari R, Hietanen JK, Glerean E (2018) Maps of subjective feelings. Proc Natl Acad Sci U S A 115:9198–9203. https://doi.org/10.1073/pnas.1807390115

O'Brien WH, Reid GJ, Jones KR (1998) Differences in heartbeat awareness among males with higher and lower levels of systolic blood pressure. Int J Psychophysiol 29:53–63. https://doi.org/10.1016/S0167-8760(98)00004-X

O'Connell KA, Hamera EK, Knapp TM et al (1984) Symptom use and self-regulation in type II diabetes. ANS Adv Nurs Sci 6:19–28. https://doi.org/10.1097/00012272-198404000-00007

O'Connell KA, Hamera EK, Schorfheide A, Guthrie D (1990) Symptom beliefs and actual blood glucose in type II diabetes. Res Nurs Health 13:145–151. https://doi.org/10.1002/nur.4770130303

Ogard WK (2011) Proprioception in sports medicine and athletic conditioning. Strength Condition J 33:111–118

Ogden J, Zoukas S (2009) Generating physical symptoms from visual cues: an experimental study. Psychol Health Med 14:695–704. https://doi.org/10.1080/13548500903311547

Olausson H, Lamarre Y, Backlund H et al (2002) Unmyelinated tactile afferents signal touch and project to insular cortex. Nat Neurosci 5:900–904. https://doi.org/10.1038/nn896

Otten S, Schötz E, Wittmann M et al (2015) Psychophysiology of duration estimation in experienced mindfulness meditators and matched controls. Front Psychol 6:1215. https://doi.org/10.3389/fpsyg.2015.01215

Paintal AS (1986) The visceral sensations—some basic mechanisms. Prog Brain Res 67:3–19. https://doi.org/10.1016/S0079-6123(08)62752-1

Palace EM, Gorzalka BB (1990) The enhancing effects of anxiety on arousal in sexually dysfunctional and functional women. J Abnorm Psychol 99:403–411

Palla A, Lenggenhager B (2014) Ways to investigate vestibular contributions to cognitive processes. Front Integr Neurosci 8:40. https://doi.org/10.3389/fnint.2014.00040

Palmer C, Ainley V, Tsakiris M (2019) Fine Tuning Your Heart: a novel method for measuring interoceptive accuracy. PsyArXiv. https://doi.org/10.31234/osf.io/qz7r9

Palser ER, Fotopoulou A, Pellicano E, Kilner JM (2018) The link between interoceptive processing and anxiety in children diagnosed with autism spectrum disorder: extending adult findings into a developmental sample. Biol Psychol 136:13–21. https://doi.org/10.1016/j.biopsycho.2018.05.003

Paluska SA, Schwenk TL (2000) Physical activity and mental health: current concepts. Sports Med 29:167–180. https://doi.org/10.2165/00007256-200029030-00003

Panerai AE (2011) Pain emotion and homeostasis. Neurol Sci 32:27–29. https://doi.org/10.1007/s10072-011-0540-5

Pánics G, Tállay A, Pavlik A, Berkes I (2008) Effect of proprioception training on knee joint position sense in female team handball players. Br J Sports Med 42:472–476. https://doi.org/10.1136/bjsm.2008.046516

Pappens M, Van den Bergh O, Vansteenwegen D et al (2013) Learning to fear obstructed breathing: comparing interoceptive and exteroceptive cues. Biol Psychol 92:36–42. https://doi.org/10.1016/j.biopsycho.2011.05.004

Parkin L, Morgan R, Rosselli A et al (2013) Exploring the relationship between mindfulness and cardiac perception. Mindfulness 5:298–313. https://doi.org/10.1007/s12671-012-0181-7

Parshall MB, Schwartzstein RM, Adams L et al (2012) An official American Thoracic Society statement: update on the mechanisms, assessment, and management of dyspnea. Am J Respir Crit Care Med 185:435–452. https://doi.org/10.1164/rccm.201111-2042ST

Pawling R, Cannon PR, McGlone FP, Walker SC (2017) C-tactile afferent stimulating touch carries a positive affective value. PLoS One 12:e0173457. https://doi.org/10.1371/journal.pone.0173457

Payne KA, Berg K, Latin RW (1997) Ankle injuries and ankle strength, flexibility, and proprioception in college basketball players. J Athl Train 32:221–225

Pennebaker JW (1981) Stimulus characteristics influencing estimation of heart rate. Psychophysiology 18:540–548. https://doi.org/10.1111/j.1469-8986.1981.tb01824.x

Pennebaker JW (1982) The psychology of physical symptoms. Springer, New York

Pennebaker JW, Epstein D (1983) Implicit psychophysiology: effects of common beliefs and idiosyncratic physiological responses on symptom reporting. J Pers 51:468–496

Pennebaker JW, Hoover CW (1984) Visceral perception versus visceral detection: disentangling methods and assumptions. Biofeedback Self Regul 9:339–352. https://doi.org/10.1007/BF00998977

Pennebaker JW, Lightner JM (1980) Competition of internal and external information in an exercise setting. J Pers Soc Psychol 39:165–174. https://doi.org/10.1037/0022-3514.39.1.165

Pennebaker JW, Roberts T-A (1992) Toward a His and Hers Theory of Emotion: gender differences in visceral perception. J Soc Clin Psychol 11:199–212. https://doi.org/10.1521/jscp.1992.11.3.199

Pennebaker JW, Skelton JA (1981) Selective monitoring of physical sensations. J Pers Soc Psychol 41:213–223

Pennebaker JW, Watson D (1988) Blood pressure estimation and beliefs among normotensives and hypertensives. Health Psychol 7:309–328. https://doi.org/10.1037/0278-6133.7.4.309

Pennebaker JW, Cox DJ, Gonder-Frederick L et al (1981) Physical symptoms related to blood glucose in insulin-dependent diabetics. Psychosom Med 43:489

Pennebaker JW, Gonder-Frederick L, Stewart H et al (1982) Physical symptoms associated with blood pressure. Psychophysiology 19:201–210. https://doi.org/10.1111/j.1469-8986.1982.tb02547.x

Perkins M, de Bruyne M, Giummarra MJ (2016) A pain in the bud? Implications of cross-modal sensitivity for pain experience. Atten Percept Psychophys 78:2348–2356. https://doi.org/10.3758/s13414-016-1217-1

Peterka RJ (2002) Sensorimotor integration in human postural control. J Neurophysiol 88:1097–1118. https://doi.org/10.1152/jn.2002.88.3.1097

Petersen S, Schroijen M, Mölders C et al (2014) Categorical interoception: perceptual organization of sensations from inside. Psychol Sci 25:1059–1066. https://doi.org/10.1177/0956797613519110

Petrie A (1967) Individuality in pain and suffering: the reducer and augmenter. University of Chicago Press, Chicago

Petrie A (1978) Individuality in pain and suffering, 23rd edn. University of Chicago Press, Chicago

Petrovic P, Dietrich T, Fransson P et al (2005) Placebo in emotional processing—induced expectations of anxiety relief activate a generalized modulatory network. Neuron 46:957–969. https://doi.org/10.1016/j.neuron.2005.05.023

Petzschner FH, Weber LA, Wellstein KV et al (2019) Focus of attention modulates the heartbeat evoked potential. NeuroImage 186:595–606. https://doi.org/10.1016/j.neuroimage.2018.11.037

Phillips GC, Jones GE, Rieger EJ, Snell JB (1999) Effects of the presentation of false heart-rate feedback on the performance of two common heartbeat-detection tasks. Psychophysiology 36:504–510

Pile V, Lau JYF, Topor M et al (2018) Interoceptive accuracy in youth with tic disorders: exploring links with premonitory urge, anxiety and quality of life. J Autism Dev Disord 48:3474–3482. https://doi.org/10.1007/s10803-018-3608-8

Pilgrim JA, Stansfeld S, Marmot M (1992) Low blood pressure, low mood? BMJ 304:75–78

Pincivero DM, Bachmeier B, Coelho AJ (2001) The effects of joint angle and reliability on knee proprioception. Med Sci Sports Exerc 33:1708–1712. https://doi.org/10.1097/00005768-200110000-00015

Pollatos O, Schandry R (2004) Accuracy of heartbeat perception is reflected in the amplitude of the heartbeat-evoked brain potential. Psychophysiology 41:476–482. https://doi.org/10.1111/1469-8986.2004.00170.x

Pollatos O, Kirsch W, Schandry R (2005) Brain structures involved in interoceptive awareness and cardioafferent signal processing: a dipole source localization study. Hum Brain Mapp 26:54–64. https://doi.org/10.1002/hbm.20121

Pollatos O, Albrecht J, Kopietz R et al (2007a) Reduced olfactory sensitivity in subjects with depressive symptoms. J Affect Disord 102:101–108. https://doi.org/10.1016/j.jad.2006.12.012

Pollatos O, Herbert BM, Kaufmann C et al (2007b) Interoceptive awareness, anxiety and cardiovascular reactivity to isometric exercise. Int J Psychophysiol 65:167–173. https://doi.org/10.1016/j.ijpsycho.2007.03.005

Pollatos O, Herbert BM, Matthias E, Schandry R (2007c) Heart rate response after emotional picture presentation is modulated by interoceptive awareness. Int J Psychophysiol 63:117–124. https://doi.org/10.1016/j.ijpsycho.2006.09.003

Pollatos O, Kopietz R, Linn J et al (2007d) Emotional stimulation alters olfactory sensitivity and odor judgment. Chem Senses 32:583–589. https://doi.org/10.1093/chemse/bjm027

Pollatos O, Traut-Mattausch E, Schroeder H, Schandry R (2007e) Interoceptive awareness mediates the relationship between anxiety and the intensity of unpleasant feelings. J Anxiety Disord 21:931–943. https://doi.org/10.1016/j.janxdis.2006.12.004

Pollatos O, Kurz A-L, Albrecht J et al (2008) Reduced perception of bodily signals in anorexia nervosa. Eat Behav 9:381–388. https://doi.org/10.1016/j.eatbeh.2008.02.001

Pollatos O, Traut-Mattausch E, Schandry R (2009) Differential effects of anxiety and depression on interoceptive accuracy. Depress Anxiety 26:167–173. https://doi.org/10.1002/da.20504

Pollatos O, Dietel A, Herbert BM et al (2011) Blunted autonomic reactivity and increased pain tolerance in somatoform patients. Pain 152:2157–2164. https://doi.org/10.1016/j.pain.2011.05.024

Pollatos O, Füstös J, Critchley HD (2012) On the generalised embodiment of pain: How interoceptive sensitivity modulates cutaneous pain perception. Pain 153:1680–1686. https://doi.org/10.1016/j.pain.2012.04.030

Pollatos O, Herbert BM, Berberich G et al (2016) Atypical self-focus effect on interoceptive accuracy in anorexia nervosa. Front Hum Neurosci 10:484. https://doi.org/10.3389/fnhum.2016.00484

Poole H, White S, Blake C et al (2009) Depression in chronic pain patients: prevalence and measurement. Pain Pract 9:173–180. https://doi.org/10.1111/j.1533-2500.2009.00274.x

Porges SW, Raskin DC (1969) Respiratory and heart rate components of attention. J Exp Psychol 81:497–503

Posserud I, Syrous A, Lindström L et al (2007) Altered rectal perception in irritable bowel syndrome is associated with symptom severity. Gastroenterology 133:1113–1123. https://doi.org/10.1053/j.gastro.2007.07.024

Powers S, Howley E (2014) Exercise physiology: theory and application to fitness and performance. McGraw-Hill, New York

Preisich P, Ádám G (1966) La discrimination nonconmiente des stimuli duodenaux: Le test de diffrentation d'habituation electroencephalographique. Acta Gastro-Enterol Belg 27:625–629

Price DD (2000) Psychological and neural mechanisms of the affective dimension of pain. Science 288:1769–1772. https://doi.org/10.1126/science.288.5472.1769

Prkachin KM (2009) Assessing pain by facial expression: facial expression as nexus. Pain Res Manag 14:53–58

Proske U, Gandevia SC (2012) The proprioceptive senses: their roles in signaling body shape, body position and movement, and muscle force. Physiol Rev 92:1651–1697. https://doi.org/10.1152/physrev.00048.2011

Pryor JB, Gibbons FX, Wicklund RA et al (1977) Self-focused attention and self-report validity. J Pers 45:513–527. https://doi.org/10.1111/j.1467-6494.1977.tb00168.x

Put C, Van den Bergh O, Van Ongeval E et al (2004) Negative affectivity and the influence of suggestion on asthma symptoms. J Psychosom Res 57:249–255. https://doi.org/10.1016/S0022-3999(03)00541-5

Racine M, Tousignant-Laflamme Y, Kloda LA et al (2012) A systematic literature review of 10 years of research on sex/gender and experimental pain perception - Part 1: Are there really differences between women and men? Pain 153:602–618. https://doi.org/10.1016/j.pain.2011.11.025

Rainville P, Duncan GH, Price DD et al (1997) Pain affect encoded in human anterior cingulate but not somatosensory cortex. Science 277:968–971. https://doi.org/10.1126/science.277.5328.968

Rainville P, Bao QVH, Chrétien P (2005) Pain-related emotions modulate experimental pain perception and autonomic responses. Pain 118:306–318. https://doi.org/10.1016/j.pain.2005.08.022

Ravn P, Frederiksen R, Skovsen AP et al (2012) Prediction of pain sensitivity in healthy volunteers. J Pain Res 5:313–326. https://doi.org/10.2147/JPR.S33925

Razon S, Basevitch I, Filho E et al (2010) Associative and dissociative imagery effects on perceived exertion and task duration. J Imagery Res Sport Phys Activity 5:1044. https://doi.org/10.2202/1932-0191.1044

Razon S, Mandler K, Arsal G et al (2014) Effects of imagery on effort perception and cycling endurance. J Imagery Res Sport Phys Activity 9:23–38. https://doi.org/10.1515/jirspa-2013-0011

Razran G (1961) The observable unconscious and the inferable conscious in current Soviet psychophysiology: interoceptive conditioning, semantic conditioning, and the orienting reflex. Psychol Rev 68:1–147

Reeder KP, Dolce JJ, Duke L et al (1990) Peak flow meters: are they monitoring tools or training devices? J Asthma 27:219–227. https://doi.org/10.3109/02770909009073330

Reek JV, Diederiks J, Philipsen H et al (1982) Subjective complaints and blood pressure. J Psychosom Res 26:155–165. https://doi.org/10.1016/0022-3999(82)90033-2

Reitz A, Schmid DM, Curt A et al (2003) Electrophysiological assessment of sensations arising from the bladder: are there objective criteria for subjective perceptions? J Urol 169:190–194. https://doi.org/10.1016/S0022-5347(05)64065-7

Rellini AH, McCall KM, Randall PK, Meston CM (2005) The relationship between women's subjective and physiological sexual arousal. Psychophysiology 42:116–124. https://doi.org/10.1111/j.1469-8986.2005.00259.x

Revel M, Minguet M, Gregoy P et al (1994) Changes in cervicocephalic kinesthesia after a proprioceptive rehabilitation program in patients with neck pain: a randomized controlled study. Arch Phys Med Rehabil 75:895–899. https://doi.org/10.1016/0003-9993(94)90115-5

Rhinewine JP, Williams OJ (2007) Holotropic Breathwork: the potential role of a prolonged, voluntary hyperventilation procedure as an adjunct to psychotherapy. J Altern Complement Med 13:771–776. https://doi.org/10.1089/acm.2006.6203

Rhudy JL, Williams AE, McCabe KM et al (2005) Affective modulation of nociception at spinal and supraspinal levels. Psychophysiology 42:579–587. https://doi.org/10.1111/j.1469-8986.2005.00313.x

Rhudy JL, Williams AE, McCabe KM et al (2006) Emotional modulation of spinal nociception and pain: the impact of predictable noxious stimulation. Pain 126:221–233. https://doi.org/10.1016/j.pain.2006.06.027

Rhudy JL, Williams AE, McCabe KM et al (2008) Emotional control of nociceptive reactions (ECON): do affective valence and arousal play a role? Pain 136:250–261. https://doi.org/10.1016/j.pain.2007.06.031

Ribeiro F, Oliveira J (2007) Aging effects on joint proprioception: the role of physical activity in proprioception preservation. Eur Rev Aging Phys Act 4:71–76. https://doi.org/10.1007/s11556-007-0026-x

Ribeiro F, Oliveira J (2011) Factors influencing proprioception: what do they reveal? In: Klika V (ed) Biomechanics in applications. InTech Open, Rijeka, pp 323–346

Richter S, Schulz A, Port J et al (2009) Cardiopulmonary baroreceptors affect reflexive startle eye blink. Physiol Behav 98:587–593. https://doi.org/10.1016/j.physbeh.2009.09.010

Richter CG, Babo-Rebelo M, Schwartz D, Tallon-Baudry C (2017) Phase-amplitude coupling at the organism level: the amplitude of spontaneous alpha rhythm fluctuations varies with the phase of the infra-slow gastric basal rhythm. NeuroImage 146:951–958. https://doi.org/10.1016/j.neuroimage.2016.08.043

Riemann BL, Myers JB, Lephart SM (2002) Sensorimotor system measurement techniques. J Athl Train 37:85–98

Rietveld S, Everaerd W, Vanbeest I (1999) Can biased symptom perception explain false-alarm choking sensations? Psychol Med 29:121–126. https://doi.org/10.1017/s0033291798007764

Rinaman L, Banihashemi L, Koehnle TJ (2011) Early life experience shapes the functional organization of stress-responsive visceral circuits. Physiol Behav 104:632–640. https://doi.org/10.1016/j.physbeh.2011.04.008

Ring C, Brener J (1996) Influence of beliefs about heart rate and actual heart rate on heartbeat counting. Psychophysiology 33:541–546. https://doi.org/10.1111/j.1469-8986.1996.tb02430.x

Ring C, Brener J (2018) Heartbeat counting is unrelated to heartbeat detection: a comparison of methods to quantify interoception. Psychophysiology 55:e13084. https://doi.org/10.1111/psyp.13084

Ring C, Liu X, Brener J (1994) Cardiac stimulus intensity and heartbeat detection: effects of tilt-induced changes in stroke volume. Psychophysiology 31:553–564

Ring C, Brener J, Knapp K, Mailloux J (2015) Effects of heartbeat feedback on beliefs about heart rate and heartbeat counting: a cautionary tale about interoceptive awareness. Biol Psychol 104:193–198. https://doi.org/10.1016/j.biopsycho.2014.12.010

Ringkamp M, Meyer R (2014) Pruriceptors. In: Carstens E, Akiyama T (eds) Itch: mechanisms and treatment. CRC Press/Taylor & Francis, Boca Raton, FL

Ringkamp M, Schepers RJ, Shimada SG et al (2011) A role for nociceptive, myelinated nerve fibers in itch sensation. J Neurosci 31:14841–14849. https://doi.org/10.1523/JNEUROSCI.3005-11.2011

Ritchie J (1973) Pain from distension of the pelvic colon by inflating a balloon in the irritable colon syndrome. Gut 14:125–132. https://doi.org/10.1136/gut.14.2.125

Ritzler B (1977) Proprioception and schizophrenia: a replication study with nonschizophrenic patient controls. J Abnorm Psychol 86:501–509. https://doi.org/10.1037/0021-843X.86.5.501

Ritzler B, Rosenbaum G (1974) Proprioception in schizophrenics and normals: effects of stimulus intensity and interstimulus interval. J Abnorm Psychol 83:106–111

Robbins TW, Granon S, Muir JL et al (1998) Neural systems underlying arousal and attention: implications for drug abusea. Ann N Y Acad Sci 846:222–237. https://doi.org/10.1111/j.1749-6632.1998.tb09740.x

Robinson DR, Gebhart GF (2008) Inside information – the unique features of visceral sensation. Mol Interv 8:242–253. https://doi.org/10.1124/mi.8.5.9

Rodríguez-Rodríguez I, Rodríguez J-V, Zamora-Izquierdo M-Á (2018) Variables to be monitored via biomedical sensors for complete type 1 diabetes mellitus management: an extension of the "on-board" concept. J Diabetes Res 2018:4826984. https://doi.org/10.1155/2018/4826984

Rollman GB, Abdel-Shaheed J, Gillespie JM, Jones KS (2004) Does past pain influence current pain: biological and psychosocial models of sex differences. Eur J Pain 8:427–433. https://doi.org/10.1016/j.ejpain.2004.03.002

Rolls ET (2015) Taste, olfactory, and food reward value processing in the brain. Prog Neurobiol 127–128:64–90. https://doi.org/10.1016/j.pneurobio.2015.03.002

Rolls BJ, Wood RJ, Rolls ET et al (1980) Thirst following water deprivation in humans. Am J Phys 239:R476–R482. https://doi.org/10.1152/ajpregu.1980.239.5.R476

Rosenbaum G, Cohen BD, Luby ED et al (1959) Comparison of sernyl with other drugs: simulation of schizophrenic performance with sernyl, LSD-25, and amobarbital (amytal) sodium. I. Attention, motor function, and proprioception. AMA Arch Gen Psychiatry 1:651–656. https://doi.org/10.1001/archpsyc.1959.03590060113013

Rosenbaum G, Flenning F, Rosen H (1965) Effects of weight intensity on discrimination thresholds of normals and schizophrenics. J Abnorm Psychol 70:446–450

Rosenberger C, Elsenbruch S, Scholle A et al (2009) Effects of psychological stress on the cerebral processing of visceral stimuli in healthy women. Neurogastroenterol Motil 21:740–e45. https://doi.org/10.1111/j.1365-2982.2009.01295.x

Ross A, Brener J (1981) Two procedures for training cardiac discrimination: a comparison of solution strategies and their relationship to heart rate control. Psychophysiology 18:62–70. https://doi.org/10.1111/j.1469-8986.1981.tb01544.x

Rothman S (1922) Beiträge zur Physiologie der Juckempfindung. Arch f Dermat 139:227–234. https://doi.org/10.1007/BF01828928

Rouillon JM, Azpiroz F, Malagelada JR (1991a) Reflex changes in intestinal tone: relationship to perception. Am J Phys 261:G280–G286. https://doi.org/10.1152/ajpgi.1991.261.2.G280

Rouillon JM, Azpiroz F, Malagelada JR (1991b) Sensorial and intestinointestinal reflex pathways in the human jejunum. Gastroenterology 101:1606–1612. https://doi.org/10.1016/0016-5085(91)90398-5

Rouse CH, Jones GE, Jones KR (1988) The effect of body composition and gender on cardiac awareness. Psychophysiology 25:400–407

Rubinfeld AR, Pain MCF (1976) Perception of asthma. Lancet 307:882–884. https://doi.org/10.1016/S0140-6736(76)92097-3

Rubinfeld AR, Pain MC (1977) Conscious perception of bronchospasm as a protective phenomenon in asthma. Chest 72:154–158. https://doi.org/10.1378/chest.72.2.154

Russo MA, Santarelli DM, O'Rourke D (2017) The physiological effects of slow breathing in the healthy human. Breathe 13:298–309. https://doi.org/10.1183/20734735.009817

Salamone PC, Sedeño L, Legaz A et al (2020) Dynamic neurocognitive changes in interoception after heart transplant. Brain Commun. https://doi.org/10.1093/braincomms/fcaa095

Sanger GJ, Andrews PLR (2006) Treatment of nausea and vomiting: gaps in our knowledge. Auton Neurosci 129:3–16. https://doi.org/10.1016/j.autneu.2006.07.009

Saoji AA, Raghavendra BR, Manjunath NK (2018) Effects of yogic breath regulation: a narrative review of scientific evidence. J Ayurveda Integr Med. https://doi.org/10.1016/j.jaim.2017.07.008

Sarnoch H, Adler F, Scholz OB (1997) Relevance of muscular sensitivity, muscular activity, and cognitive variables for pain reduction associated with EMG biofeedback in fibromyalgia. Percept Mot Skills 84:1043–1050. https://doi.org/10.2466/pms.1997.84.3.1043

Schaan L, Schulz A, Nuraydin S et al (2019) Interoceptive accuracy, emotion recognition, and emotion regulation in preschool children. Int J Psychophysiol 138:47–56. https://doi.org/10.1016/j.ijpsycho.2019.02.001

Schaefer M, Egloff B, Witthöft M (2012) Is interoceptive awareness really altered in somatoform disorders? Testing competing theories with two paradigms of heartbeat perception. J Abnorm Psychol 121:719–724. https://doi.org/10.1037/a0028509

Schaefer M, Egloff B, Gerlach AL, Witthöft M (2014) Improving heartbeat perception in patients with medically unexplained symptoms reduces symptom distress. Biol Psychol 101:69–76. https://doi.org/10.1016/j.biopsycho.2014.05.012

Schandry R (1981) Heart beat perception and emotional experience. Psychophysiology 18:483–488. https://doi.org/10.1111/j.1469-8986.1981.tb02486.x

Schandry R, Bestler M (1995) The association between parameters of cardiovascular function and heartbeat perception. In: Vaitl D, Schandry R (eds) From the heart to the brain: the psychophysiology of circulation-heart interaction. Peter Lang, Frankfurt, pp 223–250

Schandry R, Weitkunat R (1990) Enhancement of heartbeat-related brain potentials through cardiac awareness training. Int J Neurosci 53:243–253
Schandry R, Sparrer B, Weitkunat R (1986) From the heart to the brain: a study of heartbeat contingent scalp potentials. Int J Neurosci 30:261–275
Schandry R, Bestler M, Montoya P (1993) On the relation between cardiodynamics and heartbeat perception. Psychophysiology 30:467–474. https://doi.org/10.1111/j.1469-8986.1993.tb02070.x
Schandry R, Leopold C, Vogt M (1996) Symptom reporting in asthma patients and insulin-dependent diabetics. Biol Psychol 42:231–244. https://doi.org/10.1016/0301-0511(95)05157-0
Schaub N, Kano M, Farmer AD et al (2013) Su2109 the influence of the 5-HTTLPR genotype on the cerebral processing of esophageal pain - a pilot study. Gastroenterology 144:S560. https://doi.org/10.1016/S0016-5085(13)62070-3
Schauder KB, Mash LE, Bryant LK, Cascio CJ (2015) Interoceptive ability and body awareness in autism spectrum disorder. J Exp Child Psychol 131:193–200. https://doi.org/10.1016/j.jecp.2014.11.002
Schilder JD, IJzerman H, Denissen JJA (2014) Physical warmth and perceptual focus: a replication of IJzerman and Semin (2009). PLoS One 9:e112772. https://doi.org/10.1371/journal.pone.0112772
Schmid DM, Reitz A, Curt A et al (2004) Urethral evoked sympathetic skin responses and viscerosensory evoked potentials as diagnostic tools to evaluate urogenital autonomic afferent innervation in spinal cord injured patients. J Urol 171:1156–1160. https://doi.org/10.1097/01.ju.0000111809.81966.8b
Schmitz J, Blechert J, Krämer M et al (2012) Biased perception and interpretation of bodily anxiety symptoms in childhood social anxiety. J Clin Child Adolesc Psychol 41:92–102. https://doi.org/10.1080/15374416.2012.632349
Schneider TR, Ring C, Katkin ES (1998) A test of the validity of the method of constant stimuli as an index of heartbeat detection. Psychophysiology 35:86–89. https://doi.org/10.1017/S0048577298961078
Scholz OB, Ott R, Sarnoch H (2001) Proprioception in somatoform disorders. Behav Res Ther 39:1429–1438. https://doi.org/10.1016/S0005-7967(00)00108-X
Schön D, Dahme B, von Leupoldt A (2008a) Associations between the perception of dyspnea, pain, and negative affect. Psychophysiology 45:1064–1067. https://doi.org/10.1111/j.1469-8986.2008.00704.x
Schön D, Rosenkranz M, Regelsberger J et al (2008b) Reduced perception of dyspnea and pain after right insular cortex lesions. Am J Respir Crit Care Med 178:1173–1179. https://doi.org/10.1164/rccm.200805-731OC
Schönherr A, May CA (2016) Influence of caloric vestibular stimulation on body experience in healthy humans. Front Integr Neurosci 10:14. https://doi.org/10.3389/fnint.2016.00014
Schouenborg J (2002) Modular organisation and spinal somatosensory imprinting. Brain Res Rev 40:80–91. https://doi.org/10.1016/S0165-0173(02)00191-1
Schulz A, Vögele C (2015) Interoception and stress. Front Psychol 6:993. https://doi.org/10.3389/fpsyg.2015.00993
Schulz A, Lass-Hennemann J, Sütterlin S et al (2013) Cold pressor stress induces opposite effects on cardioceptive accuracy dependent on assessment paradigm. Biol Psychol 93:167–174. https://doi.org/10.1016/j.biopsycho.2013.01.007
Schulz A, Schilling TM, Vögele C et al (2016) Respiratory modulation of startle eye blink: a new approach to assess afferent signals from the respiratory system. Philos Trans R Soc Lond B Biol Sci 371:19. https://doi.org/10.1098/rstb.2016.0019
Schulz A, van Dyck Z, Lutz APC et al (2017) Gastric modulation of startle eye blink. Biol Psychol 127:25–33. https://doi.org/10.1016/j.biopsycho.2017.05.004
Schulz A, Schilling TM, Vögele C, Schächinger H (2019) Visceral-afferent signals from the cardiovascular system, but not urinary urge, affect startle eye blink. Physiol Behav 199:165–172. https://doi.org/10.1016/j.physbeh.2018.11.018

Schwabe L, Blanke O (2008) The vestibular component in out-of-body experiences: a computational approach. Front Hum Neurosci 2:17. https://doi.org/10.3389/neuro.09.017.2008
Sedeño L, Couto B, Melloni M et al (2014) How do you feel when you can't feel your body? Interoception, functional connectivity and emotional processing in depersonalization-derealization disorder. PLoS One 9:e98769. https://doi.org/10.1371/journal.pone.0098769
Seer P (1979) Psychological control of essential hypertension: review of the literature and methodological critique. Psychol Bull 86:1015–1043. https://doi.org/10.1037/0033-2909.86.5.1015
Serra J, Azpiroz F, Malagelada J-R (1995) Perception and reflex responses to intestinal distention in humans are modified by simultaneous or previous stimulation. Gastroenterology 109:1742–1749. https://doi.org/10.1016/0016-5085(95)90739-4
Serra J, Azpiroz F, Malagelada JR (1998) Modulation of gut perception in humans by spatial summation phenomena. J Physiol Lond 506(Pt 2):579–587. https://doi.org/10.1111/j.1469-7793.1998.579bw.x
Shabat-Simon M, Shuster A, Sela T, Levy DJ (2018) Objective physiological measurements but not subjective reports moderate the effect of hunger on choice behavior. Front Psychol 9:750. https://doi.org/10.3389/fpsyg.2018.00750
Shah P, Catmur C, Bird G (2017) From heart to mind: linking interoception, emotion, and theory of mind. Cortex 93:220–223. https://doi.org/10.1016/j.cortex.2017.02.010
Sharma L, Pai YC (1997) Impaired proprioception and osteoarthritis. Curr Opin Rheumatol 9:253–258
Shedivy DI, Kleinman KM (1977) Lack of correlation between frontalis EMG and either neck EMG or verbal ratings of tension. Psychophysiology 14:182–186. https://doi.org/10.1111/j.1469-8986.1977.tb03372.x
Shelley WB, Arthur RP (1957) The neurohistology and neurophysiology of the itch sensation in man. AMA Arch Derm 76:296–323. https://doi.org/10.1001/archderm.1957.01550210020004
Sherrington CS (1906) The integrative action of the nervous system. Yale University Press, New Haven
Shook NJ, Oosterhoff B, Terrizzi JA Jr, Brady KM (2017) "Dirty politics": the role of disgust sensitivity in voting. Transl Issues Psychol Sci 3:284–297. https://doi.org/10.1037/tps0000111
Sikand P, Shimada SG, Green BG, LaMotte RH (2009) Similar itch and nociceptive sensations evoked by punctate cutaneous application of capsaicin, histamine and cowhage. Pain 144:66–75. https://doi.org/10.1016/j.pain.2009.03.001
Silverman BA, Mayer D, Sabinsky R et al (1987) Training perception of air flow obstruction in asthmatics. Ann Allergy Asthma Immunol 59:350–354
Sime WE, DeGood DE (1977) Effect of EMG biofeedback and progressive muscle relaxation training on awareness of frontalis muscle tension. Psychophysiology 14:522–530. https://doi.org/10.1111/j.1469-8986.1977.tb01192.x
Simone DA, Ngeow JY, Whitehouse J et al (1987) The magnitude and duration of itch produced by intracutaneous injections of histamine. Somatosens Res 5:81–92. https://doi.org/10.3109/07367228709144620
Simone DA, Alreja M, LaMotte RH (1991) Psychophysical studies of the itch sensation and itchy skin ("alloknesis") produced by intracutaneous injection of histamine. Somatosens Mot Res 8:271–279. https://doi.org/10.3109/08990229109144750
Skinner HB, Barrack RL, Cook SD (1984) Age-related decline in proprioception. Clin Orthop Relat Res (184):208–211
Slucki H, Adam G, Porter RW (1965) Operant discrimination of an interoceptive stimulus in rhesus monkeys. J Exp Anal Behav 8:405–414. https://doi.org/10.1901/jeab.1965.8-405
Slucki H, McCoy FB, Porter RW (1969) Interoceptive Sd of the large intestine established by mechanical stimulation. Psychol Rep 24:35–42. https://doi.org/10.2466/pr0.1969.24.1.35
Smith CA, Wallston KA, Dwyer KA (1995) On babies and bathwater: disease impact and negative affectivity in the self-reports of persons with rheumatoid arthritis. Health Psychol 14:64–73. https://doi.org/10.1037//0278-6133.14.1.64

Smith KB, Oxley D, Hibbing MV et al (2011) Disgust sensitivity and the neurophysiology of left-right political orientations. PLoS One 6:e25552. https://doi.org/10.1371/journal.pone.0025552
Sommer MA, Wurtz RH (2008) Brain circuits for the internal monitoring of movements. Annu Rev Neurosci 31:317–338. https://doi.org/10.1146/annurev.neuro.31.060407.125627
Sood MR, Schwankovsky LM, Rowhani A et al (2002) Water load test in children. J Pediatr Gastroenterol Nutr 35:199–201
Soudry Y, Lemogne C, Malinvaud D et al (2011) Olfactory system and emotion: Common substrates. Eur Ann Otorhinolaryngol Head Neck Dis 128:18–23. https://doi.org/10.1016/j.anorl.2010.09.007
St Clair Gibson A, Noakes TD (2004) Evidence for complex system integration and dynamic neural regulation of skeletal muscle recruitment during exercise in humans. Br J Sports Med 38:797–806. https://doi.org/10.1136/bjsm.2003.009852
Staahl C, Drewes AM (2004) Experimental human pain models: a review of standardised methods for preclinical testing of analgesics. Basic Clin Pharmacol Toxicol 95:97–111. https://doi.org/10.1111/j.1742-7843.2004.950301.x
Stanton TR, Leake HB, Chalmers KJ, Moseley GL (2016) Evidence of impaired proprioception in chronic, idiopathic neck pain: systematic review and meta-analysis. Phys Ther 96:876–887. https://doi.org/10.2522/ptj.20150241
Staudenmayer H, Kinsman RA (1976) Awareness during electromyographic biofeedback: of signal or process? Biofeedback Self Regul 1:191–199. https://doi.org/10.1007/BF00998586
Steptoe A, Noll A (1997) The perception of bodily sensations, with special reference to hypochondriasis. Behav Res Ther 35:901–910
Steptoe A, Vögele C (1992) Individual differences in the perception of bodily sensations: the role of trait anxiety and coping style. Behav Res Ther 30:597–607
Stevens JC, Mack JD (1959) Scales of apparent force. J Exp Psychol 58:405–413. https://doi.org/10.1037/h0046906
Stilson DW, Matus I, Ball G (1980) Relaxation and subjective estimates of muscle tension: implications for a central efferent theory of muscle control. Biofeedback Self Regul 5:19–36. https://doi.org/10.1007/bf00999061
Stout C, Kotses H, Creer TL (1997) Improving perception of air flow obstruction in asthma patients. Psychosom Med 59:201–206. https://doi.org/10.1097/00006842-199703000-00013
Strid H, Norström M, Sjöberg J et al (2001) Impact of sex and psychological factors on the water loading test in functional dyspepsia. Scand J Gastroenterol 36:725–730. https://doi.org/10.1080/00365520121075
Strigo IA, Bushnell MC, Boivin M, Duncan GH (2002) Psychophysical analysis of visceral and cutaneous pain in human subjects. Pain 97:235–246. https://doi.org/10.1016/s0304-3959(02)00023-4
Stunkard AJ, Fox S (1971) The relationship of gastric motility and hunger. A summary of the evidence. Psychosom Med 33:123–134. https://doi.org/10.1097/00006842-197103000-00004
Stunkard A, Koch C (1964) The interpretation of gastric motility: I. Apparent bias in the reports of hunger by obese persons. Arch Gen Psychiatry 11:74–82. https://doi.org/10.1001/archpsyc.1964.01720250076010
Sullivan MJL, Bishop SR, Pivik J (1995) The Pain Catastrophizing Scale: development and validation. Psychol Assess 7:524–532. https://doi.org/10.1037/1040-3590.7.4.524
Sullivan MJ, Thorn B, Haythornthwaite JA et al (2001) Theoretical perspectives on the relation between catastrophizing and pain. Clin J Pain 17:52–64. https://doi.org/10.1097/00002508-200103000-00008
Sullivan CD, Garcia GJM, Frank-Ito DO et al (2014) Perception of better nasal patency correlates with increased mucosal cooling after surgery for nasal obstruction. Otolaryngol Head Neck Surg 150:139–147. https://doi.org/10.1177/0194599813509776
Suschinsky KD, Lalumière ML (2012) Is sexual concordance related to awareness of physiological states? Arch Sex Behav 41:199–208. https://doi.org/10.1007/s10508-012-9931-9

Suschinsky KD, Lalumière ML (2014) The relationship between sexual concordance and interoception in anxious and nonanxious women. J Sex Med 11:942–955. https://doi.org/10.1111/jsm.12250
Suschinsky KD, Shelley AJ, Gerritsen J et al (2015) The clitoral photoplethysmograph: a pilot study examining discriminant and convergent validity. J Sex Med 12:2324–2338. https://doi.org/10.1111/jsm.13047
Tack J, Piessevaux H, Coulie B et al (1998) Role of impaired gastric accommodation to a meal in functional dyspepsia. Gastroenterology 115:1346–1352
Takahashi T, Itoh H, Nishikawa Y et al (2015) Possible relation between olfaction and anxiety in healthy subjects. Psychiatry Clin Neurosci 69:431–438. https://doi.org/10.1111/pcn.12277
Tanner WP Jr, Swets JA (1954) A decision-making theory of visual detection. Psychol Rev 61:401–409. https://doi.org/10.1037/h0058700
Taylor DN, Lee CT (1991) Lack of correlation between frontalis electromyography and self-ratings of either frontalis tension or state anxiety. Percept Mot Skills 72:1131–1134. https://doi.org/10.2466/pms.1991.72.3c.1131
Taylor LA, Rachman SJ (1988) The effects of blood sugar level changes on cognitive function, affective state, and somatic symptoms. J Behav Med 11:279–291. https://doi.org/10.1007/BF00844433
Terasawa Y, Moriguchi Y, Tochizawa S, Umeda S (2014) Interoceptive sensitivity predicts sensitivity to the emotions of others. Cogn Emot 28:1435–1448. https://doi.org/10.1080/02699931.2014.888988
Thornton EW, Van-Toller C (1973) Operant conditioning of heart-rate changes in the functionally decorticate curarised rat. Physiol Behav 10:983–988. https://doi.org/10.1016/0031-9384(73)90176-5
Tinnermann A, Geuter S, Sprenger C et al (2017) Interactions between brain and spinal cord mediate value effects in nocebo hyperalgesia. Science 358:105–108. https://doi.org/10.1126/science.aan1221
Tracey I, Mantyh PW (2007) The cerebral signature for pain perception and its modulation. Neuron 55:377–391. https://doi.org/10.1016/j.neuron.2007.07.012
Tsai H-W, Chan P-Y, von Leupoldt A, Davenport PW (2013) The impact of emotion on the perception of graded magnitudes of respiratory resistive loads. Biol Psychol 93:220–224. https://doi.org/10.1016/j.biopsycho.2013.02.008
Tsakiris M (2010) My body in the brain: a neurocognitive model of body-ownership. Neuropsychologia 48:703–712. https://doi.org/10.1016/j.neuropsychologia.2009.09.034
Tsakiris M (2017) The multisensory basis of the self: from body to identity to others [Formula: see text]. Q J Exp Psychol (Hove) 70:597–609. https://doi.org/10.1080/17470218.2016.1181768
Tsakiris M, Tajadura-Jiménez A, Costantini M (2011) Just a heartbeat away from one's body: interoceptive sensitivity predicts malleability of body-representations. Proc Biol Sci 278:2470–2476. https://doi.org/10.1098/rspb.2010.2547
Twycross R, Greaves MW, Handwerker H et al (2003) Itch: scratching more than the surface. QJM 96:7–26. https://doi.org/10.1093/qjmed/hcg002
Tyrer P, Lee I, Alexander J (1980) Awareness of cardiac function in anxious, phobic and hypochondriacal patients. Psychol Med 10:171–174. https://doi.org/10.1017/S0033291700039726
Uher R, Treasure J, Heining M et al (2006) Cerebral processing of food-related stimuli: effects of fasting and gender. Behav Brain Res 169:111–119. https://doi.org/10.1016/j.bbr.2005.12.008
Utz KS, Dimova V, Oppenländer K, Kerkhoff G (2010) Electrified minds: transcranial direct current stimulation (tDCS) and galvanic vestibular stimulation (GVS) as methods of non-invasive brain stimulation in neuropsychology—a review of current data and future implications. Neuropsychologia 48:2789–2810. https://doi.org/10.1016/j.neuropsychologia.2010.06.002
Uzun H, Kadioglu ME, Metin NO, Akca G (2019) The association of postvoiding residual volume, uroflowmetry parameters and bladder sensation. Urol J. https://doi.org/10.22037/uj.v0i0.4368

Vallbo A, Olausson H, Wessberg J, Norrsell U (1993) A system of unmyelinated afferents for innocuous mechanoreception in the human skin. Brain Res 628:301–304. https://doi.org/10.1016/0006-8993(93)90968-s

Van den Bergh O, Winters W, Devriese S et al (2004) Accuracy of respiratory symptom perception in persons with high and low negative affectivity. Psychol Health 19:213–222. https://doi.org/10.1080/08870440410001675627

Van den Bergh O, Zacharioudakis N, Petersen S (2018) Interoception, categorization, and symptom perception. In: Tsakiris M, De Preester H (eds) The interoceptive mind. From homeostasis to awareness. Oxford University Press, Oxford, England, pp 212–226

Van Den Elzen BD, Bennink RJ, Holman R et al (2007) Impaired drinking capacity in patients with functional dyspepsia: intragastric distribution and distal stomach volume. Neurogastroenterol Motil 19:968–976. https://doi.org/10.1111/j.1365-2982.2007.00971.x

van der Schaar PJ, Lamers CBHW, Masclee AAM (1999) The role of the barostat in human research and clinical practice. Scand J Gastroenterol 34:52–63. https://doi.org/10.1080/003655299750025552

van Dyck Z (2015) The role of external and internal stimuli in the regulation of eating behaviour. Dissertation, L'Université du Luxembourg

van Dyck Z, Vögele C, Blechert J et al (2016) The water load test as a measure of gastric interoception: development of a two-stage protocol and application to a healthy female population. PLoS One 11:e0163574. https://doi.org/10.1371/journal.pone.0163574

van Laarhoven AIM, Vogelaar ML, Wilder-Smith OH et al (2011) Induction of nocebo and placebo effects on itch and pain by verbal suggestions. Pain 152:1486–1494. https://doi.org/10.1016/j.pain.2011.01.043

Van Oudenhove L, Coen S-J, Aziz Q (2007) Functional brain imaging of gastrointestinal sensation in health and disease. World J Gastroenterol 13:3438–3445. https://doi.org/10.3748/wjg.v13.i25.3438

Vidergar LJ, Lee RM, Goldman MS (1983) Discrimination of systolic blood pressure. Biofeedback Self Regul 8:45–61. https://doi.org/10.1007/BF01000536

Vierck CJ, Whitsel BL, Favorov OV et al (2013) Role of primary somatosensory cortex in the coding of pain. Pain 154:334–344. https://doi.org/10.1016/j.pain.2012.10.021

Villemure C, Bushnell MC (2002) Cognitive modulation of pain: how do attention and emotion influence pain processing? Pain 95:195–199. https://doi.org/10.1016/s0304-3959(02)00007-6

Villemure C, Slotnick BM, Bushnell MC (2003) Effects of odors on pain perception: deciphering the roles of emotion and attention. Pain 106:101–108

Villien F, Yu M, Barthélémy P, Jammes Y (2005) Training to yoga respiration selectively increases respiratory sensation in healthy man. Respir Physiol Neurobiol 146:85–96. https://doi.org/10.1016/j.resp.2004.11.010

Vlemincx E, Van Diest I, Lehrer PM et al (2010) Respiratory variability preceding and following sighs: a resetter hypothesis. Biol Psychol 84:82–87. https://doi.org/10.1016/j.biopsycho.2009.09.002

Vlemincx E, Abelson JL, Lehrer PM et al (2013) Respiratory variability and sighing: a psychophysiological reset model. Biol Psychol 93:24–32. https://doi.org/10.1016/j.biopsycho.2012.12.001

Vlemincx E, Van Diest I, Van den Bergh O (2016) A sigh of relief or a sigh to relieve: the psychological and physiological relief effect of deep breaths. Physiol Behav 165:127–135. https://doi.org/10.1016/j.physbeh.2016.07.004

von Frei M (1922) Zur Physiologie der Juckempfindung. Arch Neerl Physiol 7:142–145

von Knorring L, Monakhov K, Perris C (1978) Augmenting/reducing: an adaptive switch mechanism to cope with incoming signals in healthy subjects and psychiatric patients. Neuropsychobiology 4:150–179

von Leupoldt A, Dahme B (2005) Differentiation between the sensory and affective dimension of dyspnea during resistive load breathing in normal subjects. Chest 128:3345–3349. https://doi.org/10.1378/chest.128.5.3345

von Leupoldt A, Dahme B (2007) Experimental comparison of dyspnea and pain. Behav Res 39:137–143. https://doi.org/10.3758/BF03192852

von Leupoldt A, Dahme B (2013) The impact of emotions on symptom perception in patients with asthma and healthy controls. Psychophysiology 50:1–4. https://doi.org/10.1111/j.1469-8986.2012.01480.x

von Leupoldt A, Sommer T, Kegat S et al (2009) Down-regulation of insular cortex responses to dyspnea and pain in asthma. Am J Respir Crit Care Med 180:232–238. https://doi.org/10.1164/rccm.200902-0300OC

von Leupoldt A, Keil A, Chan P-YS et al (2010a) Cortical sources of the respiratory-related evoked potential. Respir Physiol Neurobiol 170:198. https://doi.org/10.1016/j.resp.2009.12.006

von Leupoldt A, Vovk A, Bradley MM et al (2010b) The impact of emotion on respiratory-related evoked potentials. Psychophysiology 47:579–586. https://doi.org/10.1111/j.1469-8986.2009.00956.x

von Mohr M, Fotopoulou A (2018) The cutaneous borders of interoception: Active and social inference of pain and pleasure on the skin. In: Tsakiris M, De Preester H (eds) The interoceptive mind. From homeostasis to awareness. Oxford University Press, Oxford, England, pp 102–120

von Mohr M, Krahé C, Beck B, Fotopoulou A (2018) The social buffering of pain by affective touch: a laser-evoked potential study in romantic couples. Soc Cogn Affect Neurosci 13:1121–1130. https://doi.org/10.1093/scan/nsy085

Waddington G, Adams R (1999) Ability to discriminate movements at the ankle and knee is joint specific. Percept Mot Skills 89:1037–1041

Wager TD, Atlas LY, Leotti LA, Rilling JK (2011) Predicting individual differences in placebo analgesia: contributions of brain activity during anticipation and pain experience. J Neurosci 31:439–452. https://doi.org/10.1523/JNEUROSCI.3420-10.2011

Webb J, Bolhuis DP, Cicerale S et al (2015) The relationships between common measurements of taste function. Chemosens Percept 8:11–18. https://doi.org/10.1007/s12078-015-9183-x

Weber ST, Heuberger E (2008) The impact of natural odors on affective states in humans. Chem Senses 33:441–447. https://doi.org/10.1093/chemse/bjn011

Webster KE, Colrain IM (2000) The relationship between respiratory-related evoked potentials and the perception of inspiratory resistive loads. Psychophysiology 37:831–841. https://doi.org/10.1111/1469-8986.3760831

Weck F, Esch S, Rohrmann S (2014) The role of disgust in patients with hypochondriasis. Anxiety Stress Coping 27:576–586. https://doi.org/10.1080/10615806.2013.873793

Weisenberg M, Raz T, Hener T (1998) The influence of film-induced mood on pain perception. Pain 76:365–375. https://doi.org/10.1016/S0304-3959(98)00069-4

Weiss S, Sack M, Henningsen P, Pollatos O (2014) On the interaction of self-regulation, interoception and pain perception. Psychopathology 47:377–382. https://doi.org/10.1159/000365107

Weisz J, Balázs L, Ádám G (1988) The influence of self-focused attention on heartbeat perception. Psychophysiology 25:193–199. https://doi.org/10.1111/j.1469-8986.1988.tb00987.x

Werner NS, Duschek S, Mattern M, Schandry R (2009) The relationship between pain perception and interoception. J Psychophysiol 23:35–42. https://doi.org/10.1027/0269-8803.23.1.35

Whitehead WE, Delvaux M (1997) Standardization of barostat procedures for testing smooth muscle tone and sensory thresholds in the gastrointestinal tract. The Working Team of Glaxo-Wellcome Research, UK. Dig Dis Sci 42:223–241. https://doi.org/10.1023/a:1018885028501

Whitehead WE, Drescher VM (1980) Perception of gastric contractions and self-control of gastric motility. Psychophysiology 17:552–558. https://doi.org/10.1111/j.1469-8986.1980.tb02296.x

Whitehead WE, Palsson OS (1998) Is rectal pain sensitivity a biological marker for irritable bowel syndrome: psychological influences on pain perception. Gastroenterology 115:1263–1271. https://doi.org/10.1016/S0016-5085(98)70099-X

Whitehead WE, Drescher VM, Blackwell B (1976) Lack of relationship between Autonomic Perception Questionnaire scores and actual sensitivity for perceiving one's heart beat. Psychophysiology 13:177. https://doi.org/10.1111/j.1469-8986.1984.tb00196.x

Whitehead WE, Drescher VM, Heiman P, Blackwell B (1977) Relation of heart rate control to heartbeat perception. Biofeedback Self Regul 2:371–392. https://doi.org/10.1007/BF00998623

Whitehead WE, Engel BT, Schuster MM (1980) Irritable bowel syndrome: physiological and psychological differences between diarrhea-predominant and constipation-predominant patients. Dig Dis Sci 25:404–413

Whitehead WE, Holtkotter B, Enck P et al (1990) Tolerance for rectosigmoid distention in irritable bowel syndrome. Gastroenterology 98:1187–1192

Wiebe DJ, Alderfer MA, Palmer SC et al (1994) Behavioral self-regulation in adolescents with type I diabetes: negative affectivity and blood glucose symptom perception. J Consult Clin Psychol 62:1204–1212. https://doi.org/10.1037//0022-006x.62.6.1204

Wiech K (2016) Deconstructing the sensation of pain: the influence of cognitive processes on pain perception. Science 354:584–587. https://doi.org/10.1126/science.aaf8934

Wiech K, Tracey I (2009) The influence of negative emotions on pain: behavioral effects and neural mechanisms. NeuroImage 47:987–994. https://doi.org/10.1016/j.neuroimage.2009.05.059

Wildman HE, Jones GE (1982) Consistency of heart beat discrimination scores on the Whitehead procedure in knowledge of the results trained and untrained subjects. Psychophysiology 19:592

Wiley RL, Zechman FW (1966) Perception of added airflow resistance in humans. Respir Physiol 2:73–87. https://doi.org/10.1016/0034-5687(66)90039-9

Williams LE, Bargh JA (2008) Experiencing physical warmth promotes interpersonal warmth. Science 322:606–607. https://doi.org/10.1126/science.1162548

Wilson LB, Andrew D, Craig AD (2002) Activation of spinobulbar lamina I neurons by static muscle contraction. J Neurophysiol 87:1641–1645. https://doi.org/10.1152/jn.00609.2001

Wincze JP, Hoon EF, Hoon PW (1976) Physiological responsivity of normal and sexually dysfunctional women during erotic stimulus exposure. J Psychosom Res 20:445–451. https://doi.org/10.1016/0022-3999(76)90008-8

Windmann S, Schonecke OW, Fröhlig G, Maldener G (1999) Dissociating beliefs about heart rates and actual heart rates in patients with cardiac pacemakers. Psychophysiology 36:339–342. https://doi.org/10.1017/S0048577299980381

Winters RW, McCabe PM, Schneiderman N (2002) Functional utility and neurobiology of conditioned autonomic responses. In: Moore JW (ed) A neuroscientist's guide to classical conditioning. Springer, New York, pp 46–85

Witthöft M, Bräscher A-K, Jungmann SM, Köteles F (2020) Somatic symptom perception and interoception. Z Psychol 228:100–109. https://doi.org/10.1027/2151-2604/a000403

Wolpert DM, Doya K, Kawato M (2003) A unifying computational framework for motor control and social interaction. Philos Trans R Soc Lond B Biol Sci 358:593–602. https://doi.org/10.1098/rstb.2002.1238

Woo AK (2010) Depression and anxiety in pain. Rev Pain 4:8–12. https://doi.org/10.1177/204946371000400103

Woodard TL, Diamond MP (2009) Physiologic measures of sexual function in women: a review. Fertil Steril 92:19–34. https://doi.org/10.1016/j.fertnstert.2008.04.041

Wouda JC, Hartman PM, Bakker RM et al (1998) Vaginal plethysmography in women with dyspareunia. J Sex Res 35:141–147. https://doi.org/10.1080/00224499809551927

Wright CE, Ebrecht M, Mitchell R et al (2005) The effect of psychological stress on symptom severity and perception in patients with gastro-oesophageal reflux. J Psychosom Res 59:415–424. https://doi.org/10.1016/j.jpsychores.2005.05.012

Wuyts R, Vlemincx E, Bogaerts K et al (2011) Sigh rate and respiratory variability during normal breathing and the role of negative affectivity. Int J Psychophysiol 82:175–179. https://doi.org/10.1016/j.ijpsycho.2011.07.021

Wyndaele JJ (1992) Study on the correlation between subjective perception of bladder filling and the sensory threshold towards electrical stimulation in the lower urinary tract. J Urol 147:1582–1584. https://doi.org/10.1016/s0022-5347(17)37634-6

Yamaguchi O, Honda K, Nomiya M et al (2007) Defining overactive bladder as hypersensitivity. Neurourol Urodyn 26:904–907. https://doi.org/10.1002/nau.20482

Yan S, Bates JHT (1999) Breathing responses to small inspiratory threshold loads in humans. J Appl Physiol 86:874–880. https://doi.org/10.1152/jappl.1999.86.3.874

Yang N, Waddington G, Adams R, Han J (2020) Joint position reproduction and joint position discrimination at the ankle are not related. Somatosens Mot Res 37:97–105. https://doi.org/10.1080/08990220.2020.1746638

Yao S, Becker B, Zhao W et al (2018) Oxytocin modulates attention switching between interoceptive signals and external social cues. Neuropsychopharmacology 43:294–301. https://doi.org/10.1038/npp.2017.189

Yates AJ, Jones KE, Marie GV, Hogben JH (1985) Detection of the heartbeat and events in the cardiac cycle. Psychophysiology 22:561–567. https://doi.org/10.1111/j.1469-8986.1985.tb01651.x

Yoris A, Esteves S, Couto B et al (2015) The roles of interoceptive sensitivity and metacognitive interoception in panic. Behav Brain Funct 11:14. https://doi.org/10.1186/s12993-015-0058-8

Yoris A, Garcia AM, Salamone P et al (2018) Cardiac interoception in neurological conditions and its relevance for dimensional approaches. In: Tsakiris M, De Preester H (eds) The interoceptive mind. From homeostasis to awareness. Oxford University Press, Oxford, England, pp 187–211

Yosipovitch G, Greaves MW, Schmelz M (2003) Itch. Lancet 361:690–694. https://doi.org/10.1016/S0140-6736(03)12570-6

Young HA, Gaylor CM, de Kerckhove D et al (2019) Interoceptive accuracy moderates the response to a glucose load: a test of the predictive coding framework. Proc Biol Sci 286:20190244. https://doi.org/10.1098/rspb.2019.0244

Yuan Y-Z, Tao R-J, Xu B et al (2003) Functional brain imaging in irritable bowel syndrome with rectal balloon-distention by using fMRI. World J Gastroenterol 9:1356–1360. https://doi.org/10.3748/wjg.v9.i6.1356

Yuan H, Yan H-M, Xu X-G et al (2007) Effect of heartbeat perception on heartbeat evoked potential waves. Neurosci Bull 23:357–362. https://doi.org/10.1007/s12264-007-0053-7

Zaccaro A, Piarulli A, Laurino M et al (2018) How breath-control can change your life: a systematic review on psycho-physiological correlates of slow breathing. Front Hum Neurosci 12:353. https://doi.org/10.3389/fnhum.2018.00353

Zacharioudakis N, Vlemincx E, Van den Bergh O (2020) Categorical interoception and the role of threat. Int J Psychophysiol 148:25–34. https://doi.org/10.1016/j.ijpsycho.2019.12.009

Zale EL, Lange KL, Fields SA, Ditre JW (2013) The relation between pain-related fear and disability: a meta-analysis. J Pain 14:1019–1030. https://doi.org/10.1016/j.jpain.2013.05.005

Zamariola G, Maurage P, Luminet O, Corneille O (2018) Interoceptive accuracy scores from the heartbeat counting task are problematic: evidence from simple bivariate correlations. Biol Psychol 137:12–17. https://doi.org/10.1016/j.biopsycho.2018.06.006

Zapala DA, Olsholt KF, Lundy LB (2008) A comparison of water and air caloric responses and their ability to distinguish between patients with normal and impaired ears. Ear Hear 29:585–600. https://doi.org/10.1097/AUD.0b013e3181734ed0

Zechman FW, Burki NK (1976) Tracking procedure to assess load detection threshold. Chest 70:165–168. https://doi.org/10.1378/chest.70.1_supplement.165

Zelman DC, Howland EW, Nichols SN, Cleeland CS (1991) The effects of induced mood on laboratory pain. Pain 46:105–111. https://doi.org/10.1016/0304-3959(91)90040-5

Zhao W, Martin AD, Davenport PW (2002) Detection of inspiratory resistive loads in double-lung transplant recipients. J Appl Physiol 93:1779–1785. https://doi.org/10.1152/japplphysiol.00210.2002

Zimprich D, Nusser L, Pollatos O (2020) Are interoceptive accuracy scores from the heartbeat counting task problematic? A comment on Zamariola et al. (2018). Biol Psychol 152:107868. https://doi.org/10.1016/j.biopsycho.2020.107868

Zotterman Y (1939) Touch, pain and tickling: an electro-physiological investigation on cutaneous sensory nerves. J Physiol 95:1–28 zu Eulenburg P, Baumgärtner U, Treede R-D, Dieterich M (2013) Interoceptive and multimodal functions of the operculo-insular cortex: tactile, nociceptive and vestibular representations. NeuroImage 83:75–86. https://doi.org/10.1016/j.neuroimage.2013.06.057

Zuckerman M (1971) Physiological measures of sexual arousal in the human. Psychol Bull 75:297–329. https://doi.org/10.1037/h0030923

Zweyer K, Velker B, Ruch W (2006) Do cheerfulness, exhilaration, and humor production moderate pain tolerance? A FACS study. Humor 17:85–119. https://doi.org/10.1515/humr.2004.009

Chapter 5
What Do We Perceive? Interoceptive Sensibility

5.1 Introduction

In the measurement paradigms described in the previous chapter, some sort of (spontaneous or experimentally generated) stimulus belonging to somatovisceral afference was assessed and compared to perceived (although sometimes only marginally conscious) sensations. Basically, this is how interoceptive accuracy is defined and measured (Garfinkel et al. 2015). Under everyday circumstances, however, we just feel sensations but receive no feedback about their validity or acuity. This is usually not a question at all; relying on felt sensations, we implicitly or explicitly assume the existence of corresponding physiological processes. As we will see in Chap. 6, sometimes we are right, other times we are wrong. Still, we are convinced (at least at the level of the primary reaction) that what we feel is the reality—something is clearly going on in our body. This is the tenet of realness, whose evolutionary background and functional importance were already discussed in Chaps. 1 and 2.

The dominantly verbal representation of the subjective experience, called interoceptive sensibility, is the second level of the model of interoception by Garfinkel and Critchley (Garfinkel et al. 2015, 2016). As it refers to subjective feelings and sensations, it is typically assessed via self-report. Two distinct aspects are considered at this level of the tripartite model. The acute aspect (also called confidence; we will call it IS_{acute}) refers to actual stimulation, whereas the generalized level (IS_{trait}) represents the self-reported ability to perceive internal changes generalized across modalities over time. It is important to note that the tenet of vagueness, in more detail, the difficulties with the verbalization of primary experience, is simply not taken into consideration in this model. I assume that individual differences in the ability to verbalize body sensations might substantially impact both levels of interoceptive sensibility. Of course, this caveat is difficult to avoid; still, it should be mentioned and taken into consideration when one tries to explain the dissociation between interoceptive accuracy and sensibility (Chap. 6).

F. Köteles, *Body Sensations*, https://doi.org/10.1007/978-3-030-63201-4_5

Dispositional awareness of body sensations and body-related changes, also called body awareness, is defined as sensitivity and attentiveness to internal signals (Shields et al. 1989). These two features, i.e., sensitivity and attention, largely overlap with each other but cannot be used interchangeably; the former is the ability to notice internal changes (that partly depends on the characteristics of the actual signal), whereas the latter refers to a receptive state of observation and the propensity to pay attention (Ginzburg et al. 2014). Moreover, the term attention can refer here to pure observation, i.e., preparedness to notice anything that occurs in the body, or purposeful monitoring or scanning, i.e., expecting and seeking a well-defined sensation (Ginzburg et al. 2014). Qualitative and/or quantitative characteristics of attention and the related individual differences are rarely assessed and considered in experiments measuring interoceptive accuracy as the experimental paradigms are designed to direct and maintain participants' focus on the target stimuli. Further, participants are usually explicitly asked to pay attention to the task, and it is assumed that they do their best and follow the instruction. Although stimuli with higher intensities will be perceived even in cases when attention is distracted or turned to another target, it is assumed that attention can considerably improve the detectability of exteroceptive stimuli, i.e., it lowers the detection threshold (Posner 1994; Dehaene and Changeux 2005). This rule appears to be true also for body sensations to some extent (Pennebaker and Lightner 1980; Pennebaker 1982; Nouwen et al. 2006; Matthias et al. 2009; Feldman and Friston 2010; Ainley et al. 2016). Thus, a considerable proportion of this chapter deals with various aspects of attention and their relationship with body sensations. However, it is important to see that attention is not an almighty factor and cannot make very weak stimuli perceivable. For example, in the case of so-called asymptomatic diseases or very vague signals of normal functioning, even the most perfectly focused attention is insufficient for the accurate detection of actual body changes.

First things first, in order to better understand and evaluate concepts and constructs belonging under the umbrella of interoceptive sensibility, we need to take a brief journey in the history of the research on subjective body experiences and related phenomena.

5.2 Historical Roots

Wilhelm Wundt, the first giant of the systematic research of perception, distinguished objective sensations and subjective feelings (Wundt 1903). Sensations were considered objective as they were claimed to reflect the pure sensory characteristics of the object (Wundt was primarily interested in exteroception, where sensory characteristics are less vague and easier to observe from an external point of view), whereas feelings were regarded as the subject's evaluative (affective) reactions to the object. Moreover, subjective feelings were claimed to have three distinct dimensions: pleasantness, excitement-depression, and strain-relaxation. This classification generated a heavy debate (Hayes 1906; Hollands 1906), and later only the pleasantness-unpleasantness dimension (i.e., the affective quality of the experience) was accepted (Tichtener 1926). It was also stated that affect is of a unitary quality, the result of nonconscious processes thus cannot be further divided into elements. Many researchers questioned this latter proposition, which led to a number of empirical studies inspecting the body aspects or components of feelings. For example, pleasant feelings were found to be associated with the sensation of bright pressure, whereas unpleasantness with dull pressure (Nafe 1924, 1927; Hunt 1931). Subsequent research was primarily interested in the patterns of somatic and other feelings associated with emotions (Davitz 1969); this topic will be discussed in Chap. 8. Several authors focused on nonemotive body sensations; it turned out that a wide variety of these sensations exist; however, they belong to primary consciousness, thus they are hard to verbalize and describe (Ruckmick 1936; Mason 1959, 1961). Moreover, body sensations were found to impact a number of perceptual processes (Wapner et al. 1951a, b; Werner et al. 1951, 1953; Wapner and Werner 1952; Werner and Wapner 1952a, b). It was also shown that various parts of the body are perceived to be more important than others (Schilder 1935; Fisher and Cleveland 1958, 1968; Fisher 1970, 1974). Overall, these early studies uniformly indicated the intimate bidirectional association between body sensations and affective evaluation.

Beyond basic research on emotions, psychoanalysis, an approach with a marked clinical emphasis, also focused on body sensations. Interestingly, Sigmund Freud's approach to body sensations was twofold. On the one hand, he based the so-called psycho-sexual development model on pleasant sensations (Freud 1938). According to this model, the ego, i.e., the conscious function or process of the human psychological system that is responsible for adaptive everyday functioning, primarily develops from body sensations arising from different parts of the body at different stages of development. Even the majority of stages are named after the dominant body area or erogenous zone, e.g., oral, anal, or phallic. As these sensations are pleasant, the shift from one stage to the other is often difficult for the individual. Also, it was claimed that we sometimes use these sensations in adult life to handle acute stress (this is regression, one of the basic mechanisms of psychological defense). The majority of these early ideas (particularly the expected mechanisms) are not supported by empirical findings; still, it is hard to deny that pleasant body sensations—from eating chocolate to having a yoga class—are often used to ameliorate stress and emotional upheaval.

On the other hand, unpleasant aspects of body sensations and body-related experience were also described and analyzed by Freud. Symptoms were considered manifestations of nonconscious emotional conflicts and anxiety. Conversion disorder, once also called hysteria and characterized by symptoms that impact voluntarily controllable functions, had become a diagnostic category for many decades (Brown 2004). In fact, a stressful event or trauma can often be found in the history of patients with this disorder; however, the psychological mechanism (symbolism) proposed by Freud is not accepted any more. Conversion disorder, among other conditions, such as hypochondriasis, somatoform and somatization disorder, and pain disorder, belongs to the broad category of somatization (Hurwitz 2004); the latter refers to the tendency to experience and communicate psychological distress in the form of somatic symptoms (Lipowski 1986, 1988). Somatization-related processes and models will be discussed in Chap. 7; here and now, it is enough to realize that the intimate relationship between negative affect (mostly anxiety) and body sensations is a basic experience for clinicians.

5.3 Symptom Reports

As the systematic clinical research started with the pathological aspects of body sensations, we will begin the discussion with this area. In this literature, self-reported body sensations that potentially refer to pathological processes are also called complaints or (subjective or somatic) symptoms (Pennebaker 1982). If peripheral (patho)physiological processes are present (e.g., in the case of a local inflammation or simply the urge to void the bladder), these sensations are routinely regarded as the conscious aspect of the condition (Barsky 1979). Still, the correspondence between the underlying physiology and the perceived sensation is rarely clear-cut; moreover, there are substantial individual differences in the exact shape of the curve (see Chap. 4). Perhaps the best example is sensitivity to pain, which is influenced by various dispositional and situational aspects (Gracely 1999; Wall 2000; Brooks and Tracey 2005). Moreover, there are diseases, such as stable coronary artery disease, that are not accompanied by specific body sensations or somatic symptoms (Kroenke 2003).

On the other hand, there are nonpathological and pathological conditions for which the symptoms do not have a demonstrable peripheral (patho)physiological background (Van den Bergh et al. 2017). In fact, everyone experiences such transient complaints (fatigue, various pains, etc.) on a daily basis (Pennebaker 1982); we usually do not really pay attention to them and hope that they disappear sooner and later (rather sooner). If they do not disappear and objective markers of diseases cannot be identified (which is always difficult to state), they are called medically unexplained or somatoform symptoms (Kroenke et al. 1990; Peters et al. 1998; Deary 1999). According to data from surveys, such symptoms can represent up to 70–80% of the symptoms of primary care patients (Hiller et al. 2006; Körber et al. 2011). Conditions characterized by functional impairment and suffering caused by more and/or severe somatoform symptoms are called somatization tendency, functional disorders or, above a threshold, somatoform disorders in DSM-IV (APA 1994) or somatic symptom disorder in DSM-5 (APA 2013). These conditions belong to the domain of clinical psychology and psychiatry, and the diagnosis is primarily based on the symptoms reported by the patients. Again, the common point is that the symptoms or pattern of symptoms cannot be explained by known peripheral pathophysiological mechanisms (i.e., via organic diseases). Of course, this reverse diagnostic approach (i.e., a condition is defined based on the lack of other conditions), as well as the implicit Cartesianism, i.e., the sharp distinction between mind and body, behind it received considerable criticism over the decades. Still, what is available for the diagnostic procedure is the pattern of the symptoms and their considerable impact on the functioning of the patient.

The really important point for us now is that there is a condition (we can call it somatization, as it is broad enough and not used as a diagnostic category; a more sophisticated treatment of this phenomenon comes in Chap. 7) that is typically characterized by the presence of multiple and disabling unpleasant body sensations (aka somatic symptoms) without peripheral pathophysiological background. These symptoms, such as fatigue and various pains, are nonspecific, i.e., they are no clear-cut indicators of a certain disease. Based on these features, a number of self-report measures were developed to help the diagnostic procedure of somatoform conditions. For example, the Hypochondriasis scale (Hs) of the MMPI (Hathaway and McKinley 1942), the Somatization scale of the Hopkins Symptom Checklist (SCL-90) (Derogatis et al. 1974), and the Symptoms Checklist 90-R (Derogatis 1977), or the more recent Patient Health Questionnaire Somatic Symptom Severity Scale (PHQ-15) (Kroenke et al. 2002) almost exclusively rely on somatic symptoms (Table 5.1). The idea is that if such symptoms are present over a longer period of time and they considerably impact the well-being and psychological functioning of the individual, somatization-related processes can be *suspected* (the final diagnosis always relies on clinical interviews). For these conditions, a number of overlapping terms, such as multisomatoform disorder (Kroenke et al. 1997), physical symptom disorder (Kroenke 2006), functional somatic syndromes (Barsky and Borus 1999), or the dimensional somatic symptom distress (Witthöft et al. 2016), were also proposed beyond the established diagnostic categories. Patients with somatoform disorders may represent approximately 10–15% of the primary care patients (Spitzer et al. 1994; Kroenke et al. 1998).

Table 5.1 Symptoms included in the PHQ-15 scale (Kroenke et al. 2002). Bothersomeness of the symptoms is rated with respect to the last 4 weeks on a 3-point scale

1.	Stomach pain
2.	Back pain
3.	Pain in your arms, legs, or joints (knees, hips, etc.)
4.	Menstrual cramps or other problems with your periods [Women only]
5.	Headaches
6.	Chest pain
7.	Dizziness
8.	Fainting spells
9.	Feeling your heart pound or race
10.	Shortness of breath
11.	Pain or problems during sexual intercourse
12.	Constipation, loose bowels, or diarrhea
13.	Nausea, gas, or indigestion
14.	Feeling tired or having low energy
15.	Trouble sleeping

The emotional factors (unpleasantness and negative affect) associated with these symptoms are reflected very well by the good fit of bifactor structural equation models to empirical data. For bifactor models, it is assumed that all items load on a latent factor representing what is common among them; beyond this general factor, there are specific factors representing a proportion of variance that is not accounted for by the general factor (Reise et al. 2010). For example, concerning the PHQ-15, a bifactor approach consisting of a general factor and four modality-specific factors (pain, cardiovascular, gastrointestinal, and fatigue) usually shows better explanatory power (in terms of fit indices) than the conventional solutions. Moreover, indicators of negative affect (anxiety, depression, etc.) and somatization tendency (somatosensory amplification, health anxiety) show substantially stronger associations with the general factor than the specific factors (Witthöft et al. 2013, 2016, 2020; Leonhart et al. 2018; Walentynowicz et al. 2018). This supports the idea that the experience of symptoms has a motivational-emotional and a sensory-perceptual component.

From our point of view, the bottom line is that these sensations are unpleasant and might indicate psychopathology (see Chap. 7). They are unpleasant *per se* thus they decrease subjective well-being; moreover, they are often appraised as signs of serious disease by the individual (as in the case of health anxiety) which generates further fear and anxiety. For many decades, body sensations were almost exclusively seen in this light and in medical context. However, as it was demonstrated in the early 1980s by James Pennebaker (1982) using a self-developed questionnaire consisting of nonspecific symptoms (the Pennebaker Inventory of Limbic Languidness—a telling name), such body sensations are in fact normal, i.e., they are often experienced by pretty much everyone. They simply represent an (unpleasant) aspect of body-related sensations.

5.4 Self-Focused Attention and Body Awareness

As somatic symptoms are unpleasant, they automatically catch attention and body focus. In consequence, attention and affective evaluation cannot really be separated in these cases (see below). What about evaluation-free forms of body focused attention? The research on self-focus or self-focused attention started in the first half of the twentieth century within the framework of symbolic interactionism, a sociological theory (Cooley 1902; Mead 1934). Self-focus in this context refers to the self, including the social role of the individual, and the evaluative feedback one receives from others. At about the same time, Werner Wolff played back to the participants of his experiments their own recorded voice in order to direct their attention inside (Wolff 1932). Surprisingly, most individuals found their voice alien and unpleasant from this external perspective. It took social psychologists further decades to develop the first comprehensive theory on self-focus, the objective self-awareness theory (Duval and Wicklund 1972; Wicklund 1975). This model states that internal focus automatically initializes comparison with social norms (i.e., we see ourselves from a third-person perspective, this is why the approach is called objective self-awareness theory). As nobody is able to perfectly fit the idealized norms of the society, such a comparison usually leads to unpleasant feelings, such as inadequacy and worthlessness. To put it simply, objective self-awareness typically generates (negative) self-evaluation. In contrast, subjective self-awareness, i.e., when we are involved in an activity from a first-person perspective, lacks these evaluative aspects and provides us with the more positive feelings of control and competence. Although body-related aspects were not explicitly mentioned, this view fitted very well the ideas of medicine about the maladaptivity of introspection, described as amplification tendency (Barsky 1979) and introspectiveness (Mechanic 1979, 1980). Also, a reciprocal interaction between negative affect and introspection was assumed. On the one hand, introspection mostly generates negative feelings; on the other hand, negative feelings direct attention inward as the individuals try to explain their cause (Watson and Clark 1984; Wood et al. 1990; Salovey 1992).

Being a social psychological theory, the objective self-awareness approach emphasizes the role of external cues in directing attention. The possible role of internal factors, such as dispositional characteristics, was proposed by Arnold Buss (Fenigstein et al. 1975) and led to the development of self-consciousness theory (Buss 1980). This theory describes two major dispositional aspects of self-focus, private and public self-consciousness. The first refers to the first-person perspective, whereas the second is the proneness to see ourselves through the lens of an external observer (Buss and Scheier 1976; Scheier 1976; Scheier and Carver 1977, 1985; Carver and Scheier 1978). It was assumed that more focus on internal contents (feelings, motives, emotions, attributions) leads to the better understanding and knowledge of these states. Thus, those with high levels of private self-consciousness are assumed to have a more accurate view of themselves; this is called the perceptual accuracy hypothesis or simply veridicality (aka veracity) hypothesis (Gibbons 1990; Silvia and Gendolla 2001). The theory was only partly supported by empirical findings (Gibbons 1983, 1990; Silvia and Gendolla 2001). Most importantly for our topic, it provided the theoretical basis for the development of the first questionnaire that measures the bodily aspects of self-consciousness. It is called Body Consciousness Questionnaire (BCQ), and consists of three scales: the Private and Public Body Consciousness Scale and the Body Competence Scale (Miller et al. 1981). In contrast to the symptom reporting questionnaires, these scales are not pathology-oriented; for example, the Private Body Consciousness Scale was found to be independent from the Hypochondriasis scale of MMPI (Miller et al. 1981).

Beyond the dispositional aspects, self-consciousness theory includes actual states of attention with respect to the self. It is assumed that attention focused on the self from an internal perspective, dubbed private self-awareness, enhances the perception of internal contents (Fenigstein et al. 1975). With respect to interoceptive accuracy (cardioception), however, findings of empirical studies are equivocal. Private self-awareness, induced by a mirror, improved the perception of heartbeats in a discrimination task but not in a tracking task (Weisz et al. 1988). However, only those with low level of cardiac perception showed improved performance when facing a mirror or a picture of their own face (Ainley et al. 2012; Maister and Tsakiris 2014). Finally, a photograph of own face or self-relevant words improved heartbeat tracking ability in another study (Ainley et al. 2013). Overall, these findings do not uniformly support the idea that self-focused attention automatically leads to more accurate perception of body states and changes.

Several years later, another scale, the Body Awareness Questionnaire (BAQ), was developed and published; body awareness was defined as self-reported sensitivity and attentiveness to normal (i.e., nonpathological) and nonemotive body processes (Shields et al. 1989). To clarify this definition, Stephanie Shields and her colleagues described three conceptually distinct forms of attentiveness to body sensations: the focus on (1) pathological sensations, (2) body concomitants of emotions, and (3) signs of normal body functioning. Scales measuring symptom reports described in the previous section belong to the first category. Although the authors explicitly stated the opposite (Miller et al. 1981), the Private Body Consciousness Scale of Miller and colleagues was claimed to measure the second category as four items of the five refer to sensations that accompany the emotional experience (e.g., awareness of heart activity or the rather vague "bodily tensions"). The third category, assessed by the BAQ, was assumed to be free of evaluation and emotions. This assumption was supported by empirical data: the BAQ proved to be independent of neuroticism, anxiety, self-esteem, and hypochondriasis (Shields et al. 1989). Still, the differences between the PBCS and the BAQ are way less clear-cut as originally stated in my opinion; for example, heartbeat-related sensations might equally belong to symptoms, emotions (on top of this, they can characterize both positive and negative emotions), and signs of healthy functioning. As symptoms are always characterized by negative affect, the overlap between category 1 and 2 is particularly large. Further, the BAQ assesses both direct and indirect (e.g., more abstract, such as predicted states) aspects of body experience.

Anyway, perhaps these were the first systematic attempts to define and measure the pathology-unrelated aspects of dispositional attentiveness to the body from an internal perspective. A third scale should also be mentioned here as it plays a central role in recent research on interoception. This is the Body Awareness Scale of the Body Perception Questionnaire (BPQ) by Porges (1993). Basically, it is a comprehensive list of potentially pathology-related and unrelated body sensations and perceptions, both somatosensory and visceroceptive (in fact, exteroceptive sensations, such as digestion-associated noises, are also included). Its perhaps most important advantage is the directness of the included sensations (this also holds true for the Private Body Consciousness Scale). On the other hand, the vast majority of the assessed sensations have a marked sympathetic character. In other words, they are related to stress and negative affect (Cabrera et al. 2018); thus, its applicability as the measure of trait interoceptive sensibility (see below) appears questionable to me.

It is worth noting that body sensations can also be appraised in a positive way. The BAQ showed weak to moderate positive associations with positive affectivity and subjective well-being in our studies, which can be interpreted in this way (other explanations are also possible though) (Köteles 2014; Ferentzi et al. 2019). There are scales, for example, the Body Responsiveness Questionnaire (Daubenmier 2005) or the Body Listening subscale of the Multidimensional Assessment of Interoceptive Awareness (MAIA) (Mehling et al. 2012) that are developed to measure the reliance on the body as a guide in decision-making. This form of evaluation, however, is not directly experiential—it requires conceptual knowledge, sometimes also conscious effort (see below).

5.5 Issues with Attention 1: Measurement

I am convinced that a major source of confusion with respect to body awareness is the multiple meanings of and difficulties with the measurement of attention. Let us start with the measurement part (Table 5.2, upper part). First of all, attention (at least its conscious aspect, see below), just like body sensations, is a subjective experience. Thus, it is very difficult to measure in other ways than self-report; even studies with objective measurements (e.g., EEG) usually include self-reports. Self-report represents an important means to understand human functioning for sure; however, we need to take into consideration its possible pitfalls. If someone is a member of, say, a yoga community, body awareness and body focus will easily become a must for him or her. As it is required to maintain esteem (and self-esteem) in that group, social processes will automatically and powerfully bias his or her self-perception. Moreover, body awareness is a central topic, practiced and discussed regularly, in the community. This also impacts his or her self-perception. Thus, the increase of perceived attentiveness to body changes after an intense period of mindfulness-based meditation (Parkin et al. 2013; Bornemann et al. 2015) can be explained as an actual increase (due to the amount of time spent with body focus), a systematic bias, or a blend of the two. This leads us to the second issue: attention can be acute and chronic (dispositional). As for other state-trait distinctions, it is assumed that those with higher level of the trait will show the corresponding state(s) more frequently. As mentioned, dispositional attentiveness to the body is an important component of the construct of body awareness. However, those with high level of body awareness will not necessarily be able to focus more than others in concrete situations (see below). A third related question is the way we measure self-reported attentiveness: the assessment can be retrospective or more or less real time.

Table 5.2 Various aspects of interoceptive attention discussed in this chapter

Type	Aspect
Measurement-related	Self-reported vs. objectively measured
	Dispositional (chronic) vs. acute
	Retrospective vs. real-time
Inherent	Experiential (primary) vs. secondary
	Endogenous vs. exogenous
	Unifocal vs. divided
	Receptive vs. monitoring (scanning)
	Anxious vs. neutral

Some of these issues were recently realized by Murphy and colleagues (2019). They propose the extension of the Critchley-Garfinkel model with subjective (self-reported) and objective aspects of attention. To measure the latter, they recommend the experience sampling method widely used in flow and emotion research; in those studies, participants, alarmed at random moments, are asked to characterize their actual state (Fullagar and Kelloway 2009). Such empirical studies would represent a substantial step forward; it is important to see, however, that although they can handle retrospective bias they still rely on self-report. Another option to measure the focus of attention is asking participants to verbalize their thoughts in real time (Fisher and Cleveland 1968; Geers et al. 2005); it can be used to measure actual attentiveness as well as to estimate dispositional awareness of the body. Unfortunately, this paradigm may seriously impact and distort the thought processes themselves; not to speak of the poor verbalizability of body sensations (see above).

Let us see the assumption that those with higher levels of self-reported attentiveness to the body are characterized by higher levels of the general ability to sustain attention. Unfortunately, we found no association whatsoever between four constructs tapping various aspects of dispositional body focus (body awareness, body responsiveness, somatosensory amplification, and symptom reporting) and the actual performance in a sustained attention (aka vigilance) task (Tihanyi et al. 2017b; Vig et al. 2020).

None of these issues with the measurement of attention are novel or unknown. In the literature of interoception, however, these distinctions and difficulties are rarely mentioned, or certain categories are mistakenly used in an interchangeable way. The model of Murphy and her colleagues (2019) represents perhaps the first serious step in clarifying these issues.

5.6 The Tingling Sensation

Dear reader, please find a convenient position, close your eyes, and focus your attention on a freely chosen region of your body! Your scalp, hands, or shoulders are particularly good candidates, but the choice is yours. After several seconds, you may feel that the way you sense that part of your body is somehow changed. This is called attention-related body sensation or simply the tingling sensation; it is reported by approximately 50–60% of people (Tihanyi et al. 2016, 2017a, b; Tihanyi and Köteles 2017). Such sensations can also evoke spontaneously (i.e., without attention), thus also called spontaneous sensations or SPS (Michael and Naveteur 2011; Michael et al. 2015, 2017; Salgues et al. 2021).

The phenomenon has been known for decades; for example, my older mentor, Professor György Ádám, was well aware of it and intuitively estimated that one-half of people can experience it. Although the possible role of attention in the production of tactile sensations without stimulation was described and demonstrated earlier (Schmidt et al. 1994), systematic research on the tingling sensation started with an experiment where participants expected some kind of weak cutaneous stimulation;

it was found that they definitely feel the simulation even in times when it was not delivered (Naveteur et al. 2005). It also turned out that directing gaze or only the face on the hand in the focus of attention further increases the intensity of the feeling (Naveteur et al. 2005; Michael et al. 2011); in other words, it is a perceptual phenomenon impacted by top-down processes, such as attention and multimodal integration.

The ability to perceive the tingling sensation shows a moderate temporal stability ($r = 0.47$ over 8 weeks); it is positively associated with scores of various self-reported indicators of body awareness, such as the Body Awareness Questionnaire (Shields et al. 1989), the Somatosensory Amplification Scale (Barsky et al. 1988, 1990), and the Somatic Absorption Scale (Köteles et al. 2012), and independent from positive and negative affect (Tihanyi et al. 2016, 2017b; Tihanyi and Köteles 2017). Interestingly, it was also unrelated to the ability to sustain attention, as assessed by a vigilance task, and mindfulness (Tihanyi et al. 2016, 2017b). One study reported a positive relation to cardioceptive accuracy (Schandry task) (Michael et al. 2015); however, the re-analysis of our own data from a previous study (Tihanyi et al. 2017a) showed no such association.

Medicine calls such sensations without apparent physical cause paresthesia (NINDS 2017); they belong to the specific symptoms of neuropathy caused by diabetes and transient limb ischemia (Lennertz et al. 2010). From a theoretical point of view, a particularly challenging feature of the phenomenon is that such tingling-like sensations can be evoked via both top-down and bottom-up ways (Tihanyi et al. 2018). For example, gentle stimulation of the skin (e.g., the scalp) can also elicit them. When one is sitting too long with crossed legs or a tourniquet is mounted on his arm to evoke experimental pain, ectopic neuronal activity caused by the local decrease of blood supply is accompanied by tingling too (Zhang et al. 1996; Estebe et al. 2000; St Onge 2007). Further, it can be evoked by chemical substances applied topically on the skin, e.g., hydroxy-α-sanshool, the active ingredient of Sichuan pepper, and local anesthetics (Lennertz et al. 2010; Al Luwimi et al. 2012). Similar to other body sensations, the stimulation of certain areas of the central nervous system, such as the somatosensory thalamus, the somatosensory cortices, and the insula, also leads to a tingling-like sensation (Penfield and Gage 1933; Penfield and Faulk 1955; Lenz et al. 1993; Davis et al. 2000; Ostrowsky et al. 2000, 2002; Mazzola et al. 2006; Pugnaghi et al. 2011).

In my view, this phenomenon might play a substantial role in interoception-related measurements where near-threshold stimuli are present, such as the somatosensory signal detection task by Lloyd et al. (2008) (see Chap. 4) and procedures assessing interoceptive accuracy. For example, as already mentioned in Chap. 4, focusing on the chest during the trials of a heartbeat discrimination task can evoke sensations of heartbeats, particularly if a rhythmic auditory cue is available. If the rhythm is not consistent with the actual heartbeats, this sensation can interact with (or perhaps even suppress) the weak sensations evoked by the contractions of the heart. In consistent trials, however, it can enhance the real sensation. It is important to see that the two effects are not necessarily additive, thus the consequences of attention cannot be easily eliminated. As there are individual differences in the proneness to and intensity of the tingling sensation, this factor should be taken into consideration in detection threshold measurements.

Concerning the mechanism, one possibility, already proposed by Sigmund Freud (Schilder 1935), is that attention triggers physiological changes (e.g., slight increases in blood supply or muscle tension), and these changes evoke the sensation; this is the so-called efferent model of tingling sensation (Tihanyi et al. 2018). To put it simply, attention changes something in the body that can in turn be sensed. For example, attention can change the sensitivity of muscle spindles (Hospod et al. 2007). However, we found no such changes in the temperature, skin conductance, and muscle tension in the body regions our participants focused on (Tihanyi and Köteles 2017). Although the findings of a single experiment are never conclusive, another mechanism appears more likely to me. Attention in such experiments is associated with some kind of expectation or perceptual priming with respect to the stimulus and the body area at hand (we called it monitoring or scanning in the introduction of this chapter) (Michael and Naveteur 2011; Michael et al. 2011). This expectation itself is able to shape perception if the sensory background is vague or noisy. Noise is generated by the spontaneous fluctuations of the system at multiple levels (e.g., peripheral physiology, receptors, central processing). If the signal is weak compared to the noise (in our case: there is no clear-cut change in the sensory output), top-down influences are able to shape the sensory experience. This effect was first demonstrated in the already presented seminal experiment of Pennebaker (1982); participants experienced increases and decreases in their finger temperature, depending on the information (expectation manipulation) they received before. The actual temperature showed slight fluctuations, of which mostly those sections were perceived and reported which were in accordance with participants' expectations. In the absence of specific expectations, attention directed to a body part simply leads to the lowering of the sensory threshold; in consequence, spontaneous sensory activity becomes perceived, and the body area "pops out" from the adjacent regions (Hsiao and Vega-Bermudez 2002). In line with this idea, focusing on a body part was accompanied by the simultaneous activation of the respective somatosensory area (Bauer et al. 2014). After reviewing the available evidence, we came to the conclusion that this model, called attention-disclosed model, represents the most likely mechanism as it can explain the vast majority of the features of the tingling sensation (Tihanyi et al. 2018).

The existence of spontaneous or attention-related body sensations is largely unknown both to professionals and laypeople. Again, when we feel something in our body, we are prone to accept it as the indicator of peripheral physiological changes without reservation (tenet of realness). Also, the possible impact of attention on various interoceptive processes (including pathological ones) is largely underestimated. For example, participants in the somatic signal detection task described in Chap. 4 often experienced tactile sensation in the absence of actual stimulation and those with proneness to somatoform symptoms reported such tactile illusions even more frequently (Brown et al. 2010; Katzer et al. 2011, 2012). According to our own results, the sensation itself is neutral in healthy individuals, i.e., it is not genuinely accompanied by positive or negative affect (Tihanyi et al.

2018). Thus, it can be evaluated in both directions depending on the situation. For example, tingling, originally evoked by increased body focus and selective seeking process directed by the patient information leaflet of a medicine, can be interpreted as the pharmacological side effect of a drug. This option has never been investigated empirically to date; however, it is strongly supported by the fact that tingling is among the complaints of medically unexplained symptoms (Hartman et al. 2013), nocebo symptoms (Abbott et al. 2003; Gowan and Roller 2015), and the symptoms of electromagnetic hypersensitivity (Szemerszky et al. 2010, 2015; Köteles et al. 2013) (these categories are overlapping; for more details, see Chap. 9). On the other hand, it can be the sign of vitality and the flow of energy in yoga and tai chi chuan practice (Greyson 1993a, b; Levine 2008; Chuckrow 2015), as well as the proof of the existence of healing energy in various complementary and alternative methods (Chap. 10) (Park et al. 2002; Shiflett et al. 2002; Miles and True 2003; Kong et al. 2007; Raingruber and Robinson 2007; Hui et al. 2007).

In summary, attention to the body is able to generate body sensations (we can also call them body illusions) even in the very absence of peripheral events. In the case of the tingling phenomenon, these sensations are mostly neutral. If attention is driven by positive or negative concepts (e.g., expectations of the universal energy or harmful environmental influences, respectively), however, the resulting sensations will be associated with evaluation.

5.7 Issues with Attention 2: Primary vs. Secondary Level

If one carefully reads the literature, it turns out that attentiveness to body sensations and body awareness can refer to (at least) two completely different concepts.

First, at the experiential level, there might be individual differences in the attentiveness and receptiveness to percepts from the body; this level is directly related to the topic of this book, and we primarily discuss attention from this perspective. This is in line with another accepted definition of body awareness, i.e., "the perception of bodily states, processes and actions that is presumed to originate from sensory proprioceptive and interoceptive afferents and that an individual has the capacity to be aware of" (Mehling et al. 2009, p. 4). Sometimes (actually, quite often) these percepts are automatically evaluated. As already discussed in Chap. 4, the outcome of this evaluation can be negative; for example, unpleasant symptoms characterize the majority of diseases. The valence attached to sensations can also be positive, as for affective touch or sexual excitement. Another, often raised possibility is the existence of neutral attention, i.e., the mindful observation of body percepts without any kind of evaluation (Mehling 2016, 2020). I think such an observing attitude might be possible in the case of self-directed attention, such as the tingling phenomenon. If attention is grabbed automatically by a body sensation, however, evaluation is often part and parcel of the process (see below); in conclusion, a more or less neutral observation (i.e., the separation of the evaluation from the self and seeing it as a mental process) requires mental effort and cognitive resources. In consequence, it does not belong to primary experience.

Beyond the experiential level and primary affective evaluation, higher-level processing also occurs; we can call it interpretation. For example, we can explain certain body percepts as symptoms of a disease, try to find a cause for them (attribution) or we can regard them as indicators of the unity of our body–mind system or helps in decision-making (Mehling et al. 2009, 2012). Awareness of possible long-term consequences (both positive and negative) of the perceived state also belongs here and impacts the overall evaluation and meaning of the state (Zborowski 1952). Certain authors go even further and put the preventive steps taken by the individual to avoid further injury and the use of health-care services under the umbrella of body (aka somatic) awareness (Bakal 1999; Bakal et al. 2008). From a certain point of view, the fear of future damage and its behavioral consequences also represent an aspect of attention paid to the body. For example, worrying combined with high level of conscientiousness can lead to more health behaviors and better health outcomes as emphasized by the "healthy neuroticism" theory (Weston and Jackson 2018; Weiss and Deary 2020; Graham et al. 2020). These aspects are inevitably important; however, they should be clearly distinguished from the major topic of this book, i.e., the experiential aspect of body sensations.

Wolf Mehling and his colleagues developed a questionnaire that is claimed to assess all relevant aspects of interoceptive sensations. Thus, although certain items clearly refer to proprioceptive aspects, such as posture, they prefer the term interoceptive awareness (this term is basically used as a synonym for IS_{trait} by this group), and the questionnaire is called Multidimensional Assessment of Interoceptive Awareness (MAIA) (Mehling et al. 2012, 2018). The MAIA consists of eight "distinct but related dimensions" (Mehling et al. 2012, p. 19): (1) *Noticing* (evaluation-free awareness of body sensations), (2) *Not-Distracting* (directing attention away from unpleasant sensations), (3) *Not-Worrying* (the lack of emotional distress associated with unpleasant body sensations), (4) *Attention Regulation* (voluntary direction attention to the body as a whole), (5) *Emotional Awareness* (awareness of emotion related body changes), (6) *Self-Regulation* (using the body as a refuge in stressful situations), (7) *Body Listening* (using body sensations in decision-making), and (8) *Trusting* (experiencing the body as a secure place). Out of the eight subscales, only the minority (1 and 5, perhaps partly 3 and 8) refer to direct bodily experience, the rest are associated with higher-level cognitive functioning. Empirical studies (including the original study) revealed close associations among the majority of the subscales (usually 1, 4–8), whereas the remaining two (2 and 3) were more or less independent (Mehling et al. 2013; Valenzuela-Moguillansky and Reyes-Reyes 2015; Brown et al. 2017; Reis 2017). In our own study (Ferentzi et al. 2020), the aforementioned six scales were loaded on an underlying overarching construct which was strongly associated with the BAQ. Not-worrying showed a strong negative association with pain catastrophization, somatosensory amplification, and emotionality, whereas not-distracting was not related to any of the assessed constructs. In another recently published study, those characterized by proneness to symptom reports differed from low symptom reporters exactly in these two dimensions

(Flasinski et al. 2020). To me, this indicates that the negative (unpleasant and potentially threatening) aspects of body sensations represent a distinct and important category. To avoid well-known pitfalls with negatively worded items, perhaps it would have been better to not reverse these two scales. Surprisingly enough, mindfulness, a characteristic often associated with body awareness (Baer et al. 2006; Mehling et al. 2009), showed only weak associations with any of these dimensions in our study.

According to these findings, the boundaries between primary experience (i.e., feeling an actual sensation) and more conceptual levels (i.e., treating perceived sensations as meaningful and important, or using them as guides in decision-making) appear to be vague, at least at the level of self-report. Actually, both levels rely on body sensations; however, I think the primary level is much more difficult to change and has a more profound impact on our psychological functioning.

5.8 Issues with Attention 3: Aspects of Attention

The kind of acute attention that leads to the tingling sensation is well-known for everyone. It was described by William James (1890) as a searchlight that can be voluntarily directed on a target and moved to other targets. Unfortunately, the phenomenon of attention (more accurately, a number of phenomena put under the umbrella of attention) is much more complicated. In this section, we will discuss those aspects only that are necessary for the understanding of a number of body focus-related constructs.

An important distinction between receptiveness and active monitoring or scanning was already described in the introduction of this chapter. Usually, a fundamental distinction between attentional processes with respect to control is also made. Certain authors speak of controlled vs. automatic information processing (Schneider and Shiffrin 1977; Shiffrin and Schneider 1977), whereas Posner (1980) proposed the terms endogenous vs. exogenous. Endogenous or controlled attention refers to centrally (voluntarily) controlled orienting (this is in line with the approach of William James), whereas exogenous or automatic attention is controlled by external cues. The terms top-down and bottom-up are also used for these categories (Hohwy 2014). From a neurophysiological point of view, these categories correspond with the so-called dorsal and ventral attention system, respectively (Corbetta and Shulman 2002; Vossel et al. 2014). In this section, we primarily focus on endogenous attention, whereas exogeneous (automatic) processing will be discussed later.

Strangely enough, stimuli catching attention in an exogenous way refer to internal signals originating in the body in our case; I think that this constellation characterizes the oddity of the situation very well. As automatic internal changes are assumed to indicate the salience of external stimuli (see below), this raises serious questions with respect to the applicability of concepts developed and elaborated for externally focused attention to body-related attention. Moreover, many internal sensations are vague and poorly detectable which also means that focusing on them is a difficult task (Wang et al. 2019). As shown, more attention (which often means active monitoring or scanning) not only enhances the actual signal to some extent and shortens its processing time but can generate novel body sensations too. Studies on interoceptive attention usually involve top-down (endogenous) attention; the typical task is focusing on an internal sensation (breathing, heartbeats) or an exteroceptive cue as control condition. Thus, a second question concerns the similarities of internal and external attention. Overall, do they refer to the same process? In a brain imaging study including exteroceptive (visual) and interoceptive (focus on breathing, an internal stimulus with good detectability) stimuli, substantial differences in the involved brain structures were revealed (Farb et al. 2013). External attention was characterized by the activation of the lateral fronto-parietal executive network, and increased connectivity between the visual cortex and both the inferior parietal cortex and the visual thalamus. Internal attention was associated with the activation of posterior limbic structures, the insula, and medial parietal network; enhanced connectivity with posterior insular cortical fields and putative interoceptive thalamus was also found. Moreover, the posterior part of the insula was activated during the internal attention task, whereas the anterior part was associated more with the exteroceptive task. In other studies, participants are presented with tones that are in synchrony with their heartbeats or not; moreover, sensory characteristics (frequency) of the tones varied or not. Participants are asked to make decisions with respect to these options in a number of trials, and the differences in brain activation patterns are calculated (Critchley et al. 2004; Zaki et al. 2012). Visual stimuli was also applied as a control condition for the mental tracking task (Schauder et al. 2015; Mash et al. 2017; Failla et al. 2020). Alternatively, participants are asked to count or simply focus on their own heartbeats or external tones for periods of time (Tracy et al. 2007; Ernst et al. 2013; Wiebking and Northoff 2015; Wu et al. 2019). In these studies, interoceptive attention is usually accompanied by enhanced activity of the posterior and/or anterior insula, the anterior cingulate cortex; sometimes the somatosensory and the prefrontal cortex were also involved. Differences among studies can partly be explained by differences in the tasks and interoceptive and exteroceptive modalities involved. The majority of the findings support the idea that the anterior part of the insula might be the primary coordinator of the allocation of attention, both interoceptive and exteroceptive (Critchley et al. 2004; Farb et al. 2012, 2013; Ernst et al. 2013; Avery et al. 2015; Wang et al. 2019). In a recent meta-analysis (Schulz 2016), however, focusing attention to the heartbeat was associated with the activation of the posterior insula. The author concludes that the anterior insula might be associated with the evaluative aspects of interoception rather than attention. Findings of another study show the importance of right parietal cortex to directing attention internally and focusing on a specific body signal (Tracy et al. 2007). From a functional point of view, the so-called cingulo-opercular task control network (encompassing the dorsal anterior insula and ventral anterior ACC) appears to play a role in the maintenance of interoceptive attention (Wu et al. 2019).

Another option to investigate the brain correlates of attention is electroencephalography. Heartbeat-evoked potentials (HEP) showed differences when participants were asked to conduct the Schandry task versus when they completed an auditory task in a number of studies (Schandry et al. 1986; Montoya et al. 1993; García-Cordero et al. 2017). Those with better perception performance were characterized by higher HEP amplitudes (Katkin et al. 1991; Pollatos and Schandry 2004; Pollatos et al. 2005). However, the control condition differed in many aspects from the interoceptive condition in these studies (or completely lacked), thus the results can be explained in many ways. To fill this gap, participants were continuously presented with white sound in a recent study, and their task was focusing on their heartbeat or on the sound (Petzschner et al. 2019). In this task, that posed no additional demand for participants, a pure attentional effect on HEP was found, i.e., the amplitude of HEP was larger in the interoceptive condition.

The latter study was designed to exclude any form of divided attention, a factor that was frequently criticized for forced-choice tasks (Chap. 4). However, the ability to divide attention might also be important under everyday circumstances when simultaneous internal and external cues compete for processing resources (Gregory et al. 2003). This was particularly striking in a recent study on the impact of oxytocin on interoception (Yao et al. 2018). Oxytocin, compared to placebo, did not influence heartbeat tracking performance in the absence of other stimuli. If competing social stimuli were present, however, interoceptive accuracy (to less extent for faces expressing negative emotions than neutral and positive faces) decreased in the oxytocin group. In other words, oxytocin, which improves the salience of social cues, caused a shift in the allocation of attention. In a cross-sectional study, participants with high level of interoceptive accuracy performed better in selective and divided attention tasks with exteroceptive stimuli (Matthias et al. 2009). These findings demonstrate that interoceptive and exteroceptive attention rely (at least partly) on the same cognitive resources and compete for them. Of course, this competition takes place at both levels (i.e., exogenous and endogenous) of attention.

The bottom line is that even endogenous or top-down attention is a multifaceted construct, and it is not clear which aspects are related to interoceptive sensibility and body awareness. Whereas experimental settings routinely require top-down attention, this might be relatively rare compared to bottom-up attention under everyday circumstances. In more technical terms, this seriously limits the generalizability of laboratory studies on interoceptive accuracy and IS_{acute}.

5.9 Bottom-Up Interoceptive Attention: Early Ideas

Seymour Fisher devoted his carrier to the research of the psychological representation of the body and the role of the representation on perception (Fisher and Cleveland 1958, 1968; Fisher 1970, 1974, 1986a, b). He called the representation body image; it is a vague (or very broad) term encompassing the conscious and nonconscious aspects of body experience, including the individual's attitudes toward his or her body. Most importantly for us, both external appearance (necessarily involving a third-person point of view) and internal sensations were included. Fisher and others developed several ideas (and tests) that might be interesting even today. Inspired by the psychoanalytic approach, they assumed the role of preconscious processes in the development of the representations of the body and attempted to elaborate methods that can shed light on this connection. The majority of these methods belong to the family of projective tests; participants are presented with ambiguous stimuli, and their spontaneous associations are quantitatively analyzed.

As mentioned, the ego develops from body sensations in the psycho-sexual development model of Freud (1938). A shared characteristic of these areas is that they belong to the surface of the body (i.e., the skin). Thus, using the terminology of psychoanalysis, the ego can be considered the mental projection of the surface of the body (Fisher and Cleveland 1958). As the surface represents the boundary of the body, it also functions as some kind of armor or shield protecting the organism from external influences; according to Fisher, this function should also be reflected in its internal representation and might impact general perceptual functions. As these assumed processes belong to the preconscious domain, Fisher and Cleveland (1958) developed a measurement method that was based on the Rorschach inkblot test (Rorschach 1921). This test, still used in clinical practice, has a debated position (to say the least) from a scientific point of view because of a number of validity and reliability issues (Parker 1983; Acklin et al. 1992; Gacono et al. 2002; Wood et al. 2003, 2010; Hughes et al. 2007). Fisher was well aware of the shortcomings of the Rorschach and the psychometric requirements of a good test. He did not use the inkblot test for general diagnostic purposes; he was curious only about the appearance of the representation of the body in the perceptual process. Rigorous validity and reliability analysis, as well as detailed description of the scoring, were also part of the development procedure. The basic idea was that the quantity and quality of the appearance of the body, its parts and functions in the associations of participants when interpreting the inkblots could be used as an indicator of the overall role of the body in the perceptual process. After analyzing a large number of responses, two body-related constructs emerged. They were called Barrier and Penetration of boundary. Barrier refers to the preconscious conceptualization of the body as the boundary of the self, whereas Penetration indicated the extent to which the boundary was felt as injured (Fisher and Cleveland 1958, 1968). Subsequent research focused primarily on the former because it was associated with more personality characteristics thus appeared to be easier to interpret and more salient.

It was shown that psychosomatic patients with symptoms close to the surface of the body (e.g., rheumatoid arthritis, neurodermatitis, or conversion hysteria) score high on Barrier, whereas those with internal symptoms (e.g., colitis ulcerosa or stomach disorders) score high on Penetration. High Barrier scores in normal subjects were also associated with more external than internal symptoms, higher levels of the so-called self-steering behavior (goal setting, task completion, etc.), as well as better coping with stress. Further, the Barrier score was positively related to democratic, equal, group-oriented functioning. It was concluded that the Boundary dimension is a meaningful and important characteristic with protective functions that is acquired from the family via the internalization process (Fisher and Cleveland 1968). It remained an open question, however, whether it relies on real body experience (e.g., perceived security in the body) or just projected to the boundary of the body (Fisher 1970).

Concerning validity, the Barrier score showed a positive association (1) with the number of external (i.e., related to the surface) as opposed to internal body sensations when acutely focusing on a certain part of the body or in retrospective appraisal, (2) symptom experience after taking a placebo pill, and (3) memory performance using external vs. internal related words (Fisher and Fisher 1964). Further, manipulation of the focus of attention induced changes in the Barrier score in the expected directions (Fisher and Renik 1966; Renik and Fisher 1968). These findings support the idea that this approach would be worthy of further investigation. Unfortunately, subsequent research on the Barrier score is sporadic and mostly declares the need of further studies (Siegel 1977; Jupp 1989, 2016; Zennaro and Lis 1998, 2016).

As early as in 1953, Paul F. Secord developed a homonym-based word-association test to assess body orientation (aka body cathexis, according to the psychoanalytic terminology) (Secord 1953). Body cathexis refers to the degree of feeling of satisfaction or dissatisfaction with various parts or processes of the body (Secord and Jourard 1953). These feelings were assumed to substantially contribute to the self-concept and personality. In the Secord test, words with body-related and body-unrelated meaning (e.g., tablet, colon), as well as words eliciting body-related and unrelated reactions (e.g., acid) served as stimuli. Participants were asked to write down the first word that comes into their mind after reading each stimulus word, and the number of body-related choices was considered as an indicator of body orientation. It was found that high level of body orientation is associated with either anxiety or (rarely) narcissism, whereas lower scores indicate the denial of anxiety and overcontrolling (Secord 1953). Moreover, a marked negative association with the acceptance of external characteristics of the body, but not with that of body functions, was revealed using the Body Cathexis Scale (see below). This negative association was explained by interpreting the Secord-test score as an indicator of anxiety (as the prevalence of narcissistic tendency was comparatively low, it was neglected). The negative association was replicated only for women in a subsequent study (Secord and Jourard 1953), and could not be reproduced in a second one (Weinberg 1960). Although it was not pointed out by the authors (or was overwritten by the presumed evaluative tendency), what the Secord test might primarily assess is simply the tendency to focus on the body, regardless of its evaluative background.

The Body Cathexis Scale, a self-report instrument that is designed to measure attitudes, i.e., evaluations, toward the body, represents perhaps the first questionnaire of its kind and can be considered the starting point of systematic research on the evaluative aspect of the representation of the body (Jourard and Secord 1954). Lower scores on this scale represent negative evaluation, whereas high scores indicate a positive attitude toward the body. It was assumed (Murphy 1947) and demonstrated (Secord and Jourard 1953; Johnson 1956; Weinberg 1960; Gunderson and Johnson 1965) that body-related attitudes show a considerable overlap with self-related attitudes; this finding suggests that the representation of the body might serve as the anchorage point for the self—an idea that is still accepted and intensely studied recently (see Chap. 3).

However, these results indicated the primacy of evaluative processes, whereas the present research focuses more on the sensory aspects of body-related information (see below). The assumed negative association between body cathexis and anxiety was demonstrated (Johnson 1956) using a widely used measure of anxiety, the Taylor Manifest Anxiety Scale (Taylor 1953). Also, medium-level negative correlations between body cathexis and the Maslow Test of Psychological Security-Insecurity were found (Secord and Jourard 1953; Weinberg 1960).

Another semi-projective test developed by Fisher (1970) is the Body Prominence Test. In this test, participants are simply asked to list and briefly describe 20 things that they are aware of at the time of completion. The total score is the number (or percentage) of the thoughts of bodily relevance. Psychometric properties of this test were acceptable; concerning validity, it was associated with the learning of body-related words, experienced body sensations, the score of the Secord test, and was elevated before surgery.

In summary, a number of paradigms and tests were developed to assess individual differences in attention and attitudes toward the body at the early stage of the systematic research on the representation of the body. A common characteristic of these tests is that their approach differs from the questionnaire-based self-report method widely used recently. They attempt to measure the exogenous or automatic aspect of body-focused attention; they are not suggestive, i.e., they do not direct the thinking of participants toward certain sensations or symptoms. In my opinion, particularly the Secord test and the Body Prominence Test would deserve much more research attention as their use could substantially extend our recent knowledge on the psychological representation of the body.

5.10 Bottom-Up Interoceptive Attention: Reaction Time Tests

The cognitive approach to preconscious perception emphasizes automaticity of processing (Dixon 1981; Kihlstrom 1987, 2004); this often involves affective-evaluative processes. Self-report-based measurements are not appropriate for the assessment of such automatic processes, thus new paradigms, typically based on reaction time measurements, were developed. The general idea behind these paradigms is that emotionally salient stimuli (e.g., those representing potential threat) increase reaction time as they consume attentional resources (i.e., bias attention). For example, the processing of color-related information takes longer in the Stroop paradigm (Stroop 1935) if the presented words are of emotional significance (Ray 1979; Mathews and MacLeod 1985; Watts et al. 1986; Lim and Kim 2005; Witthöft et al. 2006; Afzal et al. 2006). As these results can be explained in other ways too (e.g., as a general effect of negative affect), a novel paradigm, called dot-probe test, was developed (MacLeod et al. 1986). In this test, two words are presented simultaneously on the screen for 500 ms, then they disappear and a dot shows up almost immediately in the location of either word. Participants are asked to press a button as soon as possible after the appearance of the dot. Emotionally threatening words are characterized by an increased reaction time for those with high levels of anxiety, demonstrating that attention was grabbed (MacLeod et al. 1986; MacLeod and Mathews 1988; Vasey et al. 1995). As automatic and effortful processing might differ (Lonigan et al. 2004), presentation of the stimuli (words or pictures) was approximately 20 ms (assessing preconscious processing) or in the 300–3000 ms domain (targeting conscious processing) in later versions of the paradigm (Kroeze and van den Hout 2000; Boyer et al. 2006; Beck et al. 2011; Dehghani et al. 2018). Another method, called exogenous cueing paradigm, was also developed to assess emotional biases (Fox et al. 2001).

These paradigms were applied to body-related pathological conditions, such as chronic pain, panic disorder, and medically unexplained symptoms. Results are not without inconsistency; however, the majority of them support the existence of a body-related attentional bias under pathological conditions (Kroeze and van den Hout 2000; Lees et al. 2005; Witthöft et al. 2006; Boyer et al. 2006; Hou et al. 2008; Martin and Alexeeva 2010; Martin and Chapman 2010; Jellesma et al. 2011; Chapman and Martin 2011; Beck et al. 2011; van der Veek et al. 2014).

In these paradigms, anxiety and/or perceived threat appears to play a substantial role. In a similar vein, Fisher and others drew the conclusion from early theoretical considerations and empirical findings that the major factor behind body focus is anxiety. An intriguing question is whether other features can or cannot play a similar role. As described in Chap. 4, some body sensations (affective touch, slight to moderate physical activity, sexual excitement) are inherently accompanied by positive affect. Also, advocates of yoga and other mind–body practices emphasize the positive aspects of body-focused attention (Daubenmier 2005; Impett et al. 2006; Farb et al. 2015). Thus, the existence of a positive bias toward the body would be conceivable for certain populations; unfortunately, no study was conducted on this to date.

5.11 Top-Down Interoceptive Attention: Monitoring Tendency

As Ginzburg and colleagues point out (2014), there are conditions when individuals actively scan their body to detect changes, and this process (in accordance with Ginzburg and colleagues, we will call it monitoring) is conceptually different from the state of pure receptiveness. We already discussed an example of active monitoring: the tingling phenomenon. It was concluded that the expectation of certain body sensations can act as a self-fulfilling prophecy. Concerning physical symptoms, the phenomenon was called selective or schema-based seeking process (Pennebaker and Skelton 1981; Pennebaker 1982), which in turn can lead to the experience of more (expected) symptoms as people tend to ignore sensations that are not in accordance with their expectations or schemata and favor those that are expected. These concepts rely on laboratory experiments where participants were explicitly asked to direct their attention to certain target stimuli. However, the phenomenon is quite frequent under everyday conditions too.

I am not aware of a questionnaire that exclusively measures proneness to active monitoring. For example, in the Body Vigilance Scale developed by Schmidt and colleagues (1997), only one item of the four refers to monitoring ("On average, how much time do you spend each day 'scanning' your body for sensations (e.g., sweating, heart palpitations, dizziness)?"). Similarly, the instrument used to assess body awareness by Hansell and Mechanic (Hansell and Mechanic 1991; Hansell et al. 1991), contains only one marginally related statement ("How much do you think about how your body feels"). It turned out in two studies that (1) attentiveness to body signals (as assessed with the BAQ) and monitoring tendency (as assessed with the scanning item of Schmidt and colleagues) are related; however (2), only the latter was associated with hypochondriasis and anxiety in regression analysis after controlling for their shared variance (Ginzburg et al. 2014). Patients with hypochondriasis report more symptoms than control individuals if they are instructed to focus on body sensations; moreover, they are difficult to distract with external information (Haenen et al. 1996). Of course, some reciprocity between awareness and monitoring of internal cues should be assumed, and it cannot be stated that they represent purely bottom-up and top-down components of the perception process, respectively.

Body sensations often directly catch attention through affect (Chap. 3). Generally, monitoring tendency is often accompanied (or driven) by negative affect under everyday conditions, which can result in somatic symptom distress (for more detail, see Chap. 7). Leventhal and colleagues made a distinction between focusing on sensory and evaluative aspect of body sensations (Leventhal et al. 1979; Leventhal and Mosbach 1983; Ahles et al. 1983), and the same was proposed for pain (Melzack and Wall 1965; Melzack 1973). Automatic evaluation often gives rise to secondary processes related to the meaning and causation (attribution) of the experienced sensations. If the individual is prone to conclude that the sensations might indicate some sort of pathology, as in the case of somatosensory amplification and health anxiety, this concept might initialize the systematic seeking process which leads to the experience of more symptoms. Somatizing patients also have a proneness to such catastrophizing interpretations (Rief et al. 1998). From this point of view, both psychological and somatic attribution styles can lead to active monitoring (Robbins and Kirmayer 1991), whereas normalizing/environmental explanations (provided they refer to transient conditions) do not favor it. Environmental attribution referring to long-term harmful exposition, however, can also initialize the scanning process (Chap. 7).

Finally, absorption, the tendency to become immersed in sensory and other experiences (Tellegen and Atkinson 1974), might also interact with the scanning process. When the state of "total attention," i.e., when the available representational capacity is entirely dedicated to the experience, refers to symptoms and the assumed pathology behind them, it can further amplify body sensations. In empirical research, absorption was associated with symptom reports and various forms of idiopathic environmental intolerances (Witthöft et al. 2008, 2018; Skovbjerg et al. 2010; Simor et al. 2011; Köteles et al. 2019).

It is important to see that purely bottom-up and top-down forms of body attention occur under very special conditions only, usually in laboratory experiments. Still, considering the sensation-generating properties of the latter, the distinction appears of practical (and methodological) importance. Further positive and negative consequences of top-down monitoring and expectations are discussed in the nocebo and placebo chapters (Chaps. 9 and 10).

5.12 The Predictive Processing Approach to Attention

So far we used the exogenous vs. endogenous distinction for attention, but it was also mentioned that things are usually more complex under non-laboratory circumstances. The boundary between these two categories might be less marked, and they often interact with each other (Berger et al. 2005; MacLean et al. 2009; Hohwy 2012). To be able to understand the complexity of the phenomenon, a more sophisticated approach is needed. The predictive processing framework (for details, see Chap. 1), describing multiple, hierarchically organized levels of processing, offers such a model. Attention to a sensory channel is explained at this computational level as a factor (perhaps the factor) that increases the relative precision during hierarchical inference (Friston 2009; Feldman and Friston 2010; Hohwy 2012; Fardo et al. 2017; Petzschner et al. 2019). Higher relative precision means more relative weight, i.e., more importance in the processing chain. For example, heartbeat-evoked potentials are interpreted as indicators or brain concomitants of precision-weighted prediction errors with respect to individual heartbeats (Ainley et al. 2016). Alternatively, they might reflect predictions (Petzschner et al. 2019; Park and Blanke 2019); as the experimental manipulation of the signal at the level of heartbeats would be extremely difficult in this case, we cannot really decide.

Turning back to attention, exogenous attention can be understood as a factor that increases precision of certain signals at low (i.e., nonconscious) levels of sensory processing, whereas endogenous attention acts at the conscious (voluntary) level. Moreover, physiological (i.e., regulatory) and behavioral changes evoked by interoceptive input can also be regarded as consequences of attention paid to homeostatically relevant internal information by the organism. Monitoring tendency simply refers to the existence of strong and relatively stable high-level pre-existing hypotheses (priors) (Cioffi 1991).

Overall, attention shows substantial overlaps with salience (Parr and Friston 2019). A stimulus is considered salient exactly because it is able to catch attention (Fecteau and Munoz 2006); or, from a more evolutionary perspective, it can catch attention because it might be important for survival (Craig 2009). With respect to exteroceptive cues, e.g., the visual modality, salience primarily relies on the physical characteristics of objects, e.g., the extent to which objects stand out and contrast from other competing stimuli (Mrkva and Van Boven 2020). Salience in this context often means atypical, i.e., having a different color or being novel, which makes the object pop-up from its environment. In the literature of interoception, the terms salient and salience are often used without explicit definition (Seeley et al. 2007; Weierich et al. 2010; Menon and Uddin 2010; Uddin et al. 2014), as synonym for important and importance. Importance in this field generally refers to homeostatic significance (Damasio 2010; Barrett and Simmons 2015), with an emphasis on (possible) threats, including deviations from homeostasis, rather than pure novelty. This is in accordance with the general agreement that salience is determined by the interaction of the characteristics of the stimulus and the perceiver (Kanan et al. 2009; Blumenthal-Dramé et al. 2017; Mrkva and Van Boven 2020). Concerning external stimuli, say, a creek or a shady spot under the trees, salience is judged relying on the actual state of the organism; and the actual state is reflected by interoceptive input, such as thirst or core temperature, respectively. Now the questions are (1) what makes thirst or temperature able to serve as indicator of certain aspects of homeostasis and (2) how. The answer for the first question is obvious: these internal channels were shaped by evolution and natural selection to effectively monitor crucial aspects of homeostasis. Among other modalities, the salience of thirst and core temperature are genetically determined for mammals. As for the how question: to be able to influence regulation at the behavioral level, they have an affective-motivational aspect (Chap. 3). In other words, certain internal stimuli (more precisely, a certain intensity range of certain stimuli) are inherently associated with affect, typically negative affect. This affective quality is able to capture and maintain attention even in situations with a number of competing cues. This is not to say that complex motives, such as to eat and drink, including their behavioral aspects, are completely innate; in fact, a large proportion of them is acquired during early stages of development (Harshaw 2008). For body sensations, the classic distinction between direction (i.e., internal vs. external) and content (in this case: evaluation) of attention, made by Wicklund and Gollwitzer (1987), is less cleat-cut than it appears at first sight.

In this section, we saw that the computational background of attention can be understood through the principles of predictive processing, most importantly, hierarchical processing and weighting of top-down/bottom-up signals based on their relative (estimated) precision. Affective evaluation from this perspective simply plays a role in the estimation of precision.

5.13 Acute and Generalized Interoceptive Sensibility

In this section, we will have a closer look at the two major aspects of interoceptive sensibility. Let us start with IS_{trait}, the perceived general ability to sense internal changes. To measure this construct, Garfinkel and Critchley (2013) recommend the use of the Body Awareness Scale of the Body Perception Questionnaire (BPQ) (Porges 1993). From a certain point of view, it is an excellent choice as primary body experience is assessed with respect to a number of sensory modalities.

Unfortunately, as it was pointed out in a systematic review on questionnaires assessing body awareness (Mehling et al. 2009), we do not know too much about the validity of this instrument. A recent paper tried to fill this gap (Cabrera et al. 2018); still, more empirical evidence would definitely be needed. For example, its relation to well-established measures of body awareness, such as the BAQ or the Private Body Consciousness Scale, is unknown. It appears more problematic to me that, as clearly shown by the moderate to strong associations with the Somatosensory Amplification Scale and the Stress Reactivity Index (Cabrera et al. 2018), depression (Wiebking et al. 2010), and the somatization subscale of the SCL-90 (Wang et al. 2020), the BPQ, just like Mandler's (Mandler et al. 1958) Autonomic Perception Questionnaire (APQ; Chap. 6), focuses on modalities and feelings primarily associated with sympathetic activation (stress). In other words, negative evaluation is included in this construct. This makes the usability of BPQ for the assessment of perceived sensibility to internal cues questionable. Other studies use more neutral questionnaires, such as subscales of the MAIA or the BAQ, for the assessment of IS_{trait} (Ainley and Tsakiris 2013; Emanuelsen et al. 2015; Meessen et al. 2016; Ferentzi et al. 2017, 2018b; Stern et al. 2017; Young et al. 2017; Forkmann et al. 2019). My personal opinion is that out of the recently available questionnaires, the BAQ, although it is also far from perfect, fits best the construct of IS_{trait}.

Now let us see the construct and measurement of the ability to sense changes of an actual interoceptive stimulus, IS_{acute}. It is also called confidence, which is a bit misleading because of its multiple meanings (see below). Concerning the first option, this measurement was originally introduced by Hartl (1992) and Ehlers and her colleagues (1995) to assess aspects of cardiac perception in patients with panic disorder. For panic patients, increased or abnormal cardiac activity (arrhythmia) is of uttermost importance; it may indicate a panic attack, which is often interpreted as a heart attack (Khalsa and Feinstein 2018). In such a threatening situation, the behavioral reaction to the event can be influenced by patients' confidence in their cardiac sensations. To shed more light on this factor, participants were asked "to rate the confidence with which they had felt their heartbeats on a scale between 0 'totally uncertain' and 10 'totally certain'" (Ehlers et al. 1995, p. 71). Whereas this question is particularly relevant and meaningful for panic patients, its usability is less obvious for healthy participants who do not usually question the validity of their sensations (tenet of realness). Garfinkel and colleagues adopted this measure; they used a 10 cm visual analog scale (VAS) with the anchor points of "Total guess/No heartbeat awareness" and "Complete confidence/Full perception of heartbeat" and asked participants to rate their "confidence in their perceived accuracy of response" (Garfinkel et al. 2015, p. 67). What should be measured, however, is simply perceived accuracy of response (i.e., the subjective counterpart of objective accuracy or interoceptive accuracy shown in the test, e.g., "How well do you think you performed in this task?") and not confidence in it. The latter represents a conceptually different dimension, which is closer to metacognition (e.g., interoceptive awareness, Chap. 6) than perceived performance.

As described in Chap. 4, the perceived sensibility with respect to an actual interoceptive stimulus (IS_{acute}) is conceptually different for near-threshold and above-threshold stimuli. In the former case, when the measurement of detection is based on statistics (e.g., as in the forced-choice paradigm), the actual performance in the task is necessarily unknown to participants. Thus, if we ask them about their perceived performance, the answer will show no or only weak association with the actual performance and will be determined by other, typically top-down factors (e.g., motivation to cooperate, demand characteristics of the study, conformity, previous expectations; this will be the primary topic of Chap. 6). However, when the range of stimulation is dominantly above the sensory threshold (for example, because the sensory modality at hand usually reaches consciousness), the forced-choice approach cannot be used any more. Although the association between intensity of stimulation and the reported intensity of the sensation is not linear, participants clearly feel something; thus, the discrepancy between objective and perceived performance will be reduced to a great extent. As it was presented in Chap. 4, for dominantly conscious modalities, such as pain, itch, thermosensation, sensual touch, the fatigue of the muscles, breathlessness, the distension of the stomach, the anus, and the urinary bladder, the "acuity" of perception is necessarily measured via the comparison of some kind of self-report (see below) and the objective intensity of the sensory stimulation. In my view, the distinction between interoceptive accuracy and IS_{acute} is practically meaningless for these cases. This is not to say, of course, that no individual differences exist for these modalities or that they are not sensitive to top-down influences; still, the sensory component has comparatively more impact on the perceived sensation.

For above-threshold stimulation, self-reports do have at least one more clear advantage: they enable us to measure the stimulus–response relationship instead of simple sensory thresholds. The dose–response curve showed marked individual differences in interoception studies (Ness et al. 1990, 1998; Ádám et al. 1999); its shape is way more informative about the possible impact of the stimuli than the simple identification of the sensory threshold. The assessment of the accuracy of cardiac perception in the Schandry task represents an interesting boundary condition. Although it is generally considered an objective indicator of performance, in fact it relies on conscious processes (i.e., perceived and counted heartbeats, knowledge of heart rate, see Chap. 4), thus it should not put in the same category as forced-choice discrimination tasks. This is clearly indicated by the typically weak associations between accuracy scores obtained from the two paradigms; however, IS_{acute} scores reported by seven independent studies showed a strong pooled effect size ($r = 0.6$) (Hickman et al. 2020).

Concerning the assessment of perceived changes, a number of methods are proposed and used in the literature. There are measures that rely a bit stronger on the sensory modality at hand (i.e., the primary experience) than its verbalized representation. For example, the use of the visual analog scale and other types of graphic rating scales, such as numerical rating scales representing the entire continuum of the sensation (e.g., intensity, arousal, valence, etc.) (Hayes and Patterson 1921; Freyd 1923; Aitken 1969), as well as the so-called self-assessment manikin in the measurement of the components of the affective reaction (Bradley and Lang 1994), belong to the former category. Likert-type self-rating scales with verbal labels are closer to verbal coding, and their resolution is usually lower (5–7 options) (Likert 1932). Generally, the VAS shows excellent psychometric characteristics in interoceptive tasks, easy to administer, and requires less mental effort, thus its use is preferred over the Likert-type scales in research and clinical settings (Price et al. 1983; Wewers and Lowe 1990; Grant et al. 1999; Flint et al. 2000; Hawker et al. 2011; Adamchic et al. 2012; Klimek et al. 2017). Combined scales (e.g., a VAS with numbers and anchor words) are also recommended in interoception research (Gracely 1999; Drewes et al. 2003).

The aforementioned doubts concerning the validity of the assessment of various aspects of IS are apparently supported by the fact that they show no association with each other (Garfinkel et al. 2015; Forkmann et al. 2016). This is surprising. We perceive our body as a whole, thus some sort of integration among individual modalities at the perceptual level would be expected. This is not only a theoretical assumption, as moderate to strong associations between cardiac, respiratory, and tactile IS_{acute} were reported (Garfinkel et al. 2016). In a similar vein, association between IS_{acute} for the heartbeat tracking and the forced-choice paradigm was found (Garfinkel et al. 2015; Forkmann et al. 2016). One explanation for the lack of the association between IS_{trait} and IS_{acute} is that the perception of cardiac activity is special, another option is that the BPQ measures something else as generalized interoceptive sensibility (see above). A further possible interpretation is based on the fact that there are (at least) two conceptual differences between IS_{acute} and IS_{trait}. One concerns the sensory modality (one single modality vs. multiple modalities), another is about the concreteness of the construct (actual vs. general ability); these two levels together may considerably enlarge the gap between the two concepts. I think that the use of more valid instruments for the assessment of IS_{acute} and IS_{trait}, as well as the inclusion of a further construct that is conceptually located between the original two could shed more light on the background of the lack of association. This latter construct, which could be called $IS_{modality\text{-}specific}$, would refer to the perceived ability to detect interoceptive changes with respect to only one modality. For example, self-reported general sensitivity to pain represents such a construct. Unfortunately, it is assessed with one single item (Edwards and Fillingim 2007), which is a questionable practice from a psychometric point of view.

5.14 The Evaluative Aspects

As shown, affective evaluation represents an important aspect of body sensations. It indicates the actual salience of the sensations, i.e., the necessity of specific behavioral steps to maintain homeostasis. Taken into consideration that we primarily focus on exteroceptive stimuli and the number of stimuli usually exceeds our conscious processing capacity (Ádám 1967; Pennebaker and Lightner 1980), affect accompanied by the sensations appears a perfect solution to catch attention. For certain interoceptive modalities, a distinction between the sensory (intensity) and evaluative aspect of the stimulus is a difficult task—under everyday conditions more intense pain or more intense distension of the urinary bladder are practically equal with more discomfort and displeasure. It is important to see that novelty, a characteristic that is seemingly processed in a purely cognitive way, also triggers an affective response. Surprise and/or exploration are among the basic emotions in many models (Panksepp 1998); moreover, the response to novelty is realized within the same neural workspace as valence and arousal, i.e., the basic building stones of emotions (Weierich et al. 2010). Thus, a novel (unusual) body sensation can also catch attention.

There are constructs that refer to the primary evaluative aspect of body sensations. For example, somatosensory amplification, the proneness to experience body sensation as intense, noxious, and disturbing (Barsky et al. 1988, 1990), represents such a concept. A number of items of the Somatosensory Amplification Scale refer to the degree of affective evaluation (perceived unpleasantness) of certain stimuli. Also, the PHQ-15 (Table 5.1) rates symptoms with respect to their bothersomeness, which is basically a synonym for displeasure. These constructs are characterized by huge individual differences (Kroenke et al. 2002, 2010; Köteles and Witthöft 2017), which may partly be based on negative affect (anxiety) or the neuroticism/emotional stability dimension of personality (Witthöft et al. 2016; Köteles and Doering 2016; Ferentzi et al. 2017; Stauder et al. 2021). In other words, they partly reflect the general evaluative tendency or threat-related attentional bias (Bar-Haim et al. 2007) of the individual. Beyond this, however, body focus or body awareness is also involved (Köteles and Doering 2016), and it can be argued that interoceptive evaluation represents a distinct construct with a well-understood evolutionary background. As in the case of many dimensional characteristics, the extremities (i.e., the complete lack of evaluation and very intense evaluation) might represent maladaptive and sometimes pathological conditions which seriously threat psychological functioning (Chap. 7).

Acute affective reactions to stimuli are often measured in pain research, as well as for other modalities, e.g., distension of the bladder (urgency to void), the stomach, or for respiratory resistance or breathlessness (Stegen et al. 2000; Van den Bergh et al. 2004; Petersen et al. 2014; Hayen et al. 2015). The sensory and the evaluative aspects (intensity and unpleasantness) of the stimulus often are associated with each other; however, the strength of the association varies. The correlation between self-rated intensity and unpleasantness was found to be strong ($r = 0.80$) for the bitter taste, and moderate to strong for gastric distention ($r = 0.48$) and pain ($r = 0.51$) in our study (Ferentzi et al. 2018a). IAc, as assessed by the mental tracking task by Schandry, was independent from evaluation in a number of studies on eating disorder (Herbert and Pollatos 2018). These findings suggest that the sensory-discriminative and affective-evaluative aspects of body sensations represent two distinct dimensions that should be measured separately. The importance of the affective dimension of interoception was realized by a number of authors. The term *homeostatic emotion* (Craig 2003, 2015), *interoceptive salience* (Craig 2009), as well the concept of *interoceptive appraisal* recently proposed by Farb and Logie

(2018) clearly indicate this insight. However, interoceptive appraisal, i.e., the process of making sense, refers to both primary (affective and automatic) and secondary (dominantly cognitive, often based on logical thinking and available knowledge) forms of evaluation. Herbert and Pollatos (2018) proposed the term *interoceptive emotional evaluation* for the primary affective appraisal related to body sensations.

So far, we mainly focused on negative affect as it appears to play the dominant role in behavioral homeostatic regulation. Positive body feelings received considerably less research attention, and a part of this research focuses on secondary evaluation (Farb et al. 2015; Farb and Logie 2018). Good examples for these cognitive aspects are the Body Responsiveness Questionnaire (Daubenmier 2005; Tihanyi et al. 2017a) or the Self-Regulation and Body Listening scales of the MAIA.

A rarely raised question concerns the origin of the associations between body sensations and affect. For a number of sensations, a biologically determined association was assumed by many authors; I agree with them. Of course, additional associations can be developed by learning, but I doubt that learning can completely reverse these built-in associations. I think that the learning process primarily characterizes inherently neutral body sensations. For example, heartbeat appears to be neutral (Nummenmaa et al. 2018); thus it can be associated with positive states, such as moderate intensity physical activity, but also with negative emotions as in the case of congenital heart disease or panic attack (Karsdorp et al. 2007a, b; Khalsa and Feinstein 2018).

In the case of many body sensations, the intimate association between the sensory and the affective quality of the feeling is obvious, often overwhelming. In chronic cases, such as chronic pain, however, these two characteristics can be separated; the technique is called sensory monitoring (Leventhal et al. 1979; Ahles et al. 1983; Suls and Fletcher 1985). This separation, more precisely, focusing on the sensory aspect only, requires continuous cognitive effort, and it is never complete (Cioffi 1991). My personal opinion is that the vast majority of sensory-based body sensations are primarily associated with evaluation; the non-judgmental observation of these feelings, characterizing mindfulness, requires conscious effort. The practical consequences of this phenomenon will be discussed in Chap. 8.

5.15 Summary

In this chapter, the self-reported aspect of interoception, dubbed interoceptive sensibility, was reviewed. Of the tenets of the book, realness, vagueness, evaluation, and relevance were discussed in more detail. Although the self-reported aspect of interoception might be the most relevant dimension of the three from the viewpoint of everyday functioning, there are conceptual and measurement-related issues that substantially limit our knowledge. A number of difficulties originate in the fact that the relationship between interoceptive sensibility and attention, a poorly defined multidimensional construct, is mostly unexplored. For example, attentiveness to the body plays a marked role in the perceived proneness to be aware of body-related events and changes. However, as attentiveness is also measured via self-report, the influence of general factors biasing self-perception cannot be ruled out. On top of this, attention alone can evoke body sensations, such as tingling and warmth, thus it cannot be regarded as a factor that simply enhances perception of actual body processes. Also, the current conceptualization and assessment of interoceptive sensibility are far from perfect; for example, the latter is characterized by serious validity issues which makes the interpretation of the findings very difficult. Furthermore, the affective aspects of perceived changes are rarely measured and taken into consideration although maladaptive functioning is typically associated with altered evaluation of body sensations. A more serious issue at this level is that non-homeostatic modalities (Chap. 3), such as proprioception and the visual modality, are mostly neglected. It was mentioned that somatosensory and visual signals are integrated with those carried via the homeostatic pathway at multiple levels. Empirical research into the rubber hand illusion demonstrates integration at nonconscious levels (Chap. 3), whereas the involvement of tactile and proprioceptive information in visceral perception (Chap. 4) refers to (slightly) higher-level processes. At the conscious level, we perceive a unitary and integrated body; in other words, the distinction between homeostatic and somatosensory information is often very difficult. Thus, what we feel is much closer to body awareness than to interoceptive sensibility (i.e., sensations exclusively arising from the homeostatic pathway). This should be reflected by the content of self-report instruments. These aspects are poorly referred to in the available questionnaires, and even the status of the related items is somewhat questionable. For example, a clear-cut distinction between interoceptive and proprioceptive awareness is made by the authors of the MAIA, a questionnaire developed to assess aspects of the former category (Mehling et al. 2012, p. 2); however, several items (e.g., "When I am tense I notice where the tension is located in my body"; "I can pay attention to my breath without being distracted by things happening around me"; "When I am in conversation with someone, I can pay attention to my posture") partly or completely refer to somatosensory information. Items of the BAQ do not explicitly refer to somatosensory/proprioceptive input at all, whereas another questionnaire, the Somatic Absorption Scale by David Watson (Köteles et al. 2012), does contain such items (e.g., "I often notice how clothes feel against my skin" or "I am very aware of my body position when I go to sleep)." At the same time, awareness of somatosensory information, such as posture, is measured with a different questionnaire, the Postural Awareness Scale (Cramer et al. 2018). Such a plethora of constructs and measurement instruments appears to be disturbing rather than helpful. Even visual information should be taken into consideration as they contribute to body awareness (Mehling et al. 2009; Berntson et al. 2018).

Take-Home Messages

For laypeople: Paying attention to the body might enhance internal cues, but it can also evoke further percepts.

For clinicians: Body-focused attention and self-reported awareness of body signals are not necessarily maladaptive characteristics. They can lead to health anxiety and symptom reporting in certain cases; however, it can also be associated with positive affect and early steps to prevent certain diseases or slow down their progression.

For researchers: The measurement of acute and chronic (trait-like) aspects of self-reported sensibility to interoceptive cues is far from perfect. Whereas trait-like measurements are contaminated by negative affect, the evaluative aspect of acute sensations is rarely assessed. The lack of a unified and validated methodology (i.e., psychometrically sound self-report instruments) seriously limits the comparability and generalizability of the available findings. Also, more theoretical and empirical work exploring the associations between attention and interoceptive sensibility is needed.

References

Abbott BG, Afshar M, Berger AK, Wackers FJT (2003) Prognostic significance of ischemic electrocardiographic changes during adenosine infusion in patients with normal myocardial perfusion imaging. J Nucl Cardiol 10:9–16. https://doi.org/10.1067/mnc.2002.127625

Acklin MW, McDowell CJ, Orndoff S (1992) Statistical power and the Rorschach: 1975–1991. J Pers Assess 59:366–379. https://doi.org/10.1207/s15327752jpa5902_12

Ádám G (1967) Interoception and behavior. Akadémiai Kiadó, Budapest

Ádám G, Fent J, Kajtár M (1999) An attempt for the application of the basic law of psychophysics (Fechner-Stevenes law) in the domain of visceral stimulus in humans. Acta Biol Hung 50:19–25

Adamchic I, Langguth B, Hauptmann C, Tass PA (2012) Psychometric evaluation of visual analog scale for the assessment of chronic tinnitus. Am J Audiol 21:215–225. https://doi.org/10.1044/1059-0889(2012/12-0010)

Afzal M, Potokar JP, Probert CSJ, Munafò MR (2006) Selective processing of gastrointestinal symptom-related stimuli in irritable bowel syndrome. Psychosom Med 68:758–761. https://doi.org/10.1097/01.psy.0000232270.78071.28

Ahles TA, Blanchard EB, Leventhal H (1983) Cognitive control of pain: attention to the sensory aspects of the cold pressor stimulus. Cogn Ther Res 7:159–177. https://doi.org/10.1007/BF01190070

Ainley V, Tsakiris M (2013) Body conscious? Interoceptive awareness, measured by heartbeat perception, is negatively correlated with self-objectification. PLoS One 8:e55568. https://doi.org/10.1371/journal.pone.0055568

Ainley V, Tajadura-Jiménez A, Fotopoulou A, Tsakiris M (2012) Looking into myself: the effect of self-focused attention on interoceptive sensitivity. Psychophysiology 49:1504–1508. https://doi.org/10.1111/j.1469-8986.2012.01468.x

Ainley V, Maister L, Brokfeld J et al (2013) More of myself: manipulating interoceptive awareness by heightened attention to bodily and narrative aspects of the self. Conscious Cogn 22:1231–1238. https://doi.org/10.1016/j.concog.2013.08.004

Ainley V, Apps MAJ, Fotopoulou A, Tsakiris M (2016) "Bodily precision": a predictive coding account of individual differences in interoceptive accuracy. Philos Trans R Soc Lond B Biol Sci 371:3. https://doi.org/10.1098/rstb.2016.0003

Aitken RC (1969) Measurement of feelings using visual analogue scales. Proc R Soc Med 62:989–993

Al Luwimi I, Ammar A, Al Awami M (2012) Pathophysiology of paresthesia. Intech, Rijeka, Croatia

American Psychiatric Association (2013) Diagnostic and statistical manual of mental disorders, 5th Edition: DSM-5, 5th edn. American Psychiatric Publishing, Washington, DC
APA (1994) Diagnostic and statistical manual of mental disorders, 4th edn. American Psychiatric Association, Arlington, VA
Avery JA, Kerr KL, Ingeholm JE et al (2015) A common gustatory and interoceptive representation in the human mid-insula. Hum Brain Mapp 36:2996–3006. https://doi.org/10.1002/hbm.22823
Baer RA, Smith GT, Hopkins J et al (2006) Using self-report assessment methods to explore facets of mindfulness. Assessment 13:27–45. https://doi.org/10.1177/1073191105283504
Bakal D (1999) Minding the body: clinical uses of somatic awareness. Guilford Press, New York
Bakal D, Coll P, Schaefer J (2008) Somatic awareness in the clinical care of patients with body distress symptoms. BioPsychoSoc Med 2:6. https://doi.org/10.1186/1751-0759-2-6
Bar-Haim Y, Lamy D, Pergamin L et al (2007) Threat-related attentional bias in anxious and nonanxious individuals: a meta-analytic study. Psychol Bull 133:1–24. https://doi.org/10.1037/0033-2909.133.1.1
Barrett LF, Simmons WK (2015) Interoceptive predictions in the brain. Nat Rev Neurosci 16:419–429. https://doi.org/10.1038/nrn3950
Barsky AJ (1979) Patients who amplify bodily sensations. Ann Intern Med 91:63–70. https://doi.org/10.1059/0003-4819-91-1-63
Barsky AJ, Borus JF (1999) Functional somatic syndromes. Ann Intern Med 130:910–921
Barsky AJ, Goodson JD, Lane RS, Cleary PD (1988) The amplification of somatic symptoms. Psychosom Med 50:510–519
Barsky AJ, Wyshak G, Klerman GL (1990) The Somatosensory Amplification Scale and its relationship to hypochondriasis. J Psychiatr Res 24:323–334. https://doi.org/10.1016/0022-3956(90)90004-A
Bauer CCC, Díaz J-L, Concha L, Barrios FA (2014) Sustained attention to spontaneous thumb sensations activates brain somatosensory and other proprioceptive areas. Brain Cogn 87:86–96. https://doi.org/10.1016/j.bandc.2014.03.009
Beck JE, Lipani TA, Baber KF et al (2011) Attentional bias to pain and social threat in pediatric patients with functional abdominal pain and pain-free youth before and after performance evaluation. Pain 152:1061–1067. https://doi.org/10.1016/j.pain.2011.01.029
Berger A, Henik A, Rafal R (2005) Competition between endogenous and exogenous orienting of visual attention. J Exp Psychol Gen 134:207–221. https://doi.org/10.1037/0096-3445.134.2.207
Berntson GG, Gianaros PJ, Tsakiris M (2018) Interoception and the autonomic nervous system: bottom-up meets top-down. In: Tsakiris M, De Preester H (eds) The interoceptive mind. From homeostasis to awareness. Oxford University Press, Oxford, England, pp 3–23
Blumenthal-Dramé A, Hanulíková A, Kortmann B (2017) Editorial: Perceptual linguistic salience: modeling causes and consequences. Front Psychol 8:411. https://doi.org/10.3389/fpsyg.2017.00411
Bornemann B, Herbert BM, Mehling WE, Singer T (2015) Differential changes in self-reported aspects of interoceptive awareness through 3 months of contemplative training. Front Psychol 5:1504. https://doi.org/10.3389/fpsyg.2014.01504
Boyer MC, Compas BE, Stanger C et al (2006) Attentional biases to pain and social threat in children with recurrent abdominal pain. J Pediatr Psychol 31:209–220. https://doi.org/10.1093/jpepsy/jsj015
Bradley MM, Lang PJ (1994) Measuring emotion: the self-assessment manikin and the semantic differential. J Behav Ther Exp Psychiatry 25:49–59. https://doi.org/10.1016/0005-7916(94)90063-9
Brooks J, Tracey I (2005) From nociception to pain perception: imaging the spinal and supraspinal pathways. J Anat 207:19–33. https://doi.org/10.1111/j.1469-7580.2005.00428.x
Brown RJ (2004) Psychological mechanisms of medically unexplained symptoms: an integrative conceptual model. Psychol Bull 130:793–812. https://doi.org/10.1037/0033-2909.130.5.793
Brown RJ, Brunt N, Poliakoff E, Lloyd DM (2010) Illusory touch and tactile perception in somatoform dissociators. J Psychosom Res 69:241–248. https://doi.org/10.1016/j.jpsychores.2009.11.010

Brown TA, Berner LA, Jones MD et al (2017) Psychometric evaluation and norms for the Multidimensional Assessment of Interoceptive Awareness (MAIA) in a clinical eating disorders sample. Eur Eat Disord Rev 25:411–416. https://doi.org/10.1002/erv.2532

Buss AH (1980) Self-consciousness and social anxiety. W.H. Freeman & Co Ltd, San Francisco, CA

Buss DM, Scheier MF (1976) Self-consciousness, self-awareness, and self-attribution. J Res Pers 10:463–468. https://doi.org/10.1016/0092-6566(76)90060-X

Cabrera A, Kolacz J, Pailhez G et al (2018) Assessing body awareness and autonomic reactivity: factor structure and psychometric properties of the Body Perception Questionnaire-Short Form (BPQ-SF). Int J Methods Psychiatr Res 27:e1596. https://doi.org/10.1002/mpr.1596

Carver CS, Scheier MF (1978) Self-focusing effects of dispositional self-consciousness, mirror presence, and audience presence. J Pers Soc Psychol 36:324–332. https://doi.org/10.1037/0022-3514.36.3.324

Chapman S, Martin M (2011) Attention to pain words in irritable bowel syndrome: increased orienting and speeded engagement. Br J Health Psychol 16:47–60. https://doi.org/10.1348/135910710X505887

Chuckrow R (2015) The tai chi book. YMAA Publication Center, Wolfeboro, NH

Cioffi D (1991) Beyond attentional strategies: cognitive-perceptual model of somatic interpretation. Psychol Bull 109:25–41

Cooley CH (1902) Human nature and the social order. Charles Scribner's Sons, New York

Corbetta M, Shulman GL (2002) Control of goal-directed and stimulus-driven attention in the brain. Nat Rev Neurosci 3:201–215. https://doi.org/10.1038/nrn755

Craig AD (2003) A new view of pain as a homeostatic emotion. Trends Neurosci 26:303–307

Craig AD (2009) Emotional moments across time: a possible neural basis for time perception in the anterior insula. Philos Trans R Soc Lond B Biol Sci 364:1933–1942. https://doi.org/10.1098/rstb.2009.0008

Craig AD (2015) How do you feel? An interoceptive moment with your neurobiological self. Princeton University Press, Princeton, NJ

Cramer H, Mehling WE, Saha FJ et al (2018) Postural awareness and its relation to pain: validation of an innovative instrument measuring awareness of body posture in patients with chronic pain. BMC Musculoskelet Disord 19:109. https://doi.org/10.1186/s12891-018-2031-9

Critchley HD, Wiens S, Rotshtein P et al (2004) Neural systems supporting interoceptive awareness. Nat Neurosci 7:189–195. https://doi.org/10.1038/nn1176

Damasio A (2010) Self comes to mind: constructing the conscious brain. Pantheon, New York

Daubenmier JJ (2005) The relationship of yoga, body awareness, and body responsiveness to self-objectification and disordered eating. Psychol Women Q 29:207–219. https://doi.org/10.1111/j.1471-6402.2005.00183.x

Davis KD, Taub E, Duffner F et al (2000) Activation of the anterior cingulate cortex by thalamic stimulation in patients with chronic pain: a positron emission tomography study. J Neurosurg 92:64–69

Davitz JR (1969) The language of emotion. Academic, New York

Deary IJ (1999) A taxonomy of medically unexplained symptoms. J Psychosom Res 47:51–59

Dehaene S, Changeux J-P (2005) Ongoing spontaneous activity controls access to consciousness: a neuronal model for inattentional blindness. PLoS Biol 3:e141. https://doi.org/10.1371/journal.pbio.0030141

Dehghani M, Mohammadi S, Sharpe L, Khatibi A (2018) Attentional bias to threat-related information among individuals with dental complaints: the role of pain expectancy. Front Psychol 9:786. https://doi.org/10.3389/fpsyg.2018.00786

Derogatis LR (1977) Administration, scoring, and procedures manual for the SCL-90-R. Clinical Psycometrics Research, Baltimore

Derogatis LR, Lipman RS, Rickels K et al (1974) The Hopkins Symptom Checklist (HSCL): a self-report symptom inventory. Behav Sci 19:1–15

Dixon NF (1981) Preconscious processing. Wiley, Chichester, UK

Drewes AM, Gregersen H, Arendt-Nielsen L (2003) Experimental pain in gastroenterology: a reappraisal of human studies. Scand J Gastroenterol 38:1115–1130. https://doi.org/10.1080/00365520310004399

Duval S, Wicklund RA (1972) A theory of objective self-awareness. Academic, New York

Edwards RR, Fillingim RB (2007) Self-reported pain sensitivity. Eur J Pain 11:594–598. https://doi.org/10.1016/j.ejpain.2006.09.008

Ehlers A, Breuer P, Dohn D, Fiegenbaum W (1995) Heartbeat perception and panic disorder: possible explanations for discrepant findings. Behav Res Ther 33:69–76

Emanuelsen L, Drew R, Köteles F (2015) Interoceptive sensitivity, body image dissatisfaction, and body awareness in healthy individuals. Scand J Psychol 56:167–174. https://doi.org/10.1111/sjop.12183

Ernst J, Northoff G, Böker H et al (2013) Interoceptive awareness enhances neural activity during empathy. Hum Brain Mapp 34:1615–1624. https://doi.org/10.1002/hbm.22014

Estebe J-P, Le Naoures A, Chemaly L, Ecoffey C (2000) Tourniquet pain in a volunteer study: effect of changes in cuff width and pressure. Anaesthesia 55:21–26. https://doi.org/10.1046/j.1365-2044.2000.01128.x

Failla MD, Bryant LK, Heflin BH et al (2020) Neural correlates of cardiac interoceptive focus across development: implications for social symptoms in autism spectrum disorder. Autism Res 13:908–920. https://doi.org/10.1002/aur.2289

Farb NA, Logie K (2018) Interoceptive appraisal and mental health. In: Tsakiris M, De Preester H (eds) The interoceptive mind. From homeostasis to awareness. Oxford University Press, Oxford, England, pp 227–241

Farb NAS, Segal ZV, Anderson AK (2012) Mindfulness meditation training alters cortical representations of interoceptive attention. Soc Cogn Affect Neurosci 8(1):15–26. https://doi.org/10.1093/scan/nss066

Farb NAS, Segal ZV, Anderson AK (2013) Attentional modulation of primary interoceptive and exteroceptive cortices. Cereb Cortex 23:114–126. https://doi.org/10.1093/cercor/bhr385

Farb NA, Daubenmier JJ, Price CJ et al (2015) Interoception, contemplative practice, and health. Front Psychol 6:1–26. https://doi.org/10.3389/fpsyg.2015.00763

Fardo F, Auksztulewicz R, Allen M et al (2017) Expectation violation and attention to pain jointly modulate neural gain in somatosensory cortex. NeuroImage 153:109–121. https://doi.org/10.1016/j.neuroimage.2017.03.041

Fecteau JH, Munoz DP (2006) Salience, relevance, and firing: a priority map for target selection. Trends Cogn Sci 10:382–390. https://doi.org/10.1016/j.tics.2006.06.011

Feldman H, Friston KJ (2010) Attention, uncertainty, and free-energy. Front Hum Neurosci 4:215. https://doi.org/10.3389/fnhum.2010.00215

Fenigstein A, Scheier MF, Buss AH (1975) Public and private self-consciousness: assessment and theory. J Consult Clin Psychol 43:522–527. https://doi.org/10.1037/h0076760

Ferentzi E, Köteles F, Csala B et al (2017) What makes sense in our body? Personality and sensory correlates of body awareness and somatosensory amplification. Pers Individ Differ 104:75–81. https://doi.org/10.1016/j.paid.2016.07.034

Ferentzi E, Bogdány T, Szabolcs Z et al (2018a) Multichannel investigation of interoception: sensitivity is not a generalizable feature. Front Hum Neurosci 12:223. https://doi.org/10.3389/fnhum.2018.00223

Ferentzi E, Drew R, Tihanyi BT, Köteles F (2018b) Interoceptive accuracy and body awareness – temporal and longitudinal associations in a non-clinical sample. Physiol Behav 184:100–107. https://doi.org/10.1016/j.physbeh.2017.11.015

Ferentzi E, Horváth Á, Köteles F (2019) Do body-related sensations make feel us better? Subjective well-being is associated only with the subjective aspect of interoception. Psychophysiology 56:e13319. https://doi.org/10.1111/psyp.13319

Ferentzi E, Olaru G, Geiger M, et al (2020) Examining the factor structure and validity of the multidimensional assessment of interoceptive awareness. J Pers Assess 1–10. https://doi.org/10.1080/00223891.2020.1813147

Fisher S (1970) Body experience in fantasy and behavior. Ardent Media, New York
Fisher S (1974) Body consciousness. J. Aronson, New York
Fisher S (1986a) Development and structure of the body image, vol 1. L. Erlbaum Associates, New York
Fisher S (1986b) Development and structure of the body image, vol 2. L. Erlbaum Associates, New York
Fisher S, Cleveland SE (1958) Body image and personality, 1st edn. Van Nostrand, Princeton, NJ
Fisher S, Cleveland SE (1968) Body image and personality, 2nd rev edn. Dover Publications, New York
Fisher S, Fisher RL (1964) Body image boundaries and patterns of body perception. J Abnorm Soc Psychol 68:255–262. https://doi.org/10.1037/h0044857
Fisher S, Renik OD (1966) Induction of body image boundary changes. J Proj Tech Pers Assess 30:429–434
Flasinski T, Dierolf AM, Rost S et al (2020) Altered interoceptive awareness in high habitual symptom reporters and patients with somatoform disorders. Front Psychol 11:1859. https://doi.org/10.3389/fpsyg.2020.01859
Flint A, Raben A, Blundell JE, Astrup A (2000) Reproducibility, power and validity of visual analogue scales in assessment of appetite sensations in single test meal studies. Int J Obes 24:38–48. https://doi.org/10.1038/sj.ijo.0801083
Forkmann T, Scherer A, Meessen J et al (2016) Making sense of what you sense: disentangling interoceptive awareness, sensibility and accuracy. Int J Psychophysiol 109:71–80. https://doi.org/10.1016/j.ijpsycho.2016.09.019
Forkmann T, Volz-Sidiropoulou E, Helbing T et al (2019) Sense it and use it: interoceptive accuracy and sensibility in suicide ideators. BMC Psychiatry 19:334. https://doi.org/10.1186/s12888-019-2322-1
Fox E, Russo R, Bowles R, Dutton K (2001) Do threatening stimuli draw or hold visual attention in subclinical anxiety? J Exp Psychol Gen 130:681–700
Freud S (1938) The basic writings of Sigmund Freud. Random House, New York
Freyd M (1923) The graphic rating scale. J Educ Psychol 14:83–102. https://doi.org/10.1037/h0074329
Friston K (2009) The free-energy principle: a rough guide to the brain? Trends Cogn Sci 13:293–301. https://doi.org/10.1016/j.tics.2009.04.005
Fullagar CJ, Kelloway EK (2009) Flow at work: an experience sampling approach. J Occup Organ Psychol 82:595–615. https://doi.org/10.1348/096317908X357903
Gacono CB, Evans FB III, Viglione DJ (2002) The Rorschach in forensic practice. J Forensic Psychol Pract 2:33–53. https://doi.org/10.1300/J158v02n03_03
García-Cordero I, Esteves S, Mikulan EP et al (2017) Attention, in and out: scalp-level and intracranial EEG correlates of interoception and exteroception. Front Neurosci 11:411. https://doi.org/10.3389/fnins.2017.00411
Garfinkel SN, Critchley HD (2013) Interoception, emotion and brain: new insights link internal physiology to social behaviour. Commentary on: "Anterior insular cortex mediates bodily sensibility and social anxiety" by Terasawa et al. (2012). Soc Cogn Affect Neurosci 8:231–234. https://doi.org/10.1093/scan/nss140
Garfinkel SN, Seth AK, Barrett AB et al (2015) Knowing your own heart: distinguishing interoceptive accuracy from interoceptive awareness. Biol Psychol 104:65–74. https://doi.org/10.1016/j.biopsycho.2014.11.004
Garfinkel SN, Manassei MF, Hamilton-Fletcher G et al (2016) Interoceptive dimensions across cardiac and respiratory axes. Philos Trans R Soc Lond B Biol Sci 371:20160014. https://doi.org/10.1098/rstb.2016.0014
Geers AL, Helfer S, Kosbab K et al (2005) Reconsidering the role of personality in placebo effects: dispositional optimism, situational expectations, and the placebo response. J Psychosom Res 58:121–127. https://doi.org/10.1016/j.jpsychores.2004.08.011

Gibbons FX (1983) Self-attention and self-report: the "veridicality" hypothesis. J Pers 51:517–542. https://doi.org/10.1111/j.1467-6494.1983.tb00343.x

Gibbons FX (1990) Self-attention and behavior: a review and theoretical update. In: Zanna MP (ed) Advances in experimental social psychology. Academic, New York, pp 249–303

Ginzburg K, Tsur N, Barak-Nahum A, Defrin R (2014) Body awareness: differentiating between sensitivity to and monitoring of bodily signals. J Behav Med 37:564–575. https://doi.org/10.1007/s10865-013-9514-9

Gowan J, Roller L (2015) Disease state management: communicating the harmful effects of medication. Aust J Pharm 96:64

Gracely RH (1999) Pain measurement. Acta Anaesthesiol Scand 43:897–908. https://doi.org/10.1034/j.1399-6576.1999.430907.x

Graham EK, Weston SJ, Turiano NA et al (2020) Is healthy neuroticism associated with health behaviors? A coordinated integrative data analysis. Collabra Psychol 6:32. https://doi.org/10.1525/collabra.266

Grant S, Aitchison T, Henderson E et al (1999) A comparison of the reproducibility and the sensitivity to change of visual analogue scales, Borg scales, and Likert scales in normal subjects during submaximal exercise. Chest 116:1208–1217. https://doi.org/10.1378/chest.116.5.1208

Gregory LJ, Yágüez L, Williams SCR et al (2003) Cognitive modulation of the cerebral processing of human oesophageal sensation using functional magnetic resonance imaging. Gut 52:1671–1677

Greyson B (1993a) The physio-kundalini syndrome and mental illness. J Transpers Psychol 25:43

Greyson B (1993b) Near-death experiences and the physio-kundalini syndrome. J Relig Health 32:277–290. https://doi.org/10.1007/BF00990954

Gunderson EKE, Johnson LC (1965) Past experience, self-evaluation, and present adjustment. J Soc Psychol 66:311–321. https://doi.org/10.1080/00224545.1965.9919650

Haenen M-A, Schmidt AJ, Kroeze S, van den Hout MA (1996) Hypochondriasis and symptom reporting—the effect of attention versus distraction. Psychother Psychosom 65:43–48

Hansell S, Mechanic D (1991) Body awareness and self-assessed health among older adults. J Aging Health 3:473–492. https://doi.org/10.1177/089826439100300403

Hansell S, Sherman G, Mechanic D (1991) Body awareness and medical care utilization among older adults in an HMO. J Gerontol 46:S151–S159

Harshaw C (2008) Alimentary epigenetics: a developmental psychobiological systems view of the perception of hunger, thirst and satiety. Dev Rev 28:541–569. https://doi.org/10.1016/j.dr.2008.08.001

Hartl LA (1992) Die Panikstörung. Eine Untersuchung zur Rolle der Wahrnehmung von Körperprozessen. Peter Lang, Frankfurt

Hartman TO, Blankenstein N, Molenaar B et al (2013) NHG guideline on medically unexplained symptoms (MUS). Huisarts Wetenscap 56:222–230

Hathaway SR, McKinley JC (1942) The Minnesota Multiphasic Personality Schedule. University of Minnesota Press, Minneapolis, MN

Hawker GA, Mian S, Kendzerska T, French M (2011) Measures of adult pain: Visual Analog Scale for Pain (VAS Pain), Numeric Rating Scale for Pain (NRS Pain), McGill Pain Questionnaire (MPQ), Short-Form McGill Pain Questionnaire (SF-MPQ), Chronic Pain Grade Scale (CPGS), Short Form-36 Bodily Pain Scale (SF-36 BPS), and Measure of Intermittent and Constant Osteoarthritis Pain (ICOAP). Arthritis Care Res 63:S240–S252. https://doi.org/10.1002/acr.20543

Hayen A, Herigstad M, Wiech K, Pattinson KTS (2015) Subjective evaluation of experimental dyspnoea—effects of isocapnia and repeated exposure. Respir Physiol Neurobiol 208:21–28. https://doi.org/10.1016/j.resp.2014.12.019

Hayes SP (1906) A study of the affective qualities. I. The tridimensional theory of feeling. Am J Psychol 17:358–393. https://doi.org/10.2307/1412251

Hayes MHS, Patterson DG (1921) Experimental development of the graphic rating method. Psychol Bull 18:98–99

Herbert BM, Pollatos O (2018) The relevance of interoception for eating behavior and eating disorders. In: Tsakiris M, De Preester H (eds) The interoceptive mind. From homeostasis to awareness. Oxford University Press, Oxford, England, pp 165–186
Hickman L, Seyedsalehi A, Cook JL et al (2020) The relationship between heartbeat counting and heartbeat discrimination: a meta-analysis. Biol Psychol 156:107949
Hiller W, Rief W, Brähler E (2006) Somatization in the population: from mild bodily misperceptions to disabling symptoms. Soc Psychiatry Psychiatr Epidemiol 41:704–712. https://doi.org/10.1007/s00127-006-0082-y
Hohwy J (2012) Attention and conscious perception in the hypothesis testing brain. Front Psychol 3:96. https://doi.org/10.3389/fpsyg.2012.00096
Hohwy J (2014) The predictive mind, 1st edn. Oxford University Press, Oxford, England
Hollands EH (1906) Wundt's doctrine of psychical analysis and the psychical elements, and some recent criticism: feeling and feeling-analysis. Am J Psychol 17:206–226. https://doi.org/10.2307/1412392
Hospod V, Aimonetti J-M, Roll J-P, Ribot-Ciscar E (2007) Changes in human muscle spindle sensitivity during a proprioceptive attention task. J Neurosci 27:5172–5178. https://doi.org/10.1523/JNEUROSCI.0572-07.2007
Hou R, Moss-Morris R, Bradley BP et al (2008) Attentional bias towards health-threat information in chronic fatigue syndrome. J Psychosom Res 65:47–50. https://doi.org/10.1016/j.jpsychores.2008.03.008
Hsiao SS, Vega-Bermudez F (2002) Attention in the somatosensory system. In: Nelson RJ (ed) The somatosensory system. Deciphering the brain's own body image. CRC Press, Boca Raton, FL, pp 197–218
Hughes TL, Gacono CB, Owen PF (2007) Current status of Rorschach assessment: implications for the school psychologist. Psychol Sch 44:281–291. https://doi.org/10.1002/pits.20223
Hui KK, Nixon EE, Vangel MG et al (2007) Characterization of the "deqi" response in acupuncture. BMC Complement Altern Med 7:33. https://doi.org/10.1186/1472-6882-7-33
Hunt WA (1931) The relation of bright and dull pressure to affectivity. Am J Psychol 43:87–92. https://doi.org/10.2307/1414240
Hurwitz TA (2004) Somatization and conversion disorder. Can J Psychiatry 49:172–178. https://doi.org/10.1177/070674370404900304
Impett EA, Daubenmier JJ, Hirschman AL (2006) Minding the body: yoga, embodiment, and well-being. Sex Res Soc Policy 3:39–48. https://doi.org/10.1525/srsp.2006.3.4.39
James W (1890) The principles of psychology. Macmillan, London
Jellesma FC, Faddegon K, van der Veek SMC (2011) Children with somatic complaints: do they show a somatic attention bias? Neth J Psychol 66:78–84
Johnson LC (1956) Body cathexis as a factor in somatic complaints. J Consult Psychol 20:145–149. https://doi.org/10.1037/h0046919
Jourard SM, Secord PF (1954) Body size and body-cathexis. J Consult Psychol 18:184–184. https://doi.org/10.1037/h0063192
Jupp JJ (1989) Fisher and Cleveland barrier and penetration scores: correlations with Rorschach category scores. Percept Mot Skills 69:1011–1018. https://doi.org/10.1177/00315125890693-156
Jupp JJ (2016) Fisher and Cleveland barrier and penetration body-image scores, self-steering style, and academic achievement. Psychol Rep. https://doi.org/10.2466/pr0.1989.64.3c.1075
Kanan C, Tong MH, Zhang L, Cottrell GW (2009) SUN: top-down saliency using natural statistics. Vis Cogn 17:979–1003. https://doi.org/10.1080/13506280902771138
Karsdorp PA, Kindt M, Everaerd W, Mulder BJM (2007a) Preattentive processing of heart cues and the perception of heart symptoms in congenital heart disease. Behav Res Ther 45:1893–1902. https://doi.org/10.1016/j.brat.2007.02.010
Karsdorp PA, Kindt M, Rietveld S et al (2007b) Stress-induced heart symptoms and perceptual biases in patients with congenital heart disease. Int J Cardiol 114:352–357. https://doi.org/10.1016/j.ijcard.2006.02.004

Katkin ES, Cestaro VL, Weitkunat R (1991) Individual differences in cortical evoked potentials as a function of heartbeat detection ability. Int J Neurosci 61:269–276

Katzer A, Oberfeld D, Hiller W, Witthöft M (2011) Tactile perceptual processes and their relationship to medically unexplained symptoms and health anxiety. J Psychosom Res 71:335–341. https://doi.org/10.1016/j.jpsychores.2011.03.009

Katzer A, Oberfeld D, Hiller W et al (2012) Tactile perceptual processes and their relationship to somatoform disorders. J Abnorm Psychol 121:530–543. https://doi.org/10.1037/a0026536

Khalsa SS, Feinstein JS (2018) The somatic error hypothesis of anxiety. In: Tsakiris M, De Preester H (eds) The interoceptive mind. From homeostasis to awareness. Oxford University Press, Oxford, England, pp 144–164

Kihlstrom J (1987) The cognitive unconscious. Science 237:1445–1452. https://doi.org/10.1126/science.3629249

Kihlstrom JF (2004) Availability, accessibility, and subliminal perception. Conscious Cogn 13:92–100. https://doi.org/10.1016/j.concog.2003.09.004

Klimek L, Bergmann K-C, Biedermann T et al (2017) Visual analogue scales (VAS): measuring instruments for the documentation of symptoms and therapy monitoring in cases of allergic rhinitis in everyday health care. Allergo J Int 26:16–24. https://doi.org/10.1007/s40629-016-0006-7

Kong J, Gollub R, Huang T et al (2007) Acupuncture de qi, from qualitative history to quantitative measurement. J Altern Complement Med 13:1059–1070. https://doi.org/10.1089/acm.2007.0524

Körber S, Frieser D, Steinbrecher N, Hiller W (2011) Classification characteristics of the Patient Health Questionnaire-15 for screening somatoform disorders in a primary care setting. J Psychosom Res 71:142–147. https://doi.org/10.1016/j.jpsychores.2011.01.006

Köteles F (2014) A Testi Tudatosság Kérdőív magyar verziójának (BAQ-H) vizsgálata jógázó és fiatal felnőtt kontroll mintán. Mentálhigiéné és Pszichoszomatika 15:373–391. https://doi.org/10.1556/Mental.15.2014.4.4

Köteles F, Doering BK (2016) The many faces of somatosensory amplification: the relative contribution of body awareness, symptom labeling, and anxiety. J Health Psychol 21:2903–2911. https://doi.org/10.1177/1359105315588216

Köteles F, Witthöft M (2017) Somatosensory amplification – an old construct from a new perspective. J Psychosom Res 101:1–9. https://doi.org/10.1016/j.jpsychores.2017.07.011

Köteles F, Simor P, Tolnai N (2012) A Testi Abszorpció Skála magyar változatának pszichometriai értékelése. Mentálhigiéné és Pszichoszomatika 13:375–395. https://doi.org/10.1556/Mental.13.2012.4.2

Köteles F, Szemerszky R, Gubányi M et al (2013) Idiopathic environmental intolerance attributed to electromagnetic fields (IEI-EMF) and electrosensibility (ES) – are they connected? Int J Hyg Environ Health 216:362–370. https://doi.org/10.1016/j.ijheh.2012.05.007

Köteles F, Simor P, Szemerszky R (2019) Somatosensory amplification and absorption contribute to electrosensitivity. Ideggyógyászati Szemle 72:165–170. https://doi.org/10.18071/isz.72.0165

Kroenke K (2003) Patients presenting with somatic complaints: epidemiology, psychiatric comorbidity and management. Int J Methods Psychiatr Res 12:34–43. https://doi.org/10.1002/mpr.140

Kroenke K (2006) Physical symptom disorder: a simpler diagnostic category for somatization-spectrum conditions. J Psychosom Res 60:335–339. https://doi.org/10.1016/j.jpsychores.2006.01.022

Kroenke K, Arrington ME, Mangelsdorff AD (1990) The prevalence of symptoms in medical outpatients and the adequacy of therapy. Arch Intern Med 150:1685–1689

Kroenke K, Spitzer RL, deGruy FV et al (1997) Multisomatoform disorder. An alternative to undifferentiated somatoform disorder for the somatizing patient in primary care. Arch Gen Psychiatry 54:352–358. https://doi.org/10.1001/archpsyc.1997.01830160080011

Kroenke K, Spitzer RL, deGruy FV, Swindle R (1998) A symptom checklist to screen for somatoform disorders in primary care. Psychosomatics 39:263–272. https://doi.org/10.1016/S0033-3182(98)71343-X

Kroenke K, Spitzer RL, Williams JBW (2002) The PHQ-15: validity of a new measure for evaluating the severity of somatic symptoms. Psychosom Med 64:258–266
Kroenke K, Spitzer RL, Williams JBW, Löwe B (2010) The Patient Health Questionnaire Somatic, Anxiety, and Depressive Symptom Scales: a systematic review. Gen Hosp Psychiatry 32:345–359. https://doi.org/10.1016/j.genhosppsych.2010.03.006
Kroeze S, van den Hout MA (2000) Selective attention for cardiac information in panic patients. Behav Res Ther 38:63–72. https://doi.org/10.1016/s0005-7967(99)00023-6
Lees A, Mogg K, Bradley BP (2005) Health anxiety, anxiety sensitivity, and attentional biases for pictorial and linguistic health-threat cues. Cognit Emot 19:453–462. https://doi.org/10.1080/02699930441000184
Lennertz RC, Tsunozaki M, Bautista DM, Stucky CL (2010) Physiological basis of tingling paresthesia evoked by hydroxy-alpha-sanshool. J Neurosci 30:4353–4361. https://doi.org/10.1523/JNEUROSCI.4666-09.2010
Lenz FA, Seike M, Richardson RT et al (1993) Thermal and pain sensations evoked by microstimulation in the area of human ventrocaudal nucleus. J Neurophysiol 70:200–212
Leonhart R, de Vroege L, Zhang L et al (2018) Comparison of the factor structure of the Patient Health Questionnaire for Somatic Symptoms (PHQ-15) in Germany, the Netherlands, and China. A transcultural Structural Equation Modeling (SEM) Study. Front Psychiatry 9:240. https://doi.org/10.3389/fpsyt.2018.00240
Leventhal H, Mosbach P (1983) A perceptual-motor theory of emotion. In: Cacioppo J, Petty R (eds) Social psychophysiology. Guilford Press, New York, pp 353–388
Leventhal H, Brown D, Shacham S, Engquist G (1979) Effects of preparatory information about sensations, threat of pain, and attention on cold pressor distress. J Pers Soc Psychol 37:688–714. https://doi.org/10.1037//0022-3514.37.5.688
Levine P (2008) Trauma and spirituality. In: Measuring the immeasurable. The scientific case for spirituality. Sounds True, Inc, Boulder, pp 85–100
Likert R (1932) A technique for the measurement of attitudes. Arch Psychol 22(140):55
Lim S-L, Kim J-H (2005) Cognitive processing of emotional information in depression, panic, and somatoform disorder. J Abnorm Psychol 114:50–61. https://doi.org/10.1037/0021-843X.114.1.50
Lipowski ZJ (1986) Somatization: a borderland between medicine and psychiatry. CMAJ 135:609–614
Lipowski ZJ (1988) Somatization: the concept and its clinical application. Am J Psychiatry 145:1358–1368
Lloyd DM, Mason L, Brown RJ, Poliakoff E (2008) Development of a paradigm for measuring somatic disturbance in clinical populations with medically unexplained symptoms. J Psychosom Res 64:21–24. https://doi.org/10.1016/j.jpsychores.2007.06.004
Lonigan CJ, Vasey MW, Phillips BM, Hazen RA (2004) Temperament, anxiety, and the processing of threat-relevant stimuli. J Clin Child Adolesc Psychol 33:8–20. https://doi.org/10.1207/S15374424JCCP3301_2
MacLean KA, Aichele SR, Bridwell DA et al (2009) Interactions between endogenous and exogenous attention during vigilance. Atten Percept Psychophys 71:1042–1058. https://doi.org/10.3758/APP.71.5.1042
MacLeod C, Mathews A (1988) Anxiety and the allocation of attention to threat. Q J Exp Psychol A 40:653–670. https://doi.org/10.1080/14640748808402292
MacLeod C, Mathews A, Tata P (1986) Attentional bias in emotional disorders. J Abnorm Psychol 95:15–20. https://doi.org/10.1037//0021-843x.95.1.15
Maister L, Tsakiris M (2014) My face, my heart: cultural differences in integrated bodily self-awareness. Cogn Neurosci 5:10–16. https://doi.org/10.1080/17588928.2013.808613
Mandler G, Mandler JM, Uviller ET (1958) Autonomic feedback: the perception of autonomic activity. J Abnorm Soc Psychol 56:367–373. https://doi.org/10.1037/h0048083

Martin M, Alexeeva I (2010) Mood volatility with rumination but neither attentional nor interpretation biases in chronic fatigue syndrome. Br J Health Psychol 15:779–796. https://doi.org/10.1348/135910709X480346

Martin M, Chapman SCE (2010) Cognitive processing in putative functional gastrointestinal disorder: rumination yields orientation to social threat not pain. Eur J Pain 14:207–213. https://doi.org/10.1016/j.ejpain.2009.04.008

Mash LE, Schauder KB, Cochran C et al (2017) Associations between interoceptive cognition and age in autism spectrum disorder and typical development. J Cogn Educ Psychol 16:23–37. https://doi.org/10.1891/1945-8959.16.1.23

Mason RE (1959) Three studies relating internal sensory experiences and feeling reactions. J Gen Psychol 60:211–228. https://doi.org/10.1080/00221309.1959.9710223

Mason RE (1961) Internal perception and bodily functioning. International Universities Press, New York

Mathews A, MacLeod C (1985) Selective processing of threat cues in anxiety states. Behav Res Ther 23:563–569. https://doi.org/10.1016/0005-7967(85)90104-4

Matthias E, Schandry R, Duschek S, Pollatos O (2009) On the relationship between interoceptive awareness and the attentional processing of visual stimuli. Int J Psychophysiol 72:154–159. https://doi.org/10.1016/j.ijpsycho.2008.12.001

Mazzola L, Isnard J, Mauguière F (2006) Somatosensory and pain responses to stimulation of the second somatosensory area (SII) in humans. A comparison with SI and insular responses. Cereb Cortex 16:960–968. https://doi.org/10.1093/cercor/bhj038

Mead GH (1934) Mind, self, and society. University of Chicago Press, Chicago, IL

Mechanic D (1979) Development of psychological distress among young adults. Arch Gen Psychiatry 36:1233–1239

Mechanic D (1980) The experience and reporting of common physical complaints. J Health Soc Behav 21:146–155

Meessen J, Mainz V, Gauggel S et al (2016) The relationship between interoception and metacognition. J Psychophysiol 30:76–86. https://doi.org/10.1027/0269-8803/a000157

Mehling WE (2016) Differentiating attention styles and regulatory aspects of self-reported interoceptive sensibility. Philos Trans R Soc Lond B Biol Sci 371:13. https://doi.org/10.1098/rstb.2016.0013

Mehling WE (2020) If it all comes down to bodily awareness, how do we know? Assessing bodily awareness. Kinesiol Rev 9:254–260. https://doi.org/10.1123/kr.2020-0021

Mehling WE, Gopisetty V, Daubenmier JJ et al (2009) Body awareness: construct and self-report measures. PLoS One 4:e5614. https://doi.org/10.1371/journal.pone.0005614

Mehling WE, Price C, Daubenmier JJ et al (2012) The Multidimensional Assessment of Interoceptive Awareness (MAIA). PLoS One 7:e48230. https://doi.org/10.1371/journal.pone.0048230

Mehling WE, Daubenmier JJ, Price CJ et al (2013) Self-reported interoceptive awareness in primary care patients with past or current low back pain. J Pain Res 6:403–418. https://doi.org/10.2147/JPR.S42418

Mehling WE, Acree M, Stewart A et al (2018) The Multidimensional Assessment of Interoceptive Awareness, Version 2 (MAIA-2). PLoS One 13:e0208034. https://doi.org/10.1371/journal.pone.0208034

Melzack R (1973) The puzzle of pain. Basic Books, New York

Melzack R, Wall PD (1965) Pain mechanisms: a new theory. Science 150:971–979

Menon V, Uddin LQ (2010) Saliency, switching, attention and control: a network model of insula function. Brain Struct Funct 214:655–667. https://doi.org/10.1007/s00429-010-0262-0

Michael GA, Naveteur J (2011) The tickly homunculus and the origins of spontaneous sensations arising on the hands. Conscious Cogn 20:603–617. https://doi.org/10.1016/j.concog.2010.11.013

Michael GA, Dupuy M-A, Deleuze A et al (2011) Interacting effects of vision and attention in perceiving spontaneous sensations arising on the hands. Exp Brain Res 216:21–34. https://doi.org/10.1007/s00221-011-2901-y

Michael GA, Naveteur J, Dupuy M-A, Jacquot L (2015) My heart is in my hands: the interoceptive nature of the spontaneous sensations felt on the hands. Physiol Behav 143:113–120. https://doi.org/10.1016/j.physbeh.2015.02.030

Michael GA, Tapiero I, Gálvez-García G, Jacquot L (2017) Thoughts and sensations, twin galaxies of the inner space: the propensity to mind-wander relates to spontaneous sensations arising on the hands. Conscious Cogn 55:223–231. https://doi.org/10.1016/j.concog.2017.08.007

Miles P, True G (2003) Reiki—review of a biofield therapy history, theory, practice, and research. Altern Ther Health Med 9:62–72

Miller LC, Murphy R, Buss AH (1981) Consciousness of body: private and public. J Pers Soc Psychol 41:397–406. https://doi.org/10.1037/0022-3514.41.2.397

Montoya P, Schandry R, Müller A (1993) Heartbeat evoked potentials (HEP): topography and influence of cardiac awareness and focus of attention. Electroencephalogr Clin Neurophysiol 88:163–172

Mrkva K, Van Boven L (2020) Salience theory of mere exposure: relative exposure increases liking, extremity, and emotional intensity. J Pers Soc Psychol. https://doi.org/10.1037/pspa0000184

Murphy G (1947) Personality. Harper, New York

Murphy J, Catmur C, Bird G (2019) Classifying individual differences in interoception: implications for the measurement of interoceptive awareness. Psychon Bull Rev 26:1467–1471. https://doi.org/10.3758/s13423-019-01632-7

Nafe JP (1924) An experimental study of the affective qualities. Am J Psychol 35:507–544. https://doi.org/10.2307/1414034

Nafe JP (1927) The psychology of felt experience. Am J Psychol 39:367–389. https://doi.org/10.2307/1415424

Naveteur J, Honore J, Michael GA (2005) How to detect an electrocutaneous shock which is not delivered? Overt spatial attention influences decision. Behav Brain Res 165:254–261. https://doi.org/10.1016/j.bbr.2005.07.001

Ness TJ, Metcalf AM, Gebhart GF (1990) A psychophysiological study in humans using phasic colonic distension as a noxious visceral stimulus. Pain 43:377–386. https://doi.org/10.1016/0304-3959(90)90035-C

Ness TJ, Richter HE, Varner RE, Fillingim RB (1998) A psychophysical study of discomfort produced by repeated filling of the urinary bladder. Pain 76:61–69. https://doi.org/10.1016/s0304-3959(98)00023-2

NINDS (2017) Paresthesia Information Page. National Institute of Neurological Disorders and Stroke. https://www.ninds.nih.gov/Disorders/All-Disorders/Paresthesia-Information-Page. Accessed 7 Aug 2017

Nouwen A, Cloutier C, Kappas A et al (2006) Effects of focusing and distraction on cold pressor-induced pain in chronic back pain patients and control subjects. J Pain 7:62–71. https://doi.org/10.1016/j.jpain.2005.08.004

Nummenmaa L, Hari R, Hietanen JK, Glerean E (2018) Maps of subjective feelings. Proc Natl Acad Sci U S A 115:9198–9203. https://doi.org/10.1073/pnas.1807390115

Ostrowsky K, Isnard J, Ryvlin P et al (2000) Functional mapping of the insular cortex: clinical implication in temporal lobe epilepsy. Epilepsia 41:681–686

Ostrowsky K, Magnin M, Ryvlin P et al (2002) Representation of pain and somatic sensation in the human insula: a study of responses to direct electrical cortical stimulation. Cereb Cortex 12:376–385. https://doi.org/10.1093/cercor/12.4.376

Panksepp J (1998) Affective neuroscience: the foundations of human and animal emotions, 1st edn. Oxford University Press, New York

Park H-D, Blanke O (2019) Heartbeat-evoked cortical responses: underlying mechanisms, functional roles, and methodological considerations. NeuroImage 197:502–511. https://doi.org/10.1016/j.neuroimage.2019.04.081

Park H, Park J, Lee H, Lee H (2002) Does deqi (needle sensation) exist? Am J Chin Med 30:45–50. https://doi.org/10.1142/S0192415X02000053

Parker K (1983) A meta-analysis of the reliability and validity of the Rorschach. J Pers Assess 47:227–231. https://doi.org/10.1207/s15327752jpa4703_1

Parkin L, Morgan R, Rosselli A et al (2013) Exploring the relationship between mindfulness and cardiac perception. Mindfulness 5:298–313. https://doi.org/10.1007/s12671-012-0181-7

Parr T, Friston KJ (2019) Attention or salience? Curr Opin Psychol 29:1–5. https://doi.org/10.1016/j.copsyc.2018.10.006

Penfield W, Faulk ME (1955) The insula; further observations on its function. Brain 78:445–470. https://doi.org/10.1093/brain/78.4.445

Penfield W, Gage L (1933) Cerebral localization of epileptic manifestations. Arch NeurPsych 30:709–727. https://doi.org/10.1001/archneurpsyc.1933.02240160009001

Pennebaker JW (1982) The psychology of physical symptoms. Springer, New York

Pennebaker JW, Lightner JM (1980) Competition of internal and external information in an exercise setting. J Pers Soc Psychol 39:165–174. https://doi.org/10.1037/0022-3514.39.1.165

Pennebaker JW, Skelton JA (1981) Selective monitoring of physical sensations. J Pers Soc Psychol 41:213–223

Peters S, Stanley I, Rose M, Salmon P (1998) Patients with medically unexplained symptoms: sources of patients' authority and implications for demands on medical care. Soc Sci Med 46:559–565

Petersen S, Schroijen M, Mölders C et al (2014) Categorical interoception: perceptual organization of sensations from inside. Psychol Sci 25:1059–1066. https://doi.org/10.1177/0956797613519110

Petzschner FH, Weber LA, Wellstein KV et al (2019) Focus of attention modulates the heartbeat evoked potential. NeuroImage 186:595–606. https://doi.org/10.1016/j.neuroimage.2018.11.037

Pollatos O, Schandry R (2004) Accuracy of heartbeat perception is reflected in the amplitude of the heartbeat-evoked brain potential. Psychophysiology 41:476–482. https://doi.org/10.1111/1469-8986.2004.00170.x

Pollatos O, Kirsch W, Schandry R (2005) Brain structures involved in interoceptive awareness and cardioafferent signal processing: a dipole source localization study. Hum Brain Mapp 26:54–64. https://doi.org/10.1002/hbm.20121

Porges SW (1993) Body Perception Questionnaire. Laboratory of Developmental Assessment, University of Maryland, College Park

Posner MI (1980) Orienting of attention. Q J Exp Psychol 32:3–25. https://doi.org/10.1080/00335558008248231

Posner MI (1994) Attention: the mechanisms of consciousness. Proc Natl Acad Sci U S A 91:7398–7403

Price DD, McGrath PA, Rafii A, Buckingham B (1983) The validation of visual analogue scales as ratio scale measures for chronic and experimental pain. Pain 17:45–56. https://doi.org/10.1016/0304-3959(83)90126-4

Pugnaghi M, Meletti S, Castana L et al (2011) Features of somatosensory manifestations induced by intracranial electrical stimulations of the human insula. Clin Neurophysiol 122:2049–2058

Raingruber B, Robinson C (2007) The effectiveness of tai chi, yoga, meditation, and reiki healing sessions in promoting health and enhancing problem solving abilities of registered nurses. Issues Ment Health Nurs 28:1141–1155. https://doi.org/10.1080/01612840701581255

Ray C (1979) Examination stress and performance on a color-word interference test. Percept Mot Skills 49:400–402. https://doi.org/10.2466/pms.1979.49.2.400

Reis D (2017) Further insights into the German version of the Multidimensional Assessment of Interoceptive Awareness (MAIA): exploratory and Bayesian structural equation modeling approaches. Eur J Psychol Assess. https://doi.org/10.1027/1015-5759/a000404

Reise SP, Moore TM, Haviland MG (2010) Bifactor models and rotations: exploring the extent to which multidimensional data yield univocal scale scores. J Pers Assess 92:544–559. https://doi.org/10.1080/00223891.2010.496477

Renik OD, Fisher S (1968) Induction of body image boundary changes in male subjects. J Proj Tech Pers Assess 32:45–48. https://doi.org/10.1080/0091651X.1968.10120445
Rief W, Hiller W, Margraf J (1998) Cognitive aspects of hypochondriasis and the somatization syndrome. J Abnorm Psychol 107:587–595
Robbins JM, Kirmayer LJ (1991) Attributions of common somatic symptoms. Psychol Med 21:1029–1045. https://doi.org/10.1017/S0033291700030026
Rorschach H (1921) Psychodiagnostics: a diagnostic test based on perception. Hans Huber, Bern
Ruckmick CA (1936) The psychology of feeling and emotion. McGraw-Hill, New York
Salgues S, Plancher G, Jacquot L et al (2021) To the self and beyond: arousal and functional connectivity of the temporo-parietal junction contributes to spontaneous sensations perception. Behav Brain Res 396:112880. https://doi.org/10.1016/j.bbr.2020.112880
Salovey P (1992) Mood-induced self-focused attention. J Pers Soc Psychol 62:699–707. https://doi.org/10.1037/0022-3514.62.4.699
Schandry R, Sparrer B, Weitkunat R (1986) From the heart to the brain: a study of heartbeat contingent scalp potentials. Int J Neurosci 30:261–275
Schauder KB, Mash LE, Bryant LK, Cascio CJ (2015) Interoceptive ability and body awareness in autism spectrum disorder. J Exp Child Psychol 131:193–200. https://doi.org/10.1016/j.jecp.2014.11.002
Scheier MF (1976) Self-awareness, self-consciousness, and angry aggression. J Pers 44:627–644. https://doi.org/10.1111/j.1467-6494.1976.tb00142.x
Scheier MF, Carver CS (1977) Self-focused attention and the experience of emotion: attraction, repulsion, elation, and depression. J Pers Soc Psychol 35:625–636
Scheier MF, Carver CS (1985) Optimism, coping, and health: assessment and implications of generalized outcome expectancies. Health Psychol 4:219–247. https://doi.org/10.1037/0278-6133.4.3.219
Schilder P (1935) The image and appearance of the human body. Kegan Paul, Oxford, England
Schmidt AJ, Wolfs-Takens DJ, Oosterlaan J, van den Hout MA (1994) Psychological mechanisms in hypochondriasis: attention-induced physical symptoms without sensory stimulation. Psychother Psychosom 61:117–120
Schmidt NB, Lerew DR, Trakowski JH (1997) Body vigilance in panic disorder: evaluating attention to bodily perturbations. J Consult Clin Psychol 65:214–220. https://doi.org/10.1037/0022-006X.65.2.214
Schneider W, Shiffrin RM (1977) Controlled and automatic human information processing: I. Detection, search, and attention. Psychol Rev 84:1–66. https://doi.org/10.1037/0033-295X.84.1.1
Schulz SM (2016) Neural correlates of heart-focused interoception: a functional magnetic resonance imaging meta-analysis. Philos Trans R Soc B Biol Sci 371:20160018. https://doi.org/10.1098/rstb.2016.0018
Secord PF (1953) Objectification of word-association procedures by the use of homonyms: a measure of body cathexis. J Pers 21:479–495. https://doi.org/10.1111/j.1467-6494.1953.tb01788.x
Secord PF, Jourard SM (1953) The appraisal of body-cathexis: body-cathexis and the self. J Consult Psychol 17:343–347. https://doi.org/10.1037/h0060689
Seeley WW, Menon V, Schatzberg AF et al (2007) Dissociable intrinsic connectivity networks for salience processing and executive control. J Neurosci 27:2349–2356. https://doi.org/10.1523/JNEUROSCI.5587-06.2007
Shields SA, Mallory ME, Simon A (1989) The Body Awareness Questionnaire: reliability and validity. J Pers Assess 53:802. https://doi.org/10.1207/s15327752jpa5304_16
Shiffrin RM, Schneider W (1977) Controlled and automatic human information processing: II. Perceptual learning, automatic attending and a general theory. Psychol Rev 84:127–190. https://doi.org/10.1037/0033-295X.84.2.127
Shiflett SC, Nayak S, Bid C et al (2002) Effect of reiki treatments on functional recovery in patients in poststroke rehabilitation: a pilot study. J Altern Complement Med 8:755–763. https://doi.org/10.1089/10755530260511766

Siegel HB (1977) Body boundary and field-dependence. Percept Mot Skills 45:1097–1098. https://doi.org/10.2466/pms.1977.45.3f.1097

Silvia PJ, Gendolla GHE (2001) On introspection and self-perception: does self-focused attention enable accurate self-knowledge? Rev Gen Psychol 5:241–269. https://doi.org/10.1037/1089-2680.5.3.241

Simor P, Köteles F, Bódizs R (2011) Submersion in the experience: the examination of the Tellegen Absorption Scale in an undergraduate university sample. Mentálhigiéné és Pszichoszomatika 12:101–123. https://doi.org/10.1556/Mental.12.2011.2.1

Skovbjerg S, Zachariae R, Rasmussen A et al (2010) Attention to bodily sensations and symptom perception in individuals with idiopathic environmental intolerance. Environ Health Prev Med 15:141–150. https://doi.org/10.1007/s12199-009-0120-y

Spitzer RL, Williams JBW, Kroenke K et al (1994) Utility of a new procedure for diagnosing mental disorders in primary care: the PRIME-MD 1000 study. JAMA 272:1749–1756. https://doi.org/10.1001/jama.1994.03520220043029

St Onge PM (2007) Effect of tissue compression on the Hoffmann reflex: comparison between the ischial tuberosity and posterior thigh. Dissertation, Auburn University

Stauder A, Witthöft M, Köteles F (2021) Validation of the Hungarian PHQ-15. A latent variable approach. Ideggyógyászati Szemle/Clinical Neuroscience, in press

Stegen K, van Diest I, van de Woestijne KP, Van den Bergh O (2000) Negative affectivity and bodily sensations induced by 5.5% CO_2 enriched air inhalation: is there a bias to interpret bodily sensations negatively in persons with negative affect? Psychol Health 15:513–525. https://doi.org/10.1080/08870440008402010

Stern ER, Grimaldi SJ, Muratore A et al (2017) Neural correlates of interoception: effects of interoceptive focus and relationship to dimensional measures of body awareness. Hum Brain Mapp 38:6068–6082. https://doi.org/10.1002/hbm.23811

Stroop JR (1935) Studies of interference in serial verbal reactions. J Exp Psychol 18:643–662. https://doi.org/10.1037/h0054651

Suls J, Fletcher B (1985) The relative efficacy of avoidant and nonavoidant coping strategies: a meta-analysis. Health Psychol 4:249–288. https://doi.org/10.1037//0278-6133.4.3.249

Szemerszky R, Köteles F, Lihi R, Bárdos G (2010) Polluted places or polluted minds? An experimental sham-exposure study on background psychological factors of symptom formation in "Idiophatic Environmental Intolerance attributed to electromagnetic fields". Int J Hyg Environ Health 213:387–394. https://doi.org/10.1016/j.ijheh.2010.05.001

Szemerszky R, Gubányi M, Árvai D et al (2015) Is there a connection between electrosensitivity and electrosensibility? A replication study. Int J Behav Med 22:755–763. https://doi.org/10.1007/s12529-015-9477-z

Taylor JA (1953) A personality scale of manifest anxiety. J Abnorm Soc Psychol 48:285–290. https://doi.org/10.1037/h0056264

Tellegen A, Atkinson G (1974) Openness to absorbing and self-altering experiences ("absorption"), a trait related to hypnotic susceptibility. J Abnorm Psychol 83:268–277

Tichtener EB (1926) A textbook of psychology. Macmillan, New York

Tihanyi BT, Köteles F (2017) Physiological and psychological correlates of attention-related body sensations (tingling and warmth). Physiol Int 104:235–246. https://doi.org/10.1556/2060.104.2017.3.4

Tihanyi BT, Sági A, Csala B et al (2016) Body awareness, mindfulness and affect: does the kind of physical activity make a difference? Eur J Ment Health 11:97–111. https://doi.org/10.5708/EJMH.11.2016.1-2.6

Tihanyi BT, Ferentzi E, Daubenmier JJ et al (2017a) Body responsiveness questionnaire: validation on a European sample, mediation between body awareness and affect, connection with mindfulness, body image, and physical activity. Int Body Psychother J 16:56–73

Tihanyi BT, Ferentzi E, Köteles F (2017b) Characteristics of attention-related body sensations. Temporal stability and associations with measures of body focus, affect, sustained attention,

and heart rate variability. Somatosens Mot Res 34:179–184. https://doi.org/10.1080/08990220.2017.1384720
Tihanyi BT, Ferentzi E, Beissner F, Köteles F (2018) The neuropsychophysiology of tingling. Conscious Cogn 58:97–110. https://doi.org/10.1016/j.concog.2017.10.015
Tracy J, Goyal N, Flanders A et al (2007) Functional magnetic resonance imaging analysis of attention to one's heartbeat. Psychosom Med 69:952–960. https://doi.org/10.1097/PSY.0b013e31815b60cf
Uddin LQ, Kinnison J, Pessoa L, Anderson ML (2014) Beyond the tripartite cognition-emotion-interoception model of the human insular cortex. J Cogn Neurosci 26:16–27. https://doi.org/10.1162/jocn_a_00462
Valenzuela-Moguillansky C, Reyes-Reyes A (2015) Psychometric properties of the multidimensional assessment of interoceptive awareness (MAIA) in a Chilean population. Front Psychol 6:120. https://doi.org/10.3389/fpsyg.2015.00120
Van den Bergh O, Winters W, Devriese S et al (2004) Accuracy of respiratory symptom perception in persons with high and low negative affectivity. Psychol Health 19:213–222. https://doi.org/10.1080/0887044041000167 5627
Van den Bergh O, Witthöft M, Petersen S, Brown RJ (2017) Symptoms and the body: taking the inferential leap. Neurosci Biobehav Rev 74(Pt A):185–203. https://doi.org/10.1016/j.neubiorev.2017.01.015
van der Veek SMC, Derkx BHF, Plak RD et al (2014) Attentional bias to activity of different parts of the body in children with functional abdominal pain: an experimental study. J Pediatr Psychol. https://doi.org/10.1093/jpepsy/jsu004
Vasey MW, Daleiden EL, Williams LL, Brown LM (1995) Biased attention in childhood anxiety disorders: a preliminary study. J Abnorm Child Psychol 23:267–279. https://doi.org/10.1007/BF01447092
Vig L, Ferentzi E, Köteles F (2020) The ability to sustain attention contributes to cardioceptive accuracy but not to body awareness. Manuscript submitted for publication
Vossel S, Geng JJ, Fink GR (2014) Dorsal and ventral attention systems. Neuroscientist 20:150–159. https://doi.org/10.1177/1073858413494269
Walentynowicz M, Witthöft M, Raes F et al (2018) Sensory and affective components of symptom perception: a psychometric approach. J Exp Psychopathol. https://doi.org/10.5127/jep.059716
Wall PD (2000) Pain: the science of suffering. Columbia University Press, New York
Wang X, Wu Q, Egan L et al (2019) Anterior insular cortex plays a critical role in interoceptive attention. Elife 8:e42265. https://doi.org/10.7554/eLife.42265
Wang X, Tan Y, Van den Bergh O et al (2020) Intrinsic functional brain connectivity patterns underlying enhanced interoceptive sensibility. J Affect Disord 276:804–814. https://doi.org/10.1016/j.jad.2020.07.032
Wapner S, Werner H (1952) Experiments on sensory-tonic field theory of perception: V. Effect of body status on the kinesthetic perception of verticality. J Exp Psychol 44:126–131. https://doi.org/10.1037/h0058472
Wapner S, Werner H, Chandler KA (1951a) Experiments on sensory-tonic field theory of perception. I. Effect of extraneous stimulation on the visual perception of verticality. J Exp Psychol 42:341–344. https://doi.org/10.1037/h0053846
Wapner S, Werner H, Morant RB (1951b) Experiments on sensory-tonic field theory of perception. III. Effect of body rotation on the visual perception of verticality. J Exp Psychol 42:351–357. https://doi.org/10.1037/h0058847
Watson D, Clark LA (1984) Negative affectivity: the disposition to experience aversive emotional states. Psychol Bull 96:465–490
Watts FN, McKenna FP, Sharrock R, Trezise L (1986) Colour naming of phobia-related words. Br J Psychol 77(Pt 1):97–108. https://doi.org/10.1111/j.2044-8295.1986.tb01985.x
Weierich MR, Wright CI, Negreira A et al (2010) Novelty as a dimension in the affective brain. NeuroImage 49:2871. https://doi.org/10.1016/j.neuroimage.2009.09.047

Weinberg JR (1960) A further investigation of body-cathexis and the self. J Consult Psychol 24:277. https://doi.org/10.1037/h0038368

Weiss A, Deary IJ (2020) A new look at neuroticism: should we worry so much about worrying? Curr Dir Psychol Sci 29:92–101. https://doi.org/10.1177/0963721419887184

Weisz J, Balázs L, Ádám G (1988) The influence of self-focused attention on heartbeat perception. Psychophysiology 25:193–199. https://doi.org/10.1111/j.1469-8986.1988.tb00987.x

Werner H, Wapner S (1952a) Experiments on sensory-tonic field theory of perception: IV. Effect of initial position of a rod on apparent verticality. J Exp Psychol 43:68. https://doi.org/10.1037/h0060658

Werner H, Wapner S (1952b) Toward a general theory of perception. Psychol Rev 59:324–338. https://doi.org/10.1037/h0059673

Werner H, Wapner S, Chandler KA (1951) Experiments on sensory-tonic field theory of perception. II. Effect of supported and unsupported tilt of the body on the visual perception of verticality. J Exp Psychol 42:346–350. https://doi.org/10.1037/h0062186

Werner H, Wapner S, Bruell JH (1953) Experiments on sensory-tonic field theory of perception. VI. Effect of position of head, eyes, and of object on position of the apparent median plane. J Exp Psychol 46:293–299. https://doi.org/10.1037/h0055733

Weston SJ, Jackson JJ (2018) The role of vigilance in the relationship between neuroticism and health: a registered report. J Res Pers 73:27–34. https://doi.org/10.1016/j.jrp.2017.10.005

Wewers ME, Lowe NK (1990) A critical review of visual analogue scales in the measurement of clinical phenomena. Res Nurs Health 13:227–236. https://doi.org/10.1002/nur.4770130405

Wicklund RA (1975) Objective self-awareness. Adv Exp Soc Psychol 8:233–275

Wicklund RA, Gollwitzer PM (1987) The fallacy of the private-public self-focus distinction. J Pers 55:491–523. https://doi.org/10.1111/j.1467-6494.1987.tb00448.x

Wiebking C, Northoff G (2015) Neural activity during interoceptive awareness and its associations with alexithymia-an fMRI study in major depressive disorder and non-psychiatric controls. Front Psychol 6:589. https://doi.org/10.3389/fpsyg.2015.00589

Wiebking C, Bauer A, de Greck M et al (2010) Abnormal body perception and neural activity in the insula in depression: an fMRI study of the depressed "material me". World J Biol Psychiatry 11:538–549. https://doi.org/10.3109/15622970903563794

Witthöft M, Gerlach AL, Bailer J (2006) Selective attention, memory bias, and symptom perception in idiopathic environmental intolerance and somatoform disorders. J Abnorm Psychol 115:397–407. https://doi.org/10.1037/0021-843X.115.3.397

Witthöft M, Rist F, Bailer J (2008) Evidence for a specific link between the personality trait of absorption and idiopathic environmental intolerance. J Toxicol Environ Health A 71:795–802. https://doi.org/10.1080/15287390801985687

Witthöft M, Hiller W, Loch N, Jasper F (2013) The latent structure of medically unexplained symptoms and its relation to functional somatic syndromes. Int J Behav Med 20:172–183. https://doi.org/10.1007/s12529-012-9237-2

Witthöft M, Fischer S, Jasper F et al (2016) Clarifying the latent structure and correlates of somatic symptom distress: a bifactor model approach. Psychol Assess 28:109–115. https://doi.org/10.1037/pas0000150

Witthöft M, Freitag I, Nußbaum C et al (2018) On the origin of worries about modern health hazards: experimental evidence for a conjoint influence of media reports and personality traits. Psychol Health 33:361–380. https://doi.org/10.1080/08870446.2017.1357814

Witthöft M, Bräscher A-K, Jungmann SM, Köteles F (2020) Somatic symptom perception and interoception. Z Psychol 228:100–109. https://doi.org/10.1027/2151-2604/a000403

Wolff W (1932) Selbstbeurteilung und Fremdbeurteilung im wissentlichen und unwissentlichen Versuch: Physiognomische Untersuchungen an der Stimme, dem Profil, den Händen und einer freien Nacherzählung. Psychol Forsch 16:251–328. https://doi.org/10.1007/BF00409734

Wood JV, Saltzberg JA, Goldsamt LA (1990) Does affect induce self-focused attention? J Pers Soc Psychol 58:899–908. https://doi.org/10.1037/0022-3514.58.5.899

Wood JM, Nezworski MT, Lilienfeld SO, Garb HN (2003) What's wrong with the Rorschach?: science confronts the controversial inkblot test. Jossey-Bass, San Francisco, CA

Wood JM, Lilienfeld SO, Nezworski MT et al (2010) Validity of Rorschach Inkblot scores for discriminating psychopaths from non-psychopaths in forensic populations: a meta-analysis. Psychol Assess 22:336–349. https://doi.org/10.1037/a0018998

Wu X, Shi L, Wei D, Qiu J (2019) Brain connection pattern under interoceptive attention state predict interoceptive intensity and subjective anxiety feeling. Hum Brain Mapp 40:1760–1773. https://doi.org/10.1002/hbm.24488

Wundt W (1903) Outlines of psychology. Wilhelm Engelmann, Leipzig

Yao S, Becker B, Zhao W et al (2018) Oxytocin modulates attention switching between interoceptive signals and external social cues. Neuropsychopharmacology 43:294–301. https://doi.org/10.1038/npp.2017.189

Young HA, Williams C, Pink AE et al (2017) Getting to the heart of the matter: does aberrant interoceptive processing contribute towards emotional eating? PLoS One 12:e0186312. https://doi.org/10.1371/journal.pone.0186312

Zaki J, Davis JI, Ochsner KN (2012) Overlapping activity in anterior insula during interoception and emotional experience. NeuroImage 62:493–499. https://doi.org/10.1016/j.neuroimage.2012.05.012

Zborowski M (1952) Cultural components in responses to pain. J Soc Issues 8:16–30. https://doi.org/10.1111/j.1540-4560.1952.tb01860.x

Zennaro A, Lis A (1998) Fisher and Cleveland Barrier and Penetration scores for Italian adult and adolescent sample: comparison of two versions scored on Zulliger test. Psychol Rep 82:1411–1420. https://doi.org/10.2466/pr0.1998.82.3c.1411

Zennaro A, Lis A (2016) Fisher and Cleveland Barrier and Penetration Scores: analysis of some psychometric aspects for normal Italian adult sample. Percept Motor Skills. https://doi.org/10.2466/pms.1996.82.2.531

Zhang L, Helander MG, Drury CG (1996) Identifying factors of comfort and discomfort in sitting. Hum Factors 38:377–389

Chapter 6
Can We Trust Our Body Feelings? The Discrepancy Between Actual and Perceived Events

6.1 Dissociable and Non-dissociable Modalities

This chapter explains the tenet of dissociation and tenet of malleability, as well as their consequences, in more detail. In Chaps. 4 and 5, we explored two dimensions of (more or less) conscious aspects of interoception: the ability to detect physiological changes taking place in the body (interoceptive accuracy) and the self-reported attentiveness to and awareness of body sensations (interoceptive sensibility). What we found in Chap. 4 was a definitely poor average interoceptive accuracy with respect to a number of interoceptive modalities, particularly for those that belong to the domain of visceroception, such as heartbeat and activity of various sections of the gastrointestinal tract. Accuracy for these modalities is often measured using the forced-choice paradigm, which clearly shows that average detection ability in the normal domain of stimulus intensity is poor; in other words, most participants' performance is close to random guessing. This is why a prominent researcher of the field, Paintal, stated (1986) that we practically never sense anything from the activity of the intestinal organs. As we have no direct internal information about these events, we often rely on other (top-down) sources if we need to estimate them. This usually leads to a marked dissociation, also called discrepancy, discordance, or lack of concordance or coherence by various authors (Weinstein et al. 1968; Gannon and Haynes 1986; Suschinsky and Lalumière 2012; Farb et al. 2015), between actual and perceived events. This is particularly so in laboratory settings when the availability of additional (top-down) factors for the perception process is intentionally kept low (Pennebaker 1982; Pennebaker and Epstein 1983; Pennebaker and Roberts 1992). This dissociation will be the major topic of this chapter; the modalities that are particularly prone to such dissociation will be called *dissociable modalities*.

For other modalities, however, the picture is markedly different. Beyond pain, we have clear-cut and affectively colored conscious information on the (non-pathological) distention of the rectum, the urinary bladder and the stomach, and breathlessness. The actual rate and depth of breathing also belong here although

F. Köteles, *Body Sensations*, https://doi.org/10.1007/978-3-030-63201-4_6

somatosensory information plays a substantial role here (Parshall et al. 2012). We also saw in Chap. 4 that there are consciously accessible internal modalities and states, such as affective touch and slight to moderate physical activity that are accompanied by positive affective changes, i.e., pleasure. Also, proprioceptive and somatosensory information are in part available for conscious processing. For all of these modalities, there is a more or less monotonic relationship between the actual intensity of the stimulus and the felt state although its threshold and the steepness of the curve can show individual differences. Concerning measurement, the actual intensity of the stimulus is usually plotted against its felt intensity and/or affective characteristics. Although the shape of this function can substantially be modified by top-down influences, it cannot be stated that we are not aware of these internal events. Thus, the discrepancy between actual and perceived events will be less marked for these cases. We can call the modalities belonging here *non-dissociable modalities*.

Finally, there is a third category of sensations, which was already discussed as the tingling phenomenon in Chap. 5. This phenomenon, also called attention-related body sensations (Tihanyi et al. 2017, 2018) or spontaneous sensations (Michael and Naveteur 2011; Michael et al. 2011; Salgues et al. 2021), is characterized by the lack of specific bottom-up input. Also, placebo and nocebo sensations are often not accompanied by peripheral changes (Chaps. 9 and 10). These percepts represent body sensations (including symptoms) that do not rely on clear-cut sensory input from a certain modality. Phantom visceral and proprioceptive sensations and illusions, such as feelings localized in the limbs, the bladder, the rectum, and the womb after surgical removal of the respective organ (Head and Holmes 1911; Fisher 1970; Katz and Melzack 1990; Ramachandran and Hirstein 1998), also belong here. We will call this category *top-down sensations*.

I am not saying that these three groups represent completely distinct categories; the boundaries among them are fuzzy with possible overlaps (Fig. 6.1). For example,

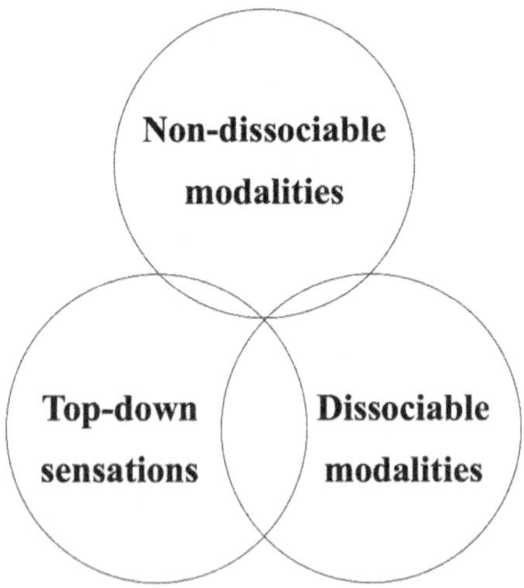

Fig. 6.1 The three major categories of body sensations

sensations belonging to non-dissociable modalities, such as pain, can be influenced by the same factors (these factors will be discussed in Chap. 7) that evoke top-down sensations and also impact dissociable modalities; also, systematic training can render an originally dissociable modality less dissociable. Still, I think this rough categorization is useful to emphasize and understand that body sensations can be evoked and maintained by very distinct factors, and the bottom-up component (sensory input) is only one of these.

It is important to note that the dissociation between actual and perceived events is not at all a rare phenomenon is experimental psychology and everyday life. For example, the manipulation of expectations combined with a conditioning paradigm led to marked perceived differences with respect to the performance in a choice reaction time task, whereas the objective performance was not impacted (Schwarz and Büchel 2015). A similar dissociation for various aspects of cognitive performance (alertness, working memory, sustained attention, and divided attention) was reported in a recent study (Winkler and Hermann 2019). In a study on the effects of essential oils, we found a similar discrepancy between objective and perceived performance in a sustained attention task (Babulka et al. 2017). Concerning body aspects, objective-subjective discrepancies were revealed for spirometry measurements (Köteles et al. 2018) and cardiovascular variables, such as heart rate and blood pressure (Köteles and Babulka 2014). Among other factors, participants' expectations considerably impacted their perceived but not objective performance in these studies; we will mention more examples for this expectation effect in Chaps. 9 and 10, discussing the nocebo and placebo phenomenon, respectively. The conclusion appears clear-cut: if we cannot accurately assess the outcome of a process (e.g., performance in a cognitive task or a body event) through internal sources, we try to estimate it using the information that is available to us. For example, perceived changes in blood pressure and heart rate were associated with changes of alertness in our study (Köteles and Babulka 2014). We speculated that participants tried to infer these consciously not accessible physiological states from their actual alertness; in consequence, mistaken inference led to a complete dissociation between physiological and perceived values.

This chapter will describe the dissociation between sensory input and perceived sensations through multiple examples, such as clinical hypochondriasis, cardiovascular reactivity, and heartbeat perception. We will also explain interoceptive awareness, i.e., the third component of the tripartite model of Garfinkel and colleagues (2015a), and critically evaluate the entire model.

6.2 Early Studies on Dissociation

In 1958, George Mandler and his colleagues published a paper that reported the development of a self-report questionnaire (Autonomic Perception Questionnaire, APQ) (Mandler et al. 1958). The APQ was meant to directly measure the general ability to sense autonomic changes, called autonomic feedback. Those with high

APQ scores showed comparatively high levels of autonomic activation as well as a tendency to overestimate their actual activation in a stressful situation, whereas individuals low on APQ were characterized by the opposite tendencies. The differences in activation (but not the over/underestimation tendency) were replicated in another study (Mandler and Kremen 1958); however, the associations between the APQ score and actual physiological changes were surprisingly weak (correlation coefficients were close to 0.2). Subsequent studies did not find a significant association between the APQ score and cardioceptive accuracy assessed under resting conditions, and it was concluded that the actual ability to perceive visceral changes cannot be measured via self-report (McFarland 1975; Whitehead et al. 1976, 1977). However, as the APQ measures the perceived general propensity to sense internal changes, these null findings are not necessarily conclusive concerning the existence of an actual dissociation.

It also turned out from two case studies in another early research by Mandler that, contrary to the expectations of the experimenters, the perception of actual changes in heart rate is not a trivial task without training (performance of both participants was not better than chance) (Mandler and Kahn 1960). These were perhaps the first findings that demonstrated the discrepancy between actual and perceived visceral changes. Later, the phenomenon was thoroughly investigated by James Pennebaker (1982), using both the between-subject and within-subject approaches. The association between actual and perceived changes was found to be nonexisting or weak for the majority of included sensory modalities (heart rate, sweatiness of hands, breathing rate). The only exception was the perceived temperature of the hands which showed a moderate association with actual temperature. Subsequent research revealed substantial discrepancies with respect to self-reported and objectively measured nasal congestion (Barsky et al. 1988), perceived palpitations and arrhythmia (Barsky 2001), and actual and perceived ability to detect heartbeats (Jones et al. 1984). Also, it turned out that, although larger deviations of blood glucose and blood pressure from the normal range can be perceived by certain individuals through various nonspecific symptoms to some extent, perceived and objective changes do not correspond for less marked fluctuations (Cox et al. 1985; Gonder-Frederick and Cox 1991).

The first researcher who considered and systematically investigated the possible pathological consequences of this dissociation (here called lack of concordance or cognitive-physiological discordance) was Linda Gannon (Gannon 1984; Gannon and Haynes 1986). In 1984, she could collect only 19 eligible studies with simultaneous physiological and self-report measurements for a systematic review. These studies investigated concordance with respect to insomnia (the association between objectively assessed and self-reported aspects of sleep), obesity (gastric motility versus felt hunger), sexual excitement (actual versus self-reported arousal), and anxiety (physiological versus perceived activation). The majority of the studies were conducted with the participation of patients (i.e., dysfunctional individuals) or healthy individuals only, some of them compared the two groups. The primary idea was that the reduced ability to sense internal changes leads to deficits in behavior, thus discordance might play an etiological and maintaining role for "psychophysiologic disorders," such as insomnia, tension headache, and sexual dysfunction (Gannon and Haynes 1986). Cognitive-physiological discordance was assumed to be caused by top-down factors, such as attention deficits, distractibility, and lack of awareness of causal relationships between internal states and negative consequences.

From a methodological point of view, an important distinction was made between true and apparent discordance, the latter is being attributable to measurement-related pitfalls and errors (Gannon and Haynes 1986). Perhaps the most important issue here is the (im)proper identification of the measured sensory channel(s), i.e., the mistaken selection of the applied stimulus. Moreover, individual differences in the physiological response patterns described by Engel (1960) and Lacey (1967) with respect to anxiety can also contribute to an apparent discordance. Also, sensitivity and range of operation of the respective receptors should be taken into consideration. For example, we can easily discriminate between temperature differences of the magnitude of 1–2 °C but not that of 0.5 °C (Beck et al. 2019). These considerations indicate very well that objective-subjective dissociations can be caused by a number of different factors beyond perceptual deficits and limitations thus always need careful theoretical evaluation. From a statistical point of view, a dissociation is a null finding that is sensitive to Type II error; this type of error is often not handled and interpreted correctly (in fact, it is often not reported at all). The risk of Type II error is particularly high for small samples, which are often used in interoception research because of the difficulties of the sensory measurements. Thus, the use of Bayesian methods and larger samples is strongly recommended for future studies.

6.3 The Perceptual Accuracy Hypothesis

Psychological approaches to self-awareness (Chap. 5), such as self-consciousness theory (Fenigstein et al. 1975; Buss 1980) and self-regulation theory (Carver and Scheier 1978, 1981; Gibbons et al. 1979; Scheier et al. 1979), proposed that self-focused attention leads to more accurate information about the self, including attitudes, motives, emotions, and body sensations. Basically, it was assumed that the consequences of introspection are comparable to that of focusing attention on an external stimulus—both lead to more detailed and clarified information on the object in question (Silvia and Gendolla 2001). The face validity of this assumption is high, thus it generated intense research in social psychology.

Unfortunately, self-related information is inherently subjective, i.e., directly accessible only via self-report, thus its accuracy is difficult to measure. In empirical studies, the hypothesized accuracy-enhancing effects of self-focused attention were typically assessed in indirect ways, for example, through changes in behavior or the impact of external information on ratings of internal states. Findings of such experiments can be interpreted in multiple ways, thus they do not necessarily prove the perceptual accuracy (also called veracity or veridicality) hypothesis (Levine and McDonald 1981; Gibbons 1990; Silvia and Gendolla 2001). Overall, the explanatory power of the hypothesis with respect to the perception of behavior, motivations, and attribution proved to be insufficient (Gibbons 1983, 1990). It was gradually realized that many aspects of the self, such as attributions and decision-making, are not accessible for conscious processing (Nisbett and Wilson 1977; Wilson et al. 1981; Gibbons 1990).

Concerning body sensations, an important difference is that objective measurement of the perceived peripheral events or their causes is often possible. Studies on the effects of alcohol and caffeine consumption revealed an overestimation tendency associated with self-focus; in other words, self-awareness led to less accurate perception in these cases (Miller et al. 1981; Hull et al. 1983). Also, reported effects of caffeine, alcohol, and other psychoactive drugs are sensitive to external influences, such as suggestions, expectations, and environmental factors (Hull and Bond Jr 1986; Kirsch and Weixel 1988; Fillmore and Vogel-Sprott 1992, 1994, 1995; Kirsch and Rosadino 1993; Fillmore et al. 1994; Walach et al. 2001, 2002; Nash et al. 2002; Harrell and Juliano 2012), which shows that the accuracy of their perception is generally low. The perception of arousal was found to be malleable to top-down factors too (Valins 1966; Barefoot and Straub 1971; Misovich and Charis 1974; Botto et al. 1974; Carver and Blaney 1977a, b; Woll and McFall 1979; Parkinson and Manstead 1986; Gray et al. 2007). Private body consciousness, i.e., the self-reported tendency to focus on internal events, was not associated with better cardiac perception in a number of studies (Gillis and Carver 1980; Pennebaker and Epstein 1983; Weisz et al. 1988). Overall, it was concluded from these studies that the perceptual accuracy hypothesis with respect to body sensations is not supported by empirical data (Silvia and Gendolla 2001). Although we are often convinced that we accurately sense our body states (this is the tenet of realness), particularly if we pay attention to them, this is not the case.

6.4 Models of Hypochondriasis

Clinical hypochondriasis, as defined in DSM-III and DSM-IV, is characterized by the interpretation of body signals as symptoms of a serious disease (APA 1980, 1987, 1994). This results in intense fear that persists despite the null findings of diagnostic procedures and medical reassurance about the lack of disease. The condition often leads to the overuse of medical services and impairment in everyday functioning (disability) (Barsky et al. 1986; Noyes et al. 1999). In DSM-5 (APA 2013), the name was changed to illness anxiety disorder, but the diagnostic criteria of the condition basically remained the same. Recently, it is considered an extreme case of a dimensional characteristic, health anxiety (Ferguson 2009; Longley et al. 2010).

Serious health anxiety is a perplexing phenomenon as patients experience and report marked body sensations (i.e., serious symptoms) and are firmly convinced of the (patho)physiological background of these symptoms. The condition has been well-known for centuries; its psychological background has been explained in various ways over time (Barsky and Klerman 1983). Analytic theorists explained it as a libidinal frustration (i.e., a denied wish) which leads to narcissistic withdrawal from the original object toward the own body, accompanied by self-punishment and guilt and/or a defense against guilt and perceived worthlessness (Fisher 1970; McCranie 1979; Barsky and Klerman 1983). Today, we could translate this as a blend of self-focus and negative affect. It is worth noting that hypochondriasis is a purposeful condition in this model, characterized by the so-called primary and secondary gains for the individual. Another option is that it is a learned behavior, usually acquired in childhood from parents and other models and maintained by various forms of reinforcement (these are similar to the gains described by psychoanalysts) (Pilowsky 1967, 1971; Wooley et al. 1978). Implicitly, both explanations assume a considerable discordance between bodily and psychological events: certain individuals perceive the same internal state as more intense than others.

It can also be speculated, however, that the condition relies on unusually intense body sensations due to low interoceptive thresholds; in this case, patients with clinical hypochondriasis would actually perceive and report internal changes accurately. In summary, those with this condition are more readily able to sense their body events than others, thus they experience very intense body sensations; perhaps this is why they interpret them as serious symptoms (Bianchi 1971; Mayou 1976; Barsky 1979). This idea, called the somatosensory model of hypochondriasis, is quite close to the concept of augmenting/reducing (Petrie 1967, 1978), which assumes the modulation of low-level processing of afferent signals by a general defense or protective mechanism (von Knorring et al. 1978). Indeed, an early paper reported elevated levels of cardiac accuracy for patients with clinical hypochondriasis (Tyrer et al. 1980), but subsequent studies found no or even a weak negative association between various measures of interoceptive accuracy and clinical hypochondriasis/high levels of health anxiety (Barsky et al. 1995; Haenen et al. 1997; Steptoe and Noll 1997; Krautwurst et al. 2014, 2016). Similarly, the existence of a general augmenting/reducing dimension was not supported by subsequent empirical research (Raine et al. 1981; Prescott et al. 1984; Stenberg et al. 1988; Carrillo-de-la-Peña 1992). Very similar findings were reported with respect to a related construct, somatosensory amplification (see Chap. 5); it was not related to higher levels of sensory sensitivity in a number of studies (Aronson et al. 2001; van der Veek et al. 2008; Ferentzi et al. 2017), even worse performance in a heartbeat detection task was reported (Mailloux and Brener 2002). Finally, an evoked potential study revealed that somatosensory amplification tendency is associated with long-latency cognitive processing rather than short-latency interoceptive sensitivity (Nakao et al. 2007). In a similar vein, a negative association (Pollatos et al. 2011; Schaefer et al. 2012; Krautwurst et al. 2014) or no association (Fairclough and Goodwin 2007; Witthöft et al. 2020) was reported between somatic symptom distress and cardiac perception. Thus, just like the functional disorders described by Gannon and Haynes (1986), hypochondriasis and symptom reporting rely on top-down processes rather than unusually intense sensory input. In accordance with this conclusion, the recently accepted cognitive model of health anxiety emphasizes the role of top-down processes, such as body focus and misinterpretation of signals of normal bodily functioning as symptoms of disease, in the generation and maintenance of symptoms (for more details, see Chap. 7).

6.5 Anxiety and Stress: From Repression to Cardiovascular Reactivity

It has been known for decades that there are substantial and temporally stable individual differences in the cardiovascular reaction (e.g., increase of heart rate and blood pressure) to stressful stimuli, including mental, physical, and social stressors (Manuck and Schaefer 1978; Manuck and Garland 1980; Light and Obrist 1983; Lovallo et al. 1986a, b; Kamarck et al. 1994; Sherwood et al. 1997; Burleson et al. 2003; Dragomir et al. 2014). This trait-like, more or less task-independent characteristic is called cardiovascular reactivity. It is a physiological phenomenon which is partly genetically determined (Wu et al. 2010; Neijts et al. 2015) and closely related to sympathetic reactivity with an assumed long-term negative impact on cardiovascular health (Sgoutas-Emch et al. 1994; Treiber et al. 2003; Matthews et al. 2003; Hassellund et al. 2010; Ginty et al. 2017). From our current perspective, the existence of this trait demonstrates that a situation evoking a certain subjective level of stress might be accompanied by physiological reactions of different intensity for different individuals; in other words, our perception of being stressed does not accurately reflect our bodily reactions.

Concerning interoceptive (cardioceptive) accuracy, acute stress improves cardiac perception as assessed by the Schandry task (Schandry and Specht 1981; Sturges and Goetsch 1996; Stevens et al. 2011; Schmitz et al. 2012; Schulz et al. 2013; Durlik et al. 2014), but decreases performance in forced-choice tasks (Fairclough and Goodwin 2007; Schulz et al. 2013). The latter finding can be explained by reduced attentional capacity due to sympathetic activation, which impacts more the forced-choice paradigms that require the division of attention between internal and external stimuli (Schulz and Vögele 2015). Moreover, those with higher cardiovascular reactivity show more accurate cardiac perception in both tasks (Eichler and Katkin 1994; Pollatos et al. 2007b; Herbert et al. 2010). The situation appears to be clear-cut by this point: stress activates the sympathetic system, which leads to increases in blood pressure and various parameters of the activity of the heart, and this in turn enhances our ability to sense cardiac changes.

However, if we compare the physiological and psychological aspects of acute stress response, we can find only weak to moderate associations in the vast majority of studies. The discrepancy between physiological and self-reported aspects of stress was described quite early (Weinstein et al. 1968). After re-analyzing the findings of six studies, Weinstein and colleagues found a close to zero average correlation between physiological and subjective indices of the stress reaction (the individual correlation coefficients ranged from −0.15 to 0.31). They partly attributed this discrepancy to a psychodynamic defense process, i.e., an indicator of coping with a threatening situation. This view, based on earlier work of Byrne on repression-sensitization (Byrne 1964; Byrne et al. 1968) and that of Lazarus (1966), demonstrates very well that the basic assumption about stress reaction was psychophysiological transparency, i.e., a high level of correspondence between physiology and phenomenology. Early studies on physiological reactions evoked by threatening thus possibly repressed materials by Jung (Ricksher and Jung 1907; Peterson and Jung 1907) supported this idea, and the assumption that nonconscious defense mechanisms might play a role here was also accepted by psychophysiologists, such as Neal A. Miller (1992). More recent studies replicated the dissociation between the emotionality of verbal reports and physiological changes when participants were asked to recall traumatic events (Pennebaker and Smyth 2016) or suppress the behavioral concomitants of emotions (Gross and Levenson 1993). Nonconscious (automatic) aspects of emotion regulation (Gross 1998; Gross and Thompson 2007) may also contribute to the discrepancy. In line with these considerations, it was supposed that the physiological stress response is partly activated by nonconscious perseverative cognition and implicit negative affect (Brosschot et al. 2010; Mossink et al. 2015; van der Ploeg et al. 2016; Brosschot 2017).

The aforementioned review paper by Gannon (1984) also reported a marked cognitive-physiological discordance in the case of experimentally evoked anxiety in anxiety patients and healthy controls. Recent models of the stress reaction and cardiovascular reactivity differentiate between psychological and physiological aspects of the stress response (Lovallo and Gerin 2003; Ursin and Eriksen 2004; Lovallo 2005, 2016; Cohen et al. 2016); the majority of more recent empirical studies support the existence of a considerable discrepancy. Whereas a moderate to strong association between subjective and endocrine stress response was reported in one study (Schlotz et al. 2008), effect sizes were small in another (Oldehinkel et al. 2011). No association between subjective and physiological stress reaction was found in a meta-analysis (Dickerson and Kemeny 2004). In another meta-analysis investigating the effect of psychosocial stress, only one-fourth of the included studies reported an association between perceived stress and heart rate or salivary cortisol, and two-third found a relation with plasma cortisol and ACTH levels (Campbell and Ehlert 2012). However, correlation coefficients between actual and perceived changes under stress with respect to five different modalities ranged from 0.47 (skin conductance) to 0.76 (heart rate) in one study (Steptoe and Vögele 1992). Beyond stress, the correspondence between physiological changes and emotional experience is generally low (Mauss et al. 2005; Mauss and Robinson 2009). For example, anxiety was more related to perceived than to actual heart rate in a recent study with psychological stress (Trotman et al. 2019). This topic will be discussed in more detail in Chap. 8.

What is the bottom line? Empirical research on nonconscious defense mechanisms is conceptually difficult. I think that the assumption that the discordance between objective and subjective aspects of the stress reaction is always the result on such processes is overly simplistic. Defense might play a role in certain cases (particularly for highly traumatic or threatening events), but the discrepancy remains there even for individuals without such challenges. Also, as interoceptive accuracy cannot be generalized across modalities (Chap. 4), it is difficult to accept that certain individuals suppress only certain interoceptive channels.

6.6 False Heart Rate Feedback Studies and Once Again on the Schandry Task

In the middle of the 1960s, Stuart Valins conducted two pioneering experiments (Valins 1966, 1967). Male participants were asked to view slides of seminude females and rate their attractiveness. A group received real-time acoustic feedback that was synchronous with their heartbeat, whereas feedback was manipulated (i.e., increased or decreased) for another group. Images associated with manipulated heartbeat were found significantly more attractive than those without perceived changes in arousal. These findings will be discussed with respect to the Schachter-Singer theory of emotion (see Chap. 8); for now, they simply demonstrate that external information on heartbeat can easily overwrite the internal input. More importantly, the findings generated a considerable debate concerning their background. Although the higher perceived attractiveness of images accompanied by bogus increased heart rate feedback was replicated by others (Barefoot and Straub 1971; Botto et al. 1974; Woll and McFall 1979; Parkinson and Manstead 1986), it was found that an actual increase in heart rate also takes place, which raises the possibility that the association was mediated by actual heart rate (Stern et al. 1972). However, this option was not supported by the findings of further studies (Goldstein et al. 1972; Thornton and Hagan 1976; Kerber and Coles 1978; Woll and McFall 1979). Also, the perceived emotional intensity of neutral faces was enhanced by false feedback of increased heart rate in a more recent study (Gray et al. 2007).

Recently, it also turned out that false feedback of actual heart rate influences participants' perceived effort during physical activity (Iodice et al. 2019). Those with a bogus feedback higher than the actual heart rate rated their actual effort higher than those with valid feedback and those with lower than actual false feedback. These results show that false feedback on heart rate can overwrite the actual sensory input, which is barely perceivable under rhythmic physical activity (Köteles et al. 2020), and can result in higher discrepancy between actual and perceived exertion.

Ehlers et al. (1988) were able to evoke marked psychological (anxiety, excitement) and physiological (heart rate, blood pressure, skin conductance) changes in patients with panic attack as opposed to controls with bogus feedback indicating a sudden increase in heart rate. Patients who suspected that the feedback did not reflect their physiological state showed no such reactions. In another study patients with congenital heart disease and high levels of trait anxiety reported disproportionately more heart-related symptoms after false feedback on the regularity of their heartbeats (Karsdorp et al. 2009). This is in line with previous findings about increased symptom reports and heart-related perceptual biases of anxious patients with congenital heart disease (Rietveld et al. 2004; Karsdorp et al. 2007a, b). In summary, those for whom heartbeats represent particularly salient stimuli are much more sensitive to top-down heart-related cues that indicate potential threat. Interestingly enough, perceived changes relying on false top-down information can trigger an affective reaction that in turn gives rise to changes in the felt and feared direction. This mechanism is quite close to the vicious circle of health anxiety (Chap. 7) and the nocebo phenomenon (Chap. 9).

These findings consistently show that cardiac activity belongs rather to the dissociable than the non-dissociable modalities, and it is very sensitive to top-down influences (Pennebaker 1981). The false heart rate feedback condition was accompanied by increased activity of the right anterior insula in a brain imaging study, which further supports the role of subjective-evaluative factors in the perception of heartbeats (Gray et al. 2007). This also means that paradigms designed to assess the accuracy of the perception of heartbeats should take careful steps to reduce the impact of top-down factors on participants' performance. The Schandry task described in Chap. 4 is particularly malleable to such factors as only the total amount of sensed heartbeats are taken into consideration, regardless of the temporal match between actual and felt heartbeats. People with a general tendency to overreport or those who are particularly sensitive to top-down factors, such as knowledge of heart rate (Ring and Brener 1996, 2018; Ring et al. 2015) or the tingling sensation, will achieve high scores in this task without actually being more accurate than those who exclusively rely on sensed heartbeats. In accordance with these considerations, false feedback on increased heart rate let to an increase of Schandry score in an experiment (Phillips et al. 1999). Overreporting tendency can be somewhat reduced by the use of more strict instruction (Ehlers et al. 1995; Desmedt et al. 2018) and by the controlling for other top-down factors (Corneille et al. 2020; Desmedt et al. 2020), but cannot completely be eliminated.

In our own study (Körmendi et al. 2020), participants were asked to rate their expectation with respect to their performance in the Schandry task. This variable was assessed between a short (15 s) training interval and the real trials. Although participants received no information about their performance in the training trial (in fact, results of the training trial were not evaluated at all), the association between their expected and actual performance, as assessed in the subsequent three real trials, was surprisingly strong ($r_s = 0.58$, $p < 0.001$). According to our (admittedly speculative) explanation, this very short period of time was enough for participants to build up a quite accurate cardioceptive accuracy-related expectation, which influenced their performance in the real trials. For example, those with high expectation might have counted vague sensations as heartbeats thus achieved higher scores, whereas lower expectations led to lower Schandry scores. This mechanism might be conceptually similar to that of attention (expectation) on the perception of gut stimulation demonstrated by Accarino et al. (1997).

As the Schandry task is perhaps the most widely used assessment of interoceptive accuracy, conclusions drawn from empirical results that utilized this method should definitely be reconsidered in the light of the novel findings. Also, new paradigms that can control and/or measure overreporting tendency should be developed.

6.7 Interoceptive Awareness

So far we primarily discussed the discrepancy between peripheral physiology and its perception; it was concluded that physiological changes in certain interoceptive modalities are not or only poorly perceived. This is why we called them dissociable modalities. In other words, we often have no cue about the processes actually taking place in our body. The left-hand side of Fig. 6.2 shows these discrepancies: there is a dissociation between actual and perceived body processes and another dissociation between actual body processes and the generalized (trait-like) ability to perceive them.

Interoceptive awareness, the third dimension of interoception proposed by Garfinkel and Critchley (Garfinkel and Critchley 2013; Garfinkel et al. 2015a), however, indicates the correspondence between actual detection ability (i.e., interoceptive accuracy) and people's confidence in their judgements (Fig. 6.2, right-hand side). By definition, interoceptive awareness refers to people's metacognitive awareness of their interoceptive accuracy (Garfinkel et al. 2015a) "that quantifies individuals' explicit knowledge of (and confidence in) their interoceptive accuracy" (Garfinkel et al. 2015a, p. 67).

The term itself is a bit misleading for at least two reasons. First, interoceptive awareness has been defined and operationalized in different ways in previous research. For example, it is a synonym for interoceptive sensibility, i.e., the self-reported aspects of interoception for some (Mehling et al. 2012; Ceunen et al. 2013), whereas it generally referred to both behavioral and self-reported aspects of interoception before the development of the tripartite model (Critchley et al. 2004; Pollatos et al. 2007a; Ceunen et al. 2013). Second, it is questionable to what extent it measures explicit knowledge of interoceptive accuracy. Metacognition, i.e., cognition about cognition usually refers to higher level representation of self-related materials and processes, such as the experiential aspects of emotions (Flavell 1979;

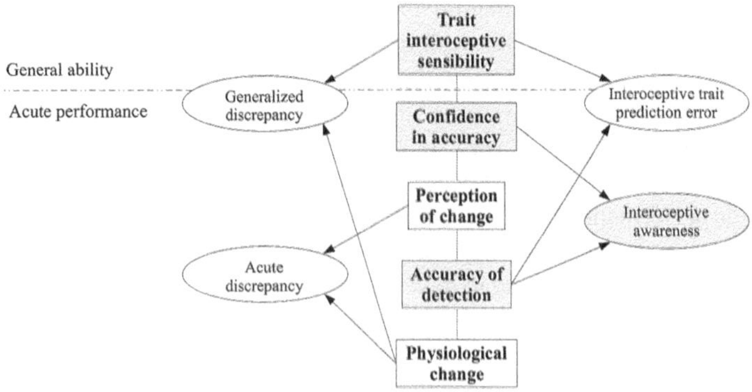

Fig. 6.2 Conceptual levels of central processing of bodily changes (in the middle) and various measures of the discrepancy between objective and self-reported levels. The major categories of the tripartite model of Garfinkel and Critchley (2015a) are marked with gray background

Teasdale et al. 2002; Mehling 2016). The term metacognitive accuracy is used to characterize the dissociation between objective and subjective performance in various tasks with an outcome that is unknown to the individual (e.g., near-threshold stimulation) (Kunimoto et al. 2001; Washburn et al. 2005; Fleming et al. 2010). Whereas type-1 performance refers to the accuracy of discrimination, type-2 performance means the ability to discriminate between correct and incorrect responses (Barttfeld et al. 2013). Interoceptive awareness is operationalized as follows. Participants are asked to rate their confidence in their perceived accuracy of response using a visual analog scale with the anchor points "Total guess/No heartbeat awareness" and "Complete confidence/Full perception of heartbeat" after each interoceptive accuracy trial (Garfinkel et al. 2015a). Concerning the Schandry score, accuracy scores achieved in the individual trials are correlated with confidence judgements and the resulting within-subject correlation coefficient is used as the measure of interoceptive awareness, also called coherence between confidence ratings and accuracy by others (Parkin et al. 2013). As an alternative measure, the difference between accuracy and confidence scores following a linear transformation into a 0–100 scale was proposed (Weineck et al. 2019). What these indices show is a mixed bag as intuitive or explicit knowledge of heart rate can substantially influence confidence ratings (see above). If one perceives and counts one heartbeat in approximately each second, he or she can be sure that the majority of heartbeats were counted. If sensed and counted heartbeats are substantially rarer, this fact may influence the confidence judgments. In this case, explicit knowledge of accuracy in the task at hand is assessed, but it partly relies on top-down information. For forced-choice discrimination tasks, correct judgments with high confidence (i.e., hits) are plotted against incorrect judgments with high confidence (i.e., false alarms) for all possible detection thresholds. The area under the receiver operating characteristics curve (Garfinkel et al. 2015a, 2016; Ewing et al. 2017; Rae et al. 2019) or alternatively the sum of the squares of the difference between accuracy scores and the respective confidence judgments divided by the number of trials (Meessen et al. 2016) are used to characterize individual interoceptive awareness. It is important to see that these measures reflect neither explicit knowledge of interoceptive accuracy nor confidence in interoceptive ability. The former should be based on feedback on actual performance, whereas the latter is a belief that does not need reality check. What is measured here is the coherence between actual detection ability (which is typically poor) and its perception (Parkin et al. 2013), thus the term "interoceptive insight" (Khalsa et al. 2018) appears more appropriate to me. Unfortunately, descriptive statistics of indicators of interoceptive awareness are rarely reported, thus little is known about their mean values and variance. Most importantly, these two approaches (i.e., tracking and discrimination-related indices) should not put in the same category (interoceptive awareness) and used as alternatives.

Whatever is assessed with these indices, perhaps it plays a substantial role in human functioning. Let us see the empirical findings! First of all, concerning the discrimination (forced-choice) approach, both cardiac and respiratory interoceptive awareness is unrelated to accuracy (Garfinkel et al. 2015a, 2016). The same was reported for the tracking-based approach (Garfinkel et al. 2015a; Meessen et al.

2016). As interoceptive awareness shows individual differences, its associations with other characteristics are conceivable. A medium-level association between anxiety and forced choice-based cardiac but not respiratory interoceptive awareness was found in one study (Garfinkel et al. 2016). Cardiac interoceptive awareness calculated from the tracking task showed a medium-level negative association with only one of the three categories of symptoms (tic impairment) of Tourette syndrome (and even this association became nonsignificant after correcting for multiple comparisons); none of the three categories were associated with interoceptive awareness calculated from a discrimination task (Rae et al. 2019). Cardiac tracking-related interoceptive awareness was negatively associated with mixed anxiety and depressive disorder and the interaction between sleep difficulties and depression/mixed anxiety and depressive disorder, and positively associated with the interaction between sleep effectiveness and mixed anxiety and depressive disorder (the strength of associations was in the moderate domain); however, no significant associations for discrimination-related awareness were revealed (Ewing et al. 2017). In another study, interoceptive awareness calculated from a heartbeat tracking task showed a moderate negative association with emotional eating and restrained eating but not with external eating and the body mass index (Young et al. 2017). Tracking-based interoceptive awareness decreased after a power posing intervention mainly because of the increase of accuracy (Weineck et al. 2019). Patients with frontotemporal dementia and Alzheimer's disease showed impaired interoceptive awareness due to the overestimation of performance; this might be related to general cognitive disabilities characterizing dementia (García-Cordero et al. 2016). Finally, no association between discrimination-related cardiac awareness score and autism spectrum disorder was revealed (Garfinkel et al. 2015b). In summary, tracking-related interoceptive awareness score often shows a weak to moderate association with various constructs (although the statistical significance of these associations is often spurious because of the high number of statistical tests carried out), whereas discrimination-related score does not. In the light of the validity issues of the mental tracking task, these already not too convincing findings appear even more spurious to me. Thus, the available empirical evidence is scarce and only partially supports the idea that interoceptive awareness plays a role in healthy psychological functioning. Just like cardioceptive accuracy, cardioceptive awareness did not improve after intense mindfulness-based meditation training; whereas trait mindfulness was weakly associated with coherence, the association was negative for the observe facet of mindfulness (Parkin et al. 2013).

Beyond interoceptive awareness, we can also estimate the average extent of "general interoceptive awareness" from the correlation between actual performance in an interoceptive accuracy task and a questionnaire assessing general interoceptive sensibility, i.e., the perceived awareness of bodily changes with respect to a number of interoceptive modalities. As mentioned in Chap. 5, the Body Perception Questionnaire (Critchley et al. 2004; Garfinkel et al. 2015a), the Body Awareness Questionnaire (Ainley and Tsakiris 2013; Emanuelsen et al. 2015; Ferentzi et al. 2017, 2018), or various scales of the MAIA (Meessen et al. 2016; Young et al. 2017) are used as measures of general interoceptive sensibility ($IS_{general}$). The calculated correlations are typically nonsignificant and close to zero; a weak positive association was reported in one study only (Zamariola et al. 2018). Thus, people's actual interoceptive accuracy is not related to their perceived general awareness of body changes at the level of the group. Just like the nonsignificant association between physiological changes and body awareness, this null finding does not necessarily mean that there is no association with respect to single modalities. To demonstrate the existence of such an association, a trait-like modality-specific indicator (e.g., assessing the generalized ability to perceive heartbeats) would be required (see Chap. 5).

A weak point of the approach discussed in the present paragraph is that it characterizes the group-level (between-subject) association but cannot provide us with individual scores. To fill this gap, a novel measure of the discrepancy between actual accuracy and general sensibility, called interoceptive trait prediction error, was proposed. It refers to the difference between the standardized values of the two constructs (Garfinkel et al. 2015b). Cardiac interoceptive trait prediction error was associated with autism spectrum disorder, trait and state anxiety, abnormal skin sensation, and emotional eating in several studies (Garfinkel et al. 2015b; Eccles et al. 2015; Young et al. 2017); the pathology-related conditions were consistently associated with higher levels of prediction error. However, no association between cardiac trait prediction error (both tracking and discrimination-based) and any indices of the Tourette syndrome was found (Rae et al. 2019).

Overall, the two measures of the discrepancy between actual and self-reported ability to sense interoceptive events (interoceptive awareness, interoceptive trait prediction error) show associations with various indicators of psychological functioning, but the available evidence is scarce and might be confounded by methodological issues.

6.8 Philosophical Speculations

At this point, I cannot resist the temptation to take a quick and highly speculative detour. As it was presented in Chaps. 4 and 5 as well as in the present chapter, we have only very limited conscious access to the functioning of our internal organs (interoceptive accuracy with respect to the visceroceptive modalities is generally low). At the same time we have marked (mis)beliefs about our ability to detect internal changes and are not aware of the discrepancy between the two. To put it simply, we think and feel that we directly sense our body, but in fact we rely on its internal representation; the background of this very strong conviction was discussed in Chap. 2 as the background of the tenet of realness. As a result, we can easily implement mental operations (voluntarily and nonvoluntarily) on the internal representation of our body (this is the tenet of malleability) and experience these changes as real. Fortunately, the actual body is not malleable to such mental operations; however, we rarely receive feedback on the actual changes, thus our beliefs often remain intact. This apparent mind-over-body constellation clearly favors the animistic form of dualistic thinking or even idealism, thus it might contribute to the general proneness of humankind to develop magical and supernatural beliefs and religions, including healing practices.

This lack of interoceptive insight has multiple practical consequences. On the one hand, it makes us sensitive to placebo interventions (including magical healing practices and many complementary and alternative techniques; Chap. 10) and nocebo influences (Chap. 9). On the other hand, if the actual state of our body is not reflected by its conscious representation, as in the case of asymptomatic or silent killer diseases, we will not feel sick, thus we will be less willing to seek medical help and adhere to treatment prescriptions. These aspects sometimes overlap which usually leads to negative consequences (e.g., cessation of medication caused by perceived improvement; Chap. 10).

6.9 On the Background of the Dissociation

Our limited ability to sense visceral changes and the discordance between actual peripheral events and their perception has been well known for many decades. Different authors attempted to explain it in different, often multiple ways. Certain explanations take an evolutionary perspective, others simply rely on biological and/or psychological facts. Thus, these ideas do not necessarily exclude each other. In this section, I try to summarize the possible reasons for and existing explanations of the actual-perceived dissociation.

From a biological (evolutionary) point of view, perhaps the most straightforward explanation is that we cannot sense the majority of internal changes because we are designed so. This is supported by the small number of visceroceptors compared to exteroceptors. As the amount of visceral input is clearly sufficient for the maintenance of homeostasis via low-level (nonconscious) regulatory loops, the scarcity of conscious information simply reflects the fact that our brainstem does as excellent job and behavioral interventions are rarely required. The fact that sensory innervation is significantly greater in visceral regions which are involved in behavioral regulation, e.g., that of the distal part of the colon and the rectum compared to the midproximal colon (Kyloh et al. 2011; Keszthelyi et al. 2012), also supports this idea. Even if behavioral change is mandated, gross overall changes in the homeostatic status may be more important than minor fluctuations in certain modalities, thus only the former reach consciousness (Pennebaker 1982; Ferentzi et al. 2018). The evolutionary approach of Craig (2003, 2009, 2011, 2015) also emphasizes the adaptive value of the overall evaluation of the homeostatic state of the body. In summary, natural selection was able to shape the hardware of body sensations in cases which pay off; in other cases, the additional metabolic and other costs would have been too high compared to the benefits, thus conscious access was not implemented (Metzinger 2009).

From a cultural point of view, a substantial problem with internal events, including body sensations, is that they cannot be cross-verified by others and socially reinforced, i.e., the experience of different individuals cannot be directly compared (Mandler et al. 1958; Hohwy 2014). Moreover, interoceptive channels show less overlap and interaction with each other than exteroceptive channels; in consequence, incoming information cannot be cross-verified by other channels during development thus necessarily remains vague and difficult to identify and verbalize (Bahrick and Lickliter 2012). This leads to a shortage of words describing body sensations (Ruckmick 1936; Mason 1959, 1961; Davitz 1969; Ádám 1998), which in turn impacts their perception and reporting. Another related process is the change from interoception to exteroception during development in early childhood, which may reflect the evolutionary priority of exteroceptive information over interoception. This shift partly relies on learning processes but has a firm biological background (Ádám 1967). The limitedness of cognitive (attentional and processing) capacity also supports this process, which results in the preference of exteroceptive over interoceptive information (Pennebaker and Lightner 1980; Hohwy 2014).

The most important characteristic of perception, i.e., its reliance on top-down information, also contributes to the dissociation (Pennebaker 1982). In the absence of clear-cut bottom-up information (using the terminology of the predictive processing approach, the majority of sensory input are "explained away" in low levels of processing), expectations (priors) can shape conscious experience more than in the case of exteroception (Barrett and Simmons 2015). This idea is not novel either; Ádám (1998) described it as the consequence of life-long "overlearning" of interoceptive information.

Finally, visceral and interoceptive illusions suggest that issues in multisensory integration can also contribute to the dissociation; for example, visual and tactile cues are able to modify interoceptive information, such as pain and thermosensation in the rubber hand illusion (Kammers et al. 2011; Hänsel et al. 2011; Hegedüs et al. 2014; Siedlecka et al. 2014, 2018; Giummarra et al. 2015).

6.10 What About Non-dissociable Modalities?

As discussed in Chap. 4, the forced-choice paradigm is not really applicable to modalities that are readily accessible for conscious processing under normal circumstances. Sometimes it can be adapted to these modalities, and individual differences in the detection of slight differences between stimuli can be measured. What these tasks assess is the discrimination threshold, i.e., the ability to sense very small differences in respiratory resistance, skin temperature, or nociceptive stimulation. However, they do not inform us about the exact association between actual and perceived stimulation, which might be more relevant from a practical point of view. Interoceptive accuracy with respect to non-dissociable modalities is typically assessed by comparing self-reported and actual intensity of stimuli, i.e., interoceptive sensibility and physiology. For example, verbal reports are required for the assessment of the perception of pain, respiratory parameters (breathlessness, frequency, and depth of breathing), gastric distension, and affective touch (for details, see Chap. 4). Thus, the distinction between actual accuracy and sensibility is not as clear-cut here as for the dissociable modalities. Also, the aforementioned indicators of interoceptive awareness cannot be calculated for these modalities. From a conceptual point of view, although the stimulus–response relation may show marked individual differences, these differences cannot be regarded as dissociation, even less as differences in confidence.

However, in the case of pain, the sensation that perhaps most readily accesses conscious awareness, a discrepancy between self-reported general and actual measures of sensitivity (e.g., threshold) was found (Edwards and Fillingim 2007; Meiselles et al. 2017). Moreover, only a medium-level association between actual pain threshold and self-reported severity of pain ($r = -0.41$) and between discomfort threshold and self-reported bloating ($r = 0.35$) was found in patients with irritable bowel syndrome; the associations between thresholds and other self-reported aspects of the condition were even lower (Posserud et al. 2007). Also, substantial individual differences in thermosensation-related and pain-related metacognition (metacognitive efficiency and bias) were found in a recent study using a forced-choice discrimination task (Beck et al. 2019). Thus, top-down biases play a role even for non-dissociable modalities.

6.11 Summary

In this chapter, evidence from multiple lines of research was presented for the existence of a more or less marked discrepancy (1) between perceived body sensations and changes that actually take place in the body, and (2) between the actual and perceived ability to sense bodily changes (tenet of dissociation).

Concerning the first case, it was concluded that there is a near complete dissociation for certain interoceptive (mainly visceroceptive) modalities. As mentioned in Chap. 3, low-level physiological regulation can effectively control these functions and activities, thus their appearance in conscious awareness is not required, perhaps even maladaptive, for healthy functioning. As we have no clear-cut conscious information about the actual state of these organs and activities, there is a place for "body hallucinations," caused by top-down factors, that can have harmful consequences (see Chaps. 7, 9, and 10). The modalities we called non-dissociable, however, are accessible for conscious awareness because they play a role in the regulation of behavior, including motor control. Still, there are marked individual differences in the sensitivity to these modalities (e.g., pain, distension of the stomach) which can also lead to a negative outcome in extreme cases (Chap. 4).

Interoceptive awareness, i.e., the concordance between actual and believed ability to detect internal changes, is a bit different. Detection ability is rarely perfect, thus we usually detect only a proportion of the actual events (e.g., heartbeats). If the actual ability (interoceptive accuracy) is low, i.e., detection performance is close to random guessing, which is the case for non-dissociable modalities, confidence is difficult to understand. Moreover, confidence ratings might be influenced by a number of general (i.e., non-interoceptive) cognitive-emotional factors which impact other perceptual processes too. Therefore, an association between interoceptive awareness and other characteristics might represent the classic third variable problem. As the number of reported significant results is quite low, both in absolute terms and relatively, compared to the overall number of tested associations in the studies, one cannot state that interoceptive awareness plays a role in healthy psychological functioning.

Concerning tripartite model of Garfinkel and Critchley, the three major categories (i.e., interoceptive accuracy, sensibility, and awareness) are not related to each other (Garfinkel et al. 2015a, 2016; Forkmann et al. 2016). In my view, this represents a clear limitation of the model beyond the validity issues discussed in Chaps. 5 and 6. Moreover, the approach is purely descriptive: what we have is three distinct categories with an unknown relationship. In consequence, this model does not enable us to create testable hypotheses in order to better understand and predict the flow of information from the periphery to the cortex and the other way around. Beyond the three major dimensions, other categories influencing or representing different levels of processing of interoceptive information were proposed recently. For example, the overall physiological state of the body might be such a category (Forkmann et al. 2019). Critchley and Garfinkel (2017) also included two additional dimensions (preconscious processing and executive aspects) in the most recent version of the model. Still, it remains descriptive and static even with these additions. Extensions with respect to aspects of attentional (Murphy et al. 2019a, b) and affective-evaluative processing (Farb and Logie 2018; Herbert and Pollatos 2018) were also proposed. I think these aspects are more promising because they might explain the dynamics of human psycho(physio)logical functioning. This will be the primary topic of the next chapter.

Take-Home Messages

For laypeople: Although we often experience body sensations, these feelings do not always reflect actual body processes. In other cases, we do not sense actual body processes.

For clinicians: Body sensations reported by patients are real and very convincing experiences, but they are not necessarily associated with peripheral physiological events. This dissociation is automatic and normal; it does not indicate malfunctioning or the deliberate fabrication or faking of symptoms.

For researchers: The discrepancy between actual and perceived body events is well-documented in the literature. Less known are the mechanisms of dissociation and the possible interactions between the physiological and psychological levels.

References

Accarino AM, Azpiroz F, Malagelada JR (1997) Attention and distraction: effects on gut perception. Gastroenterology 113:415–422. https://doi.org/10.1053/gast.1997.v113.pm9247458

Ádám G (1967) Interoception and behavior. Akadémiai Kiadó, Budapest

Ádám G (1998) Visceral perception: understanding internal cognition. Plenum Press, New York

Ainley V, Tsakiris M (2013) Body conscious? Interoceptive awareness, measured by heartbeat perception, is negatively correlated with self-objectification. PLoS One 8:e55568. https://doi.org/10.1371/journal.pone.0055568

APA (1980) Diagnostic and statistical manual of mental disorders, 3rd edn. American Psychiatric Association, Arlington, VA

APA (1987) Diagnostic and statistical manual of mental disorders, 3rd Rev edn. American Psychiatric Association, Arlington, VA

APA (1994) Diagnostic and statistical manual of mental disorders, 4th edn. American Psychiatric Association, Arlington, VA

APA (2013) Diagnostic and statistical manual of mental disorders, 5th edn. American Psychiatric Association, Arlington, VA

Aronson KR, Barrett LF, Quigley KS (2001) Feeling your body or feeling badly: evidence for the limited validity of the Somatosensory Amplification Scale as an index of somatic sensitivity. J Psychosom Res 51:387–394

Babulka P, Berkes T, Szemerszky R, Köteles F (2017) No effects of rosemary and lavender essential oil and a placebo pill on sustained attention, alertness, and heart rate. Flavour Fragr J 32:305–311. https://doi.org/10.1002/ffj.3392

Bahrick LE, Lickliter R (2012) The role of intersensory redundancy in early perceptual, cognitive, and social development. In: Bremner AJ, Lewkowicz DJ, Spence C (eds) Multisensory development. Oxford University Press, Oxford, UK, pp 183–206

Barefoot JC, Straub RB (1971) Opportunity for information search and the effect of false heart-rate feedback. J Pers Soc Psychol 17:154–157. https://doi.org/10.1037/h0030383

Barrett LF, Simmons WK (2015) Interoceptive predictions in the brain. Nat Rev Neurosci 16:419–429. https://doi.org/10.1038/nrn3950

Barsky AJ (1979) Patients who amplify bodily sensations. Ann Intern Med 91:63–70. https://doi.org/10.1059/0003-4819-91-1-63

Barsky AJ (2001) Palpitations, arrhythmias, and awareness of cardiac activity. Ann Intern Med 134:832–837

Barsky AJ, Klerman GL (1983) Overview: hypochondriasis, bodily complaints, and somatic styles. Am J Psychiatry 140:273–283. https://doi.org/10.1176/ajp.140.3.273

Barsky AJ, Wyshak G, Klerman GL (1986) Hypochondriasis. An evaluation of the DSM-III criteria in medical outpatients. Arch Gen Psychiatry 43:493–500

Barsky AJ, Goodson JD, Lane RS, Cleary PD (1988) The amplification of somatic symptoms. Psychosom Med 50:510–519

Barsky AJ, Brener J, Coeytaux RR, Cleary PD (1995) Accurate awareness of heartbeat in hypochondriacal and non-hypochondriacal patients. J Psychosom Res 39:489–497

Barttfeld P, Wicker B, McAleer P et al (2013) Distinct patterns of functional brain connectivity correlate with objective performance and subjective beliefs. Proc Natl Acad Sci U S A 110:11577–11582. https://doi.org/10.1073/pnas.1301353110

Beck B, Peña-Vivas V, Fleming S, Haggard P (2019) Metacognition across sensory modalities: vision, warmth, and nociceptive pain. Cognition 186:32–41. https://doi.org/10.1016/j.cognition.2019.01.018

Bianchi GN (1971) Origins of disease phobia. Aust N Z J Psychiatry 5:241–257. https://doi.org/10.1080/00048677109159654

Botto RW, Galbraith GG, Stern RM (1974) Effects of false heart rate feedback and sex-guilt upon attitudes toward sexual stimuli. Psychol Rep 35:267–274. https://doi.org/10.2466/pr0.1974.35.1.267

Brosschot JF (2017) Ever at the ready for events that never happen. Eur J Psychotraumatol 8:1309934. https://doi.org/10.1080/20008198.2017.1309934
Brosschot JF, Verkuil B, Thayer JF (2010) Conscious and unconscious perseverative cognition: is a large part of prolonged physiological activity due to unconscious stress? J Psychosom Res 69:407–416. https://doi.org/10.1016/j.jpsychores.2010.02.002
Burleson MH, Poehlmann KM, Hawkley LC et al (2003) Neuroendocrine and cardiovascular reactivity to stress in mid-aged and older women: long-term temporal consistency of individual differences. Psychophysiology 40:358–369. https://doi.org/10.1111/1469-8986.00039
Buss AH (1980) Self-consciousness and social anxiety. W.H. Freeman & Co Ltd, San Francisco
Byrne D (1964) Repression-sensitization as a dimension of personality. Prog Exp Pers Res 72:169–220
Byrne D, Steinberg MA, Schwartz MS (1968) Relationship between repression-sensitization and physical illness. J Abnorm Psychol 73:154–155
Campbell J, Ehlert U (2012) Acute psychosocial stress: does the emotional stress response correspond with physiological responses? Psychoneuroendocrinology 37:1111–1134. https://doi.org/10.1016/j.psyneuen.2011.12.010
Carrillo-de-la-Peña MT (1992) ERP augmenting/reducing and sensation seeking: a critical review. Int J Psychophysiol 12:211–220. https://doi.org/10.1016/0167-8760(92)90059-K
Carver CS, Blaney PH (1977a) Perceived arousal, focus of attention, and avoidance behavior. J Abnorm Psychol 86:154–162
Carver CS, Blaney PH (1977b) Avoidance behavior and perceived arousal. Motiv Emot 1:61–73. https://doi.org/10.1007/BF00997581
Carver CS, Scheier MF (1978) Self-focusing effects of dispositional self-consciousness, mirror presence, and audience presence. J Pers Soc Psychol 36:324–332. https://doi.org/10.1037/0022-3514.36.3.324
Carver CS, Scheier MF (1981) Attention and self-regulation: a control-theory approach to human behavior. Springer, New York
Ceunen E, Van Diest I, Vlaeyen JWS (2013) Accuracy and awareness of perception: related, yet distinct (commentary on Herbert et al., 2012). Biol Psychol 92:426–427. https://doi.org/10.1016/j.biopsycho.2012.09.012
Cohen S, Gianaros PJ, Manuck SB (2016) A stage model of stress and disease. Perspect Psychol Sci 11:456–463. https://doi.org/10.1177/1745691616646305
Corneille O, Desmedt O, Zamariola G et al (2020) A heartfelt response to Zimprich et al. (2020), and Ainley et al. (2020)'s commentaries: acknowledging issues with the HCT would benefit interoception research. Biol Psychol 152:107869. https://doi.org/10.1016/j.biopsycho.2020.107869
Cox DJ, Clarke WL, Gonder-Frederick L et al (1985) Accuracy of perceiving blood glucose in IDDM. Diabetes Care 8:529–536. https://doi.org/10.2337/diacare.8.6.529
Craig AD (2003) Interoception: the sense of the physiological condition of the body. Curr Opin Neurobiol 13:500–505. https://doi.org/10.1016/S0959-4388(03)00090-4
Craig AD (2009) How do you feel—now? The anterior insula and human awareness. Nat Rev Neurosci 10:59–70. https://doi.org/10.1038/nrn2555
Craig AD (2011) Significance of the insula for the evolution of human awareness of feelings from the body. Ann N Y Acad Sci 1225:72–82. https://doi.org/10.1111/j.1749-6632.2011.05990.x
Craig AD (2015) How do you feel? An interoceptive moment with your neurobiological self. Princeton University Press, Princeton, NJ
Critchley HD, Garfinkel SN (2017) Interoception and emotion. Curr Opin Psychol 17:7–14. https://doi.org/10.1016/j.copsyc.2017.04.020
Critchley HD, Wiens S, Rotshtein P et al (2004) Neural systems supporting interoceptive awareness. Nat Neurosci 7:189–195. https://doi.org/10.1038/nn1176
Davitz JR (1969) The language of emotion. Academic, New York
Desmedt O, Luminet O, Corneille O (2018) The heartbeat counting task largely involves non-interoceptive processes: evidence from both the original and an adapted counting task. Biol Psychol 138:185–188. https://doi.org/10.1016/j.biopsycho.2018.09.004

Desmedt O, Corneille O, Luminet O et al (2020) Contribution of time estimation and knowledge to heartbeat counting task performance under original and adapted instructions. Biol Psychol 154:107904. https://doi.org/10.1016/j.biopsycho.2020.107904

Dickerson SS, Kemeny ME (2004) Acute stressors and cortisol responses: a theoretical integration and synthesis of laboratory research. Psychol Bull 130:355–391. https://doi.org/10.1037/0033-2909.130.3.355

Dragomir AI, Gentile C, Nolan RP, D'Antono B (2014) Three-year stability of cardiovascular and autonomic nervous system responses to psychological stress. Psychophysiology 51:921–931. https://doi.org/10.1111/psyp.12231

Durlik C, Brown G, Tsakiris M (2014) Enhanced interoceptive awareness during anticipation of public speaking is associated with fear of negative evaluation. Cognit Emot 28:530–540. https://doi.org/10.1080/02699931.2013.832654

Eccles JA, Garfinkel SN, Harrison NA et al (2015) Sensations of skin infestation linked to abnormal frontolimbic brain reactivity and differences in self-representation. Neuropsychologia 77:90–96. https://doi.org/10.1016/j.neuropsychologia.2015.08.006

Edwards RR, Fillingim RB (2007) Self-reported pain sensitivity. Eur J Pain 11:594–598. https://doi.org/10.1016/j.ejpain.2006.09.008

Ehlers A, Margraf J, Roth WT et al (1988) Anxiety induced by false heart rate feedback in patients with panic disorder. Behav Res Ther 26:1–11. https://doi.org/10.1016/0005-7967(88)90028-9

Ehlers A, Breuer P, Dohn D, Fiegenbaum W (1995) Heartbeat perception and panic disorder: possible explanations for discrepant findings. Behav Res Ther 33:69–76

Eichler S, Katkin ES (1994) The relationship between cardiovascular reactivity and heartbeat detection. Psychophysiology 31:229–234

Emanuelsen L, Drew R, Köteles F (2015) Interoceptive sensitivity, body image dissatisfaction, and body awareness in healthy individuals. Scand J Psychol 56:167–174. https://doi.org/10.1111/sjop.12183

Engel BT (1960) Stimulus-response and individual-response specificity. JAMA Psychiat 2:305–313. https://doi.org/10.1001/archpsyc.1960.03590090061010

Ewing DL, Manassei M, Gould van Praag C et al (2017) Sleep and the heart: interoceptive differences linked to poor experiential sleep quality in anxiety and depression. Biol Psychol 127:163–172. https://doi.org/10.1016/j.biopsycho.2017.05.011

Fairclough SH, Goodwin L (2007) The effect of psychological stress and relaxation on interoceptive accuracy: implications for symptom perception. J Psychosom Res 62:289–295. https://doi.org/10.1016/j.jpsychores.2006.10.017

Farb NA, Logie K (2018) Interoceptive appraisal and mental health. In: Tsakiris M, De Preester H (eds) The interoceptive mind. From homeostasis to awareness. Oxford University Press, Oxford, England, pp 227–241

Farb NA, Daubenmier JJ, Price CJ et al (2015) Interoception, contemplative practice, and health. Front Psychol 6:1–26. https://doi.org/10.3389/fpsyg.2015.00763

Fenigstein A, Scheier MF, Buss AH (1975) Public and private self-consciousness: assessment and theory. J Consult Clin Psychol 43:522–527. https://doi.org/10.1037/h0076760

Ferentzi E, Köteles F, Csala B et al (2017) What makes sense in our body? Personality and sensory correlates of body awareness and somatosensory amplification. Personal Individ Differ 104:75–81. https://doi.org/10.1016/j.paid.2016.07.034

Ferentzi E, Drew R, Tihanyi BT, Köteles F (2018) Interoceptive accuracy and body awareness – temporal and longitudinal associations in a non-clinical sample. Physiol Behav 184:100–107. https://doi.org/10.1016/j.physbeh.2017.11.015

Ferguson E (2009) A taxometric analysis of health anxiety. Psychol Med 39:277–285. https://doi.org/10.1017/S0033291708003322

Fillmore MT, Vogel-Sprott M (1992) Expected effect of caffeine on motor performance predicts the type of response to placebo. Psychopharmacology 106:209–214

Fillmore MT, Vogel-Sprott M (1994) Psychomotor performance under alcohol and under caffeine: expectancy and pharmacological effects. Exp Clin Psychopharmacol 2:319–327

Fillmore MT, Vogel-Sprott M (1995) Behavioral effects of alcohol in novice and experienced drinkers: alcohol expectancies and impairment. Psychopharmacology 122:175–181. https://doi.org/10.1007/BF02246092

Fillmore MT, Mulvihill LE, Vogel-Sprott M (1994) The expected drug and its expected effect interact to determine placebo responses to alcohol and caffeine. Psychopharmacology 115:383–388

Fisher S (1970) Body experience in fantasy and behavior. Ardent Media, New York

Flavell JH (1979) Metacognition and cognitive monitoring: a new area of cognitive–developmental inquiry. Am Psychol 34:906–911. https://doi.org/10.1037/0003-066X.34.10.906

Fleming SM, Weil RS, Nagy Z et al (2010) Relating introspective accuracy to individual differences in brain structure. Science 329:1541–1543. https://doi.org/10.1126/science.1191883

Forkmann T, Scherer A, Meessen J et al (2016) Making sense of what you sense: disentangling interoceptive awareness, sensibility and accuracy. Int J Psychophysiol 109:71–80. https://doi.org/10.1016/j.ijpsycho.2016.09.019

Forkmann T, Volz-Sidiropoulou E, Helbing T et al (2019) Sense it and use it: interoceptive accuracy and sensibility in suicide ideators. BMC Psychiatry 19:334. https://doi.org/10.1186/s12888-019-2322-1

Gannon L (1984) Awareness of internal cues and concordance among verbal behavioral, and physiological systems in dysfunction. Psychol Rep 54:631–650. https://doi.org/10.2466/pr0.1984.54.2.631

Gannon LR, Haynes SN (1986) Cognitive-physiological discordance as an etiological factor in psychophysiologic disorders. Adv Behav Res Ther 8:223–236. https://doi.org/10.1016/0146-6402(86)90006-8

García-Cordero I, Sedeño L, de la Fuente L et al (2016) Feeling, learning from and being aware of inner states: interoceptive dimensions in neurodegeneration and stroke. Philos Trans R Soc Lond B Biol Sci 371:20160006. https://doi.org/10.1098/rstb.2016.0006

Garfinkel SN, Critchley HD (2013) Interoception, emotion and brain: new insights link internal physiology to social behaviour. Commentary on: "Anterior insular cortex mediates bodily sensibility and social anxiety" by Terasawa et al. (2012). Soc Cogn Affect Neurosci 8:231–234. https://doi.org/10.1093/scan/nss140

Garfinkel SN, Seth AK, Barrett AB et al (2015a) Knowing your own heart: distinguishing interoceptive accuracy from interoceptive awareness. Biol Psychol 104:65–74. https://doi.org/10.1016/j.biopsycho.2014.11.004

Garfinkel SN, Tiley C, O'Keeffe S et al (2015b) Discrepancies between dimensions of interoception in autism: implications for emotion and anxiety. Biol Psychol 114:117–126. https://doi.org/10.1016/j.biopsycho.2015.12.003

Garfinkel SN, Manassei MF, Hamilton-Fletcher G et al (2016) Interoceptive dimensions across cardiac and respiratory axes. Philos Trans R Soc Lond B Biol Sci 371:20160014. https://doi.org/10.1098/rstb.2016.0014

Gibbons FX (1983) Self-attention and self-report: the "veridicality" hypothesis. J Pers 51:517–542. https://doi.org/10.1111/j.1467-6494.1983.tb00343.x

Gibbons FX (1990) Self-attention and behavior: a review and theoretical update. In: Zanna MP (ed) Advances in experimental social psychology. Academic, New York, pp 249–303

Gibbons FX, Carver CS, Scheier MF, Hormuth SE (1979) Self-focused attention and the placebo effect: fooling some of the people some of the time. J Exp Soc Psychol 15:263–274. https://doi.org/10.1016/0022-1031(79)90037-4

Gillis R, Carver CS (1980) Self-focus and estimation of heart rate following physical exertion. Bull Psychon Soc 15:118–120. https://doi.org/10.3758/BF03334484

Ginty AT, Kraynak TE, Fisher JP, Gianaros PJ (2017) Cardiovascular and autonomic reactivity to psychological stress: neurophysiological substrates and links to cardiovascular disease. Auton Neurosci 207:2–9. https://doi.org/10.1016/j.autneu.2017.03.003

Giummarra MJ, Georgiou-Karistianis N, Verdejo-Garcia A, Gibson SJ (2015) Feeling the burn: when it looks like it hurts, and belongs to me, it really does hurt more. Conscious Cogn 36:314–326. https://doi.org/10.1016/j.concog.2015.07.010

Goldstein D, Fink D, Mettee DR (1972) Cognition of arousal and actual arousal as determinants of emotion. J Pers Soc Psychol 21:41–51. https://doi.org/10.1037/h0031873

Gonder-Frederick LA, Cox DJ (1991) Symptom perception, symptom beliefs, and blood glucose discrimination in the self-treatment of insulin-dependent diabetes. In: Skelton JA, Croyle RT (eds) Mental representation in health and illness. Springer, New York, pp 220–246

Gray MA, Harrison NA, Wiens S, Critchley HD (2007) Modulation of emotional appraisal by false physiological feedback during fMRI. PLoS One 2:e546. https://doi.org/10.1371/journal.pone.0000546

Gross JJ (1998) The emerging field of emotion regulation: an integrative review. Rev Gen Psychol 2:271–299. https://doi.org/10.1037/1089-2680.2.3.271

Gross JJ, Levenson RW (1993) Emotional suppression: physiology, self-report, and expressive behavior. J Pers Soc Psychol 64:970–986. https://doi.org/10.1037/0022-3514.64.6.970

Gross JJ, Thompson RA (2007) Emotion regulation: conceptual foundations. In: Gross JJ (ed) Handbook of emotion regulation. Guilford Press, New York, pp 3–24

Haenen MA, Schmidi AJ, Schoenmakers M, van den Hout MA (1997) Tactual sensitivity in hypochondriasis. Psychother Psychosom 66:128–132. https://doi.org/10.1159/000289122

Hänsel A, Lenggenhager B, von Känel R et al (2011) Seeing and identifying with a virtual body decreases pain perception. Eur J Pain 15:874–879. https://doi.org/10.1016/j.ejpain.2011.03.013

Harrell PT, Juliano LM (2012) A direct test of the influence of nicotine response expectancies on the subjective and cognitive effects of smoking. Exp Clin Psychopharmacol 20:278–286. https://doi.org/10.1037/a0028652

Hassellund SS, Arnljot F, Leiv S et al (2010) Long-term stability of cardiovascular and catecholamine responses to stress tests. Hypertension 55:131–136. https://doi.org/10.1161/HYPERTENSIONAHA.109.143164

Head H, Holmes G (1911) Sensory disturbances from cerebral lesions. Brain 34:102–254. https://doi.org/10.1093/brain/34.2-3.102

Hegedüs G, Darnai G, Szolcsányi T et al (2014) The rubber hand illusion increases heat pain threshold. Eur J Pain 18:1173–1181. https://doi.org/10.1002/j.1532-2149.2014.00466.x

Herbert BM, Pollatos O (2018) The relevance of interoception for eating behavior and eating disorders. In: Tsakiris M, De Preester H (eds) The interoceptive mind. From homeostasis to awareness. Oxford University Press, Oxford, England, pp 165–186

Herbert BM, Pollatos O, Flor H et al (2010) Cardiac awareness and autonomic cardiac reactivity during emotional picture viewing and mental stress. Psychophysiology 47:342–354. https://doi.org/10.1111/j.1469-8986.2009.00931.x

Hohwy J (2014) The predictive mind, 1st edition. Oxford University Press, Oxford, UK

Hull JG, Bond CF Jr (1986) Social and behavioral consequences of alcohol consumption and expectancy: a meta-analysis. Psychol Bull 99:347–360

Hull JG, Levenson RW, Young RD, Sher KJ (1983) Self-awareness-reducing effects of alcohol consumption. J Pers Soc Psychol 44:461–473. https://doi.org/10.1037/0022-3514.44.3.461

Iodice P, Porciello G, Bufalari I et al (2019) An interoceptive illusion of effort induced by false heart-rate feedback. Proc Natl Acad Sci U S A 116:13897–13902. https://doi.org/10.1073/pnas.1821032116

Jones GE, O'Leary RT, Pipkin BL (1984) Comparison of the Brener-Jones and Whitehead procedures for assessing cardiac awareness. Psychophysiology 21:143–148. https://doi.org/10.1111/j.1469-8986.1984.tb00196.x

Kamarck TW, Jennings JR, Pogue-Geile M, Manuck SB (1994) A multidimensional measurement model for cardiovascular reactivity: stability and cross-validation in two adult samples. Health Psychol 13:471–478

Kammers MPM, Rose K, Haggard P (2011) Feeling numb: temperature, but not thermal pain, modulates feeling of body ownership. Neuropsychologia 49:1316–1321. https://doi.org/10.1016/j.neuropsychologia.2011.02.039

Karsdorp PA, Kindt M, Everaerd W, Mulder BJM (2007a) Preattentive processing of heart cues and the perception of heart symptoms in congenital heart disease. Behav Res Ther 45:1893–1902. https://doi.org/10.1016/j.brat.2007.02.010

Karsdorp PA, Kindt M, Rietveld S et al (2007b) Stress-induced heart symptoms and perceptual biases in patients with congenital heart disease. Int J Cardiol 114:352–357. https://doi.org/10.1016/j.ijcard.2006.02.004

Karsdorp PA, Kindt M, Rietveld S et al (2009) False heart rate feedback and the perception of heart symptoms in patients with congenital heart disease and anxiety. Int J Behav Med 16:81–88. https://doi.org/10.1007/s12529-008-9001-9

Katz J, Melzack R (1990) Pain "memories" in phantom limbs: review and clinical observations. Pain 43:319–336

Kerber KW, Coles MG (1978) The role of perceived physiological activity in affective judgments. J Exp Soc Psychol 14:419–433. https://doi.org/10.1016/0022-1031(78)90039-2

Keszthelyi D, Troost FJ, Masclee AA (2012) Irritable bowel syndrome: methods, mechanisms, and pathophysiology. Methods to assess visceral hypersensitivity in irritable bowel syndrome. Am J Physiol Gastrointest Liver Physiol 303:G141–G154. https://doi.org/10.1152/ajpgi.00060.2012

Khalsa SS, Adolphs R, Cameron OG et al (2018) Interoception and mental health: a roadmap. Biol Psychiatry Cogn Neurosci Neuroimaging 3:501–513. https://doi.org/10.1016/j.bpsc.2017.12.004

Kirsch I, Rosadino MJ (1993) Do double-blind studies with informed consent yield externally valid results? An empirical test. Psychopharmacology 110:437–442

Kirsch I, Weixel LJ (1988) Double-blind versus deceptive administration of a placebo. Behav Neurosci 102:319–323

Körmendi J, Ferentzi E, Köteles F (2020) Expectation predicts performance in the mental heartbeat tracking task. Manuscript submitted for publication

Köteles F, Babulka P (2014) Role of expectations and pleasantness of essential oils in their acute effects. Acta Physiol Hung 101:329–340. https://doi.org/10.1556/APhysiol.101.2014.3.8

Köteles F, Babulka P, Szemerszky R et al (2018) Inhaled peppermint, rosemary and eucalyptus essential oils do not change spirometry in healthy individuals. Physiol Behav 194:319–323. https://doi.org/10.1016/j.physbeh.2018.06.022

Köteles F, Éliás I, Szabolcs Z et al (2020) Accuracy of reproduction of physical training load is not associated with resting heartbeat perception in healthy individuals. Biol Psychol 150:107831. https://doi.org/10.1016/j.biopsycho.2019.107831

Krautwurst S, Gerlach AL, Gomille L et al (2014) Health anxiety—an indicator of higher interoceptive sensitivity? J Behav Ther Exp Psychiatry 45:303–309. https://doi.org/10.1016/j.jbtep.2014.02.001

Krautwurst S, Gerlach AL, Witthöft M (2016) Interoception in pathological health anxiety. J Abnorm Psychol 125:1179–1184. https://doi.org/10.1037/abn0000210

Kunimoto C, Miller J, Pashler H (2001) Confidence and accuracy of near-threshold discrimination responses. Conscious Cogn 10:294–340. https://doi.org/10.1006/ccog.2000.0494

Kyloh M, Nicholas S, Zagorodnyuk VP et al (2011) Identification of the visceral pain pathway activated by noxious colorectal distension in mice. Front Neurosci 5:16. https://doi.org/10.3389/fnins.2011.00016

Lacey JI (1967) Somatic response patterning and stress: some revisions of activation theory. In: Appley MH, Thumbull R (eds) Psychological stress: issues in research. Appletin-Century-Crofts, New York, pp 14–42

Lazarus RS (1966) Psychological stress and the coping process. McGraw-Hill, New York

Levine DW, McDonald PJ (1981) Self-awareness and the veracity hypothesis. Personal Soc Psychol Bull 7:655–660. https://doi.org/10.1177/014616728174023

Light KC, Obrist PA (1983) Task difficulty, heart rate reactivity, and cardiovascular responses to an appetitive reaction time task. Psychophysiology 20:301–312. https://doi.org/10.1111/j.1469-8986.1983.tb02158.x

Longley SL, Broman-Fulks JJ, Calamari JE et al (2010) A taxometric study of hypochondriasis symptoms. Behav Ther 41:505–514. https://doi.org/10.1016/j.beth.2010.02.002
Lovallo WR (2005) Cardiovascular reactivity: mechanisms and pathways to cardiovascular disease. Int J Psychophysiol 58:119–132. https://doi.org/10.1016/j.ijpsycho.2004.11.007
Lovallo WR (2016) Stress and health. Biological and psychological interactions. Sage Publications, Inc, London
Lovallo W, Gerin W (2003) Psychophysiological reactivity: mechanisms and pathways to cardiovascular disease. Psychosom Med 65:36–45. https://doi.org/10.1097/01.PSY.0000033128.44101.C1
Lovallo WR, Pincomb GA, Wilson MF (1986a) Heart rate reactivity and type A behavior as modifiers of physiological response to active and passive coping. Psychophysiology 23:105–112. https://doi.org/10.1111/j.1469-8986.1986.tb00603.x
Lovallo WR, Pincomb GA, Wilson MF (1986b) Predicting response to a reaction time task: heart rate reactivity compared with type A behavior. Psychophysiology 23:648–656. https://doi.org/10.1111/j.1469-8986.1986.tb00686.x
Mailloux J, Brener J (2002) Somatosensory amplification and its relationship to heartbeat detection ability. Psychosom Med 64:353–357
Mandler G, Kahn M (1960) Discrimination of changes in heart rate: two unsuccessful attempts. J Exp Anal Behav 3:21–25. https://doi.org/10.1901/jeab.1960.3-21
Mandler G, Kremen I (1958) Autonomic feedback: a correlational study. J Pers 26:388–399. https://doi.org/10.1111/j.1467-6494.1958.tb01594.x
Mandler G, Mandler JM, Uviller ET (1958) Autonomic feedback: the perception of autonomic activity. J Abnorm Soc Psychol 56:367–373. https://doi.org/10.1037/h0048083
Manuck SB, Garland FN (1980) Stability of individual differences in cardiovascular reactivity: a thirteen month follow-up. Physiol Behav 24:621–624. https://doi.org/10.1016/0031-9384(80)90261-9
Manuck SB, Schaefer DC (1978) Stability of individual differences in cardiovascular reactivity. Physiol Behav 21:675–678. https://doi.org/10.1016/0031-9384(78)90150-6
Mason RE (1959) Three studies relating internal sensory experiences and feeling reactions. J Gen Psychol 60:211–228. https://doi.org/10.1080/00221309.1959.9710223
Mason RE (1961) Internal perception and bodily functioning. International Universities Press, New York
Matthews KA, Salomon K, Brady SS, Allen MT (2003) Cardiovascular reactivity to stress predicts future blood pressure in adolescence. Psychosom Med 65:410. https://doi.org/10.1097/01.PSY.0000057612.94797.5F
Mauss IB, Robinson MD (2009) Measures of emotion: a review. Cognit Emot 23:209–237. https://doi.org/10.1080/02699930802204677
Mauss IB, Levenson RW, McCarter L et al (2005) The tie that binds? Coherence among emotion experience, behavior, and physiology. Emotion 5:175–190. https://doi.org/10.1037/1528-3542.5.2.175
Mayou R (1976) The nature of bodily symptoms. Br J Psychiatry 129:55–60
McCranie EJ (1979) Hypochondriacal neurosis. Psychosomatics 20:11–15. https://doi.org/10.1016/S0033-3182(79)70879-6
McFarland RA (1975) Heart rate perception and heart rate control. Psychophysiology 12:402–405. https://doi.org/10.1111/j.1469-8986.1975.tb00011.x
Meessen J, Mainz V, Gauggel S et al (2016) The relationship between interoception and metacognition. J Psychophysiol 30:76–86. https://doi.org/10.1027/0269-8803/a000157
Mehling WE (2016) Differentiating attention styles and regulatory aspects of self-reported interoceptive sensibility. Philos Trans R Soc Lond B Biol Sci 371:20160013. https://doi.org/10.1098/rstb.2016.0013
Mehling WE, Price C, Daubenmier JJ et al (2012) The multidimensional assessment of interoceptive awareness (MAIA). PLoS One 7:e48230. https://doi.org/10.1371/journal.pone.0048230

Meiselles D, Aviram J, Suzan E et al (2017) Does self-perception of sensitivity to pain correlate with actual sensitivity to experimental pain? J Pain Res 10:2657–2663. https://doi.org/10.2147/JPR.S149663

Metzinger T (2009) The ego tunnel. The science of the mind and the myth of the self. Basic Books, New York

Michael GA, Naveteur J (2011) The tickly homunculus and the origins of spontaneous sensations arising on the hands. Conscious Cogn 20:603–617. https://doi.org/10.1016/j.concog.2010.11.013

Michael GA, Dupuy M-A, Deleuze A et al (2011) Interacting effects of vision and attention in perceiving spontaneous sensations arising on the hands. Exp Brain Res 216:21–34. https://doi.org/10.1007/s00221-011-2901-y

Miller NE (1992) Some examples of psychophysiology and the unconscious. Biofeedback Self Regul 17:3–16. https://doi.org/10.1007/bf01000088

Miller LC, Murphy R, Buss AH (1981) Consciousness of body: private and public. J Pers Soc Psychol 41:397–406. https://doi.org/10.1037/0022-3514.41.2.397

Misovich S, Charis PC (1974) Information need, affect, and congnition of autonomic activity. J Exp Soc Psychol 10:274–283

Mossink JCL, Verkuil B, Burger AM et al (2015) Ambulatory assessed implicit affect is associated with salivary cortisol. Front Psychol 6:111. https://doi.org/10.3389/fpsyg.2015.00111

Murphy J, Brewer R, Plans D et al (2019a) Testing the independence of self-reported interoceptive accuracy and attention. Q J Exp Psychol. https://doi.org/10.1177/1747021819879826

Murphy J, Catmur C, Bird G (2019b) Classifying individual differences in interoception: implications for the measurement of interoceptive awareness. Psychon Bull Rev 26:1467–1471. https://doi.org/10.3758/s13423-019-01632-7

Nakao M, Barsky AJ, Nishikitani M et al (2007) Somatosensory amplification and its relationship to somatosensory, auditory, and visual evoked and event-related potentials (P300). Neurosci Lett 415:185–189. https://doi.org/10.1016/j.neulet.2007.01.021

Nash JM, Holroyd KA, Rokicki LA et al (2002) The influence of placebo awareness on stimulant drug response in a double-blind trial. Psychopharmacology 161:213–221. https://doi.org/10.1007/s00213-002-1050-1

Neijts M, van Lien R, Kupper N et al (2015) Heritability and temporal stability of ambulatory autonomic stress reactivity in unstructured 24-hour recordings. Psychosom Med 77:870–881. https://doi.org/10.1097/PSY.0000000000000227

Nisbett RE, Wilson TD (1977) Telling more than we can know: verbal reports on mental processes. Psychol Rev 84:231–259. https://doi.org/10.1037/0033-295X.84.3.231

Noyes R, Happel RL, Yagla SJ (1999) Correlates of hypochondriasis in a nonclinical population. Psychosomatics 40:461–469

Oldehinkel AJ, Ormel J, Bosch NM et al (2011) Stressed out? Associations between perceived and physiological stress responses in adolescents: the TRAILS study. Psychophysiology 48:441–452. https://doi.org/10.1111/j.1469-8986.2010.01118.x

Paintal AS (1986) The visceral sensations—some basic mechanisms. Prog Brain Res 67:3–19. https://doi.org/10.1016/S0079-6123(08)62752-1

Parkin L, Morgan R, Rosselli A et al (2013) Exploring the relationship between mindfulness and cardiac perception. Mindfulness 5:298–313. https://doi.org/10.1007/s12671-012-0181-7

Parkinson B, Manstead ASR (1986) False autonomic feedback: effects of attention to feedback on ratings of erotic stimuli. Motiv Emot 10:11–24. https://doi.org/10.1007/BF00992147

Parshall MB, Schwartzstein RM, Adams L et al (2012) An official American Thoracic Society statement: update on the mechanisms, assessment, and management of dyspnea. Am J Respir Crit Care Med 185:435–452. https://doi.org/10.1164/rccm.201111-2042ST

Pennebaker JW (1981) Stimulus characteristics influencing estimation of heart rate. Psychophysiology 18:540–548. https://doi.org/10.1111/j.1469-8986.1981.tb01824.x

Pennebaker JW (1982) The psychology of physical symptoms. Springer, New York

Pennebaker JW, Epstein D (1983) Implicit psychophysiology: effects of common beliefs and idiosyncratic physiological responses on symptom reporting. J Pers 51:468–496
Pennebaker JW, Lightner JM (1980) Competition of internal and external information in an exercise setting. J Pers Soc Psychol 39:165–174. https://doi.org/10.1037/0022-3514.39.1.165
Pennebaker JW, Roberts T-A (1992) Toward a His and Hers Theory of Emotion: gender differences in visceral perception. J Soc Clin Psychol 11:199–212. https://doi.org/10.1521/jscp.1992.11.3.199
Pennebaker JW, Smyth JM (2016) Opening up by writing it down. How expressive writing improves health and eases emotional pain, 3rd edn. The Guilford Press, New York
Peterson F, Jung CG (1907) Psycho-physical investigations with the galvanometer and pneumograph in normal and insane individuals. Brain 30:153–218. https://doi.org/10.1093/brain/30.2.153
Petrie A (1967) Individuality in pain and suffering: the reducer and augmenter. University of Chicago Press, Chicago
Petrie A (1978) Individuality in pain and suffering, 23rd edn. University of Chicago Press, Chicago
Phillips GC, Jones GE, Rieger EJ, Snell JB (1999) Effects of the presentation of false heart-rate feedback on the performance of two common heartbeat-detection tasks. Psychophysiology 36:504–510
Pilowsky I (1967) Dimensions of hypochondriasis. Br J Psychiatry 113:89–93. https://doi.org/10.1192/bjp.113.494.89
Pilowsky I (1971) The diagnosis of abnormal illness behaviour. Aust N Z J Psychiatry 5:136–138. https://doi.org/10.3109/00048677109159637
Pollatos O, Gramann K, Schandry R (2007a) Neural systems connecting interoceptive awareness and feelings. Hum Brain Mapp 28:9–18. https://doi.org/10.1002/hbm.20258
Pollatos O, Herbert BM, Kaufmann C et al (2007b) Interoceptive awareness, anxiety and cardiovascular reactivity to isometric exercise. Int J Psychophysiol 65:167–173. https://doi.org/10.1016/j.ijpsycho.2007.03.005
Pollatos O, Herbert BM, Wankner S et al (2011) Autonomic imbalance is associated with reduced facial recognition in somatoform disorders. J Psychosom Res 71:232–239. https://doi.org/10.1016/j.jpsychores.2011.03.012
Posserud I, Syrous A, Lindström L et al (2007) Altered rectal perception in irritable bowel syndrome is associated with symptom severity. Gastroenterology 133:1113–1123. https://doi.org/10.1053/j.gastro.2007.07.024
Prescott J, Connolly JF, Gruzelier JH (1984) The augmenting/reducing phenomenon in the auditory evoked potential. Biol Psychol 19:31–44
Rae CL, Larsson DEO, Garfinkel SN, Critchley HD (2019) Dimensions of interoception predict premonitory urges and tic severity in Tourette syndrome. Psychiatry Res 271:469–475. https://doi.org/10.1016/j.psychres.2018.12.036
Raine A, Mitchell DA, Venables PH (1981) Cortical augmenting-reducing—modality specific? Psychophysiology 18:700–708. https://doi.org/10.1111/j.1469-8986.1981.tb01847.x
Ramachandran VS, Hirstein W (1998) The perception of phantom limbs. The D. O. Hebb lecture. Brain 121(Pt 9):1603–1630
Ricksher C, Jung CG (1907) Further investigations on the galvanic phenomenon and respiration in normal and insane individuals. J Abnorm Psychol 2:189–217. https://doi.org/10.1037/h0073786
Rietveld S, Karsdorp PA, Mulder BJM (2004) Heartbeat sensitivity in adults with congenital heart disease. Int J Behav Med 11:203–211. https://doi.org/10.1207/s15327558ijbm1104_3
Ring C, Brener J (1996) Influence of beliefs about heart rate and actual heart rate on heartbeat counting. Psychophysiology 33:541–546. https://doi.org/10.1111/j.1469-8986.1996.tb02430.x
Ring C, Brener J (2018) Heartbeat counting is unrelated to heartbeat detection: a comparison of methods to quantify interoception. Psychophysiology 55:e13084. https://doi.org/10.1111/psyp.13084

Ring C, Brener J, Knapp K, Mailloux J (2015) Effects of heartbeat feedback on beliefs about heart rate and heartbeat counting: a cautionary tale about interoceptive awareness. Biol Psychol 104:193–198. https://doi.org/10.1016/j.biopsycho.2014.12.010

Ruckmick CA (1936) The psychology of feeling and emotion. McGraw-Hill, New York

Salgues S, Plancher G, Jacquot L et al (2021) To the self and beyond: arousal and functional connectivity of the temporo-parietal junction contributes to spontaneous sensations perception. Behav Brain Res 396:112880. https://doi.org/10.1016/j.bbr.2020.112880

Schaefer M, Egloff B, Witthöft M (2012) Is interoceptive awareness really altered in somatoform disorders? Testing competing theories with two paradigms of heartbeat perception. J Abnorm Psychol 121:719–724. https://doi.org/10.1037/a0028509

Schandry R, Specht G (1981) The influence of psychological and physical stress on the perception of heartbeats. [Abstract]. Psychophysiology 18:54. https://doi.org/10.1111/j.1469-8986.1981.tb02928.x

Scheier MF, Carver CS, Gibbons FX (1979) Self-directed attention, awareness of bodily states, and suggestibility. J Pers Soc Psychol 37:1576–1588

Schlotz W, Kumsta R, Layes I et al (2008) Covariance between psychological and endocrine responses to pharmacological challenge and psychosocial stress: a question of timing. Psychosom Med 70:787–796. https://doi.org/10.1097/PSY.0b013e3181810658

Schmitz J, Blechert J, Krämer M et al (2012) Biased perception and interpretation of bodily anxiety symptoms in childhood social anxiety. J Clin Child Adolesc Psychol 41:92–102. https://doi.org/10.1080/15374416.2012.632349

Schulz A, Vögele C (2015) Interoception and stress. Front Psychol 6:993. https://doi.org/10.3389/fpsyg.2015.00993

Schulz A, Lass-Hennemann J, Sütterlin S et al (2013) Cold pressor stress induces opposite effects on cardioceptive accuracy dependent on assessment paradigm. Biol Psychol 93:167–174. https://doi.org/10.1016/j.biopsycho.2013.01.007

Schwarz KA, Büchel C (2015) Cognition and the placebo effect - dissociating subjective perception and actual performance. PLoS One 10:e0130492. https://doi.org/10.1371/journal.pone.0130492

Sgoutas-Emch SA, Cacioppo JT, Uchino BN et al (1994) The effects of an acute psychological stressor on cardiovascular, endocrine, and cellular immune response: a prospective study of individuals high and low in heart rate reactivity. Psychophysiology 31:264–271. https://doi.org/10.1111/j.1469-8986.1994.tb02215.x

Sherwood A, Girdler SS, Bragdon EE et al (1997) Ten-year stability of cardiovascular responses to laboratory stressors. Psychophysiology 34:185–191. https://doi.org/10.1111/j.1469-8986.1997.tb02130.x

Siedlecka M, Klimza A, Łukowska M, Wierzchoń M (2014) Rubber hand illusion reduces discomfort caused by cold stimulus. PLoS One 9:e109909. https://doi.org/10.1371/journal.pone.0109909

Siedlecka M, Spychała N, Łukowska M et al (2018) Rubber hand illusion increases pain caused by electric stimuli. J Pain 19:35–45. https://doi.org/10.1016/j.jpain.2017.08.005

Silvia PJ, Gendolla GHE (2001) On introspection and self-perception: does self-focused attention enable accurate self-knowledge? Rev Gen Psychol 5:241–269. https://doi.org/10.1037/1089-2680.5.3.241

Stenberg G, Rosén I, Risberg J (1988) Personality and augmenting/reducing in visual and auditory evoked potentials. Personal Individ Differ 9:571–579. https://doi.org/10.1016/0191-8869(88)90155-9

Steptoe A, Noll A (1997) The perception of bodily sensations, with special reference to hypochondriasis. Behav Res Ther 35:901–910

Steptoe A, Vögele C (1992) Individual differences in the perception of bodily sensations: the role of trait anxiety and coping style. Behav Res Ther 30:597–607

Stern RM, Botto RW, Herrick CD (1972) Behavioral and physiological effects of false heart rate feedback: a replication and extension. Psychophysiology 9:21–29. https://doi.org/10.1111/j.1469-8986.1972.tb00739.x

Stevens S, Gerlach AL, Cludius B et al (2011) Heartbeat perception in social anxiety before and during speech anticipation. Behav Res Ther 49:138–143. https://doi.org/10.1016/j.brat.2010.11.009

Sturges LV, Goetsch VL (1996) Psychophysiological reactivity and heartbeat awareness in anxiety sensitivity. J Anxiety Disord 10:283–294. https://doi.org/10.1016/0887-6185(96)00012-6

Suschinsky KD, Lalumière ML (2012) Is sexual concordance related to awareness of physiological states? Arch Sex Behav 41:199–208. https://doi.org/10.1007/s10508-012-9931-9

Teasdale JD, Moore RG, Hayhurst H et al (2002) Metacognitive awareness and prevention of relapse in depression: empirical evidence. J Consult Clin Psychol 70:275–287. https://doi.org/10.1037//0022-006x.70.2.275

Thornton EW, Hagan PJ (1976) A failure to explain the effects of false heart-rate feedback on affect by induced changes in physiological response. Br J Psychol 67:359–365. https://doi.org/10.1111/j.2044-8295.1976.tb01522.x

Tihanyi BT, Ferentzi E, Köteles F (2017) Characteristics of attention-related body sensations. Temporal stability and associations with measures of body focus, affect, sustained attention, and heart rate variability. Somatosens Mot Res 34:179–184. https://doi.org/10.1080/08990220.2017.1384720

Tihanyi BT, Ferentzi E, Beissner F, Köteles F (2018) The neuropsychophysiology of tingling. Conscious Cogn 58:97–110. https://doi.org/10.1016/j.concog.2017.10.015

Treiber FA, Kamarck T, Schneiderman N, Sheffield D, Kapuku G, Taylor T (2003) Cardiovascular reactivity and development of preclinical and clinical disease states. Psychosom Med 65(1):46–62. https://doi.org/10.1097/00006842-200301000-00007

Trotman GP, van Zanten JJCSV, Davies J et al (2019) Associations between heart rate, perceived heart rate, and anxiety during acute psychological stress. Anxiety Stress Coping 32:711–727. https://doi.org/10.1080/10615806.2019.1648794

Tyrer P, Lee I, Alexander J (1980) Awareness of cardiac function in anxious, phobic and hypochondriacal patients. Psychol Med 10:171–174. https://doi.org/10.1017/S0033291700039726

Ursin H, Eriksen HR (2004) The cognitive activation theory of stress. Psychoneuroendocrinology 29:567–592. https://doi.org/10.1016/S0306-4530(03)00091-X

Valins S (1966) Cognitive effects of false heart-rate feedback. J Pers Soc Psychol 4:400–408. https://doi.org/10.1037/h0023791

Valins S (1967) Emotionality and information concerning internal reactions. J Pers Soc Psychol 6:458–463. https://doi.org/10.1037/h0024842

van der Ploeg MM, Brosschot JF, Thayer JF, Verkuil B (2016) The implicit positive and negative affect test: validity and relationship with cardiovascular stress-responses. Front Psychol 7:425. https://doi.org/10.3389/fpsyg.2016.00425

van der Veek PPJ, van Rood YR, Masclee AAM (2008) Symptom severity but not psychopathology predicts visceral hypersensitivity in irritable bowel syndrome. Clin Gastroenterol Hepatol 6:321–328. https://doi.org/10.1016/j.cgh.2007.12.005

von Knorring L, Monakhov K, Perris C (1978) Augmenting/reducing: an adaptive switch mechanism to cope with incoming signals in healthy subjects and psychiatric patients. Neuropsychobiology 4:150–179

Walach H, Schimdt S, Bihr Y-M, Wiesch S (2001) The effects of a caffeine placebo and experimenter expectation on blood pressure, heart rate, well-being, and cognitive performance. Eur Psychol 6:15–25. https://doi.org/10.1027//1016-9040.6.1.15

Walach H, Schmidt S, Dirhold T, Nosch S (2002) The effects of a caffeine placebo and suggestion on blood pressure, heart rate, well-being and cognitive performance. Int J Psychophysiol 43:247–260. https://doi.org/10.1016/S0167-8760(01)00188-X

Washburn DA, Smith JD, Taglialatela LA (2005) Individual differences in metacognitive responsiveness: cognitive and personality correlates. J Gen Psychol 132:446–461. https://doi.org/10.3200/GENP.132.4.446-461
Weineck F, Messner M, Hauke G, Pollatos O (2019) Improving interoceptive ability through the practice of power posing: a pilot study. PLoS One 14:e0211453. https://doi.org/10.1371/journal.pone.0211453
Weinstein J, Averill JR, Opton EM Jr, Lazarus RS (1968) Defensive style and discrepancy between self-report and physiological indexes of stress. J Pers Soc Psychol 10:406–413. https://doi.org/10.1037/h0026829
Weisz J, Balázs L, Ádám G (1988) The influence of self-focused attention on heartbeat perception. Psychophysiology 25:193–199. https://doi.org/10.1111/j.1469-8986.1988.tb00987.x
Whitehead WE, Drescher VM, Blackwell B (1976) Lack of relationship between Autonomic Perception Questionnaire scores and actual sensitivity for perceiving one's heart beat. Psychophysiology 13:177. https://doi.org/10.1111/j.1469-8986.1984.tb00196.x
Whitehead WE, Drescher VM, Heiman P, Blackwell B (1977) Relation of heart rate control to heartbeat perception. Biofeedback Self Regul 2:371–392. https://doi.org/10.1007/BF00998623
Wilson TD, Hull JG, Johnson J (1981) Awareness and self-perception: verbal reports on internal states. J Pers Soc Psychol 40:53–71. https://doi.org/10.1037/0022-3514.40.1.53
Winkler A, Hermann C (2019) Placebo- and nocebo-effects in cognitive neuroenhancement: when expectation shapes perception. Front Psych 10:498. https://doi.org/10.3389/fpsyt.2019.00498
Witthöft M, Bräscher A-K, Jungmann SM, Köteles F (2020) Somatic symptom perception and interoception. Z Psychol 228:100–109. https://doi.org/10.1027/2151-2604/a000403
Woll SB, McFall ME (1979) The effects of false feedback on attributed arousal and rated attractiveness in female subjects. J Pers 47:214–229. https://doi.org/10.1111/j.1467-6494.1979.tb00200.x
Wooley SC, Blackwell B, Winget C (1978) A learning theory model of chronic illness behavior: theory, treatment, and research. Psychosom Med 40:379–401. https://doi.org/10.1097/00006842-197808000-00003
Wu T, Snieder H, de Geus E (2010) Genetic influences on cardiovascular stress reactivity. Neurosci Biobehav Rev 35:58–68. https://doi.org/10.1016/j.neubiorev.2009.12.001
Young HA, Williams C, Pink AE et al (2017) Getting to the heart of the matter: does aberrant interoceptive processing contribute towards emotional eating? PLoS One 12:e0186312. https://doi.org/10.1371/journal.pone.0186312
Zamariola G, Vlemincx E, Corneille O, Luminet O (2018) Relationship between interoceptive accuracy, interoceptive sensibility, and alexithymia. Personal Individ Differ 125:14–20. https://doi.org/10.1016/j.paid.2017.12.024

Chapter 7
Putting Everything Together: Integrated Models and Pathological Aspects of Symptom Perception

7.1 Introduction

In the previous chapters, all important factors that influence body sensations were introduced and described. Sometimes there exists a weaker or stronger sensory input from the body which is modified by higher-level processing. In other times, sensations are of dominantly top-down origin, primarily determined by attentional and/or affective processes. So far, we followed the tripartite model of Garfinkel and Critchley (Garfinkel and Critchley 2013; Garfinkel et al. 2015), and discussed the three major dimensions of interoception, i.e., interoceptive accuracy, sensibility, and awareness in Chaps. 4, 5, and 6, respectively. However, as mentioned in the previous chapters, the distinction between interoceptive accuracy and sensibility is not clear-cut for a number of interoceptive modalities that readily reach conscious awareness, such as pain or gastric distension. Moreover, the model is static, i.e., it does not describe the interactions and dynamics among the major dimensions. These difficult relationships can be explained only if we apply a systemic approach, i.e., put all factors and their interplay in a model. Fortunately, a number of such attempts were made over time. In this chapter we will discuss these models in a more or less temporal order as changes and modifications very well reflect the huge progress that was made in this field in the last two decades.

Although the majority of the models were originally developed in order to explain symptom perception, i.e., the pathological aspect of body sensations, they often include non-pathological processes and explanations thereof. The boundary between these categories is not sharp; pathological conditions can often be understood as extremities of dimensional characteristics, such as health anxiety or symptom reporting tendency. Moreover, the knowledge of the negative (i.e., the clinical and subclinical) aspects can be helpful in the understanding of the underlying non-pathological processes. In line with this, a brief summary on the psychopathological aspects of body sensations (e.g., somatization) appears to be appropriate. Finally, I would like to illustrate the phenomena described in this chapter with our own

F. Köteles, *Body Sensations*, https://doi.org/10.1007/978-3-030-63201-4_7

findings on electromagnetic hypersensitivity, a condition dominantly characterized by misattributed symptoms. Concerning the major tenets of this work, the present chapter discusses aspects of the tenet of evaluation, dissociation, relevance, and malleability, with a special emphasis on the latter.

7.2 The Bottom-Up Approach 1: Naïve Realism

Let us start with an approach to body sensations that represents one extremity of the bottom-up—top-down dimension. It describes a clear-cut monotonic relationship between the sensory input and perceived sensations; here, perception is assumed to accurately and objectively represent the world. From a scientific point of view, this approach has clearly been unacceptable for many decades; in fact, Wundt himself was very interested in the top-down elements of perception in the second half of the nineteenth century, and the New Look movement, starting with the works of Jerome Bruner (Bruner and Goodman 1947; Bruner and Postman 1948), realized and described the importance of internal states and values in the process of perception.

This conceptual progress appears to be considerably slower in the field of internal perception for multiple reasons. First, the implicit or explicit assumption of underlying (typically peripheral) pathology still prevails in the medical field (Van den Bergh et al. 2018). This is also called biomedical model (Cioffi 1991) or simply naïve realism (Costa and McCrae 1985). Symptoms do have a somatic background for many organic diseases; this seemingly supports this idea and easily becomes implicitly generalized to all cases. From a methodological point of view, the option that there is some sort of peripheral change behind sensations is way more difficult to rule out for interoception than for external senses. This is particularly so in the clinical field, where the experimental method is available only with restrictions. The exclusion of peripheral changes is extremely difficult in pathological cases, i.e., for individuals with multiple symptoms who obviously show signs of disturbed homeostasis and functioning, as well as elevated levels of arousal. Second, as scholars in the field of interoception research were primarily interested in stimulus–reaction relationships (i.e., the detectability of weak interoceptive stimuli), they simply regarded false alarms in forced-choice measurements as mistakes. As these experiments applied near threshold stimuli, this approach was partly justified. But what about people who report clear-cut, well-defined internal sensations or "bodily hallucinations"? For many decades, the first thing that appeared in the mind of clinicians was serious psychopathology, typically schizophrenic and paranoid conditions (Fisher and Cleveland 1968; Fisher 1970). Phantom limb and analog phantom organ experiences after surgical removal of the bladder, the rectum, or the womb were described quite early (Head and Holmes 1911; Fisher 1970; Katz and Melzack 1990; Ramachandran and Hirstein 1998), but these sensations also represent a condition with pathological background. What about healthy people? Today we know very well that bodily hallucinations may occur for physically and psychologically healthy people too; a perfect example is the tingling sensation discussed in Chap. 4.

Also, it is accepted that there are somatic symptoms without underlying pathological processes (Van den Bergh et al. 2017a, b, 2018).

7.3 The Bottom-Up Approach 2: Linear (Unidirectional) Models of Symptom Perception

The first generation of models of symptom perception developed from the late 1970s relied on the explicit or implicit assumption of the existence of underlying physiological processes. Although the possibility that symptoms can be the products of purely top-down mechanisms was not excluded (Gijsbers van Wijk and Kolk 1997), this option was not handled by the models. Moreover, these models assumed a linear processing of information, which starts in peripheral receptors and ends at various centers of the cortex.

Perhaps the first attempt to systematically include a top-down component in the approach to internal perception was the mechanism of *schema-guided selective search* by Pennebaker (see Chap. 4) (Pennebaker and Skelton 1981; Pennebaker 1982). The process relies on bottom-up sensory input; however, this input is filtered and moderated by top-down cognitive processes called schemata or expectations. In consequence, only those components of the sensory input are allowed to reach consciousness that are in accordance with the expectations, whereas the rest will be ignored. Another top-down factor emphasized by Pennebaker is attention. It was stated that conscious attentional capacity is limited, thus a constant *competition between external and internal stimuli* takes place; in conclusion, we tend to experience more internal input in the relative absence of external cues (Pennebaker and Lightner 1980; Pennebaker 1982). Pennebaker also described and emphasized the importance of a number of additional factors that impact symptom perception, such as gender, negative affect, beliefs, and previous experiences (Pennebaker 1982, 1994; Pennebaker and Epstein 1983; Watson and Pennebaker 1989; Pennebaker and Roberts 1992; Roberts and Pennebaker 1995) but did not included them in a comprehensive model.

The first really comprehensive approach to symptom perception was developed and tested in the 1990s (Gijsbers van Wijk and Kolk 1996). The model was slightly changed one year later (e.g., the behavioral aspect was elaborated) (Gijsbers van Wijk and Kolk 1997), but the overall structure remained the same. Although the authors mention the possibility that body sensations can be experienced without peripheral background, again, the model does not reflect this option and primarily relies on bottom-up information input. As we can see (Fig. 7.1), it integrates all the elements described by Pennebaker, and includes one more factor: causal attribution. Attribution is a secondary (although not necessarily conscious) cognitive step that relies on the presence of some sort of somatic sensation.

The most widely used model of symptom attribution describes three major options: environmental/normalizing, somatic, and psychological attribution

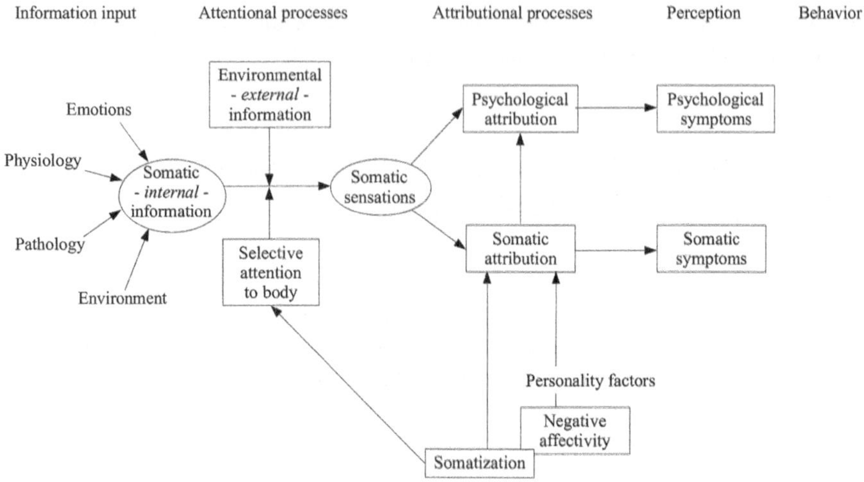

Fig. 7.1 The symptom perception model of Gijsbers van Wijk and Kolk (1996, p. 56)

(Robbins and Kirmayer 1991). Normalizing attribution ascribes the sensations to situational (transient) internal or external factors. As this does not usually refer to pathology and symptoms, this option was not included in the model. Sensations with somatic attribution will be recognized as indicators of pathophysiological processes (i.e., disease), whereas psychological attribution refers to an assumed psychological background (i.e., emotions). Another categorization attempts to apply the classic dimensions of attribution theory (Heider 1958; Weiner et al. 1972; Kelley 1973; Weiner 1979) to symptoms: they can be (1) caused by physical or non-physical factors, (2) which can be stable or transient, and (3) controllable or not controllable by the individual (Swartzman and Lees 1996). The point is that these implicit decisions can have serious behavioral consequences with respect to illness behavior, utilization of medical services, and subjective health (Mechanic 1962, 1972; Robbins and Kirmayer 1991; Gijsbers van Wijk and Kolk 1997); these aspect will be briefly discussed later.

Kolk et al. (2003) modified the model at several points (Fig. 7.2), and Rief and Barsky (2005) attempted to elaborate the generation of body sensations (the primary process depicted in the left-hand side of Figs. 7.1 and 7.2) in more detail (Fig. 7.3). Still, the key assumptions remained the same: (1) bottom-up input from the body is required for the perception of symptoms and (2) the peripheral input is processed in a linear fashion. For example, these models cannot explain the impact of perceived symptoms on the direction of attention and negative affect, or monitoring tendencies, which in turn influence filtering and higher-level processing.

An interesting aspect of these models is the transition from body sensations to symptoms. Symptoms are characterized by negative affective evaluation (i.e., they are unpleasant). As there are many unpleasant body sensations that are not considered pathological (e.g., the urge to defecate or urinate) and the same percept can be considered symptom or simple sensation, this decision cannot be explained by the

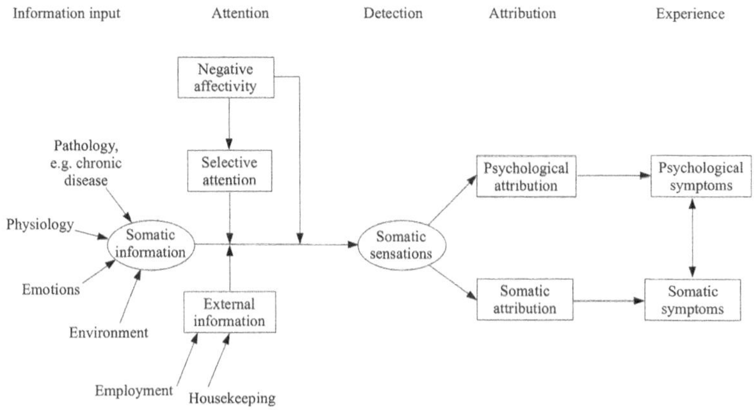

Fig. 7.2 The symptom perception model of Kolk et al. (2003, p. 2344)

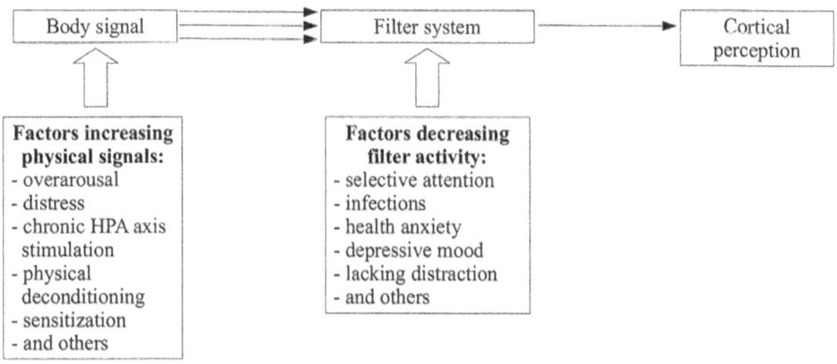

Fig. 7.3 The symptom perception model of Rief and Barsky (2005, p. 1000)

affective processes alone. A crucial difference is that symptoms as opposed to body sensations negatively impact everyday functioning and decrease subjective well-being. This impact can be direct: the experience of multiple symptoms over a period of time, also called somatic symptom distress (Witthöft et al. 2012), is a disabling condition (see below). Moreover, symptoms can indicate threat, i.e., they can be interpreted as signs of actual disease or of processes that can lead to disease, as in the case of somatosensory amplification and health anxiety. This assumes the operation of higher (cognitive) processes.

In a series of elegantly designed experiments, it was shown that categorization affects perceived differences in respiratory loads in both the sensory (perceived intensity) and evaluative (perceived unpleasantness) domains (Petersen et al. 2014, 2015a; Van den Bergh et al. 2018; Zacharioudakis et al. 2020). In some of these experiments, neutral categories (simply called A and B, representing lower and higher levels of respiratory resistance) were used; a more meaningful grouping (e.g., normal sensation or symptom; indicating threat or not) was applied only recently. In line with this, negative affect (see below) can be considered as a basis for such primary categorization as it may act as a precise prior (using the terminology of predictive processing, see below) that strongly influences ascending information (Petersen et al. 2015b). Such categorization effects primarily impact vague and/or unpredictable stimuli (Petersen et al. 2015b; Van den Bergh et al. 2018) because uncertainty favors top-down processes. At this point, a brief summary on negative affect and its possible roles in the perception of body sensations appears to be appropriate.

7.4 Negative Affect

In Chap. 2, the term core affect was introduced as the most elementary form of affective feelings, i.e., subjectivity; it is assessed along two more or less independent dimensions, valence and activation. A related term, negative affect, refers to a general dimension of subjective distress, including a variety of mood states, such as anger, guilt, anxiety, sadness, and self-dissatisfaction (Watson and Clark 1984; Watson and Pennebaker 1989). The common characteristic of these states is their unpleasantness or aversivity, i.e., the high level of negative evaluation. There are large individual differences in the frequency of occurrence of states belonging to the umbrella of negative affect; this mood-dispositional dimension is called negative affectivity and shows a substantial overlap with the personality trait neuroticism (Watson and Clark 1984). Individuals high on negative affectivity are more likely to experience distress and dissatisfaction, even in the absence of overt stress. Negative affect, as well as the conceptually related Behavioral Inhibition System, do not represent a purely reactive mood state: they include the active monitoring of the environment for possible threats (Grey 1985; Tellegen 1985; Watson and Pennebaker 1989). In consequence, negative affect (and its perhaps most typical representative, anxiety) increases sensitivity to negative external stimuli and bias ambiguous external stimuli in a threatening manner (Watson and Clark 1984; Eysenck 1992; Byrne and Eysenck 1995; Bar-Haim et al. 2007; Cisler and Koster 2010). A further feature of negative affect is introspection and self-scrutiny as the individual tries to find an explanation for the aversive state (Watson and Clark 1984; Mor and Winquist 2002; Weston and Jackson 2018). If we take these characteristics together, we can picture an individual who expects and experiences comparatively more unpleasant and stressful internal and external events. Concerning our topic, i.e., internal body-related events, this may lead to increased symptom reporting as persons with high negative affectivity pay more attention to body processes and tend to interpret them as symptoms (Barsky 1979; Barsky et al. 1988; Watson and Pennebaker 1989; Stegen et al. 2000).

Empirical findings support this idea; negative affectivity usually shows a moderate to strong association with symptom reports, i.e., the experience of unpleasant body sensations (Watson and Pennebaker 1989; Wessely et al. 1999; De Peuter et al. 2007; Witthöft and Hiller 2010; Jasper et al. 2012; Witthöft et al. 2012, 2013; Walentynowicz et al. 2018). Interestingly, whereas body sensations are related to objective indicators of physical health too, negative affectivity is more or less independent from the latter (Costa and McCrae 1985, 1987; Watson and Pennebaker 1989). Thus, the latter dominantly operates at the experiential level.

The association between negative affect and symptoms can be explained at least in three different but not completely unrelated ways. One option, what can be called *bias approach*, is that negative affective states (and consequently negative affectivity too) represent a systematic bias or a "general nuisance factor" that should be excluded or controlled for in health research (Watson and Pennebaker 1989). With respect to interoceptive accuracy research, this means that negative affect may seriously interfere with sensory-discriminative measurements, and this is particularly so for vague stimuli (De Peuter et al. 2010). In doing so, it distorts objective detection performance, i.e., the stimulus–response relationship. For example, somatosensory amplification tendency was negatively associated with cardiac IAw in a study which was interpreted as a cognitive-emotional bias (Mailloux and Brener 2002). Similarly, the accuracy of perception of air hunger was reversely associated with negative affectivity (Van den Bergh et al. 2004; Bogaerts et al. 2005). However, no association between cardioceptive accuracy and somatosensory amplification was found in other studies with the participation of university students (Aronson et al. 2001; Ferentzi et al. 2017), and state anxiety is usually unrelated to, or sometimes even positively associated with, performance in cardioceptive accuracy tasks (Schandry 1981; Pollatos et al. 2007; Garfinkel et al. 2016). On top of this, trait anxiety and anxiety sensitivity (both are strongly related to negative affectivity) showed a positive association with cardiac accuracy in a meta-analysis (Domschke et al. 2010). In summary, negative affect in certain cases is associated with improved sensory-discriminative performance.

A second option, dubbed *systemic approach*, relies on the idea that perception should not be accurate in objective terms; it reflects a more holistic and subjective evaluation of the actual homeostatic state, including the affective state (Hohwy 2014; Clark 2015). Thus, a perceptual bias can be considered an adaptive characteristic in many situations (Lynn and Barrett 2014; Petersen et al. 2015a). From this point of view, objective detection performance is way less important than the perception of those stimuli that really matter to the organism as the latter takes into consideration its actual condition and need states (accuracy of detection is utterly important at the level of physiological regulation though). Using the terms of the predictive processing approach, always the entire representation is optimized which can lead to less precise representations at certain levels (see below). As mentioned, negative affect might facilitate categorization processes for uncertain stimuli which also leads to the suppression of single stimuli (Petersen et al. 2015b). The idea that negative affect is characterized by an increased attentional focus on bodily sensations supports this idea (Stegen et al. 2001; Kolk et al. 2003), Also, sensations accompanied (or marked in the course of natural selection as they carry important message) by unpleasantness are able to effectively grab attention which shows that the perceptual system judges certain stimuli more salient than others. Consequently, what symptom reports accurately indicate is not the raw sensory input but the overall state of the individual. The relatively higher symptom reporting tendency of more anxious and stressed individuals (Barsky et al. 1998), and the fact that experimentally induced negative affect can evoke more symptoms in high symptom reporters (Bogaerts et al. 2010; Constantinou et al. 2013) also support this view. In a similar vein, the intensity of pain was increased by the induction of depressive mood state (Berna et al. 2010).

A third possibility, aka *confusion approach*, is that the distinction between body sensations and mood states is often vague, thus sometimes we simply confuse one with the other or tend to favor one over the other (this is another example of implicit categorization). As the majority of body sensations are accompanied by negative affective evaluation and body sensations represent important components of emotions, these two categories overlap to a large extent and might interact with each other. Thus, the distinction between "feeling your body or feeling badly" (Aronson et al. 2001) is not always clear-cut. In line with this, Watson and Pennebaker propose the term somatopsychic distress, and note that "the distinction between psychological and physical complaining is clearly arbitrary and inadequate" (Watson and Pennebaker 1989, p. 248). The German term "Gemeingefühl" (as well as its Hungarian counterpart, "közérzet") refers to the appraisal of the overall condition of the individual, primarily relying on the felt state of the body. On the one hand, it is clearly evaluative and unitary which shows its affective nature; on the other, it clearly refers to the body. Perhaps this feeling reflects the embodiedness of our affects or the mind–body unity; alternatively, it may indicate the difficulties with the verbalization of contents of primary consciousness (tenet of vagueness). The point is that certain feelings can be interpreted (categorized) as body sensation or affective sensation, and the boundary between the two may vary among individuals—this step is not included in any of the symptom perception models described above. This idea, although it does not assume pathological processes, is quite close to the construct of alexithymia (Sifneos 1973). Alexithymia refers to difficulties in the expression of affective states because of the lack of appropriate words. As no other means are available to them, individuals with high levels of alexithymia are assumed to try to express their internal states (most importantly unpleasant and stressful states) in terms of body sensations and symptoms (Apfel and Sifneos 1979). The condition has a neurophysiological background (Jørgensen et al. 2007; Hogeveen et al. 2016). As the findings of empirical studies on the association between alexithymia and somatic symptoms are equivocal, and there are conceptual issues with the self-report measurement of the construct, the assumed causal role of alexithymia in the generation of symptoms is still questioned (Cohen et al. 1994; Rief et al. 1996; De Gucht and Heiser 2003; Rief and Broadbent 2007). Still, there might be merit in the idea that body sensations and affective feelings can be interchangeable to some extent. We will discuss the associations between interoception and alexithymia in more detail in Chap. 8.

As I mentioned, these three approaches, particularly options 1 (the bias approach) and 2 (the systemic approach), and options 2 and 3 (the confusion approach), are clearly related to each other, thus are not mutually exclusive. A further source of confusion concerning the role of negative affect might be that body sensations that rely on sensory input and are associated with unpleasantness differ from perceived sensations that are primarily produced by top-down mechanisms. In the former case, for example, for pain or rectal distension, negative affect makes an already present and salient stimulus, which can be solved through behavioral means only, even more marked (Smith et al. 1995; Mora et al. 2007). This is basically in accordance with the systemic approach. In the latter case, however, there is no sensory input to be enhanced or it is very weak; thus, the sensation generated by the affective state and other top-down factors can interfere with sensory input as described by the bias and confusion approach.

I think that the most widely used and mentioned interoceptive modality, the perception of the heartbeat, is special from this point of view, and this special position can partly explain seemingly inconsistent findings. For the majority of people, heartbeats represent a very weak and inherently neutral stimulus, which is reflected very well in their poor detectability under resting conditions. In other words, heartbeat belong to the dissociable modalities (Chap. 6) in these cases, which is more or less independent from affective evaluation. For those with high levels of anxiety, frequent panic attacks or cardiac pathology, however, heartbeats are the indicators of a salient negative state, thus their position shifts from the dissociable toward the non-dissociable category to some extent, and evaluation plays an important role in their perception. From the available cross-sectional and quasi-experimental studies we cannot draw a conclusion on the direction of causality (Schulz and Vögele 2015). Although the wording of some authors often suggests a causal relationship from cardioceptive accuracy to psychopathology, I do not think this is supported by empirical data. The reverse direction appears equally likely to me, for example, heartbeat may become associated with negative affect over time which somewhat enhances their detectability.

As we can see, the position of negative affect is not as clear-cut as suggested by symptom perception models; perhaps the model of Kolk and colleagues (Fig. 7.2) represents this uncertainty to some extent. Still, it cannot be satisfactorily explained without assuming interactions and feed-back loops in the process of the generation of symptoms.

7.5 The Top-Down Approach 1: Nonlinear Models

The most important problem with the symptom perception models presented above is that they are linear, i.e., assume a unidirectional flow of information from the periphery to the cortex; in consequence, (1) the assumed role of top-down influences is simply the filtering (moderating) of the available sensory input and (2) circularity in not allowed or handled. A number of authors realized and attempted to handle these issues in their models. For example, the model of Kirmayer and Taillefer (1997) allows an interaction between affective and physiological processes and self-enhancing cycles between affect, body attention, sensations attributed to illness, and their cognitive and behavioral consequences, including illness behavior. Although it was developed to explain somatization (see below), it regards processes leading to pathology as extreme cases of normal processes that characterize everyone. Still, the body sensations rely on more or less modified sensory input in this model too.

Body sensations, including symptoms, can also be generated by top-down processes. We are already familiar with the tingling phenomenon (Chap. 5), which was basically explained by an attentional process acting on sensory noise (Tihanyi et al. 2018). Richard Brown (2004, 2006) went even further and mentioned conditions and cases which do not even require sensory noise. He argues that materials can reach consciousness from multiple directions, and the only thing what really matters is that a representation appears in conscious awareness. Once it is there, one cannot distinguish between sensations of actual sensory origin and sensations originating from other sources (this is a consequence of tenet of realness). For example, bodily hallucinations and illusions or body sensations elicited via hypnotic suggestions represent such phenomena. In these cases, top-down processes, such as memories and external information, are at work. The system (called primary attentional system) automatically selects among possible explanations (hypotheses) of the actual sensory pattern and favors one over the others. The favored model then reaches consciousness and serves as the basis for control of behavior. The selection is usually partly determined by the actual sensory input, and partly based on previous experiences and memories. So far the model describes healthy functioning; in fact, it described and incorporated many components of the predictive processing framework before the latter was elaborated for interoceptive processes. Pathology (i.e., somatoform or medically unexplained symptoms and related disorders) comes into the picture when the system chooses and applies a representation which dominantly (sometimes perhaps exclusively) relies on top-down processes (e.g., memories of own and other's previous pathological somatic states, sociocultural representations, suggestion) and ignores bottom-up input in the long run. This representation is called rogue representation by Brown; it is maintained by selective attention, thus it further distorts perceptual and higher processes (rumination, illness behavior) in a self-reinforcing manner (Fig. 7.4). The assumed role of negative affect is complex: emotional arousal contributes to the representation of symptoms and enhances body focus (this can also be a conversion-like coping strategy under recurrent traumatizing conditions).

Rief and Broadbent (2007) also realized the importance of top-down processes in symptom perception and completed the Rief-Barsky model (Fig. 7.3) with factors

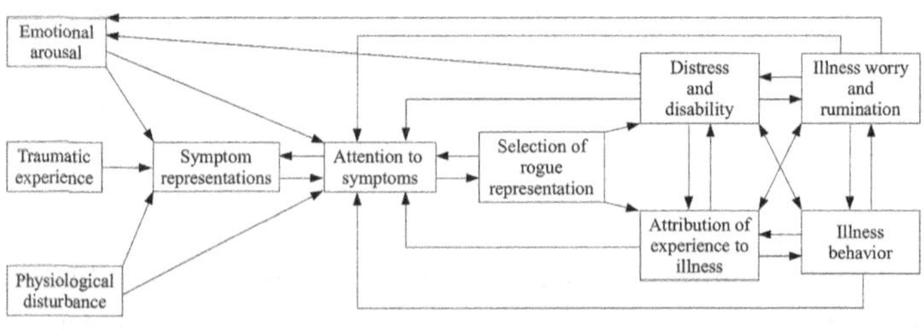

Fig. 7.4 The symptom perception model of Brown (2004, p. 798)

influencing "cortical perception", i.e., sensations that reach conscious awareness (Fig. 7.5). Beyond the importance of expectations (which will be the topic of the next section), the actual state of the central nervous system is also included here. It reflects the fact that some individuals process information, for example, body-related information, in a different way because of altered neurophysiological functioning. Sensitization theories focus on this aspect from both psychological and biological point of view, and highlight the importance of neuroendocrine and immunological changes in pathological symptom reporting (Ursin 1997; Ursin and Eriksen 2001, 2004, 2010; Yunus 2007, 2008). Of course, the physiological and the psychological level might reflect the same phenomenon from different viewpoints, but the fact that an immunological change caused by an infection can modify perceptual processes shows that the assumption that interoception and symptoms perception are purely determined by psychological factors (top-down attention, defense, etc.) is an oversimplification.

7.6 The Top-Down Approach 2: Expectations

The term expectation refers to a belief about the future outcome of an event. It is rarely defined in the literature, which leads to a number of issues and questions concerning its meaning and applicability. For example, conditioned responses and allostatic regulation clearly represent some kind of future-related representations (Winters et al. 2002; McEwen and Wingfield 2003; Sterling 2012); however, they rely on nonconscious processes, thus the majority of researchers in the placebo field (an area where expectations are intensely studied) do not include them in the category of expectations (Chap. 10). Similarly, although the anticipation of threat is part and parcel of the concept of negative affect (Watson and Pennebaker 1989), the latter is not considered a representative of expectations in the field of nocebo research; also, conditioned negative responses (e.g., nausea before chemotherapy evoked by the medical environment) are not regarded as consequences of expectations (Chap. 9). In fact, responses based on conditioning and expectations are often considered two alternative, sometimes competing mechanisms (Voudouris et al.

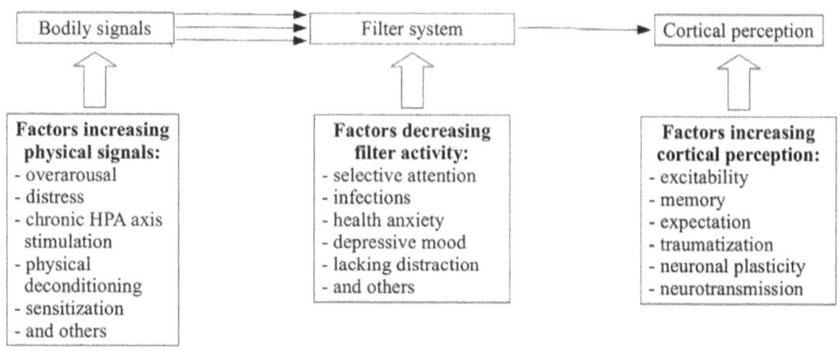

Fig. 7.5 The symptom perception model of Rief and Broadbent (2007, p. 825)

1990; de Jong and Arntz 1993; de Jong et al. 1996; Montgomery and Kirsch 1996; Stewart-Williams and Podd 2004; Stewart-Williams 2004). Expectations are typically measured with self-report, which assumes that they are accessible for conscious awareness. Some authors allow the existence of nonconscious expectations (Hahn 1997; Clark 2015), others exclude this possibility (Kirsch 1985, 1997a, b). A related term, often used as a synonym for expectation, is expectancy. Expectancy is a *terminus technicus* of social learning theory (Rotter 1953; Palenzuela 1987); it refers to the assumed outcome of an action and to the perceived ability to perform an action (self-efficacy). In this context, action means voluntary behavior (Rotter 1953; Bandura 1969; Bootzin 1985). In the field of placebo research, its meaning was extended: the term *response expectancy* was proposed for the predicted outcome of voluntarily not controllable (e.g., visceral or emotional) changes (Kirsch 1985). Response expectancies are expectancies that *somehow* manifest themselves. Considering the large overlap between the placebo phenomenon (Chap. 10) and body sensations, this concept appears to be very well usable at first sight. However, the picture is much more complex in the field of nocebo reactions and unpleasant body sensations (symptoms). We usually do not want these sensations to happen and even less want to cause them; we just anticipate them. Conceptually, this issue can be solved by assuming an unconscious desire to self-punishment (this is the classic analytic explanation, as described in Chap. 5) or by allowing the existence of automatic (nonconscious) processes that are independent from our wishes. Now we will follow the second path.

The schema-guided seeking process by Pennebaker (1982) mentioned above illustrates very well a possible cognitive mechanism. Most importantly, the schema simply represents an automatically acting representation that is independent of our preferences. We might be aware of this representation (belief), as in many experimental situations that include explicit verbal manipulation, or not, like under everyday circumstances where the overall meaning of the situation often processed automatically (Moerman 2002a, b, 2011). Even the regular completion of a questionnaire assessing nonspecific symptoms can activate the symptom schema, which in turn leads to the increase of experienced symptoms (Ferrari and Russell 2010; Ferrari 2015; MacKrill et al. 2020). In the latter cases, we cannot directly measure expectations with self-report. In the experiments of Pennebaker, no novel bottom-up input was generated, only those components of the signal that are in line with the active schema (expectation) were selected, and the non-matching parts were filtered out. What is needed for this mechanism to take place is spontaneous fluctuations. The tingling sensation, evoked by attention, can generate very concrete and well-localized body sensations thus can represent a mechanism which does not even need fluctuations, only sensory noise or a low signal-to-noise ratio (Tihanyi et al. 2018). If we combine the tingling phenomenon with active monitoring evoked by an alarming message and add some negative affect to this blend to increase arousal, sensory noise, and introspection, a wide variety of symptoms will be explainable through nonconscious or only partly conscious expectations.

The next logical question is whether beyond these processes, which uniformly rely on already existing patterns of events, can an expectation actually change sensory input? Yes, it can. We have convincing empirical evidence (including the neurobiological mechanisms) from pain and itch research that expectations are able to modify these sensations to both directions (Benedetti 2009, 2011; Bartels et al. 2014, 2015; Meeuwis et al. 2019). For pain, the major relay stations of the hierarchical mechanism are well described; it even involves the spinal cord, the lowest central relay station of pain processing (Eippert et al. 2009). The extent of such modifications can be considerable: complete analgesia, as well as the conversion of non-painful sensations to pain, was reported (Colloca et al. 2007). Taking into consideration that many nonspecific symptoms are pain-related (for example, almost half of the items of the PHQ-15 scale presented in Chap. 5 refers to pain experience), this process alone can explain a large proportion of symptom reports.

At a much higher level, expectations can also "evoke" symptoms through attribution processes. For example, worries about the harmful effects of a pesticide spraying increased the number of symptoms attributed to the intervention but not the overall symptom score in a more natural (if one can use this adjective for pesticide spraying) setting (Petrie et al. 2005). In such cases, expectations change the interpretation of minor fluctuations of homeostasis and negative sensations that were originally unnoticed or not focused on. This leads to a perceived increase of symptoms that are attributed to the predicted factor.

I think the definition of expectation, as well as the operationalization of its measurement, is among the weakest points of the expectation-based approach to body sensations, including placebo and nocebo responses. In the recent literature, the concept is often used as an umbrella construct for a number of conceptually very different automatic processes, from perceptual biases to conditioned responses and attribution. If one assesses conscious expectations at a given point of time, then measures the change of the respective condition, and finds an association between the two, it is tempting to state that expectations played a role in the change (Fig. 7.6, left-hand side). In fact, a third variable may lie behind both measured variables, and the actual value of expectation simply indicates the extent the individual is able to assess (or estimate) the process represented by the third variable (Fig. 7.6, right-hand side). The bottom line is that the dominant proportion of expectations may act at nonconscious level(s), thus self-report is simply not an appropriate tool to assess them.

I also have the feeling that the importance of affect is underestimated in this field of research. The schema-based mechanism is a "cold" cognitive process which is usually associated with (or driven by) fear, anxiety, and the perception of threat in real-life situations (we will further discuss these ideas in Chap. 9). Moreover, as

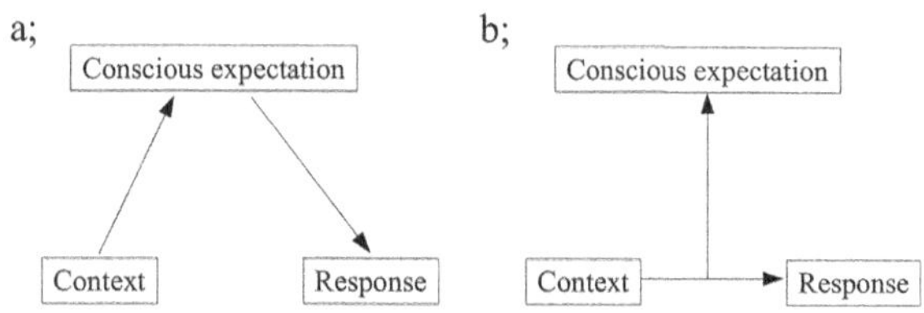

Fig. 7.6 Possible associations between conscious expectations and change (response) (Köteles 2013, p. 308). (**a**) Causal relationship. (**b**) Lack of causality

pain research shows, the transition from a conscious content (expectation) to neurophysiological changes is a multi-step (hierarchical) process that might involve bidirectional interactions between different levels of hierarchy. Exactly such a sophisticated mechanism characterizes the most recent model of interoception, the predictive processing approach.

7.7 The Top-Down Approach 3: Predictive Processing

The general principles of the predictive processing framework were already presented in Chap. 1. In this section, its application to interoception, also called interoceptive inference (Seth 2013), will be discussed. In this context, prediction refers to the prediction of body states via active inference that is tested against incoming body signals (Barrett and Simmons 2015). In fact, not the incoming body signals themselves but error signals, generated by comparison with the respective priors, climb upwards trough the levels of processing hierarchy. Prediction error can be minimized via modifying the (posterior) priors (perception) or changing the sampling of the incoming signals (action), e.g., via initiating regulatory steps supposed to decrease the discrepancy. However, unlike exteroceptive perception, the trajectories (i.e., in the long-term sustainable states) of interoception are limited (Gu and FitzGerald 2014). These homeostatic or allostatic necessities are above all other priorities of a living organism (Friston 2009, 2010); therefore, the representations of these can be considered first priors (Allen and Tsakiris 2018). These first priors, which are shaped by natural selection and determined by genetic information (Friston 2010; Fotopoulou 2013), are necessarily very rigid. In the language of predictive processing, as the number of sustainable physiological states of an organism is far from infinite, its primary motive is the avoidance of "surprising" states, i.e., those that are out of the acceptable range (Friston 2010). From this point of view, homeostatic regulation is simply the continuous reduction of discrepancies between predicted (perhaps expected or preferred are better terms here) and actual physiological states (Gu and FitzGerald 2014). As the first priors are resistant to change, action-based solutions (both physiological and behavioral) necessarily play the dominant role here. To provide the reader with an everyday example, removal of the noxious stimulus in a painful situation (i.e., the withdrawal of the affected limb) can be considered an action that eliminates "surprising" (i.e., unexpected from an adaptivity point of view) sensory input (Friston 2009). Similarly, keeping the level of CO_2 within the physiological range in the blood via respiratory and cardiovascular maneuvers is simply a pattern of autonomous actions in order to reduce "surprise" and maintain homeostasis. This process is online, i.e., perception and action are in continuous interaction at all levels of the hierarchy (see Fig. 7.7) (Clark 2015).

In order to further optimize regulation, the brain anticipates potential future deviations from the homeostatic range and takes steps to prevent them. This idea was recognized decades ago, for example, with respect to the homeostatic regulation of food intake: the motive to eat evokes well before the depletion of mobilizable energy

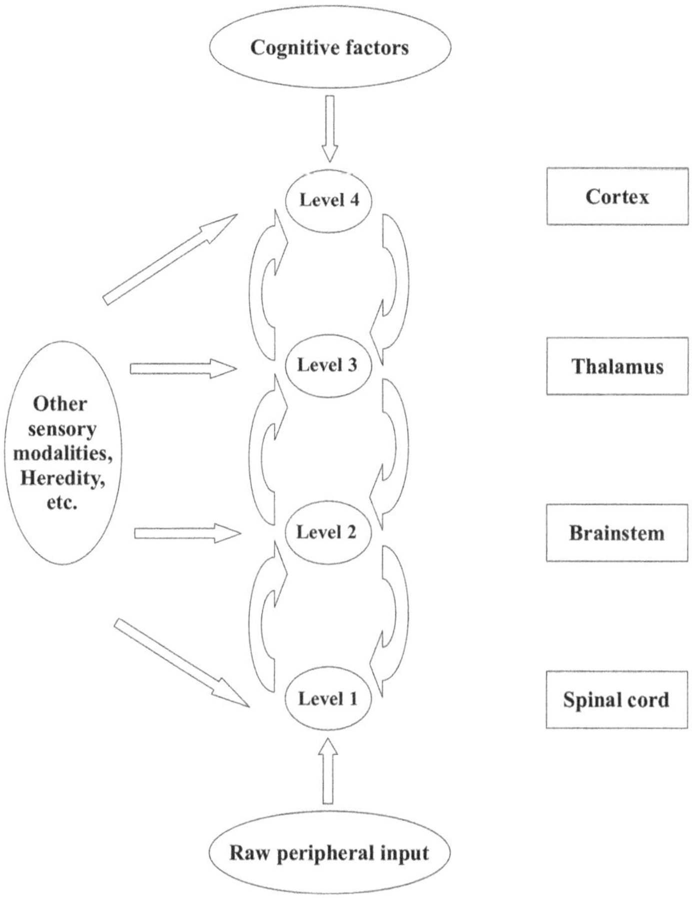

Fig. 7.7 A schematic presentation of the predictive processing approach to the perception of body symptoms. Ascending and descending information are compared at multiple levels of hierarchical processing (the listed brain regions at the right-hand side serve only as examples). The ascending signal is the result of the comparison at the respective level. The descending signal represents the system's expectations (priors) with respect to the ascending signal. Processing is impacted by information from other modalities (multisensory integration) and genetically determined factors (e.g., homeostatic set points). Cognitive factors, such as knowledge and social influences, impact perception through high-level priors

resources, and the gastrointestinal system is prepared in a step-by-step manner for the arrival of food. The sight or smell of food can increase the insulin level of the blood as the brain anticipates the following changes (this is called the cephalic phase of digestion) (Teff 2011). Allostasis, a more dynamic concept, emphasizes even more the importance of predictive regulation (Sterling and Eyer 1988; McEwen 1998; Sterling 2012); it was already reformulated using the terms of predictive processing (Pezzulo et al. 2015; Barrett 2017; Kleckner et al. 2017; Peters et al. 2017; Corcoran and Hohwy 2018). A key element of the novel approaches to allostatic (and homeostatic) regulation is interoceptive inference. In summary, the ultimate goal of interoceptive predictive processing is the identification of those hidden causes, both internal and external, that can result in serious deviations from the sustainable ("expected") physiological range and prevent them via physiological and behavioral regulatory steps. It is a complicated process, for example, a (meta-) representation of the brain's regulatory capacity, called allostatic self-efficacy, is also required (Stephan et al. 2016).

As it was shown in Chap. 3, the interoceptive neural hierarchy ranges from the spinal cord to the insula, the anterior cingulate cortex, and the orbitofrontal cortex (Seth et al. 2012; Barrett and Simmons 2015). The dominant proportion of homeostatic regulation takes place at lower (physiological) levels; using the terminology of the predictive coding approach, forward propagated (ascending) interoceptive error signals are attenuated (explained away) at early states of processing, thus they do not change priors at higher (i.e., conscious) levels (Seth et al. 2012; Ainley et al. 2016). In consequence, no clear-cut body sensations arise from the majority of these operations; however, random fluctuations and noise are still present and might be processed in the light of top-down factors in certain cases (see below).

Higher (i.e., conscious) levels of processing become impacted (1) if the error signal cannot be handled at lower levels, thus the reduction of surprise requires behavioral steps, (2) processing of the error signal is "aberrant", for example, in the case of anxiety (Paulus and Stein 2006, 2010; Khalsa and Feinstein 2018), or (3) unusually strong and rigid higher-level priors determine perceptual hallucinations. In all cases, attention, conceptualized as the optimization of expected precision (or uncertainty) of the signals, plays a central role (Edwards et al. 2012; Ainley et al. 2016). However, precision of error signals (i.e., the reliability of the input) cannot be directly measured just estimated by the brain (Hohwy 2014; Clark 2015); this estimation mostly relies on high-level priors. For those modalities that are normally characterized by conscious sensations (for example, the urge to urinate, pain, or sensual touch; see Chap. 4), the high precision of these priors is partly determined by heredity and evolution, thus they cannot be substantially changed. This also means that the salience of these priors, i.e., their value for the organism, is very high, at both the subjective and the physiological levels. Once we experience a body feeling, pleasant or unpleasant, it is very hard to suppress, and it changes our cognition and behavior (tenet of relevance) (Allen and Tsakiris 2018).

Abnormal weighting of (in other words, attention focused on) certain modalities can be a consequence or remainder of past experiences, e.g., infections or inflammations; alternatively, it can be determined by external influences, such as suggestions. Due to aberrant processing, certain signals acquire higher than normal precision, whereas prior-driven hallucinatory experiences rely on the inherently high noise present in interoceptive channels (Paulus and Stein 2010; Van den Bergh et al. 2017b, 2018). The noisiness of the system makes the process particularly malleable to external influences that change precision-related expectations. Thus, just like flames, clouds, or the inkblots of the Rorschach test, it can provide an excellent projection surface for conscious and nonconscious expectations (priors; with alternative terms: schemas, beliefs, biases), particularly for strong ones (Clark 2015;

Leder 2018). This point was perhaps best illustrated by attention-related body sensations in Chap. 5 (Tihanyi et al. 2018); medically unexplained symptoms (aka body illusions, see below) and the placebo/nocebo phenomena are also nice examples.

In the predictive processing framework, interoceptive accuracy refers to the (actual) optimization of the precision of error signals at high (conscious) levels in the interoceptive system, and individual differences are explained by differences in the ability to do so (Fotopoulou 2013; Ainley et al. 2016). In other words, attention is assumed to play the dominant role here, which is an oversimplification in my opinion (see Chap. 5). For example, there are large differences among individuals with respect to the sheer strength of the bottom-up signal. As mentioned in Chap. 4, aerobic fitness (to put it simply: the strength of individual heartbeats) and body fat percent can impact the ascending signal which makes the tasks measuring cardioceptive accuracy more difficult for certain individuals than for others. Second, it is important to mention once again that the main preference of predictive processing is not perfect accuracy but optimal overall predictive power (Hohwy 2012; Pezzulo et al. 2015; Clark 2015; Seth and Friston 2016). Optimal means here a good balance between complexity (expensiveness) of the model (higher complexity requires more metabolic resources) and accuracy with respect to allostatic regulation. Also, as discussed in Chap. 5 as well as in Sect. 7.4 of the present chapter, evaluation of the condition and the overall affective state of the organism can change the optimal set point to a great extent.

In summary, the hierarchical simultaneous processing of top-down and bottom-up information as described by the predictive processing approach is theoretically able to handle all the issues described by the previous models of symptom perception. As expectations (predictions) are conceptualized at multiple (conscious and nonconscious) levels, it can also solve the problems I mentioned with respect to the expectation-based approach in the previous section. Still, I think that the objection I mentioned in Chap. 1 is valid: what we see in the majority of papers adopting this perspective is the translation of well-known ideas and phenomena, such as homeostasis, affect, and attention, to the computational language of the predictive processing framework. This understanding certainly represents an important step toward the description of the way the brain operates, but its practical (e.g., clinical) applicability is yet to be demonstrated. Measuring the neural processes in order to obtain actual and realistic values for simulations is an extremely difficult task. Today, even the role of certain brain centers and their activity is poorly understood. For example, heartbeat-evoked potentials might represent the neural concomitant of predictions (Petzschner et al. 2019; Park and Blanke 2019) or prediction errors (Ainley et al. 2016). Beyond methodological issues, it is not written in stone that one single mechanism should exclusively prevail such a complex machinery as the human brain. Still, the predictive processing framework appears the most promising tool to understand the development and maintenance of body sensations, including their pathological aspect.

7.8 The Pathological Aspect: Dissociation, Conversion, Somatization

After reviewing the comprehensive models describing the generation of body sensations and symptoms, we will focus on the latter. Symptoms (i.e., unpleasant body sensations that might indicate pathology) without clear-cut pathophysiological background come and go for everyone (Pennebaker 1982; Rosenzweig et al. 1993); however, they remain and cause severe suffering for some. Historically, these persistent symptoms or patterns of symptoms were explained in various ways. As certain aspects of these explanations are also accepted today and other aspects, although not accepted scientifically any more, still influence the thinking of clinicians, a brief summary appears to be useful. The interested reader can find an excellent and very thorough review on these concepts in the paper of Brown (2004).

The dissociation theory of Janet (1889, 1907) conceptualized the phenomenon that was called hysteria that time as an attentional deficit. It was proposed that patients with hysteria cannot distract their attention from irrelevant sensory information in order to focus on important pieces of the input (Horvath et al. 1980). As attention dominantly focuses on certain sensory channels, input from non-attended channels cannot become integrated with that from other modalities and becomes dissociated (i.e., the individual will not be aware of it). In the long run, as a proportion of information cannot be integrated with the conscious representation of the body, these dissociated memories can be unexpectedly triggered at any time and are regarded by the individual as actual input (perception) instead of memory. A major factor that often leads to dissociation is trauma, in childhood or later (Janet 1889, 1907; Pennebaker 1982; Brown et al. 2005); beyond external events, people with hysteria were assumed by Janet to be characterized by constitutional weakness which makes them particularly prone to dissociation.

Breuer and Freud (1893) proposed a different mechanism: they thought that repression, a nonconscious defense mechanism, plays the major role in dissociation instead of simple attentional deficit. In this model, emotionally threatening material, such as a traumatic memory, becomes repressed in order to prevent the psychic system from serious overloading. As the maintenance of repression requires continuous effort which consumes mental resources, the system tries to discharge the repressed energy in a modified, less threatening way—via symptoms. This process is called conversion. In conclusion, in this approach symptoms (1) represent an effective way to reduce internal tension (anxiety) and maintain the functioning of the system (this is called primary gain), and (2) are symbolic expressions of a hidden content. Although this concept is not accepted any more, it is still implicitly and sometimes even explicitly active in the clinical tradition (Brown 2004). As presented above, symptoms are often accompanied by high levels of negative affect; thus, the conversion approach offers a way (although a logically incongruent way, as repressed emotions should not be present) to explain this association.

A third term, somatization, was proposed by Lipowski (1968, p. 413) for the "tendency to experience, conceptualize, and/or communicate psychological states or contents as bodily sensations, functional changes, or somatic metaphors." This definition is very broad, thus it can encompass a number of more or less interrelated phenomena from the perception of body changes in affective states to symptoms of symbolic and metaphoric origin (including conversion), hypochondriasis, and body delusions. As the approach is so broad and a number of different psychological mechanisms can be involved, somatization does not represent a diagnostic category. It simply refers to the experience of multiple symptoms over a period of time that cannot be explained by an organic disease (aka somatoform symptoms). An important aspect of somatization is the seeking of medical help; individuals adopt the "sick role" and extensively use health-care services in their quest to find an explanation for their condition (Lipowski 1986, 1988; Kirmayer and Taillefer 1997).

From a learning perspective, symptoms can be regarded as products of learning processes (including classic and social conditioning) (Mechanic 1972; Barsky 1979; Barsky and Klerman 1983; Kellner 1986; Van den Bergh et al. 1995, 1997). This is in line with the observations that the complains are not necessarily the products or consequences of affective states; however, they often evoke negative affective reactions. A core component of models describing the development and maintenance of symptoms characterizing various pathological states under the umbrella of somatization is a self-enhancing circle, in which originally weak and harmless body sensations are interpreted as harmful and threatening (catastrophization); this in turn leads (directly or through behavioral changes) to the further enhancement of sensations (see Fig. 7.8) (Barsky and Wyshak 1990; Rief et al. 1998; Abramowitz et al. 2007; Witthöft and Hiller 2010). The schema was adopted from models explaining various anxiety disorders, such as panic disorder and obsessive–compulsive disorder (Salkovskis 1985; Clark 1986).

As the term somatization assumes psychological distress at work in the background, and this is not always the case, more neutral terms, such as functional somatic symptoms/syndromes and medically unexplained symptoms, were proposed (Mayou 1993; Kellner 1994; Barsky and Borus 1999; Sirri and Grandi 2012).

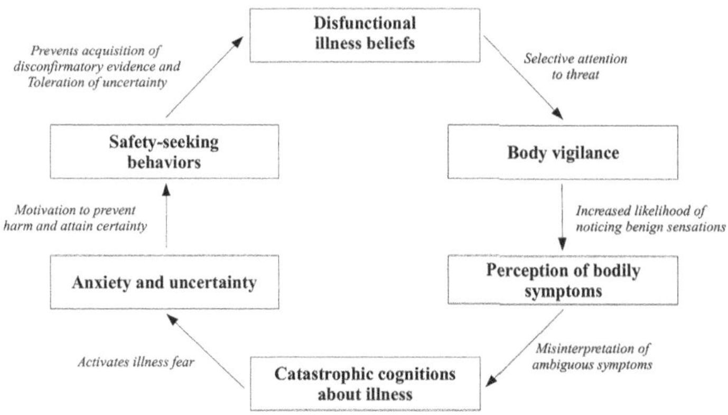

Fig. 7.8 The cognitive-behavioral model of hypochondriasis (Abramowitz et al. 2007, p. 873)

The individual's response to this condition (including the personal, social, and even sociological aspects) is often called abnormal illness behavior (Pilowsky 1969, 1971, 1986); it is determined by the overall perception or cognitive representation of the condition by the sufferer, including assumed cause (attribution), consequences, and controllability (Leventhal et al. 1984, 1997; Weinman et al. 1996). As this does not belong to the primary topic of this book, we will not go into the details. For us, the bottom line is that certain unpleasant body sensations can have a profound impact on the functioning of the organism (this is the tenet of relevance), even if they do not indicate disease or possibly threatening peripheral changes.

Of course, attempts were made to adopt the predictive processing approach to symptom perception and somatoform disorders (Edwards et al. 2012; Van den Bergh et al. 2017a, b; Henningsen et al. 2018; Pezzulo et al. 2019). A very appealing feature of this approach is that it allows the top-down generation of symptoms, even in the very absence of bottom-up input; this is a process that was suspected for decades but could not be theoretically explained in a convincing way before. It is very frustrating for physicians and clinical psychologists to sit in front of a patient who lists his or her complaints and obviously suffers from them without having a satisfying scientific explanation for this condition; frustrating from both a practical and a theoretical point of view. Concerning the latter, the inability of scientific theories to explain such symptoms might challenge the entire worldview and the professional self-esteem of the clinician. In response, professionals sometimes are prone to adopt (of course non-conscious) defense mechanisms, such as minimizing the symptoms, blaming the patient, or accepting a symbolic explanation. Simply ascribing symptoms to expectations or beliefs does not really help as expectations and beliefs are typically considered conscious contents (see above). Thus, referring to expectations often means that the patient should be able to modify these contents. In fact, there are patients who accept the lack of an organic diagnosis (i.e., do not expect symptoms any more) and their symptoms just persist. The predictive processing model is a real revelation for these cases as it offers scientific explanation and a possible mechanism; to put it simply, it can answer the *How*? question (Henningsen et al. 2018). As for the practical consequences, it can also be helpful for developing new interventions. Concerning the *Why*? question, however, things are less rosy. In a nutshell, somatoform symptoms are understood by the predictive processing approach as a failure of inference because of the overall domination of top-down processes over weak bottom-up input or a particularly stubborn (i.e., very precise thus self-maintaining) prior in the processing chain that is difficult to modify (Henningsen et al. 2018). As the entire approach is based on the idea of learning and the modification of existing priors, such a lack of flexibility is hard to explain from the predictive processing perspective. The difficulties with somatization-type conditions will be illustrated via the example of electromagnetic hypersensitivity.

7.9 An Example: Electromagnetic Hypersensitivity

A very important thing that cannot be emphasized enough is that there is no experiential difference between symptoms with and without known pathophysiological background (tenet of realness). Their unpleasantness and the suffering they cause are the same, and we cannot (and should not) differentiate between them. I would like to illustrate this point and also many others discussed in this chapter by electromagnetic hypersensitivity, a condition we have thoroughly studied in the last two decades.

Electromagnetic hypersensitivity, also called Idiopathic Environmental Intolerance attributed to Electromagnetic Fields (IEI-EMF) (WHO 2006), refers to a self-diagnosed condition characterized by symptoms experienced in the proximity of electric devices and/or electromagnetic fields. These symptoms cause considerable suffering and functional impairment for those with severe IEI-EMF. As in other cases of IEI, such as multiple chemical sensitivity and noise sensitivity, the same exposure (which is usually well below the allowed security thresholds) does not cause complaints for the majority of the population (Staudenmayer 2006; Kjellqvist et al. 2016). A further common point is that the symptoms reported by individuals with IEI-EMF are not associated with actual exposure under double-blind conditions (Frick et al. 2005; Rubin et al. 2005, 2008, 2010; Seitz et al. 2005; Röösli 2008). For example, participants with and without IEI-EMF were asked to put their hand in a coil system generating a homogeneous magnetic field of 50 Hz in one of our own laboratory studies (Szemerszky et al. 2015). The coil system was turned on and off for short periods of time in a randomized fashion, and participants had to guess its actual state and report experienced symptoms after each period. Guessing performance was not better than chance; however, the statistical analysis showed that the reported symptoms were associated with the assumed but not with the actual presence of the exposure, and electrosensitives reported more symptoms than controls. In other studies, we were able to demonstrate that the expectation of symptoms can lead to the experience of symptoms even in the absence of actual electromagnetic exposure, both in healthy participants and individuals with IEI-EMF (Szemerszky et al. 2010; Dömötör et al. 2016). Concerning individual differences, the electrosensitive group was consistently characterized by higher levels of somatosensory amplification (see Chap. 5) and more worries about harmful effects of modern technologies than controls in a number of studies (Köteles et al. 2013, 2019; Szemerszky et al. 2015; Dömötör et al. 2016, 2017). Overall, as the claim that symptoms are caused by electromagnetic radiation is usually not supported by objective tests, it was proposed that they are of top-down (aka psychogenic or nocebogenic) origin (Göthe et al. 1995; Röösli 2008; Rubin et al. 2010; Szemerszky et al. 2010; Köteles et al. 2013). Findings of experiments, in which expectations and worry were triggered by materials from media reports and advertisements and the experienced symptoms were attributed to exposure to sham electromagnetic exposure by participants, also support this view (Witthöft and Rubin 2013; Köteles et al.

2016; Bräscher et al. 2017, 2020; Witthöft et al. 2018). In summary, negative expectations and beliefs manifest themselves, and this substantially impacts the condition.

These findings appear really convincing; the lack of association between symptoms and actual exposure was even shown by systematic reviews (Rubin et al. 2005, 2010). However, the widely used double-blind provocation studies are not without flaws either. For example, only one frequency domain can be used at a time (the choice usually relies on the self-report of the participant), and, as the latency between exposure and the appearance of symptoms is often unknown and might vary among participants, carry-over effects from one trial to the subsequent ones are hard to prevent (Hocking 1998; Röösli et al. 2004; Rubin et al. 2010; Verrender et al. 2018). Even the cause–effect relation can disappear because of the latency. A paradigm that can handle these issues and has substantially higher external validity is ecological momentary assessment (EMA) (Stone and Shiffman 1994). In EMA studies, the simultaneous assessment of a number of variables is carried out under everyday conditions for longer periods of time (days or even weeks). In our own study (Szemerszky et al. 2020a), three individuals with severe electrosensitivity carried a dosimeter continuously registering EMF exposure in multiple frequency domains, weared a mobile ECG-device, and completed a brief questionnaire assessing their actual symptoms, mood state, assumed EMF exposure, and activity at random points of time for 3 weeks. Data sets obtained from the three participants were analyzed individually using time series analysis, allowing delays between actual exposure and other (physiological and psychological) variables. It turned out that the reported symptoms were not predicted by the exposure to the suspected EMF domain for any of the participants. In fact, a negative association between symptoms and another EMF domains was found for two of them, i.e., higher exposures in these domains were associated with less symptoms. A positive association between complaints and a non-suspected EMF domain was also revealed for the third participant; however, she was not aware of it. Indicators of physiological activation (heart rate, respiratory rate, heart rate variability) did not show a clear-cut negative effect of exposure; again, higher exposure at certain frequency domains was associated with lower heart rate and respiratory rate in certain cases, whereas heart rate variability-related results were equivocal. To sum it up, we could demonstrate (case studies are always about demonstrating the existence of a phenomenon) using a paradigm with high ecological validity (1) that the claimed association between symptoms and actual exposure does not exist for certain cases, (2) that there are associations that were not realized by our participants, and (3) that there are considerable individual differences in the overall psychophysiological pattern. Similar findings and conclusions were reported in other recently published studies (Bogers et al. 2018; Bolte et al. 2019).

Medical and psychosocial assessment-related (including a clinical interview and the completion of the MMPI) findings of the study are also telling (Szemerszky et al. 2020b). The symptoms reported by our participants (most frequently: tinnitus) could medically not be explained. Each individual had a (partner and/or job-related) traumatic experience before they started to consider themselves electrosensitives.

The psychological profile of two of them was strikingly similar with a marked (but still subclinical) paranoid tendency, whereas the third participant did not show any deviation from the average scores whatsoever. Again, it was demonstrated that people with IEI-EMF does not represent thus should not be regarded as a homogeneous group.

Further findings from recent questionnaire studies shed more light on the IEI-EMF phenomena. Interestingly, it turned out that more than one-fifth of individuals categorizing themselves as electrosensitives do not experience any symptoms and negative impact on daily functioning (Szemerszky et al. 2019). They simply apply this label to themselves (e.g., holistic thinking or New Age-related influences can lie in the background), but in fact do not experience any negative effects (Dömötör et al. 2017). Paranoid thinking was associated with IEI-EMF only for those with symptoms and functional impairment in another, still unpublished, study, which might indicate that paranoid tendencies are associated with the symptom attribution process.

Like other individuals with IEI (Staudenmayer 1998), our participants categorically excluded and rejected the idea that there is no biophysical and/or physiological background behind their condition, and a psychological approach (e.g., psychotherapy) would be helpful for them. In fact, when they learned the null findings of the EMA, all of them refused further cooperation with us. The two with paranoid tendencies started to accuse us for their symptoms or claimed that the study was paid by telecommunication companies, and we deceived them with the results. Such attacks are very difficult to handle personally (sometimes we still recall and discuss these bad experiences); a more important point is that these tendencies render the treatment and even the research of such conditions very hard if not impossible in many cases.

IEIs have two core components: symptoms and a specific attribution. As the former is heavily based on top-down processes and not caused by the suspected environmental factor (in the medical or pathophysiological sense), the neutral term "symptoms associated with environmental factors" was proposed recently (Haanes et al. 2020). The latter component, which often contains paranoid elements (e.g., evil companies that do not publish findings that indicate the harmfulness of certain factors), is very strong and resistant to change or contrary information. It is called an overvalued idea (Henningsen and Priebe 2003), the origin of which usually remains unclear. Such a firm idea heavily influences the perception of symptoms and behavior thus can contribute to the maintenance of the condition as it is illustrated in Fig. 7.7. Still, I think that the presence of symptoms is the primary factor in the development of the condition; if the symptoms disappear (for example, because of the improvement of the stressful life situation), there remains nothing to attribute and the patient recovers spontaneously. Exactly this happened to one of our participants. The fact that the condition is not necessarily stable over time is promising. As we learn more on the background of IEI-EMF, perhaps we will be able to ameliorate the suffering of these individuals.

7.10 Summary

In this chapter, systematic attempts made to understand the primary and secondary aspects of perception of symptoms were summarized. We saw that the boundaries among body sensations, symptoms, and negative affective states are often vague, and they are dominantly determined by nonconscious processes. The output of these processes is very convincing (tenet of realness); thus, it is difficult to modify the individuals' beliefs about the origin and meaning of their sensations, particularly if they are categorized as symptoms and persist over longer periods of time. The newest generation of symptom perception models is able to handle the fact which was discussed in the previous chapters, i.e., that body sensations can arise even in the absence of sensory input, as well as interactions between factors impacting the process. As nonspecific somatic symptoms can be explained and understood in scientific terms, they are not that mysterious any more.

Take-Home Messages

For laypeople: A number of somatic symptoms can develop without clear-cut bodily background, whereas others indicate disturbances in the functioning of the body. We cannot differentiate between them without diagnostic procedures. The possibility that certain sensations does not indicate bodily processes is often hard to accept as they are clearly felt in and related to the body. Fortunately, somatic complaints typically disappear within hours. If they persist, the best thing one can do is visiting his or her general practitioner and follow the doctor's instructions.

For clinicians: The experience of symptoms, both those with and without known pathophysiological background, is the consequence of a sophisticated multi-step process. Symptom reports that involve a number of different complaints belonging to multiple modalities reflect rather the overall psychophysiological state of the patient than the output of independent interoceptive processes. Unfortunately, patients with somatization-type conditions typically are convinced that their symptoms are caused by well-defined peripheral mechanisms and are not willing to accept a psychological (top-down) explanation and therapy.

For researchers: Prediction and causality are among the weakest points of models of symptom perception. Conscious expectations can account for a comparatively small proportion of overall variance at best, whereas their nonconscious counterparts are difficult to measure, both behaviorally and neurophysiologically.

References

Abramowitz JS, Deacon BJ, Valentiner DP (2007) The Short Health Anxiety Inventory: psychometric properties and construct validity in a non-clinical sample. Cogn Ther Res 31:871–883. https://doi.org/10.1007/s10608-006-9058-1

Ainley V, Apps MAJ, Fotopoulou A, Tsakiris M (2016) "Bodily precision": a predictive coding account of individual differences in interoceptive accuracy. Philos Trans R Soc Lond B Biol Sci 371:20160003. https://doi.org/10.1098/rstb.2016.0003

Allen M, Tsakiris M (2018) The body as first prior: interoceptive predictive processing and the primacy of self-models. In: Tsakiris M, De Preester H (eds) The interoceptive mind. From homeostasis to awareness. Oxford University Press, Oxford, England, pp 27–45

Apfel RJ, Sifneos PE (1979) Alexithymia: concept and measurement. Psychother Psychosom 32:180–190

Aronson KR, Barrett LF, Quigley KS (2001) Feeling your body or feeling badly: evidence for the limited validity of the Somatosensory Amplification Scale as an index of somatic sensitivity. J Psychosom Res 51:387–394

Bandura A (1969) Principles of behavior modification. Holt, Rinehart & Winston, New York

Bar-Haim Y, Lamy D, Pergamin L et al (2007) Threat-related attentional bias in anxious and nonanxious individuals: a meta-analytic study. Psychol Bull 133:1–24. https://doi.org/10.1037/0033-2909.133.1.1

Barrett LF (2017) The theory of constructed emotion: an active inference account of interoception and categorization. Soc Cogn Affect Neurosci 12:1–23. https://doi.org/10.1093/scan/nsw154

Barrett LF, Simmons WK (2015) Interoceptive predictions in the brain. Nat Rev Neurosci 16:419–429. https://doi.org/10.1038/nrn3950

Barsky AJ (1979) Patients who amplify bodily sensations. Ann Intern Med 91:63–70. https://doi.org/10.1059/0003-4819-91-1-63

Barsky AJ, Borus JF (1999) Functional somatic syndromes. Ann Intern Med 130:910–921

Barsky AJ, Klerman GL (1983) Overview: hypochondriasis, bodily complaints, and somatic styles. Am J Psychiatry 140:273–283. https://doi.org/10.1176/ajp.140.3.273

Barsky AJ, Wyshak G (1990) Hypochondriasis and somatosensory amplification. Br J Psychiatry 157:404–409. https://doi.org/10.1192/bjp.157.3.404

Barsky AJ, Goodson JD, Lane RS, Cleary PD (1988) The amplification of somatic symptoms. Psychosom Med 50:510–519

Barsky AJ, Orav JE, Delamater BA et al (1998) Cardiorespiratory symptoms in response to physiological arousal. Psychosom Med 60:604–609. https://doi.org/10.1097/00006842-199809000-00016

Bartels DJP, van Laarhoven AIM, Haverkamp EA et al (2014) Role of conditioning and verbal suggestion in placebo and nocebo effects on itch. PLoS One 9:e91727. https://doi.org/10.1371/journal.pone.0091727

Bartels DJP, van Laarhoven AIM, van de Kerkhof PCM, Evers AWM (2015) Placebo and nocebo effects on itch: effects, mechanisms, and predictors. Eur J Pain 20:8. https://doi.org/10.1002/ejp.750

Benedetti F (2009) Placebo effects: understanding the mechanisms in health and disease. Oxford University Press, New York

Benedetti F (2011) The patient's brain. The neuroscience behind the doctor-patient relationship. New York, Oxford

Berna C, Leknes S, Holmes EA et al (2010) Induction of depressed mood disrupts emotion regulation neurocircuitry and enhances pain unpleasantness. Biol Psychiatry 67:1083–1090. https://doi.org/10.1016/j.biopsych.2010.01.014

Bogaerts K, Notebaert K, Van Diest I et al (2005) Accuracy of respiratory symptom perception in different affective contexts. J Psychosom Res 58:537–543. https://doi.org/10.1016/j.jpsychores.2004.12.005

Bogaerts K, Janssens T, De Peuter S et al (2010) Negative affective pictures can elicit physical symptoms in high habitual symptom reporters. Psychol Health 25:685–698. https://doi.org/10.1080/08870440902814639

Bogers RP, van Gils A, Clahsen SCS et al (2018) Individual variation in temporal relationships between exposure to radiofrequency electromagnetic fields and non-specific physical symptoms: a new approach in studying "electrosensitivity". Environ Int 121:297–307. https://doi.org/10.1016/j.envint.2018.08.064

Bolte JFB, Clahsen S, Vercruijsse W et al (2019) Ecological momentary assessment study of exposure to radiofrequency electromagnetic fields and non-specific physical symptoms with self-declared electrosensitives. Environ Int 131:104948. https://doi.org/10.1016/j.envint.2019.104948

Bootzin RR (1985) The role of expectancy in behavior change. In: White L, Thursky B, Schwartz GE (eds) Placebo. Theory, research, and mechanisms. The Guilford Press, New York, pp 196–210

Bräscher A-K, Raymaekers K, Van den Bergh O, Witthöft M (2017) Are media reports able to cause somatic symptoms attributed to WiFi radiation? An experimental test of the negative expectation hypothesis. Environ Res 156:265–271. https://doi.org/10.1016/j.envres.2017.03.040

Bräscher A-K, Schulz SM, Van den Bergh O, Witthöft M (2020) Prospective study of nocebo effects related to symptoms of idiopathic environmental intolerance attributed to electromagnetic fields (IEI-EMF). Environ Res 190:110019. https://doi.org/10.1016/j.envres.2020.110019

Breuer J, Freud S (1893) Studies on hysteria. In: Strachey J, Strachey A (eds) The Penguin Freud library, vol 3). (Original work published 1893–1895). Penguin, London

Brown RJ (2004) Psychological mechanisms of medically unexplained symptoms: an integrative conceptual model. Psychol Bull 130:793–812. https://doi.org/10.1037/0033-2909.130.5.793

Brown RJ (2006) Medically unexplained symptoms: a new model. Psychiatry 5:43–47

Brown RJ, Schrag A, Trimble MR (2005) Dissociation, childhood interpersonal trauma, and family functioning in patients with somatization disorder. Am J Psychiatry 162:899–905. https://doi.org/10.1176/appi.ajp.162.5.899

Bruner JS, Goodman CC (1947) Value and need as organizing factors in perception. J Abnorm Psychol 42:33. https://doi.org/10.1037/h0058484

Bruner JS, Postman L (1948) Symbolic value as an organizing factor in perception. J Soc Psychol 27:203–208. https://doi.org/10.1080/00224545.1948.9918925

Byrne A, Eysenck MW (1995) Trait anxiety, anxious mood, and threat detection. Cognit Emot 9:549–562. https://doi.org/10.1080/02699939508408982

Cioffi D (1991) Beyond attentional strategies: cognitive-perceptual model of somatic interpretation. Psychol Bull 109:25–41

Cisler JM, Koster EHW (2010) Mechanisms of attentional biases towards threat in anxiety disorders: an integrative review. Clin Psychol Rev 30:203–216. https://doi.org/10.1016/j.cpr.2009.11.003

Clark DM (1986) A cognitive approach to panic. Behav Res Ther 24:461–470. https://doi.org/10.1016/0005-7967(86)90011-2

Clark A (2015) Surfing uncertainty: prediction, action, and the embodied mind, 1st edn. Oxford University Press, Oxford

Cohen K, Auld F, Brooker H (1994) Is alexithymia related to psychosomatic disorder and somatizing? J Psychosom Res 38:119–127. https://doi.org/10.1016/0022-3999(94)90085-x

Colloca L, Vighetti S, Sigaudo M, Benedetti F (2007) Hyperalgesic and allodynic effects of nocebo suggestions. Eur J Pain 11:S198. https://doi.org/10.1016/j.ejpain.2007.03.462

Constantinou E, Bogaerts K, Van Diest I, Van den Bergh O (2013) Inducing symptoms in high symptom reporters via emotional pictures: the interactive effects of valence and arousal. J Psychosom Res 74:191–196. https://doi.org/10.1016/j.jpsychores.2012.12.015

Corcoran AW, Hohwy J (2018) Allostasis, interoception, and the free energy principle: feeling our way forward. In: Tsakiris M, De Preester H (eds) The interoceptive mind. From homeostasis to awareness. Oxford University Press, Oxford, England, pp 272–292

Costa PT, McCrae RR (1985) Hypochondriasis, neuroticism, and aging. When are somatic complaints unfounded? Am Psychol 40:19–28

Costa PT, McCrae RR (1987) Neuroticism, somatic complaints, and disease: is the bark worse than the bite? J Pers 55:299–316. https://doi.org/10.1111/j.1467-6494.1987.tb00438.x

De Gucht V, Heiser W (2003) Alexithymia and somatisation: quantitative review of the literature. J Psychosom Res 54:425–434. https://doi.org/10.1016/s0022-3999(02)00467-1

de Jong PJ, Arntz A (1993) The role of conditioning in the placebo response: a comment on Voudouris et al., Pain, 43 (1990) 121–128. Pain 54:232–235
de Jong PJ, van Baast R, Arntz A, Merckelbach H (1996) The placebo effect in pain reduction: the influence of conditioning experiences and response expectancies. Int J Behav Med 3:14–29. https://doi.org/10.1207/s15327558ijbm0301_2
De Peuter S, Put C, Lemaigre V et al (2007) Context-evoked overperception in asthma. Psychol Health 22:737–748. https://doi.org/10.1080/14768320601151702
De Peuter S, Van den Bergh O, Vlaeyen JW (2010) Breathtaking! About the comparison of the subjective sensations of pain and dyspnea. Pain 149:411–412. https://doi.org/10.1016/j.pain.2010.03.021
Dömötör Z, Doering BK, Köteles F (2016) Dispositional aspects of body focus and idiopathic environmental intolerance attributed to electromagnetic fields (IEI-EMF). Scand J Psychol 57:136–143. https://doi.org/10.1111/sjop.12271
Dömötör Z, Szemerszky R, Köteles F (2017) Nature relatedness is connected with modern health worries and electromagnetic hypersensitivity. J Health Psychol 24:1359105317699681. https://doi.org/10.1177/1359105317699681
Domschke K, Stevens S, Pfleiderer B, Gerlach AL (2010) Interoceptive sensitivity in anxiety and anxiety disorders: an overview and integration of neurobiological findings. Clin Psychol Rev 30:1–11. https://doi.org/10.1016/j.cpr.2009.08.008
Edwards MJ, Adams RA, Brown H et al (2012) A Bayesian account of "hysteria". Brain 135:3495–3512. https://doi.org/10.1093/brain/aws129
Eippert F, Finsterbusch J, Bingel U, Büchel C (2009) Direct evidence for spinal cord involvement in placebo analgesia. Science 326:404. https://doi.org/10.1126/science.1180142
Eysenck MW (1992) Anxiety: the cognitive perspective. Erlbaum, Hove, UK
Ferentzi E, Köteles F, Csala B et al (2017) What makes sense in our body? Personality and sensory correlates of body awareness and somatosensory amplification. Personal Individ Differ 104:75–81. https://doi.org/10.1016/j.paid.2016.07.034
Ferrari R (2015) Effect of a pain diary use on recovery from acute low back (lumbar) sprain. Rheumatol Int 35:55–59. https://doi.org/10.1007/s00296-014-3082-3
Ferrari R, Russell AS (2010) Effect of a symptom diary on symptom frequency and intensity in healthy subjects. J Rheumatol 37:2387–2389. https://doi.org/10.3899/jrheum.100513
Fisher S (1970) Body experience in fantasy and behavior. Ardent Media, New York
Fisher S, Cleveland SE (1968) Body image and personality, 2nd rev edn. Dover Publications, New York
Fotopoulou A (2013) Beyond the reward principle: consciousness as precision seeking. Neuropsychoanalysis 15:33–38. https://doi.org/10.1080/15294145.2013.10773715
Frick U, Kharraz A, Hauser S et al (2005) Comparison perception of singular transcranial magnetic stimuli by subjectively electrosensitive subjects and general population controls. Bioelectromagnetics 26:287–298. https://doi.org/10.1002/bem.20085
Friston K (2009) The free-energy principle: a rough guide to the brain? Trends Cogn Sci 13:293–301. https://doi.org/10.1016/j.tics.2009.04.005
Friston K (2010) The free-energy principle: a unified brain theory? Nat Rev Neurosci 11:127–138. https://doi.org/10.1038/nrn2787
Garfinkel SN, Critchley HD (2013) Interoception, emotion and brain: new insights link internal physiology to social behaviour. Commentary on: "Anterior insular cortex mediates bodily sensibility and social anxiety" by Terasawa et al. (2012). Soc Cogn Affect Neurosci 8:231–234. https://doi.org/10.1093/scan/nss140
Garfinkel SN, Seth AK, Barrett AB et al (2015) Knowing your own heart: distinguishing interoceptive accuracy from interoceptive awareness. Biol Psychol 104:65–74. https://doi.org/10.1016/j.biopsycho.2014.11.004
Garfinkel SN, Manassei MF, Hamilton-Fletcher G et al (2016) Interoceptive dimensions across cardiac and respiratory axes. Philos Trans R Soc Lond B Biol Sci 371:20160014. https://doi.org/10.1098/rstb.2016.0014

Gijsbers van Wijk CM, Kolk AM (1996) Psychometric evaluation of symptom perception related measures. Personal Individ Differ 20:55–70. https://doi.org/10.1016/0191-8869(95)90023-N
Gijsbers van Wijk CM, Kolk AM (1997) Sex differences in physical symptoms: the contribution of symptom perception theory. Soc Sci Med 45:231–246
Göthe CJ, Molin C, Nilsson CG (1995) The environmental somatization syndrome. Psychosomatics 36:1–11
Grey JA (1985) Issues in the neuropsychology of anxiety. In: Tuma AH, Maser JD (eds) Anxiety and the anxiety disorders. Erlbaum, Hillsdale, NJ, pp 5–25
Gu X, FitzGerald THB (2014) Interoceptive inference: homeostasis and decision-making. Trends Cogn Sci 18:269–270. https://doi.org/10.1016/j.tics.2014.02.001
Haanes JV, Nordin S, Hillert L et al (2020) "Symptoms associated with environmental factors" (SAEF) – towards a paradigm shift regarding "idiopathic environmental intolerance" and related phenomena. J Psychosom Res 131:109955. https://doi.org/10.1016/j.jpsychores.2020.109955
Hahn RA (1997) The nocebo phenomenon: scope and foundations. In: Harrington A (ed) The placebo effect. An interdisciplinary exploration. Harvard University Press, Cambridge, pp 56–76
Head H, Holmes G (1911) Sensory disturbances from cerebral lesions. Brain 34:102–254. https://doi.org/10.1093/brain/34.2-3.102
Heider F (1958) The psychology of interpersonal relations. Wiley, New York
Henningsen P, Priebe S (2003) New environmental illnesses: what are their characteristics? Psychother Psychosom 72:231–234
Henningsen P, Gündel H, Kop WJ et al (2018) Persistent physical symptoms as perceptual dysregulation: a neuropsychobehavioral model and its clinical implications. Psychosom Med 80:422–431. https://doi.org/10.1097/PSY.0000000000000588
Hocking B (1998) Preliminary report: symptoms associated with mobile phone use. Occup Med (Lond) 48:357–360. https://doi.org/10.1093/occmed/48.6.357
Hogeveen J, Bird G, Chau A et al (2016) Acquired alexithymia following damage to the anterior insula. Neuropsychologia 82:142–148. https://doi.org/10.1016/j.neuropsychologia.2016.01.021
Hohwy J (2012) Attention and conscious perception in the hypothesis testing brain. Front Psychol 3:96. https://doi.org/10.3389/fpsyg.2012.00096
Hohwy J (2014) The predictive mind, 1st edn. Oxford University Press, Oxford, England
Horvath T, Friedman J, Meares R (1980) Attention in hysteria: a study of Janet's hypothesis by means of habituation and arousal measures. Am J Psychiatry 137:217–220. https://doi.org/10.1176/ajp.137.2.217
Janet P (1889) L'automatisme psychologique. Alcan, Paris
Janet P (1907) The major symptoms of hysteria. Macmillan, New York
Jasper F, Hiller W, Rist F et al (2012) Somatic symptom reporting has a dimensional latent structure: results from taxometric analyses. J Abnorm Psychol 121:725–738. https://doi.org/10.1037/a0028407
Jørgensen MM, Zachariae R, Skytthe A, Kyvik K (2007) Genetic and environmental factors in alexithymia: a population-based study of 8,785 Danish twin pairs. Psychother Psychosom 76:369–375. https://doi.org/10.1159/000107565
Katz J, Melzack R (1990) Pain "memories" in phantom limbs: review and clinical observations. Pain 43:319–336
Kelley HH (1973) The processes of causal attribution. Am Psychol 28:107–128. https://doi.org/10.1037/h0034225
Kellner R (1986) Somatization and hypochondriasis. Praeger, New York
Kellner R (1994) Psychosomatic syndromes, somatization and somatoform disorders. Psychother Psychosom 61:4–24. https://doi.org/10.1159/000288868
Khalsa SS, Feinstein JS (2018) The somatic error hypothesis of anxiety. In: Tsakiris M, De Preester H (eds) The interoceptive mind. From homeostasis to awareness. Oxford University Press, Oxford, England, pp 144–164

Kirmayer LJ, Taillefer S (1997) Somatoform disorders. In: Turner SM, Hersen M (eds) Adult psychopathology and diagnosis, 3rd edn. Wiley, New York, pp 333–383
Kirsch I (1985) Response expectancy as a determinant of experience and behavior. Am Psychol 40:1189–1202
Kirsch I (1997a) Response expectancy theory and application: a decennial review. Appl Prev Psychol 6:69–79. https://doi.org/10.1016/S0962-1849(05)80012-5
Kirsch I (1997b) Specifying nonspecifics: psychological mechanisms of placebo effects. In: Harrington A (ed) The placebo effect. An interdisciplinary exploration. Harvard University Press, Cambridge, pp 166–186
Kjellqvist A, Palmquist E, Nordin S (2016) Psychological symptoms and health-related quality of life in idiopathic environmental intolerance attributed to electromagnetic fields. J Psychosom Res 84:8–12. https://doi.org/10.1016/j.jpsychores.2016.03.006
Kleckner IR, Zhang J, Touroutoglou A et al (2017) Evidence for a large-scale brain system supporting allostasis and interoception in humans. Nat Hum Behav 1:s41562-017-0069-017. https://doi.org/10.1038/s41562-017-0069
Kolk AM, Hanewald GJFP, Schagen S, Gijsbers van Wijk CMT (2003) A symptom perception approach to common physical symptoms. Soc Sci Med 57:2343–2354
Köteles F (2013) A placebo-válasz. Medicina Könyvkiadó Zrt, Budapest
Köteles F, Szemerszky R, Gubányi M et al (2013) Idiopathic environmental intolerance attributed to electromagnetic fields (IEI-EMF) and electrosensibility (ES) – are they connected? Int J Hyg Environ Health 216:362–370. https://doi.org/10.1016/j.ijheh.2012.05.007
Köteles F, Tarján E, Berkes T (2016) Artificial concerns. Effects of a commercial advertisement on modern health worries and sympathetic activation. Mentálhigiéné és Pszichoszomatika 17:61–79. https://doi.org/10.1556/0406.17.2016.1.4
Köteles F, Simor P, Szemerszky R (2019) Somatosensory amplification and absorption contribute to electrosensitivity. Ideggyógyászati Szemle 72:165–170. https://doi.org/10.18071/isz.72.0165
Leder D (2018) Inside insights: a phenomenology of interoception. In: Tsakiris M, De Preester H (eds) The interoceptive mind. From homeostasis to awareness. Oxford University Press, Oxford, England, pp 307–322
Leventhal H, Nerenz DR, Steele DS (1984) Illness representations and coping with health threats. In: Baum A, Taylor SE, Singer JE (eds) Handbook of psychology and health. Erlbaum, Hillsdale, NJ, pp 219–252
Leventhal H, Benyamini Y, Brownlee S et al (1997) Illness representations: theoretical foundations. In: Petrie KJ, Weinman JA (eds) Perceptions of health and illness. Harwood, Amsterdam, pp 19–45
Lipowski ZJ (1968) Review of consultation psychiatry and psychosomatic medicine. 3. Theoretical issues. Psychosom Med 30:395–422. https://doi.org/10.1097/00006842-196807000-00005
Lipowski ZJ (1986) Somatization: a borderland between medicine and psychiatry. CMAJ 135:609–614
Lipowski ZJ (1988) Somatization: the concept and its clinical application. Am J Psychiatry 145:1358–1368
Lynn SK, Barrett LF (2014) "Utilizing" signal detection theory. Psychol Sci 25:1663–1673. https://doi.org/10.1177/0956797614541991
MacKrill K, Groom KM, Petrie KJ (2020) The effect of symptom-tracking apps on symptom reporting. Br J Health Psychol 25:1074. https://doi.org/10.1111/bjhp.12459
Mailloux J, Brener J (2002) Somatosensory amplification and its relationship to heartbeat detection ability. Psychosom Med 64:353–357
Mayou R (1993) Somatization. Psychother Psychosom 59:69–83. https://doi.org/10.1159/000288650
McEwen BS (1998) Stress, adaptation, and disease. Allostasis and allostatic load. Ann N Y Acad Sci 840:33–44
McEwen BS, Wingfield JC (2003) The concept of allostasis in biology and biomedicine. Horm Behav 43:2–15

Mechanic D (1962) The concept of illness behavior. J Chronic Dis 15:189–194. https://doi.org/10.1016/0021-9681(62)90068-1
Mechanic D (1972) Social psychologic factors affecting the presentation of bodily complaints. N Engl J Med 286:1132–1139. https://doi.org/10.1056/NEJM197205252862105
Meeuwis SH, van Middendorp H, van Laarhoven AIM et al (2019) Effects of open- and closed-label nocebo and placebo suggestions on itch and itch expectations. Front Psych 10:436. https://doi.org/10.3389/fpsyt.2019.00436
Moerman DE (2002a) Explanatory mechanisms for placebo effects: cultural influences and the meaning response. In: Guess HA, Kleinman A, Kusek JW, Engel LW (eds) The science of the placebo. Toward an interdisciplinary research agenda. BMJ Books, London, pp 77–107
Moerman DE (2002b) Meaning, medicine and the "placebo effect". Cambridge University Press, Cambridge
Moerman DE (2011) Examining a powerful healing effect through a cultural lens, and finding meaning. J Mind Body Regul 1:63–72
Montgomery GH, Kirsch I (1996) Mechanisms of placebo pain reduction: an empirical investigation. Psychol Sci 7:174–176. https://doi.org/10.1111/j.1467-9280.1996.tb00352.x
Mor N, Winquist J (2002) Self-focused attention and negative affect: a meta-analysis. Psychol Bull 128:638–662. https://doi.org/10.1037/0033-2909.128.4.638
Mora PA, Halm E, Leventhal H, Ceric F (2007) Elucidating the relationship between negative affectivity and symptoms: the role of illness-specific affective responses. Ann Behav Med 34:77–86. https://doi.org/10.1007/bf02879923
Palenzuela DL (1987) The expectancy construct within the social learning theories of Rotter and Bandura: a reply to Krisch's approach. J Soc Behav Pers 2:437–452
Park H-D, Blanke O (2019) Heartbeat-evoked cortical responses: underlying mechanisms, functional roles, and methodological considerations. NeuroImage 197:502–511. https://doi.org/10.1016/j.neuroimage.2019.04.081
Paulus MP, Stein MB (2006) An insular view of anxiety. Biol Psychiatry 60:383–387. https://doi.org/10.1016/j.biopsych.2006.03.042
Paulus MP, Stein MB (2010) Interoception in anxiety and depression. Brain Struct Funct 214:451–463. https://doi.org/10.1007/s00429-010-0258-9
Pennebaker JW (1982) The psychology of physical symptoms. Springer, New York
Pennebaker JW (1994) Psychological bases of symptom reporting: perceptual and emotional aspects of chemical sensitivity. Toxicol Ind Health 10:497–511
Pennebaker JW, Epstein D (1983) Implicit psychophysiology: effects of common beliefs and idiosyncratic physiological responses on symptom reporting. J Pers 51:468–496
Pennebaker JW, Lightner JM (1980) Competition of internal and external information in an exercise setting. J Pers Soc Psychol 39:165–174. https://doi.org/10.1037/0022-3514.39.1.165
Pennebaker JW, Roberts T-A (1992) Toward a His and Hers Theory of Emotion: gender differences in visceral perception. J Soc Clin Psychol 11:199–212. https://doi.org/10.1521/jscp.1992.11.3.199
Pennebaker JW, Skelton JA (1981) Selective monitoring of physical sensations. J Pers Soc Psychol 41:213–223
Peters A, McEwen BS, Friston K (2017) Uncertainty and stress: why it causes diseases and how it is mastered by the brain. Prog Neurobiol 156:164–188. https://doi.org/10.1016/j.pneurobio.2017.05.004
Petersen S, Schroijen M, Mölders C et al (2014) Categorical interoception: perceptual organization of sensations from inside. Psychol Sci 25:1059–1066. https://doi.org/10.1177/0956797613519110
Petersen S, Van Staeyen K, Vögele C et al (2015a) Interoception and symptom reporting: disentangling accuracy and bias. Front Psychol 6:732. https://doi.org/10.3389/fpsyg.2015.00732
Petersen S, von Leupoldt A, Van den Bergh O (2015b) Interoception and the uneasiness of the mind: affect as perceptual style. Front Psychol 6:1408. https://doi.org/10.3389/fpsyg.2015.01408
Petrie KJ, Broadbent E, Kley N et al (2005) Worries about modernity predict symptom complaints after environmental pesticide spraying. Psychosom Med 67:778–782. https://doi.org/10.1097/01.psy.0000181277.48575.a4

Petzschner FH, Weber LA, Wellstein KV et al (2019) Focus of attention modulates the heartbeat evoked potential. NeuroImage 186:595–606. https://doi.org/10.1016/j.neuroimage.2018.11.037
Pezzulo G, Rigoli F, Friston K (2015) Active inference, homeostatic regulation and adaptive behavioural control. Prog Neurobiol 134:17–35. https://doi.org/10.1016/j.pneurobio.2015.09.001
Pezzulo G, Maisto D, Barca L, Van den Bergh O (2019) Symptom perception from a predictive processing perspective. Clin Psychol Eur 1(4):e35952. https://doi.org/10.32872/cpe.v1i4.35952
Pilowsky I (1969) Abnormal illness behaviour. Br J Med Psychol 42:347–351. https://doi.org/10.1111/j.2044-8341.1969.tb02089.x
Pilowsky I (1971) The diagnosis of abnormal illness behaviour. Aust N Z J Psychiatry 5:136–138. https://doi.org/10.3109/00048677109159637
Pilowsky I (1986) Abnormal illness behaviour: a review of the concept and its implications. In: McHugh S, Vallis TM (eds) Illness behavior: a multidisciplinary model. Springer, Boston, MA, pp 391–395
Pollatos O, Herbert BM, Kaufmann C et al (2007) Interoceptive awareness, anxiety and cardiovascular reactivity to isometric exercise. Int J Psychophysiol 65:167–173. https://doi.org/10.1016/j.ijpsycho.2007.03.005
Ramachandran VS, Hirstein W (1998) The perception of phantom limbs. The D. O. Hebb lecture. Brain 121(Pt 9):1603–1630
Rief W, Barsky AJ (2005) Psychobiological perspectives on somatoform disorders. Psychoneuroendocrinology 30:996–1002. https://doi.org/10.1016/j.psyneuen.2005.03.018
Rief W, Broadbent E (2007) Explaining medically unexplained symptoms-models and mechanisms. Clin Psychol Rev 27:821–841. https://doi.org/10.1016/j.cpr.2007.07.005
Rief W, Heuser J, Fichter MM (1996) What does the Toronto Alexithymia Scale TAS-R measure? J Clin Psychol 52:423–429. https://doi.org/10.1002/(SICI)1097-4679(199607)52:4<423::AID-JCLP6>3.0.CO;2-Q
Rief W, Hiller W, Margraf J (1998) Cognitive aspects of hypochondriasis and the somatization syndrome. J Abnorm Psychol 107:587–595
Robbins JM, Kirmayer LJ (1991) Attributions of common somatic symptoms. Psychol Med 21:1029–1045. https://doi.org/10.1017/S0033291700030026
Roberts T-A, Pennebaker JW (1995) Gender differences in perceiving internal state: toward a His-And-Hers Model of perceptual cue use (Ch. 4). In: Advances in experimental social psychology. Academic Press INC Ltd, New York, pp 143–175
Röösli M (2008) Radiofrequency electromagnetic field exposure and non-specific symptoms of ill health: a systematic review. Environ Res 107:277–287. https://doi.org/10.1016/j.envres.2008.02.003
Röösli M, Moser M, Baldinini Y et al (2004) Symptoms of ill health ascribed to electromagnetic field exposure – a questionnaire survey. Int J Hyg Environ Health 207:141–150. https://doi.org/10.1078/1438-4639-00269
Rosenzweig P, Brohier S, Zipfel A (1993) The placebo effect in healthy volunteers: influence of experimental conditions on the adverse events profile during phase I studies. Clin Pharmacol Ther 54:578–583. https://doi.org/10.1038/clpt.1993.190
Rotter JB (1953) Social learning and clinical psychology. Prentice-Hall, Englewood Cliffs, NJ
Rubin GJ, Das Munshi J, Wessely S (2005) Electromagnetic hypersensitivity: a systematic review of provocation studies. Psychosom Med 67:224–232. https://doi.org/10.1097/01.psy.0000155664.13300.64
Rubin GJ, Cleare AJ, Wessely S (2008) Psychological factors associated with self-reported sensitivity to mobile phones. J Psychosom Res 64:1–12. https://doi.org/10.1016/j.jpsychores.2007.05.006
Rubin GJ, Nieto-Hernandez R, Wessely S (2010) Idiopathic environmental intolerance attributed to electromagnetic fields (formerly 'electromagnetic hypersensitivity'): an updated systematic review of provocation studies. Bioelectromagnetics 31:1–11. https://doi.org/10.1002/bem.20536

Salkovskis PM (1985) Obsessional-compulsive problems: a cognitive-behavioural analysis. Behav Res Ther 23:571–583. https://doi.org/10.1016/0005-7967(85)90105-6
Schandry R (1981) Heart beat perception and emotional experience. Psychophysiology 18:483–488. https://doi.org/10.1111/j.1469-8986.1981.tb02486.x
Schulz A, Vögele C (2015) Interoception and stress. Front Psychol 6:993. https://doi.org/10.3389/fpsyg.2015.00993
Seitz H, Stinner D, Eikmann T et al (2005) Electromagnetic hypersensitivity (EHS) and subjective health complaints associated with electromagnetic fields of mobile phone communication—a literature review published between 2000 and 2004. Sci Total Environ 349:45–55. https://doi.org/10.1016/j.scitotenv.2005.05.009
Seth AK (2013) Interoceptive inference, emotion, and the embodied self. Trends Cogn Sci 17:565–573. https://doi.org/10.1016/j.tics.2013.09.007
Seth AK, Friston KJ (2016) Active interoceptive inference and the emotional brain. Philos Trans R Soc Lond B Biol Sci 371:20160007. https://doi.org/10.1098/rstb.2016.0007
Seth AK, Suzuki K, Critchley HD (2012) An interoceptive predictive coding model of conscious presence. Front Psychol 2:395. https://doi.org/10.3389/fpsyg.2011.00395
Sifneos PE (1973) The prevalence of 'alexithymic' characteristics in psychosomatic patients. Psychother Psychosom 22:255–262. https://doi.org/10.1159/000286529
Sirri L, Grandi S (2012) Illness behavior. Adv Psychosom Med 32:160–181. https://doi.org/10.1159/000330015
Smith CA, Wallston KA, Dwyer KA (1995) On babies and bathwater: disease impact and negative affectivity in the self-reports of persons with rheumatoid arthritis. Health Psychol 14:64–73. https://doi.org/10.1037//0278-6133.14.1.64
Staudenmayer H (1998) Environmental illness: myth & reality. CRC Press, Boca Raton, FL
Staudenmayer H (2006) Idiopathic environmental intolerance (IEI): a causation analysis. In: Hansson Mild K, Repacholi M, van Deventer E, Ravazzani P (eds) Electromagnetic hypersensitivity. Proceedings: International workshop on EMF hypersensitivity. WHO, Prague, pp 39–53
Stegen K, van Diest I, van de Woestijne KP, Van den Bergh O (2000) Negative affectivity and bodily sensations induced by 5.5% CO_2 enriched air inhalation: is there a bias to interpret bodily sensations negatively in persons with negative affect? Psychol Health 15:513–525. https://doi.org/10.1080/08870440008402010
Stegen K, Diest IV, Woestijne KPVD, Van den Bergh O (2001) Do persons with negative affect have an attentional bias to bodily sensations? Cognit Emot 15:813–829. https://doi.org/10.1080/02699930143000211
Stephan KE, Manjaly ZM, Mathys CD et al (2016) Allostatic self-efficacy: a metacognitive theory of dyshomeostasis-induced fatigue and depression. Front Hum Neurosci 10:550. https://doi.org/10.3389/fnhum.2016.00550
Sterling P (2012) Allostasis: a model of predictive regulation. Physiol Behav 106:5–15. https://doi.org/10.1016/j.physbeh.2011.06.004
Sterling P, Eyer J (1988) Allostasis: a new paradigm to explain arousal pathology. In: Handbook of life stress, cognition and health. Wiley, Oxford, England, pp 629–649
Stewart-Williams S (2004) The placebo puzzle: putting together the pieces. Health Psychol 23:198–206. https://doi.org/10.1037/0278-6133.23.2.198
Stewart-Williams S, Podd J (2004) The placebo effect: dissolving the expectancy versus conditioning debate. Psychol Bull 130:324–340. https://doi.org/10.1037/0033-2909.130.2.324
Stone AA, Shiffman S (1994) Ecological momentary assessment (EMA) in behavorial medicine. Ann Behav Med 16:199–202
Swartzman LC, Lees MC (1996) Causal dimensions of college students' perceptions of physical symptoms. J Behav Med 19:95–110. https://doi.org/10.1007/BF01857601
Szemerszky R, Köteles F, Lihi R, Bárdos G (2010) Polluted places or polluted minds? An experimental sham-exposure study on background psychological factors of symptom formation in "Idiophatic Environmental Intolerance attributed to electromagnetic fields". Int J Hyg Environ Health 213:387–394. https://doi.org/10.1016/j.ijheh.2010.05.001

Szemerszky R, Gubányi M, Árvai D et al (2015) Is there a connection between electrosensitivity and electrosensibility? A replication study. Int J Behav Med 22:755–763. https://doi.org/10.1007/s12529-015-9477-z

Szemerszky R, Dömötör Z, Köteles F (2019) One single question is not sufficient to identify individuals with electromagnetic hypersensitivity. Clin Psychol Eur 1(4):e35668. https://doi.org/10.32872/cpe.v1i4.35668

Szemerszky R, Dömötör Z, Ruzsa G, et al (2020a) An idiographic approach to Idiopathic Environmental Intolerance attributed to Electromagnetic Fields (IEI-EMF) Part I. Ecological momentary assessment. Manuscript submitted for publication

Szemerszky R, Dömötör Z, Szabolcs Z, et al (2020b) An idiographic approach to Idiopathic Environmental Intolerance attributed to Electromagnetic Fields (IEI-EMF) Part II. Environmental, medical and psychosocial assessment. Manuscript submitted for publication

Teff KL (2011) How neural mediation of anticipatory and compensatory insulin release helps us tolerate food. Physiol Behav 103:44–50. https://doi.org/10.1016/j.physbeh.2011.01.012

Tellegen A (1985) Structures of mood and personality and their relevance to assessing anxiety, with an emphasis on self-report. In: Tuma AH, Maser JD (eds) Anxiety and the anxiety disorders. Erlbaum, Hillsdale, NJ, pp 681–706

Tihanyi BT, Ferentzi E, Beissner F, Köteles F (2018) The neuropsychophysiology of tingling. Conscious Cogn 58:97–110. https://doi.org/10.1016/j.concog.2017.10.015

Ursin H (1997) Sensitization, somatization, and subjective health complaints. Int J Behav Med 4:105–116. https://doi.org/10.1207/s15327558ijbm0402_1

Ursin H, Eriksen HR (2001) Sensitization, subjective health complaints, and sustained arousal. Ann N Y Acad Sci 933:119–129. https://doi.org/10.1111/j.1749-6632.2001.tb05819.x

Ursin H, Eriksen HR (2004) The cognitive activation theory of stress. Psychoneuroendocrinology 29:567–592. https://doi.org/10.1016/S0306-4530(03)00091-X

Ursin H, Eriksen HR (2010) Cognitive activation theory of stress (CATS). Neurosci Biobehav Rev 34:877–881. https://doi.org/10.1016/j.neubiorev.2009.03.001

Van den Bergh O, Kempynck PJ, van de Woestijne KP et al (1995) Respiratory learning and somatic complaints: a conditioning approach using CO_2-enriched air inhalation. Behav Res Ther 33:517–527

Van den Bergh O, Stegen K, Van de Woestijne KP (1997) Learning to have psychosomatic complaints: conditioning of respiratory behavior and somatic complaints in psychosomatic patients. Psychosom Med 59:13–23

Van den Bergh O, Winters W, Devriese S et al (2004) Accuracy of respiratory symptom perception in persons with high and low negative affectivity. Psychol Health 19:213–222. https://doi.org/10.1080/08870440410001675627

Van den Bergh O, Brown RJ, Petersen S, Witthöft M (2017a) Idiopathic environmental intolerance: a comprehensive model. Clin Psychol Sci 5:551–567. https://doi.org/10.1177/2167702617693327

Van den Bergh O, Witthöft M, Petersen S, Brown RJ (2017b) Symptoms and the body: taking the inferential leap. Neurosci Biobehav Rev 74(Pt A):185–203. https://doi.org/10.1016/j.neubiorev.2017.01.015

Van den Bergh O, Zacharioudakis N, Petersen S (2018) Interoception, categorization, and symptom perception. In: Tsakiris M, De Preester H (eds) The interoceptive mind. From homeostasis to awareness. Oxford University Press, Oxford, England, pp 212–226

Verrender A, Loughran SP, Anderson V et al (2018) IEI-EMF provocation case studies: a novel approach to testing sensitive individuals. Bioelectromagnetics 39:132–143. https://doi.org/10.1002/bem.22095

Voudouris NJ, Peck CL, Coleman G (1990) The role of conditioning and verbal expectancy in the placebo response. Pain 43:121–128. https://doi.org/10.1016/0304-3959(90)90057-K

Walentynowicz M, Witthöft M, Raes F et al (2018) Sensory and affective components of symptom perception: a psychometric approach. J Exp Psychopathol 9:jep.059716. https://doi.org/10.5127/jep.059716

Watson D, Clark LA (1984) Negative affectivity: the disposition to experience aversive emotional states. Psychol Bull 96:465–490

Watson D, Pennebaker JW (1989) Health complaints, stress, and distress: exploring the central role of negative affectivity. Psychol Rev 96:234–254

Weiner B (1979) A theory of motivation for some classroom experiences. J Educ Psychol 71:3–25

Weiner B, Frieze I, Kukla A et al (1972) Perceiving the causes of success and failure. In: Jones EE, Kanouse DE, Kelley HH et al (eds) Attribution: perceiving the causes of behaviour. General Learning Press, Morristown, NJ

Weinman J, Petrie KJ, Moss-morris R, Horne R (1996) The illness perception questionnaire: a new method for assessing the cognitive representation of illness. Psychol Health 11:431–445. https://doi.org/10.1080/08870449608400270

Wessely S, Nimnuan C, Sharpe M (1999) Functional somatic syndromes: one or many? Lancet 354:936–939. https://doi.org/10.1016/S0140-6736(98)08320-2

Weston SJ, Jackson JJ (2018) The role of vigilance in the relationship between neuroticism and health: a registered report. J Res Pers 73:27–34. https://doi.org/10.1016/j.jrp.2017.10.005

WHO (2006) Electromagnetic hypersensitivity. In: Mild KH, Repacholi M, van Deventer E, Ravazzani P (eds) Proceedings, International workshop on electromagnetic field hypersensitivity, Prague, Czech Republic. WHO, Geneva

Winters RW, McCabe PM, Schneiderman N (2002) Functional utility and neurobiology of conditioned autonomic responses. In: Moore JW (ed) A neuroscientist's guide to classical conditioning. Springer, New York, pp 46–85

Witthöft M, Hiller W (2010) Psychological approaches to origins and treatments of somatoform disorders. Annu Rev Clin Psychol 6:257–283. https://doi.org/10.1146/annurev.clinpsy.121208.131505

Witthöft M, Rubin GJ (2013) Are media warnings about the adverse health effects of modern life self-fulfilling? An experimental study on idiopathic environmental intolerance attributed to electromagnetic fields (IEI-EMF). J Psychosom Res 74:206–212. https://doi.org/10.1016/j.jpsychores.2012.12.002

Witthöft M, Basfeld C, Steinhoff M, Gerlach AL (2012) Can't suppress this feeling: automatic negative evaluations of somatosensory stimuli are related to the experience of somatic symptom distress. Emotion 12:640–649. https://doi.org/10.1037/a0024924

Witthöft M, Hiller W, Loch N, Jasper F (2013) The latent structure of medically unexplained symptoms and its relation to functional somatic syndromes. Int J Behav Med 20:172–183. https://doi.org/10.1007/s12529-012-9237-2

Witthöft M, Freitag I, Nußbaum C et al (2018) On the origin of worries about modern health hazards: experimental evidence for a conjoint influence of media reports and personality traits. Psychol Health 33:361–380. https://doi.org/10.1080/08870446.2017.1357814

Yunus MB (2007) Fibromyalgia and overlapping disorders: the unifying concept of central sensitivity syndromes. Semin Arthritis Rheum 36:339–356. https://doi.org/10.1016/j.semarthrit.2006.12.009

Yunus MB (2008) Central sensitivity syndromes: a new paradigm and group nosology for fibromyalgia and overlapping conditions, and the related issue of disease versus illness. Semin Arthritis Rheum 37:339–352. https://doi.org/10.1016/j.semarthrit.2007.09.003

Zacharioudakis N, Vlemincx E, Van den Bergh O (2020) Categorical interoception and the role of threat. Int J Psychophysiol 148:25–34. https://doi.org/10.1016/j.ijpsycho.2019.12.009

Chapter 8
Body Sensations and Emotions

8.1 Introduction

This chapter aims to summarize theories on emotions from the viewpoint of body sensations. This area of research is a really challenging field; I do not intend to describe and evaluate all issues in detail. For example, I will not present the often cited criticism of the Jamesian theory by Cannon (1927, 1931) or the famous debate of Zajonc (1980, 1984) and Lazarus and Smith (1988) on the relationship between cognition and affect. Also, the intriguing idea that emotions do not represent a natural class, i.e., conceptually distinct phenomena are put under this umbrella term (Prinz 2004), will not be discussed. This chapter focuses on the body sensations-related aspects of emotions only and largely neglect other aspects.

Body feelings, i.e., conscious body-related experiences, are involved in the majority of emotion theories, including naïve theories. In fact, only a small number of theories, such as the early model of Cannon, excluded the subjective dimension from the concept of emotions as this was not objectively measurable (Dror 2014). However, those who accept the importance of the experiential component of emotions could not reach an agreement on the role of body percepts. Whereas a number of theories on emotions assume the *causal role* of body sensations in the generation of the emotional experience (i.e., the affective feeling), other approaches regard them as *concomitants* or even *consequences* of the process. There are conceptual differences between affective and body feelings (see Chap. 2): the former term refers to non-isomorphic evaluative representations, whereas body sensations are localized (Feinberg and Mallatt 2016). As we will see, this issue can be resolved, at least theoretically, by assuming distinct patterns of body sensations behind each emotion. To make the situation even more complex, the association is bidirectional: body sensations can contribute to affective feelings and at the same time can be subjects to affective evaluation too (tenet of evaluation). This state of affairs can easily lead to a chicken-and-egg dilemma. On top of this, both affective feelings and body feelings are difficult to verbalize (tenet of vagueness), thus the overlap between

F. Köteles, *Body Sensations*, https://doi.org/10.1007/978-3-030-63201-4_8

the two shown by self-reports may simply be the consequence of ineffability (i.e., an artifact).

Perhaps the most interesting area discussed in this chapter is the relationship between body sensations and the emotional experience. As presented in Chaps. 6 and 7, the peripheral physiological background of body sensations is not always clear-cut; we have a good reason to assume that there are body sensations without such background. The major question is how can we apply this knowledge to emotions.

8.2 The Background of Emotional Experience

Emotions are proposed to have two major adaptive functions: a survival-related, i.e., organizing an instant adaptive response to typical situations and a communicative, i.e., informing others about the internal state of the organism (Plutchik 1982). The objectively observable starting and end points of the process are not questioned: emotions are evoked by a stimulus or pattern of (usually external) stimuli which results in a well-defined behavioral pattern (including visceral and somatic changes). In the middle of the process, some sort of evaluation necessarily occurs as only certain stimuli evoke an emotional response, and there is a specificity in the stimulus–response relationship. Thus, emotions are considered stereotypic adaptations and reactions to typical challenges of the life of social mammals (Nesse 1990; LeDoux 1998). The visceral component (e.g., increased heart rate and blood pressure) prepares the organism for action, whereas the somatic component (e.g., increased muscle tone or a typical pose) represents the behavioral reaction in the strict sense. Beyond this, informing other members of the group about the actual internal state of the individual is also adaptive (Darwin 1872). All these factors are objectively measurable and the assumption that something else, i.e., subjective experience, also takes place appears unnecessary thus unimportant at first sight. As mentioned in Chap. 2, however, we have a good reason to think that affective feelings are present in higher animals, and they play a marked role in the course of adaptation (Panksepp 1998; Metzinger 2009). The major debate is about the causal background of this emotional experience. In this section, I will try to summarize the standpoint of various theories on the temporal sequence of three components of emotions: primary evaluation, feeling (subjective experience), and response (including visceral and motor responses).

Central theories of emotions (Cannon 1927, 1931; Papez 1937; MacLean 1949; Olds and Olds 1961; Olds 1967; Oatley and Johnson-laird 1987; Davis 1989; LeDoux 1990) assume the primary role of the central nervous system, most precisely, subcortical structures in the development of the emotional response (both the behavioral and experiential components although many scholars did not pay much attention to the latter). As mentioned, the available environmental stimuli are subject to evaluation. This evaluation is called appraisal (Arnold 1960) and often divided into a primary (automatic) and secondary component (Lazarus and Folkman

1984). Many theorists think that primary appraisal already gives rise to a subjective feeling although there is a disagreement concerning its quality and/or complexity; it can simply be a judgment along the dimension of pleasantness–unpleasantness (Cabanac 1971), a blend of valence and arousal (Tichtener 1908, 1926; Osgood et al. 1957; Russell 1980), well-defined primary emotions (Panksepp 1982, 1998), or a decision between harm, threat, or challenge (Lazarus and Folkman 1984). Subsequent steps make the experience more differentiated and complex; still, the existence of such a first step is not questioned. Bodily response is considered a consequence of primary evaluation (and can run simultaneously with secondary evaluation) by certain theorists (Zajonc 1980, 1984; Ellsworth 1994).

However, according to the interpretation of most textbooks, the James-Lange theory of emotions proposes that salient stimuli elicit bodily changes which in turn are perceived by the organism and the emotional experience is based on those felt changes (James 1884, 1890; Lange 1885). Actually, James (1890) wrote about "the perception of the exciting fact," which necessarily refers to a primary appraisal process; not all facts are exciting (Ellsworth 1994). Further, he described the existence of a mechanism that is close to primary appraisal; he called it instinct and assumed that it relies on reflex-like reactions to certain external stimuli (Lang 1994). The subjective aspect of this primary appraisal was simply not regarded by James as a full-featured emotional experience; he assumed that the experience of bodily changes is necessary for the latter. Concerning the bodily changes, although James was not clear in this respect, he probably believed that visceral changes and the related sensations play a more prominent role than somatosensory events (Leventhal 1984; Ellsworth 1994). The experience of body feelings is then called emotion by James and its quantity and quality are assumed to be determined by the extent and specific pattern of physiological activation, respectively. Lange, however, highlighted the importance of cardiovascular changes in the generation of emotions and was interested in the behavioral rather than the experiential aspect of the process (Lang 1994). Thus, although the theory is named after both authors, perhaps the only thing he agreed upon was that different emotions are caused by different physiological activation patterns. This point, i.e., the specificity of the physiological patterns behind various emotions, was already questioned by Cannon (1927) and still debated (see below). More recently, a similar philosophical approach to emotions was proposed by Prinz (2004). This theory puts less emphasis on the conscious experience than the adaptive behavior; still, it argues that the basic emotions are encoded as specific body patterns.

A further issue with this approach already mentioned by Cannon (1927) concerns timing: it takes physiological changes seconds to build up, and this step represents only the medium component of the entire process. Before that, primary appraisal takes place, and the processing of evoked physiological feedback also takes time. Thus, the entire process appears to be the way to slow to explain the immediateness of emotional response (including the feeling aspect) we often experience. For example, in reaction time-based measurements, the delay caused by emotional processing is usually in the magnitude of tens of milliseconds, and this delay includes additional steps, such as attentional processes beyond the pure response. Central simulation of the somatic part of the process, such as the "as-if loop" proposed by Damasio (1994, 1996, 1999), can theoretically shorten the processing time. If this is the case, however, then the role of peripheral physiological activation (emotion-specific or general, see below) in the generation of the emotional experience remains unknown. In other words, once we accept the existence of central simulation, the theory loses its most characteristic component, i.e., the causal sequence of peripheral change and affective feeling.

Another feedback theory that primarily relies on somatic feedback, such as facial expression, was proposed by Darwin (1872) and further elaborated by Tomkins (1962, 1981) and Izard (1971, 1990). Although a proportion of the empirical results supporting the facial feedback theory was questioned recently (Noah et al. 2018), the possibility that somatosensory information from receptors located in the skin and the motor system can contribute to the emotional experience cannot be excluded. For example, recent findings on power posing support the idea that postures can influence emotional experience (Carney et al. 2010; Ranehill et al. 2015; Weineck et al. 2019), and basic emotions were accompanied by specific movement patterns in another study (Shafir et al. 2015). Also, certain arm movements were found to be associated with biases in effective evaluation (flexing was related to more positive, whereas extending with more negative evaluation) (Cacioppo et al. 1993b; Neumann and Strack 2000); in a similar vein, breathing patterns can evoke emotional states (Philippot et al. 2002). Unfortunately, the feed-forward mechanism of muscle control (Chap. 3) puts these instances in a similar position as centrally simulated visceral changes: as the output of the motor command is predicted by the brain (even the motor command can be considered a prediction), the slow peripheral loop is not required to explain the phenomenon.

Regardless the causation issue, a (seemingly, see below) necessary prerequisite for the James-Lange theory is that different emotions are characterized by different physiological activation patterns. This idea had received marked research attention in the heydays of psychophysiology and was supported by empirical findings from different laboratories (Wolf and Wolff 1943; Arnold 1945; Malmo et al. 1950; Ax 1953; Davis et al. 1955, 1957; Funkenstein 1956; Davis and Buchwald 1957; Schachter 1957; Graham et al. 1958; Dykman et al. 1959; Hess and Polt 1960). To explain the revealed substantial inter- and individual differences in the physiological reaction pattern, Lacey developed the concepts of situational, intra-stressor, and inter-stressor stereotypy (Lacey 1950, 1956; Lacey and Lacey 1958). This idea, also called stimulus–response and individual–response specificity (Wenger et al. 1957; Engel 1960; Engel and Bickford 1961; Roessler and Engel 1974), was then applied to psychosomatic medicine to explain individual differences in pathological responses (Alexander 1950). Later studies supported the existence of differences among emotions with respect to distinct physiological activation patterns (Roberts and Weerts 1982; Ekman et al. 1983; Miller et al. 1987; Levenson et al. 1990; Lang et al. 1993; Bandler et al. 2000; Folkow 2000; Friedman 2010; Stephens et al. 2010), and a recent review also drew this conclusion (Kreibig 2010). However, these patterns show individual differences and explain only a small proportion of the total variance (Hsü 1952; Cacioppo et al. 1993a; Lang 1994; Christie and Friedman

2004; Moors 2009). Moreover, recent studies show that impaired peripheral feedback is not accompanied by considerably limited emotional capacity (Heims et al. 2004; Cobos et al. 2004); the central simulation of bodily states can explain this apparent controversy (Wiens 2005; Smith and Lane 2015). These and other conceptual issues do not allow a final decision (Barrett 2006a; Levenson 2014; Smith and Lane 2015). Overall, considering our limited ability to detect visceral events, the direct causal link between specific visceral changes and the emotional experience for emotions evoked by external stimuli appears unlikely to me.

Schachter's theory (Schachter and Singer 1962; Schachter 1964) basically kept the primary evaluation—peripheral response—emotional experience sequence and tried to resolve the specificity issue by reducing the somatic part to nonspecific activation (physiological arousal) that elicits the (conscious) cognitive evaluation of the entire setting in a later step. This approach is rooted in early experimental work that reported that adrenaline evoked only "cold" emotions in the absence of objects and/or valid reasons for genuine emotions (Maranon 1924; Cantril and Hunt 1932; Landis and Hunt 1932). Consequently, cognitive evaluation, primarily relying on situational stimuli and knowledge, determines the quality of emotional experience. Again, it is not clear in this theory how a stimulus leads to an aroused state (Moors 2009); in other words, some kind of automatic evaluation necessarily takes place before the increase of arousal. However, this evaluation is assumed to be less specific (this feature is shared with a number of central theories), as it simply judges whether the stimulus at hand is worthy of attention and further processing (in other word, salient) or not.

Subsequent theories of appraisal partly replaced conscious evaluation with non-conscious or automatic processing (Lazarus 1966, 1991; Frijda 1986; Scherer 1999, 2004). Important components of this processing are goal relevance, goal congruence, certainty, etc. (Moors 2009). Evaluation of stimuli can be rule-based or associative (i.e., non-cognitive). The former is a cognitive computation that relies on the aforementioned components, whereas the latter is based on association developed by previous experiences. In these models, primary (mostly associative) evaluation leads to the emotional experience and the peripheral changes, but the latter two does not directly impact each other. An important question here is the origin of emotional (affective) feelings that become associated with new and new stimuli over time. A number of prominent authors argue that there must be primary (inherited) feelings that are genetically linked to certain stimuli and settings (Leventhal 1984; Leventhal and Scherer 1987; Panksepp 1998; Damasio 2004). For example, according to affective neuroscience, these primary affective feelings cover the major categories of situations that are salient for mammals, and they have well-defined functions (i.e., they are not simply emergent phenomena): they can be associated with novel stimuli and can be combined with each other; this leads to behavioral flexibility and a plethora of higher-order emotions (Panksepp 1998). Alternatively, basic emotions are not natural entities but rely on the two independent dimensions (i.e., arousal and valence) of core affect and are the results of automatic categorization processes (Russell 1980, 2003; Barrett and Russell 1999; Barrett 2006a, b). Still, the existence of inborn associations that serve as a starting point for further learning is a logical necessity here too. Network theories (Bower 1981; Leventhal 1984), including connectionist theories (Lewis 2005), also emphasize the role of evaluative conditioning (Martin and Levey 1978) thus implicitly or explicitly assume the existence of primary feelings. In summary, appraisal-based and related theories do not necessarily assume peripheral body processes behind the generation of the emotional experience. Evaluation takes place in the central nervous system (the exact localization varies from upper brainstem structures to the thalamus), and the output of this primary evaluation triggers the subsequent components of emotional processing, including affective experience and emotion-specific peripheral changes, i.e., action tendencies (Ekman 1992; Izard 1993; Panksepp 1998; Scherer 2004). Perhaps the most important brain center involved in the emotional experience is the anterior part of the insula (Gu et al. 2013; Barrett and Simmons 2015). The direct stimulation and lesion of the anterior insula impacted the recognition of emotions in a number of studies (Terasawa et al. 2015; Motomura et al. 2019).

8.3 Possible Consequences of the Actual-Perceived Discrepancy

Despite the disagreement on causation and timing, most theories on emotions accept that the emotional experience is embodied, i.e., accompanied by bodily changes. Thus, perception of bodily events should be associated with the emotional experience as proposed by Darwin, James, Lange, Schachter, and others (this does not necessarily mean causation though). A number of early (Yokoyama 1921; Nafe 1924; Young 1927, 1932; Hoisington 1928; Hunt 1931, 1939; Gemelli 1949; Mason 1959; Davitz 1969) and more recent (Shields 1984; Nummenmaa et al. 2014, 2018; Jung et al. 2017; Novembre et al. 2019; Torregrossa et al. 2019) phenomenological studies documented the existence of such a connection. However, self-reported attentiveness to and awareness of nonemotional body processes (assessed with the BAQ) and the perception of emotion-related body changes (measured with the APQ) show only a weak association (Shields and Simon 1991), which indicates that the emotional experience cannot be exclusively explained by perceived somatic changes. Moreover, as presented in Chaps. 4 and 6, our ability to detect and sense visceral events is very limited, and we often sense things without peripheral changes. These facts raise a number of questions and options with respect to theories that emphasize the causal role of body changes in the emotional experience. Logically, there are four possibilities, summarized in Table 8.1; just like the kinds of body sensations in Chap. 6, these categories exclusively serve demonstrative purposes. I do not think that things are so simple and there are clear-cut boundaries between categories (see below). Still, such a classification can help us to systematically consider all possibilities.

Table 8.1 The contribution of possible combinations of actual and perceived body changes to the emotional experience (for details, see the text)

The background of emotional experience		Perceived bodily change	
		Present	Absent
Actual bodily change	Present	A	B
	Absent	C	D

First, one can assume that well-defined body changes takes place and are perceived during the emotional experience (Table 8.1, case A); this is the standpoint of the classic peripheral theories.

Second, an option is that the emotional response is accompanied by peripheral changes; however, they remain unnoticed for the individual (Table 8.1, case B). Classic peripheral theories would not assume any change in the affective experience in this case; in fact, they would not accept this as emotion. Approaches that emphasize the behavioral aspect and rather neglect the experiential aspect of emotions would consider this pattern as an emotion if there are changes in behavior (including visceral changes). Also, evaluation of actual body changes may occur at nonconscious levels, leading to affective feelings but not to body sensations. Finally, as in the case of cardiovascular reactivity (Chap. 6), sometimes there is a marked physiological reaction to a situation, but the individual does not experience a similar activation at the subjective level at all. These cases thus can be called implicit or nonconscious emotions (see below).

Third (Table 8.1, case C), as demonstrated by the tingling phenomenon in Chap. 5, we can experience body changes in the very absence of actual changes. William James most probably did not differentiate actual and perceived body changes as this discrepancy was not known that time. More precisely, he must have assumed that felt changes that lead to the emotional experience always reflect actual changes. In consequence, empirical studies designed to test this model should rely on self-reported sensations, regardless of their peripheral physiological background. Early empirical studies on emotions (Chap. 6) reported a marked discordance between feelings and the physiological aspects (Mandler and Kremen 1958; Mandler et al. 1958, 1961; Katkin 1965); subsequent research replicated this dissociation in different settings for various emotions (Marks and Huson 1973; Rachman and Hodgson 1974; Cook et al. 1988; Lang et al. 1993). It is possible, then, that the experience of emotions is associated with (and perhaps caused by) perceived body events, i.e., body sensations.

Concerning the James-Lange theory, James Pennebaker devoted an entire chapter to this problem, called perceived specificity hypothesis, in his excellent monograph (1982). In a series of studies, it was shown that (1) there are associations between emotional feelings and self-reported body sensations using both between-subject and within-subject designs, i.e., at the level of the group as well as the individual; (2) participants asked to induce an emotion reported body feelings that showed specific associations with certain emotions; (3) participants asked to produce certain patterns of body feelings reported emotions that were specifically associated with certain constellations of body sensations; (4) there was no association between actual and reported body events (the warmth of hands and gastric contractions) in situations that evoked emotions. Taken together, these findings show that the link between body sensations and affective feelings is not necessarily one-directional, and perceived body feelings are associated with emotions regardless of their physiological background. Of course, it is indeed possible that certain peripheral changes are also perceived to some degree. This option, which is in essence an overlap between category A and C (thus clearly shows the pitfalls of my rigid categorization), will be discussed in the subsequent section.

In a number of qualitative and quantitative studies, participants were asked to mark their body feelings by indicating the areas that were activated in a body map (Nummenmaa et al. 2014). Activation patterns characterizing basic (e.g., fear, anger, happiness, surprise) and nonbasic or complex (e.g., anxiety, love, shame, envy) emotions could be differentiated from each other and from the neutral state above chance level. It does not mean perfect discrimination though. In a subsequent study (Nummenmaa et al. 2018), participants were asked to characterize 100 subjective states (e.g., heartburn, sexual arousal, panic, reasoning), including their topography and perceived similarity. All of them were emotionally valenced; the saliency of the accompanying body sensations was related to the saliency of the feeling states. Finally, a sophisticated dimension reduction procedure resulted in five distinct clusters: negative and positive emotions, illnesses, homeostasis, and cognition. Interestingly enough, the feeling of heartbeat did not belong to any of these dimensions, it is located between the major clusters (Fig. 8.1). At a higher (i.e., more verbalized) level, a systematic attempt was made to describe the emotions as unique

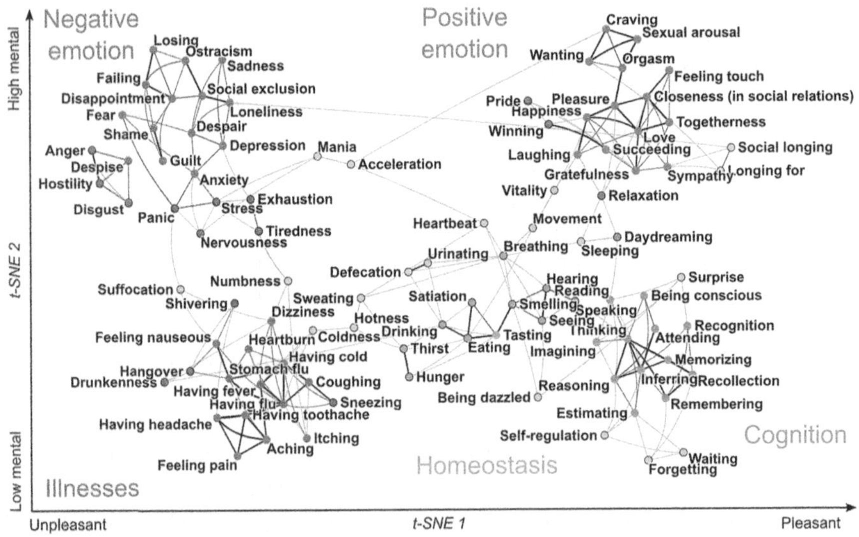

Fig. 8.1 Two-dimensional map of cognitive and affective aspects of affective and body feelings (Nummenmaa et al. 2018, p. 3)

patterns of more or less elementary feelings (Davitz 1969). It was found that there are considerable similarities among people in the emotion-specific patterns; moreover, body sensations were part and parcel of these patterns. These studies, following the phenomenological tradition, neatly indicate that body sensations represent an important component of almost all affective states. In other words, no clear-cut distinction between purely affective and physical states can be made (see also Chap. 7).

The findings of Pennebaker and Nummenmaa do not really support the idea of William James on the causation of the emotional experience by body sensations. Also, they highlight the importance of perceived as opposed to actual events. It is important to see at this point, however, that it is almost impossible to demonstrate that there are no actual changes whatsoever behind felt changes. To do this, a large number of interoceptive channels, from thermosensation to the actual state of the muscles, should be monitored over large body areas to exclude all possible peripheral influences (not to speak of referred sensations). All we can state is that the known empirical results, monitoring a limited number of interoceptive channels only, make this option very unlikely. The Schachter-Singer theory is more promising in this respect as it relies on one single interoceptive "channel," i.e., arousal. It can be argued that overall interoceptive activation, as a summary of activation belonging to different modalities, becomes integrated or summarized in the brainstem and this is what we perceive (Chap. 4) (Ádám 1967, 1998). Unfortunately, findings do not favor the expected causal link between physiological activation and emotion either. The theory and the classic experiments were criticized from many points of view (Plutchik and Ax 1967). Most importantly for our topic, arousal is not neutral as originally assumed; it was found to be associated with valence, more precisely negative affect (Rogers and Deckner 1975; Marshall and Zimbardo 1979; Maslach 1979). Overall, the entire arousal concept was insufficiently elaborated (Manstead and Wagner 1981; Reisenzein 1983). Even if we forget the fact that arousal itself is not a unitary construct (Aston-Jones et al. 1996; Robbins et al. 1998; Berntson and Cacioppo 2007) and focus on sympathetic arousal only, empirical findings demonstrate that (1) there is a marked discrepancy between the actual and perceived arousal (Colombetti and Harrison 2018), and (2) the latter is enough to evoke attributions and emotions (Gerdes 1979) (see also the false heart rate feedback studies presented in Chap. 6).

Finally (Table 8.1, case D), if neither actual nor perceived body change occurs, the vast majority of theorists will agree that the condition in question does not represent an emotion.

8.4 Heartbeat Perception and the Emotional Experience

Although the actual level of arousal is not perceived accurately, it can be argued that its perception is mediated by the individual's ability to detect visceral changes, i.e., interoceptive accuracy. Cardioceptive accuracy, the acuity of perception of heartbeats, appears a conceptually appropriate measure of interoception in this case. Thus, it can be assumed that those with higher level of cardioceptive accuracy experience more intense emotions than poor perceivers. This assumption is in concert with the Schachter-Singer theory of emotion and could neatly fill the gap between the discordance between actual and felt arousal. However, as cardioception represents one single interoceptive channel, I do not think it can support or disprove the James-Lange theory or the somatic marker hypothesis that assume specific patterns of body sensations. The lack of association between cardioceptive accuracy and the emotional experience does not exclude possible associations with other modalities, whereas the existence of such an association simply proves that cardiac activity is related to the emotional experience but says nothing about the specificity of body patterns for different emotions. Recently, however, a weak evidence on the association between cardioceptive accuracy and the specificity of self-reported topographic changes associated with various emotion was found (Jung et al. 2017).

The first scholar who attempted to systematically test the mediation idea was Rainer Schandry (1981). The mental tracking task was originally designed exactly for this purpose which partly explains its shortcomings that received so much criticism later (see Chaps. 4 and 6). Accepting the reasoning of James regarding the impact of perceived events on the emotional experience, the Schandry task focuses on sensible events (i.e., perceived heartbeats) and compares them to actual events. From this point of view, the use of external cues and available knowledge to estimate heart rate and heartbeats appears to be a less serious issue as the same estimation might take place with respect to the emotional experience in everyday situations (Pennebaker 1982). The problem is that the same performance (score) can be achieved in different ways, e.g., relying exclusively on internal information or estimating internal events from other sources of information. Thus, the Schandry paradigm appears to be more appropriate for the assessment of the ability to estimate cardiac activity from whatever sources (internal and/or external, depending on their availability) than for the measurement of pure interoceptive accuracy. Following the original research of Schandry (1981), a number of studies were conducted to test the association between performance in the Schandry task and the emotional experience. These studies usually induced emotional states via the presentation of emotionally intense material (e.g., pictures or film clips) and assumed a positive association between the Schandry score and features of the emotional experience. It was found in all studies but one (Pollatos et al. 2007b) that heartbeat perception is associated with the intensity (arousal) of actual emotional experience but not with its quality (valence) (Schandry 1981; Pollatos et al. 2005, 2007a, c, e; Herbert et al. 2007, 2010; Dunn et al. 2010a). We found a similar pattern for emotional states induced by physical activity: increasing physical exertion was positively associated with the intensity (arousal) of felt affective state and those with better cardioceptive performance consistently reported higher levels of activation (Köteles et al. 2020). Certain findings obtained with the use of the forced-choice paradigm also support the association between arousal and cardioceptive accuracy (Hantas et al. 1982; Wiens et al. 2000) although there are contradictory results (Ferguson and Katkin 1996). Also, performance in the mental tracking task was related to measures of trait-like emotionality, such as emotional lability/neuroticism or trait anxiety (Schandry 1981; Pollatos et al. 2007b, d, e). The association between cardioceptive accuracy and anxiety and anxiety-related conditions is also supported by a meta-analysis (Domschke et al. 2010). In social context, things appear less clear-cut though. Oxytocin, a neuropeptide that shifts attention toward emotionally relevant external stimuli (e.g., faces) (Domes et al. 2013; Tollenaar et al. 2013) and increases the salience of external social cues (Shamay-Tsoory and Abu-Akel 2016), decreased cardioceptive accuracy in the presence of emotional faces (Yao et al. 2018). This finding shows that the accuracy of the perception of body states during emotions can actually be worse under certain circumstances.

Beyond experiencing emotions, the ability to regulate them might also be associated with cardioceptive accuracy. It was shown in an early study that better heartbeat tracking ability is marginally associated with more actual anxiety and less emotional expressiveness (Ludwick-Rosenthal and Neufeld 1985); this combination can be regarded as an instance of self-regulation. A similar conclusion with respect to both emotional and body feelings (pain, hunger) was drawn from more recent experimental and questionnaire studies (Füstös et al. 2013; Herbert et al. 2013; Weiss et al. 2014; Kever et al. 2015). Concerning the mechanism, it is assumed that more intense emotions associated with more accurately perceived body changes enable people to better differentiate among various emotions. This enhanced differentiation ability in turn improves regulation (Pollatos and Ferentzi 2018). On the other hand, more intense emotions are more difficult to regulate; thus, the relationship between interoceptive accuracy and emotion regulation appears more complex to me.

8.5 Somatic Markers and Decision-Making

The somatic marker hypothesis by Damasio (Damasio et al. 1991; Damasio 1994, 1996; Bechara and Damasio 2005) assumes a learned link between certain stimuli/situations and previous homeostatic body reactions, including emotions. Thus, the occurrence of a stimulus also evokes the reaction pattern it is associated with (the ventromedial prefrontal cortex is assumed to play the primary role in this process). The activated reaction pattern is claimed to implicitly influence conscious functions, most importantly reasoning and decision-making. The theory is based on case studies with patients with ventromedial lesion whose intellectual abilities remained intact; however, their decision-making and social functioning became severely impaired. The idea is that emotions and body feelings are necessary for logical reasoning, particularly in the case of complex problems, and the subjective aspects of previous experiences are stored in homeostatic-emotional body patterns. In other words, the theory highlights the importance of emotions and the differences between emotion-based and reason-based functioning, just like the theories of Zajonc (1980), Epstein (Epstein et al. 1996; Epstein 2014), or Kahneman (Tversky and Kahneman 1983; Kahneman 2011). In summary, this theory emphasizes the influence of implicit and affective factors (the relation of the two is somewhat unclear though) on overt reasoning and the adaptivity of such unconscious biases in everyday functioning (Bechara et al. 1997; Bechara 2004).

This theory is often referred to as a modern form of the Jamesian model, although the overlap between the two approaches is smaller than it appears at first sight. For example, the conscious experience of the body reaction pattern (i.e., body sensations) is not needed; this factor remains implicit and manifest itself through images or affective feelings only. Moreover, no peripheral reaction (dubbed body loop) is assumed: there is a so-called as-if loop through which the brain can simulate the body reaction (see above). Finally, the theory is about learned associations between behaviors and reward or punishment; it does not state that certain emotions are characterized by specific body patterns. In concert with this, the card gambling task (Iowa Gambling Task) designed and used to explain the differences between patients with ventromedial lesion and healthy controls simply measures the ability to implicitly learn contingencies through reward and punishment and apply the learned rules in decision-making (Bechara et al. 1994, 1996).

Of course, this very popular theory received criticism from many directions. Most importantly for our topic, some kind of primary appraisal connecting certain body patterns with certain stimuli is necessary at least at the first instance (Dunn et al. 2006). If such a primary appraisal (i.e., a direct link between stimulus and response) exists, however, it is not clear why a body loop or an as-if loop is needed (Rolls et al. 1996). Also, there are issues with the electrodermal response that was used to support the somatic aspect of the theory: there are findings that demonstrate that the response occurs after and not before decision-making (Amiez et al. 2003; Simonovic et al. 2019). Moreover, it represents only one single channel, not a pattern of body responses; this is not sufficient to draw conclusions about the validity of the idea. Finally, as shown in Chap. 4, the electrodermal response cannot be accurately perceived; thus, a question is whether actual or perceived changes influence decision-making.

Social psychological experiments on arousal and attribution raise further questions about the somatic marker hypothesis. For example, what if we mistakenly attribute our perceived bodily response? It turned out quite early that if students attribute their arousal to a drug (which was in fact a placebo pill), they cheat more than if the activation is ascribed to the act of cheating (Dienstbier and Munter 1971). In the same vein, (mis)attribution of arousal impacted smoking behavior (Schwarz et al. 1985). Even more intriguingly, similar effects could be achieved by false feedback on arousal (Valins 1966; Batson et al. 1995). In a direct test of the hypothesis, false feedback on electrodermal activity impacted participants' attitudes toward equality and freedom (Batson et al. 1999). Overall, these findings suggest that perceived rather than actual arousal plays the major role in the decision-making process.

Concerning the adaptivity of intuitive decisions, it is tempting to consider them as the natural wisdom of the body that should be followed at any cost. The fact is that intuition is a double-edged sword which can be beneficial as well as harmful (Myers 2004). As intuitive thinking prefers short-term benefits, thus it is often inferior to rational thinking in the long run (Kahneman 2011; Epstein 2014). It is important to see that the gambling task represents a quite simple situation, in which intuitive decision-making, based on implicitly learned contingencies, can be beneficial. Still, it is not clear whether the organism predicts reward or the level of risk associated with reward forecasts (Preuschoff et al. 2008). Cardioceptive accuracy (as assessed with the Schandry task) moderated performance in the Iowa Gambling Task in one study; those with higher accuracy earned more money in the game (Werner et al. 2009). However, cardiac perception hindered adaptive intuitive decision-making if the rules of the game were changed (Dunn et al. 2010a). Also, enhanced cardiac perception was associated with worse performance in the Iowa Gambling Task for patients with panic disorder who experience cardiac cues as threatening (Wölk et al. 2014). Thus, the usability of internal signals in decision-making situations is partly determined by the general meaning of the signals for the individual.

8.6 Empathy and Altruism

Emotions have a communicative role, thus the ability to decode messages about others' internal state impacts social functioning. Empathy refers to experiencing an affective or sensory state similar to that of an observed individual and being aware at the same time that the state was evoked by the other (Eisenberg and Fabes 1990; de Vignemont and Singer 2006). The state is accompanied with the activation of a pattern of brain regions (anterior insula, anterior cingulate cortex, thalamus, medial prefrontal cortex) that shows considerable similarities with the interoceptive network (Singer and Lamm 2009; Fan et al. 2011; Lamm et al. 2011). In line with these findings, interoceptive as opposed to exteroceptive attention enhanced the neural activity during a subsequent empathy task (Ernst et al. 2013), and empathy was associated with the amplitude of heartbeat-evoked potential, an indicator of interoceptive processing (Fukushima et al. 2011). Heartbeat tracking ability was positively associated with the ability to decode emotions from facial expressions (Terasawa et al. 2014), automatic imitation tendency (Ainley et al. 2014), emotional aspects of theory of mind (Shah et al. 2017), and affective and cognitive empathy evoked by painful pictures (Grynberg and Pollatos 2015). Case studies with patients with insular lesion support these findings (Terasawa et al. 2015). However, no associations with emotional sensitivity and questionnaire measures of various aspects of empathy were reported in other studies (Ainley et al. 2015; Piech et al. 2017). Moreover, cardioceptive accuracy was only marginally related to better self/other distinction (Tajadura-Jiménez and Tsakiris 2014).

Overall, the available data supports the existence of a positive association between cardioceptive accuracy and empathy; however, the exact mechanism that links these two characteristics is yet to be explored. An option is that, similar to emotion regulation, the perceived intensity (i.e., the arousal component) of emotions can be a mediating factor. In the light of the uncertainty with respect to causation, the proposal that cardioceptive accuracy training would improve empathy

(Gao et al. 2019) appears premature to me. For example, altruism, a related (Eisenberg and Miller 1987; Eisenberg and Fabes 1990) social characteristics, was also associated with cardioceptive accuracy; however, the improvement of the latter by training did not lead to the increase of altruism (Piech et al. 2017).

8.7 Alexithymia

Another insight into the complex relationship between interoception and emotions can be gained through alexithymia, a construct characterized by impaired emotional awareness. Individuals with high level of alexithymia have difficulties in identifying and describing their affective feelings, show a constricted phantasy life and a reality-based cognitive style (Nemiah and Sifneos 1970; Sifneos 1973). It is also proposed that emotional states are explained and understood as somatic states by these individuals; thus, people with alexithymia are prone to misinterpret emotions as somatic symptoms, more precisely, ascribe emotional arousal to somatic distress (Lane and Schwartz 1987). This confusion is claimed to make them prone to somatization and somatic symptom distress (Bagby et al. 1988; Taylor et al. 1992; De Gucht and Heiser 2003). Based on these ideas, the stress-alexithymia hypothesis was developed, which states that alexithymic characteristics lead to an impairment in identifying (primarily social) stressors and coping with them thus further increase stress (Martin and Pihl 1985).

Brain imaging studies indicate a considerable overlap between regions processing interoceptive information and those activated in individuals with high levels of alexithymia. Most importantly, the left anterior insula and the anterior cingulate cortex are involved (Silani et al. 2008; Bird et al. 2010; Herbert and Pollatos 2012; Ernst et al. 2014; Wiebking and Northoff 2015). From a theoretical point of view, it can be assumed that alexithymia is associated with an impaired accuracy of perception of body signals (Lane et al. 1997, 1998; Herbert et al. 2011; Herbert and Pollatos 2012). Empirical evidence, however, is equivocal. Alexithymia was accompanied by inferior estimation of heart rate (Näring and van der Staak 1995), worse perception of palpitations (Chiaie et al. 1996), lower scores in the Schandry task in a number of studies (Herbert et al. 2011; Shah et al. 2016; Bornemann and Singer 2017; Murphy et al. 2018b) (although not in all of them; Zamariola et al. 2018; Nicholson et al. 2018), lower concordance between electrodermal activity and self-rated arousal (Gaigg et al. 2018), and other interoceptive modalities (Borhani et al. 2017; Murphy et al. 2018a). It was also shown that the accuracy of cardiac perception, improved by contemplative mental training, predicts changes in alexithymia (Bornemann and Singer 2017). However, alexithymia, particularly the difficulties in identifying feelings aspect, was associated with higher sensitivity to touch, visceral stimulation, and acute and chronic pain in several studies (Sivik 1993; Nyklíček and Vingerhoets 2000; Lumley et al. 2002; Glaros and Lumley 2005; Celikel and Saatcioglu 2006; Kano et al. 2007; Porcelli et al. 2007; Huber et al. 2009; Katz et al. 2009; Shibata et al. 2014; Pollatos et al. 2015). Null findings (de Zwaan et al. 1996) and a positive association between difficulties in describing feelings and pain sensitivity (Pollatos et al. 2015) were also reported. As pain is considered an interoceptive modality, these findings are in contrast with those with other modalities and raise questions about the generalizability of the association between alexithymia and interoception.

The relationship between alexithymia and various aspects of the subjective dimension of interoception, i.e., interoceptive sensibility, is also complex. Positive associations with questionnaires assessing stress-related aspects of body awareness, such as the Body Perception Questionnaire (Porges 1993), the Illness Attitude Scale (Kellner 1987), and the Somatosensory Amplification Scale (Barsky et al. 1990), were reported (Wise and Mann 1994, 1995; Kosturek et al. 1998; Nakao et al. 2002, 2007; De Berardis et al. 2007; Tominaga et al. 2014; Ernst et al. 2014; Gulpek et al. 2014; Longarzo et al. 2015; Taycan et al. 2015; Chio and Zaroff 2015; Palser et al. 2018), although null findings are also known (Duddu et al. 2003; Jones et al. 2004; Valdés et al. 2008). Concerning the nonevaluative aspects of interoceptive sensibility, positive (Longarzo et al. 2015) as well as negative associations were reported (Zamariola et al. 2018; Berenguer et al. 2019). Overall, this pattern is in accordance with the stress-alexithymia hypothesis. Also, alexithymia was positively related to difficulties in interpreting interoceptive states (aka interoceptive confusion) and discriminate emotional and nonemotional internal states (Brewer et al. 2016).

8.8 Autism

Autism spectrum disorder, a developmental disease, is characterized by impaired social communication and restricted repetitive behavior and interest (American Psychiatric Association 2013). Unusual sensitivity to exteroceptive stimuli, such as light, sounds, or touch, is also a diagnostic criterion (DuBois et al. 2016). Individuals with autism have difficulties with the understanding of beliefs and intentions of others, aka theory of mind or mentalizing (Premack and Woodruff 1978; Frith and Frith 2006). This impairment is explained mainly by altered cognitive functioning (Frith 2012); models based on the predictive processing approach emphasize the tendency to overweight sensory information compared to priors in updating representations of the environment (Pellicano and Burr 2012; Palmer et al. 2015, 2017). The possibility that affective processes requiring body and emotional awareness, such as empathy, might also be altered received marked research attention recently. Altered connectivity between the insula and somatosensory and limbic regions (Ebisch et al. 2011) and an aberrant oxytocin system impacting interoception (Quattrocki and Friston 2014) was assumed to play a role in the development of the condition. It was also proposed that empathy deficits in autism can be explained by the large comorbidity with alexithymia (Bird et al. 2010; Brewer et al. 2015; Shah et al. 2016). Individuals with autism show decreased cardioceptive accuracy as assessed with the Schandry paradigm but not with a forced-choice test in two studies (Garfinkel et al. 2015; Mul et al. 2018); however, no difference was found (Schandry test) in other studies (Schauder et al. 2015; Shah et al. 2016; Failla et al. 2020). In a research involving children and adults, only the former group showed worse cardioceptive accuracy than controls; it was speculated that adults learn to compensate for their impairment (Nicholson et al. 2019). Also, interoceptive accuracy was associated with the repetitive restricted behavior component of the condition only, whereas interoceptive sensibility was related to social-affective aspects (Palser et al. 2020). A qualitative study revealed a general hyposensitivity to body-related stimuli (Elwin et al. 2012), which is supported by empirical data on impaired tactile sensitivity (Failla et al. 2017). Difficulty in integrating cardiac and visual stimuli in patients with autism was also reported (Noel et al. 2018). Data on interoceptive sensibility is equivocal; some studies using the Body Awareness Questionnaire and the MAIA report lower level of body awareness (Fiene and Brownlow 2015; Mul et al. 2018), whereas people with autism score higher on the Porges Body Perception Questionnaire (Garfinkel et al. 2015).

8.9 Anxiety, Depression, and Other Psychopathologies

Concerning the associations between psychopathology and interoceptive accuracy, the most established findings refer to anxiety-related states. Perceivable elevations of heart rate (e.g., evoked by isoproterenol, a nonselective beta adrenergic agonist) lead to anxiety even in healthy individuals (Cameron and Minoshima 2002; Khalsa et al. 2009). Accuracy of cardiac perception shows a moderate to strong association with anxiety sensitivity, a weak to medium association with trait anxiety, and a medium association with panic disorder (Van der Does et al. 2000; Domschke et al. 2010). This means that better perception of heartbeats might be a risk factor of these conditions, particularly if it is accompanied by catastrophization, e.g., the misinterpretation of heartbeats as indicators of heart attack (Chambless et al. 2000). It is important to note, however, that (1) only a proportion of individuals with higher levels of anxiety was characterized by better cardioceptive accuracy, and (2) the vast majority of the studies applied the Schandry task, thus biases in perception might have impacted the results. It is also worth mentioning that heartbeat might be associated with a negative evaluation in these conditions as it refers to increased levels of anxiety. This evaluative aspect clearly differentiates people with and without anxiety and might also represent a factor that moderates cardioception (see Chap. 7). Also, processing of fear-related signals is modulated by the cardiac cycle and mediated by cardioceptive accuracy (usually assessed using a tracking paradigm) (Garfinkel et al. 2014; Critchley and Garfinkel 2015; Garfinkel and Critchley 2016; Pfeifer et al. 2017), which suggests a specific link between anxiety and rhythmic cardiac activity beyond arousal.

The presented anxiety-related findings appear to support the notion that above average cardioceptive accuracy is maladaptive. As heart-related sensations represent a substantial component of anxiety, this association appears logical. In other pathologies and aversive states, however, a negative association with cardioception was revealed. Before we briefly summarize the results, it is important to mention that cardiac perception was regarded as a measure of general interoceptive accuracy in these studies. As we know very well, this is not the case, thus the presented results should be interpreted with caution. For example, patients with depression showed a negative association with heartbeat perception in the majority of the studies (Van der Does et al. 1997; Pollatos et al. 2009; Herbert et al. 2011; Terhaar et al. 2012; Furman et al. 2013), albeit not without exceptions (Mussgay et al. 1999; Dunn et al. 2007, 2010b). Self-reported depression showed a moderate to strong positive association with cardioceptive accuracy for healthy participants (Critchley et al. 2004). Overall, these findings support the notion that depression is characterized by dysfunctional (impaired) interoception (Harshaw 2015); however, the methodological issues with the Schandry task somewhat temper this conclusion. Individuals with various forms of substance use disorders showed worse cardioceptive accuracy than controls in several studies (Ateş Çöl et al. 2016; Sönmez et al. 2017), whereas no association was reported in another paper (Schmidt et al. 2013). The majority of the studies on eating disorders such as anorexia nervosa and bulimia nervosa also revealed a negative association (Pollatos et al. 2008; Klabunde et al. 2013; Herbert and Pollatos 2014); again, null findings were also reported (Eshkevari et al. 2014; Pollatos and Georgiou 2016). On the other hand, both intuitive eating (Herbert et al. 2013) and emotional eating (Koch and Pollatos 2014; Young et al. 2017) showed a positive association with cardioception. Finally, findings concerning the association between somatization tendency in the broad sense and cardiac interoception are also equivocal (Chaps. 6 and 7). Overall, these conditions were accompanied by altered insular processing of interoceptive signals in brain imaging studies (Wiebking et al. 2010, 2015; Mata et al. 2015; Kerr et al. 2016).

To me, the most important message of these findings is that altered interoception might be involved in pathological conditions, however, causality cannot be established. Abnormal (both enhanced and impaired) interoceptive accuracy might be the consequence of more general perceptual and cognitive-emotional changes characterizing these conditions.

8.10 Nonconscious Emotions

So far we followed the path of peripheral theories of emotion and (well, sometimes implicitly) accepted that the emotional experience is determined by (or at least associated with) body sensations. James even claimed (1890) that if we remove body sensations from the experience, the rest cannot be called emotion. The same conclusion is often drawn by the authors of recent studies investigating the associations between interoception and affective functioning. This approach implicitly assumes that emotions necessarily reach conscious awareness.

But what about nonconscious emotions, i.e., emotions with well-measurable physiological and behavioral changes that are not accompanied by affective feelings or any form of conscious experience? The first theory that comes into one's mind with respect to this topic is most likely psychoanalysis. For example, certain mechanisms of defense, such as denial or intellectualization, are characterized by the complete suppression of the emotional experience (Freud 1937). Actually, William James was present at Sigmund Freud's lecture at Clark University in 1909 and found the idea of unconscious interesting (Rosenzweig 1993); thus, his approach to emotions appears open and flexible enough to accept such special cases. Since then, many authors emphasized the fact that affect impacts conscious functioning (including reasoning) in multiple, partly nonconscious ways. Just to mention a few, theories describing the two basic ways of information processing, such as the first and second signaling system of Pavlov, or the dual-process theories of Paivio, Kahneman, and Epstein, follow this path (Epstein 1994). The somatic marker hypothesis also allows the existence of nonconscious perception of body states, i.e., nonconscious emotions (Bechara et al. 1997). Research on subliminal perception (Bower 1981; Bornstein et al. 1987; Kihlstrom 2004) and the existence of implicit attitudes (Banaji and Greenwald 1995) also supports the idea that affective processing taking place at a nonconscious level impacts our conscious functioning. Sophisticated experiments demonstrate that implicitly generated emotional states impact our behavior without reaching conscious awareness (Winkielman and Berridge 2004). Recently, the Implicit Positive and Negative Affect Test was developed to assess nonconscious affective processes (Quirin et al. 2009a). Empirical findings with this test show that implicit affect predict physiological reactions to stress better than explicit ratings (Quirin et al. 2009b; Brosschot et al. 2014; van der Ploeg et al. 2016). These ideas, usually supported by empirical results, clearly demonstrate that a proportion of emotional functioning does not reach consciousness. Still, this is not necessarily challenging the peripheral theories as they do not state that all emotions are conscious. Perhaps the aforementioned ideas are based on cases where no sensible body changes occur thus no emotional experience develops. On the other hand, it does not appear logical to assume that the very same body signals (most importantly, heartbeat) sometimes play a role in the affective experience and sometimes do not.

It was also mentioned that primary evaluation, preceding body reactions, is a necessary component of the affective response; however, there is no agreement on whether this primary evaluation results in conscious feelings or not (Lambie and Marcel 2002; Wiens 2005). As mentioned, affective neuroscience assumes the existence of the so-called primary affective feelings, which are conscious and represent the building stones of more complex emotions. Interestingly enough, however, Panksepp makes a clear-cut distinction between affective and body feelings (Panksepp and Biven 2012). In this case, the emotional experience is conscious but not directly impacted by body events. Also, emotional experience can be generated by top-down mechanisms, such as hypnosis (Bower 1981). In these cases, low-level primary evaluation is replaced by high-level processes. This idea leads us to the most recent approach to emotions.

8.11 The Predictive Processing Approach

As the reader might suspect, the predictive processing approach (Chap. 1) abandons the assumption of the primacy of bottom-up processes and emphasizes the importance of top-down factors in the generation of emotions. More precisely, it assumes an ongoing, real-time interaction between ascending and descending signals along a chain of brain centers (Seth 2013). These centers are mostly identical with the centers described by the EPIC model (Barrett and Simmons 2015) presented in Chap. 3. Of course, the most important structure here is the anterior part of the insula that integrates bottom-up and top-down (from the anterior cingulate the prefrontal cortex) information and gives rise to conscious emotional feelings (Singer et al. 2009; Seth et al. 2012; Xiang et al. 2013; Gu et al. 2013; Pace-Schott et al. 2019).

The view of emotions that emerge from this approach is based on active inference, relying on prediction and categorization processes mostly determined by previous experiences (Barrett 2017).

As the optimization of predictions (i.e., explaining away of the prediction errors) occurs simultaneously at each level of the hierarchy, the integrative power of this model is enormous. It can easily handle enactivist models at lower levels, whereas in-brain simulation of peripheral changes (as-if loops) might also be included. Also, it elegantly avoids possible timing issues since the interactions between various levels of hierarchy occur simultaneously in a fast timescale (Seth and Critchley 2013).

Beyond the conceptual level, however, the complexity of emotions poses serious challenges even for the extremely flexible predictive processing model; perhaps this is why Andy Clark, the doyen of the predictive processing approach, also feels that this is a particularly difficult area for the model (Clark 2015). Emotions reorganize our entire functioning, including attention, memory, information processing, physiology, and overt behavior (Smith and Lane 2015). A computational model aimed to describe such complicated effects should be able to handle not only the interaction but also the coexistence of and sometimes the conflict between rational and nonrational factors, genetically determined connections, even the co-occurrence of different emotions.

8.12 Summary

In this chapter, associations between the emotional experience, including pathological conditions, and dimensions of interoception were discussed. Contrary to the assumptions of a variety of theories, the association between actual body changes and the emotional experience does not appear either obvious or unidirectional. Emotions are accompanied by body changes and *vice versa*; body sensations are typically subjects of affective evaluation. This is why the latter category is called "homeostatic emotions" by Craig (2003). Interestingly, the emotional experience is associated with perceived body changes (i.e., body sensations) rather than actual peripheral events in many cases; in other words, the dissociation between actual and perceived body changes is present in this area too. In these cases, perceived changes are evaluated, and the outcome of the evaluation can impact behavior (see Chaps. 9 and 10).

Take-Home Messages

For laypeople: Many body changes we perceive and associate with and experience in emotional states do not actually take place. Moreover, our gut feelings do not originate in the visceral organs and does not represent the superior wisdom of the body. They are products of an ancient learning mechanism that biases our behavior in favor of short-term benefits.

For clinicians: The association between body states and emotions is less clear-cut than it appears to be. The peripheral changes accompanying many emotions are below the perception threshold, and perceived changes do not necessarily reflect actual changes.

For researchers: The majority of available findings on the associations between the emotional experience and interoceptive accuracy rely on the Schandry paradigm (see Chap. 4). This is problematic from at least two viewpoints. First, cardioceptive accuracy cannot be generalized to acuity shown in other interoceptive modalities (Garfinkel et al. 2016; Ferentzi et al. 2017, 2018; Horváth et al. 2020). Second, the Schandry paradigm does not represent a measure of pure (i.e., bottom-up-based) interoception (Ehlers et al. 1995; Ring and Brener 1996; Ring et al. 2015; Brener and Ring 2016; Desmedt et al. 2018, 2020). On the other hand, cardiac activity is a prominent (perhaps the most prominent) peripheral concomitant of emotions, and the contribution of top-down factors to the emotional experience is well-known. Keeping in mind these points, the available evidence should systematically be revised, and new conclusions on the associations between emotional feelings and body sensations should be drawn.

References

Ádám G (1967) Interoception and behavior. Akadémiai Kiadó, Budapest

Ádám G (1998) Visceral perception: understanding internal cognition. Plenum Press, New York

Ainley V, Brass M, Tsakiris M (2014) Heartfelt imitation: high interoceptive awareness is linked to greater automatic imitation. Neuropsychologia 60C:21–28. https://doi.org/10.1016/j.neuropsychologia.2014.05.010

Ainley V, Maister L, Tsakiris M (2015) Heartfelt empathy? No association between interoceptive awareness, questionnaire measures of empathy, reading the mind in the eyes task or the director task. Front Psychol 6:554. https://doi.org/10.3389/fpsyg.2015.00554

Alexander F (1950) Psychosomatic medicine, its principles and applications. W. W. Norton & Company, New York

American Psychiatric Association (2013) Diagnostic and statistical manual of mental disorders, 5th Edition: DSM-5, 5th edn. American Psychiatric Publishing, Washington, DC

Amiez C, Procyk E, Honoré J et al (2003) Reward anticipation, cognition, and electrodermal activity in the conditioned monkey. Exp Brain Res 149:267–275. https://doi.org/10.1007/s00221-002-1353-9

Arnold MB (1945) Physiological differentiation of emotional states. Psychol Rev 52:35–48. https://doi.org/10.1037/h0061746

Arnold MB (1960) Emotion and personality. Columbia University Press, New York

Aston-Jones G, Rajkowski J, Kubiak P et al (1996) Role of the locus coeruleus in emotional activation. Prog Brain Res 107:379–402

Ateş Çöl I, Sönmez MB, Vardar ME (2016) Evaluation of interoceptive awareness in alcohol-addicted patients. Noro Psikiyatr Ars 53:17–22. https://doi.org/10.5152/npa.2015.9898

Ax AF (1953) The physiological differentiation between fear and anger in humans. Psychosom Med 15:433–442. https://doi.org/10.1097/00006842-195309000-00007

Bagby RM, Taylor GJ, Atkinson L (1988) Alexithymia: a comparative study of three self-report measures. J Psychosom Res 32:107–116. https://doi.org/10.1016/0022-3999(88)90094-3

Banaji MR, Greenwald AG (1995) Implicit gender stereotyping in judgments of fame. J Pers Soc Psychol 68:181–198. https://doi.org/10.1037//0022-3514.68.2.181

Bandler R, Keay KA, Floyd N, Price J (2000) Central circuits mediating patterned autonomic activity during active vs. passive emotional coping. Brain Res Bull 53:95–104. https://doi.org/10.1016/S0361-9230(00)00313-0

Barrett LF (2006a) Are emotions natural kinds? Perspect Psychol Sci 1:28–58. https://doi.org/10.1111/j.1745-6916.2006.00003.x

Barrett LF (2006b) Solving the emotion paradox: categorization and the experience of emotion. Personal Soc Psychol Rev 10:20–46. https://doi.org/10.1207/s15327957pspr1001_2

Barrett LF (2017) The theory of constructed emotion: an active inference account of interoception and categorization. Soc Cogn Affect Neurosci 12:1–23. https://doi.org/10.1093/scan/nsw154

Barrett LF, Russell JA (1999) The structure of current affect: controversies and emerging consensus. Curr Dir Psychol Sci 8:10–14. https://doi.org/10.1111/1467-8721.00003

Barrett LF, Simmons WK (2015) Interoceptive predictions in the brain. Nat Rev Neurosci 16:419–429. https://doi.org/10.1038/nrn3950

Barsky AJ, Wyshak G, Klerman GL (1990) The Somatosensory Amplification Scale and its relationship to hypochondriasis. J Psychiatr Res 24:323–334. https://doi.org/10.1016/0022-3956(90)90004-A

Batson CD, Turk CL, Shaw LL, Klein TR (1995) Information function of empathic emotion: learning that we value the other's welfare. J Pers Soc Psychol 68:300–313. https://doi.org/10.1037/0022-3514.68.2.300

Batson CD, Engel CL, Fridell SR (1999) Value judgments: testing the somatic-marker hypothesis using false physiological feedback. Pers Soc Psychol Bull 25:1021–1032. https://doi.org/10.1177/01461672992511009

Bechara A (2004) The role of emotion in decision-making: evidence from neurological patients with orbitofrontal damage. Brain Cogn 55:30–40. https://doi.org/10.1016/j.bandc.2003.04.001

Bechara A, Damasio AR (2005) The somatic marker hypothesis: a neural theory of economic decision. Games Econ Behav 52:336–372. https://doi.org/10.1016/j.geb.2004.06.010

Bechara A, Damasio AR, Damasio H, Anderson SW (1994) Insensitivity to future consequences following damage to human prefrontal cortex. Cognition 50:7–15. https://doi.org/10.1016/0010-0277(94)90018-3

Bechara A, Tranel D, Damasio H, Damasio AR (1996) Failure to respond autonomically to anticipated future outcomes following damage to prefrontal cortex. Cereb Cortex 6:215–225. https://doi.org/10.1093/cercor/6.2.215

Bechara A, Damasio H, Tranel D, Damasio AR (1997) Deciding advantageously before knowing the advantageous strategy. Science 275:1293–1295. https://doi.org/10.1126/science.275.5304.1293

Berenguer C, Rebôlo C, Costa RM (2019) Interoceptive awareness, alexithymia, and sexual function. J Sex Marital Ther 45:729–738. https://doi.org/10.1080/0092623X.2019.1610128

Berntson GG, Cacioppo JT (2007) Integrative physiology: homeostasis, allostasis, and the orchestration of systemic physiology. In: Handbook of psychophysiology, 3rd edn. Cambridge University Press, New York, pp 433–452

Bird G, Silani G, Brindley R et al (2010) Empathic brain responses in insula are modulated by levels of alexithymia but not autism. Brain 133:1515–1525. https://doi.org/10.1093/brain/awq060

Borhani K, Làdavas E, Fotopoulou A, Haggard P (2017) "Lacking warmth": alexithymia trait is related to warm-specific thermal somatosensory processing. Biol Psychol 128:132–140. https://doi.org/10.1016/j.biopsycho.2017.07.012

Bornemann B, Singer T (2017) Taking time to feel our body: steady increases in heartbeat perception accuracy and decreases in alexithymia over 9 months of contemplative mental training. Psychophysiology 54:469–482. https://doi.org/10.1111/psyp.12790

Bornstein RF, Leone DR, Galley DJ (1987) The generalizability of subliminal mere exposure effects: influence of stimuli perceived without awareness on social behavior. J Pers Soc Psychol 53:1070–1079. https://doi.org/10.1037/0022-3514.53.6.1070

Bower GH (1981) Mood and memory. Am Psychol 36:129–148. https://doi.org/10.1037/0003-066X.36.2.129

Brener J, Ring C (2016) Towards a psychophysics of interoceptive processes: the measurement of heartbeat detection. Philos Trans R Soc Lond B Biol Sci 371:20160015. https://doi.org/10.1098/rstb.2016.0015

Brewer R, Happé F, Cook R, Bird G (2015) Commentary on "Autism, oxytocin and interoception": Alexithymia, not Autism Spectrum Disorders, is the consequence of interoceptive failure. Neurosci Biobehav Rev 56:348–353. https://doi.org/10.1016/j.neubiorev.2015.07.006

Brewer R, Cook R, Bird G (2016) Alexithymia: a general deficit of interoception. R Soc Open Sci 3:150664. https://doi.org/10.1098/rsos.150664

Brosschot JF, Geurts SAE, Kruizinga I et al (2014) Does unconscious stress play a role in prolonged cardiovascular stress recovery? Stress Health 30:179–187. https://doi.org/10.1002/smi.2590

Cabanac M (1971) Physiological role of pleasure. Science 173:1103–1107. https://doi.org/10.1126/science.173.4002.1103

Cacioppo JT, Berntson GG, Larsen JT et al (1993a) The psychophysiology of emotion. In: Lewis R, Haviland-Jones JM (eds) The handbook of emotion, 2nd edn. Guilford Press, New York, pp 173–191

Cacioppo JT, Priester JR, Berntson GG (1993b) Rudimentary determinants of attitudes: II. Arm flexion and extension have differential effects on attitudes. J Pers Soc Psychol 65:5–17. https://doi.org/10.1037/0022-3514.65.1.5

Cameron OG, Minoshima S (2002) Regional brain activation due to pharmacologically induced adrenergic interoceptive stimulation in humans. Psychosom Med 64:851–861

Cannon WB (1927) The James-Lange theory of emotions: a critical examination and an alternative theory. Am J Psychol 39:106–124. https://doi.org/10.2307/1415404

Cannon WB (1931) Again the James-Lange and the thalamic theories of emotion. Psychol Rev 38:281–295. https://doi.org/10.1037/h0072957

Cantril H, Hunt WA (1932) Emotional effects produced by the injection of adrenalin. Am J Psychol 44:300–307. https://doi.org/10.2307/1414829

Carney DR, Cuddy AJC, Yap AJ (2010) Power posing: brief nonverbal displays affect neuroendocrine levels and risk tolerance. Psychol Sci 21:1363–1368. https://doi.org/10.1177/0956797610383437

Celikel FC, Saatcioglu O (2006) Alexithymia and anxiety in female chronic pain patients. Ann General Psychiatry 5:13. https://doi.org/10.1186/1744-859X-5-13

Chambless DL, Beck AT, Gracely EJ, Grisham JR (2000) Relationship of cognitions to fear of somatic symptoms: a test of the cognitive theory of panic. Depress Anxiety 11:1–9

Chiaie RD, Baciarello G, Villani M et al (1996) Cardiovascular reactivity of mitral valve prolapse patients during experimental stress exposure: evidence for a functional nature of cardiovascular

symptoms. Acta Psychiatr Scand 93:434–441. https://doi.org/10.1111/j.1600-0447.1996.tb10674.x
Chio PH, Zaroff CM (2015) Traditional Chinese medicinal herbal tea consumption, self-reported somatization, and alexithymia. Asia Pac Psychiatry 7:127–134. https://doi.org/10.1111/appy.12161
Christie IC, Friedman BH (2004) Autonomic specificity of discrete emotion and dimensions of affective space: a multivariate approach. Int J Psychophysiol 51:143–153. https://doi.org/10.1016/j.ijpsycho.2003.08.002
Clark A (2015) Surfing uncertainty: prediction, action, and the embodied mind, 1st edn. Oxford University Press, Oxford, England
Cobos P, Sánchez M, Pérez N, Vila J (2004) Effects of spinal cord injuries on the subjective component of emotions. Cognit Emot 18:281–287. https://doi.org/10.1080/02699930244000471
Colombetti G, Harrison N (2018) From physiology to experience: enriching existing conceptions of "arousal" in affective science. In: Tsakiris M, De Preester H (eds) The interoceptive mind. From homeostasis to awareness. Oxford University Press, Oxford, England, pp 245–258
Cook EW, Melamed BG, Cuthbert BN et al (1988) Emotional imagery and the differential diagnosis of anxiety. J Consult Clin Psychol 56:734–740. https://doi.org/10.1037/0022-006X.56.5.734
Craig AD (2003) A new view of pain as a homeostatic emotion. Trends Neurosci 26:303–307
Critchley HD, Garfinkel SN (2015) Interactions between visceral afferent signaling and stimulus processing. Front Neurosci 9:286. https://doi.org/10.3389/fnins.2015.00286
Critchley HD, Wiens S, Rotshtein P et al (2004) Neural systems supporting interoceptive awareness. Nat Neurosci 7:189–195. https://doi.org/10.1038/nn1176
Damasio A (1994) Descartes's error: emotion, reason, and the human brain. Penguin Books, New York
Damasio A (1996) The somatic marker hypothesis and the possible functions of the prefrontal cortex. Philos Trans R Soc Lond B Biol Sci 351:1413–1420. https://doi.org/10.1098/rstb.1996.0125
Damasio A (1999) The feeling of what happens: body and emotion in the making of consciousness. Harcourt Brace and Co, San Diego, CA
Damasio A (2004) Emotions and feelings: a neurobiological perspective. In: Manstead ASR, Frijda N, Fischer A (eds) Feelings and emotions. The Amsterdam Symposium. Cambridge University Press, Cambridge, UK, pp 49–57
Damasio AR, Tranel D, Damasio HC (1991) Somatic markers and the guidance of behavior: theory and preliminary testing. In: Frontal lobe function and dysfunction. Oxford University Press, New York, pp 217–229
Darwin C (1872) The expression of the emotions in man and animals [Original work published in 1872]. CreateSpace Independent Publishing Platform, Scotts Valley, CA
Davis M (1989) Neural systems involved in fear-potentiated startle. Ann N Y Acad Sci 563:165–183. https://doi.org/10.1111/j.1749-6632.1989.tb42197.x
Davis RC, Buchwald AM (1957) An exploration of somatic response patterns: stimulus and sex differences. J Comp Physiol Psychol 50:44–52. https://doi.org/10.1037/h0046428
Davis RC, Buchwald AM, Frankmann RW (1955) Autonomic and muscular responses, and their relation to simple stimuli. Psychol Monogr Gen Appl 69:1–71. https://doi.org/10.1037/h0093734
Davis RC, Lundervold A, Miller JD (1957) The pattern of somatic response during a repetitive motor task and its modification by visual stimuli. J Comp Physiol Psychol 50:53–60. https://doi.org/10.1037/h0045021
Davitz JR (1969) The language of emotion. Academic, New York
De Berardis D, Campanella D, Gambi F et al (2007) Alexithymia, fear of bodily sensations, and somatosensory amplification in young outpatients with panic disorder. Psychosomatics 48:239–246. https://doi.org/10.1176/appi.psy.48.3.239
De Gucht V, Heiser W (2003) Alexithymia and somatisation: quantitative review of the literature. J Psychosom Res 54:425–434. https://doi.org/10.1016/s0022-3999(02)00467-1

de Vignemont F, Singer T (2006) The empathic brain: how, when and why? Trends Cogn Sci 10:435–441. https://doi.org/10.1016/j.tics.2006.08.008
de Zwaan M, Biener D, Bach M et al (1996) Pain sensitivity, alexithymia, and depression in patients with eating disorders: are they related? J Psychosom Res 41:65–70. https://doi.org/10.1016/0022-3999(96)00088-8
Desmedt O, Luminet O, Corneille O (2018) The heartbeat counting task largely involves non-interoceptive processes: evidence from both the original and an adapted counting task. Biol Psychol 138:185–188. https://doi.org/10.1016/j.biopsycho.2018.09.004
Desmedt O, Corneille O, Luminet O et al (2020) Contribution of time estimation and knowledge to heartbeat counting task performance under original and adapted instructions. Biol Psychol 154:107904. https://doi.org/10.1016/j.biopsycho.2020.107904
Dienstbier RA, Munter PO (1971) Cheating as a function of the labeling of natural arousal. J Pers Soc Psychol 17:208–213. https://doi.org/10.1037/h0030384
Domes G, Sibold M, Schulze L et al (2013) Intranasal oxytocin increases covert attention to positive social cues. Psychol Med 43:1747–1753. https://doi.org/10.1017/S0033291712002565
Domschke K, Stevens S, Pfleiderer B, Gerlach AL (2010) Interoceptive sensitivity in anxiety and anxiety disorders: an overview and integration of neurobiological findings. Clin Psychol Rev 30:1–11. https://doi.org/10.1016/j.cpr.2009.08.008
Dror OE (2014) The Cannon–Bard thalamic theory of emotions: a brief genealogy and reappraisal. Emot Rev 6:13–20. https://doi.org/10.1177/1754073913494898
DuBois D, Ameis SH, Lai M-C et al (2016) Interoception in autism spectrum disorder: a review. Int J Dev Neurosci 52:104–111. https://doi.org/10.1016/j.ijdevneu.2016.05.001
Duddu V, Isaac MK, Chaturvedi SK (2003) Alexithymia in somatoform and depressive disorders. J Psychosom Res 54:435–438
Dunn BD, Dalgleish T, Lawrence AD (2006) The somatic marker hypothesis: a critical evaluation. Neurosci Biobehav Rev 30:239–271. https://doi.org/10.1016/j.neubiorev.2005.07.001
Dunn BD, Dalgleish T, Ogilvie AD, Lawrence AD (2007) Heartbeat perception in depression. Behav Res Ther 45:1921–1930. https://doi.org/10.1016/j.brat.2006.09.008
Dunn BD, Galton HC, Morgan R et al (2010a) Listening to your heart. How interoception shapes emotion experience and intuitive decision making. Psychol Sci 21:1835–1844. https://doi.org/10.1177/0956797610389191
Dunn BD, Stefanovitch I, Evans D et al (2010b) Can you feel the beat? Interoceptive awareness is an interactive function of anxiety- and depression-specific symptom dimensions. Behav Res Ther 48:1133–1138. https://doi.org/10.1016/j.brat.2010.07.006
Dykman RA, Reese WG, Galbrecht CR, Thomasson PJ (1959) Psychophysiological reactions to novel stimuli: measurement, adaptation, and relationship of psychological and physiological variables in the normal human. Ann N Y Acad Sci 79:43–107. https://doi.org/10.1111/j.1749-6632.1959.tb42777.x
Ebisch SJH, Gallese V, Willems RM et al (2011) Altered intrinsic functional connectivity of anterior and posterior insula regions in high-functioning participants with autism spectrum disorder. Hum Brain Mapp 32:1013–1028. https://doi.org/10.1002/hbm.21085
Ehlers A, Breuer P, Dohn D, Fiegenbaum W (1995) Heartbeat perception and panic disorder: possible explanations for discrepant findings. Behav Res Ther 33:69–76
Eisenberg N, Fabes RA (1990) Empathy: conceptualization, measurement, and relation to prosocial behavior. Motiv Emot 14:131–149. https://doi.org/10.1007/BF00991640
Eisenberg N, Miller PA (1987) The relation of empathy to prosocial and related behaviors. Psychol Bull 101:91–119. https://doi.org/10.1037/0033-2909.101.1.91
Ekman P (1992) An argument for basic emotions. Cognit Emot 6:169–200. https://doi.org/10.1080/02699939208411068
Ekman P, Levenson RW, Friesen WV (1983) Autonomic nervous system activity distinguishes among emotions. Science 221:1208–1210. https://doi.org/10.1126/science.6612338
Ellsworth PC (1994) William James and emotion: is a century of fame worth a century of misunderstanding? Psychol Rev 101:222–229. https://doi.org/10.1037/0033-295X.101.2.222

Elwin M, Ek L, Schröder A, Kjellin L (2012) Autobiographical accounts of sensing in Asperger syndrome and high-functioning autism. Arch Psychiatr Nurs 26:420–429. https://doi.org/10.1016/j.apnu.2011.10.003

Engel BT (1960) Stimulus-response and individual-response specificity. JAMA Psychiatry 2:305–313. https://doi.org/10.1001/archpsyc.1960.03590090061010

Engel BT, Bickford AF (1961) Response specificity: stimulus-response and individual-response specificity in essential hypertensives. Arch Gen Psychiatry 5:478–489. https://doi.org/10.1001/archpsyc.1961.01710170056007

Epstein S (1994) Integration of the cognitive and the psychodynamic unconscious. Am Psychol 49:709–724

Epstein S (2014) Cognitive-experiential theory. Oxford University Press, Oxford, England

Epstein S, Pacini R, Denes-Raj V, Heier H (1996) Individual differences in intuitive-experiential and analytical-rational thinking styles. J Pers Soc Psychol 71:390–405

Ernst J, Northoff G, Böker H et al (2013) Interoceptive awareness enhances neural activity during empathy. Hum Brain Mapp 34:1615–1624. https://doi.org/10.1002/hbm.22014

Ernst J, Böker H, Hättenschwiler J et al (2014) The association of interoceptive awareness and alexithymia with neurotransmitter concentrations in insula and anterior cingulate. Soc Cogn Affect Neurosci 9:857–863. https://doi.org/10.1093/scan/nst058

Eshkevari E, Rieger E, Musiat P, Treasure J (2014) An investigation of interoceptive sensitivity in eating disorders using a heartbeat detection task and a self-report measure. Eur Eat Disord Rev 22:383–388. https://doi.org/10.1002/erv.2305

Failla MD, Peters BR, Karbasforoushan H et al (2017) Intrainsular connectivity and somatosensory responsiveness in young children with ASD. Mol Autism 8:25. https://doi.org/10.1186/s13229-017-0143-y

Failla MD, Bryant LK, Heflin BH et al (2020) Neural correlates of cardiac interoceptive focus across development: implications for social symptoms in autism spectrum disorder. Autism Res 13:908–920. https://doi.org/10.1002/aur.2289

Fan Y, Duncan NW, de Greck M, Northoff G (2011) Is there a core neural network in empathy? An fMRI based quantitative meta-analysis. Neurosci Biobehav Rev 35:903–911. https://doi.org/10.1016/j.neubiorev.2010.10.009

Feinberg TE, Mallatt JM (2016) The ancient origins of consciousness. How the brain created experience. The MIT Press, Cambridge, MA

Ferentzi E, Köteles F, Csala B et al (2017) What makes sense in our body? Personality and sensory correlates of body awareness and somatosensory amplification. Pers Individ Differ 104:75–81. https://doi.org/10.1016/j.paid.2016.07.034

Ferentzi E, Bogdány T, Szabolcs Z et al (2018) Multichannel investigation of interoception: sensitivity is not a generalizable feature. Front Hum Neurosci 12:223. https://doi.org/10.3389/fnhum.2018.00223

Ferguson ML, Katkin ES (1996) Visceral perception, anhedonia, and emotion. Biol Psychol 42:131–145

Fiene L, Brownlow C (2015) Investigating interoception and body awareness in adults with and without autism spectrum disorder. Autism Res 8:709–716. https://doi.org/10.1002/aur.1486

Folkow B (2000) Perspectives on the integrative functions of the "sympatho-adrenomedullary system". Auton Neurosci 83:101–115. https://doi.org/10.1016/S1566-0702(00)00171-5

Freud A (1937) The ego and the mechanisms of defence. Hogarth Press and Institute of Psycho-Analysis, London

Friedman BH (2010) Feelings and the body: the Jamesian perspective on autonomic specificity of emotion. Biol Psychol 84:383–393. https://doi.org/10.1016/j.biopsycho.2009.10.006

Frijda NH (1986) The emotions. Cambridge University Press, New York

Frith U (2012) Why we need cognitive explanations of autism. Q J Exp Psychol (Hove) 65:2073–2092. https://doi.org/10.1080/17470218.2012.697178

Frith CD, Frith U (2006) The neural basis of mentalizing. Neuron 50:531–534. https://doi.org/10.1016/j.neuron.2006.05.001

Fukushima H, Terasawa Y, Umeda S (2011) Association between interoception and empathy: evidence from heartbeat-evoked brain potential. Int J Psychophysiol 79:259–265. https://doi.org/10.1016/j.ijpsycho.2010.10.015

Funkenstein DH (1956) Nor-epinephrine-like and epinephrine-like substances in relation to human behavior. J Nerv Ment Dis 124:58–68. https://doi.org/10.1097/00005053-195607000-00009

Furman DJ, Waugh CE, Bhattacharjee K et al (2013) Interoceptive awareness, positive affect, and decision making in major depressive disorder. J Affect Disord 151:780–785. https://doi.org/10.1016/j.jad.2013.06.044

Füstös J, Gramann K, Herbert BM, Pollatos O (2013) On the embodiment of emotion regulation: interoceptive awareness facilitates reappraisal. Soc Cogn Affect Neurosci 8:911–917. https://doi.org/10.1093/scan/nss089

Gaigg SB, Cornell AS, Bird G (2018) The psychophysiological mechanisms of alexithymia in autism spectrum disorder. Autism 22:227–231. https://doi.org/10.1177/1362361316667062

Gao Q, Ping X, Chen W (2019) Body influences on social cognition through interoception. Front Psychol 10:2066. https://doi.org/10.3389/fpsyg.2019.02066

Garfinkel SN, Critchley HD (2016) Threat and the body: how the heart supports fear processing. Trends Cogn Sci 20:34–46. https://doi.org/10.1016/j.tics.2015.10.005

Garfinkel SN, Minati L, Gray MA et al (2014) Fear from the heart: sensitivity to fear stimuli depends on individual heartbeats. J Neurosci 34:6573–6582. https://doi.org/10.1523/JNEUROSCI.3507-13.2014

Garfinkel SN, Tiley C, O'Keeffe S et al (2015) Discrepancies between dimensions of interoception in autism: implications for emotion and anxiety. Biol Psychol 114:117–126. https://doi.org/10.1016/j.biopsycho.2015.12.003

Garfinkel SN, Manassei MF, Hamilton-Fletcher G et al (2016) Interoceptive dimensions across cardiac and respiratory axes. Philos Trans R Soc Lond B Biol Sci 371:20160014. https://doi.org/10.1098/rstb.2016.0014

Gemelli A (1949) Orienting concepts in the study of effective states (transl.). J Nerv Ment Dis 110:198; passim

Gerdes EP (1979) Autonomic arousal as a cognitive cue in stressful situations. J Pers 47:677–711. https://doi.org/10.1111/j.1467-6494.1979.tb00216.x

Glaros AG, Lumley MA (2005) Alexithymia and pain in temporomandibular disorder. J Psychosom Res 59:85–88. https://doi.org/10.1016/j.jpsychores.2005.05.007

Graham DT, Stern JA, Winokur G (1958) Experimental investigation of the specificity of attitude hypothesis in psychosomatic disease. Psychosom Med 20:446–457. https://doi.org/10.1097/00006842-195811000-00002

Grynberg D, Pollatos O (2015) Perceiving one's body shapes empathy. Physiol Behav 140:54–60. https://doi.org/10.1016/j.physbeh.2014.12.026

Gu X, Hof PR, Friston KJ, Fan J (2013) Anterior insular cortex and emotional awareness. J Comp Neurol 521:3371–3388. https://doi.org/10.1002/cne.23368

Gulpek D, Kelemence Kaplan F, Kesebir S, Bora O (2014) Alexithymia in patients with conversion disorder. Nord J Psychiatry 68:300–305. https://doi.org/10.3109/08039488.2013.814711

Hantas MN, Katkin ES, Blascovich J (1982) Relationship between heartbeat discrimination and subjective experience of affective state. Psychophysiology 19:563

Harshaw C (2015) Interoceptive dysfunction: toward an integrated framework for understanding somatic and affective disturbance in depression. Psychol Bull 141:311–363. https://doi.org/10.1037/a0038101

Heims HC, Critchley HD, Dolan R et al (2004) Social and motivational functioning is not critically dependent on feedback of autonomic responses: neuropsychological evidence from patients with pure autonomic failure. Neuropsychologia 42:1979–1988. https://doi.org/10.1016/j.neuropsychologia.2004.06.001

Herbert BM, Pollatos O (2012) The body in the mind: on the relationship between interoception and embodiment. Top Cogn Sci 4:692–704. https://doi.org/10.1111/j.1756-8765.2012.01189.x

Herbert BM, Pollatos O (2014) Attenuated interoceptive sensitivity in overweight and obese individuals. Eat Behav 15:445–448. https://doi.org/10.1016/j.eatbeh.2014.06.002
Herbert BM, Pollatos O, Schandry R (2007) Interoceptive sensitivity and emotion processing: an EEG study. Int J Psychophysiol 65:214–227. https://doi.org/10.1016/j.ijpsycho.2007.04.007
Herbert BM, Pollatos O, Flor H et al (2010) Cardiac awareness and autonomic cardiac reactivity during emotional picture viewing and mental stress. Psychophysiology 47:342–354. https://doi.org/10.1111/j.1469-8986.2009.00931.x
Herbert BM, Herbert C, Pollatos O (2011) On the relationship between interoceptive awareness and alexithymia: is interoceptive awareness related to emotional awareness? J Pers 79:1149–1175. https://doi.org/10.1111/j.1467-6494.2011.00717.x
Herbert BM, Blechert J, Hautzinger M et al (2013) Intuitive eating is associated with interoceptive sensitivity. Effects on body mass index. Appetite 70:22–30. https://doi.org/10.1016/j.appet.2013.06.082
Hess EH, Polt JM (1960) Pupil size as related to interest value of visual stimuli. Science 132:349–350. https://doi.org/10.1126/science.132.3423.349
Hoisington LB (1928) Pleasantness and unpleasantness as modes of bodily experience. In: Reymert M (ed) Feelings and emotions: the Wittenberg symposium. Clark University Press, Worcester, MA
Horváth Á, Vig L, Ferentzi E, Köteles F (2021) Cardiac and proprioceptive accuracy are not related to body awareness, perceived body competence, and affect. Frontiers in Psychology in Press: https://doi.org/10.3389/fpsyg.2020.575574
Hsü EH (1952) Comparative study of factor patterns, physiologically and psychologically determined. J Gen Psychol 47:105–128. https://doi.org/10.1080/00221309.1952.9918295
Huber A, Suman AL, Biasi G, Carli G (2009) Alexithymia in fibromyalgia syndrome: associations with ongoing pain, experimental pain sensitivity and illness behavior. J Psychosom Res 66:425–433. https://doi.org/10.1016/j.jpsychores.2008.11.009
Hunt WA (1931) The relation of bright and dull pressure to affectivity. Am J Psychol 43:87–92. https://doi.org/10.2307/1414240
Hunt WA (1939) A critical review of current approaches to affectivity. Psychol Bull 36:807–828. https://doi.org/10.1037/h0054060
Izard CE (1971) The face of emotion. Appletin-Century-Crofts, New York
Izard CE (1990) Facial expressions and the regulation of emotions. J Pers Soc Psychol 58:487–498. https://doi.org/10.1037/0022-3514.58.3.487
Izard CE (1993) Four systems for emotion activation: cognitive and noncognitive processes. Psychol Rev 100:68–90. https://doi.org/10.1037/0033-295x.100.1.68
James W (1884) What is an emotion? Mind 9:188–205
James W (1890) The principles of psychology. Macmillan, London
Jones MP, Schettler A, Olden K, Crowell MD (2004) Alexithymia and somatosensory amplification in functional dyspepsia. Psychosomatics 45:508–516. https://doi.org/10.1176/appi.psy.45.6.508
Jung W-M, Ryu Y, Lee Y-S et al (2017) Role of interoceptive accuracy in topographical changes in emotion-induced bodily sensations. PLoS One 12:e0183211. https://doi.org/10.1371/journal.pone.0183211
Kahneman D (2011) Thinking, fast and slow. Farrar, Straus and Giroux, New York
Kano M, Hamaguchi T, Itoh M et al (2007) Correlation between alexithymia and hypersensitivity to visceral stimulation in human. Pain 132:252–263. https://doi.org/10.1016/j.pain.2007.01.032
Katkin ES (1965) Relationship between manifest anxiety and two indices of autonomic response to stress. J Pers Soc Psychol 2:324–333. https://doi.org/10.1037/h0022303
Katz J, Martin AL, Pagé MG, Calleri V (2009) Alexithymia and fear of pain independently predict heat pain intensity ratings among undergraduate university students. Pain Res Manag 14:299–305. https://doi.org/10.1155/2009/468321
Kellner R (1987) Abridged manual of the Illness Attitudes Scale. Department of Psychiatry, School of Medicine, University of New Mexico, Albuquerque

Kerr KL, Moseman SE, Avery JA et al (2016) Altered insula activity during visceral interoception in weight-restored patients with anorexia nervosa. Neuropsychopharmacology 41:521–528. https://doi.org/10.1038/npp.2015.174

Kever A, Pollatos O, Vermeulen N, Grynberg D (2015) Interoceptive sensitivity facilitates both antecedent- and response-focused emotion regulation strategies. Pers Individ Differ 87:20–23. https://doi.org/10.1016/j.paid.2015.07.014

Khalsa SS, Rudrauf D, Sandesara C et al (2009) Bolus isoproterenol infusions provide a reliable method for assessing interoceptive awareness. Int J Psychophysiol 72:34–45. https://doi.org/10.1016/j.ijpsycho.2008.08.010

Kihlstrom JF (2004) Availability, accessibility, and subliminal perception. Conscious Cogn 13:92–100. https://doi.org/10.1016/j.concog.2003.09.004

Klabunde M, Acheson DT, Boutelle KN et al (2013) Interoceptive sensitivity deficits in women recovered from bulimia nervosa. Eat Behav 14:488–492. https://doi.org/10.1016/j.eatbeh.2013.08.002

Koch A, Pollatos O (2014) Interoceptive sensitivity, body weight and eating behavior in children: a prospective study. Front Psychol 5:1003. https://doi.org/10.3389/fpsyg.2014.01003

Kosturek A, Gregory RJ, Sousou AJ, Trief P (1998) Alexithymia and somatic amplification in chronic pain. Psychosomatics 39:399–404. https://doi.org/10.1016/S0033-3182(98)71298-8

Köteles F, Teufel B, Körmendi J et al (2020) Cardioceptive accuracy is associated with arousal but not with valence and perceived exertion under physical load. Psychophysiology 57:e13620. https://doi.org/10.1111/psyp.13620

Kreibig SD (2010) Autonomic nervous system activity in emotion: a review. Biol Psychol 84:394–421. https://doi.org/10.1016/j.biopsycho.2010.03.010

Lacey JI (1950) Individual differences in somatic response patterns. J Comp Physiol Psychol 43:338–350. https://doi.org/10.1037/h0057436

Lacey JI (1956) The evaluation of autonomic responses: toward a general solution. Ann N Y Acad Sci 67:125–163. https://doi.org/10.1111/j.1749-6632.1956.tb46040.x

Lacey JI, Lacey BC (1958) Verification and extension of the principle of autonomic response-stereotypy. Am J Psychol 71:50–73

Lambie JA, Marcel AJ (2002) Consciousness and the varieties of emotion experience: a theoretical framework. Psychol Rev 109:219–259. https://doi.org/10.1037/0033-295x.109.2.219

Lamm C, Decety J, Singer T (2011) Meta-analytic evidence for common and distinct neural networks associated with directly experienced pain and empathy for pain. NeuroImage 54:2492–2502. https://doi.org/10.1016/j.neuroimage.2010.10.014

Landis C, Hunt WA (1932) Adrenalin and emotion. Psychol Rev 39:467–485. https://doi.org/10.1037/h0074176

Lane RD, Schwartz GE (1987) Levels of emotional awareness: a cognitive-developmental theory and its application to psychopathology. Am J Psychiatr 144:133–143. https://doi.org/10.1176/ajp.144.2.133

Lane RD, Ahern GL, Schwartz GE, Kaszniak AW (1997) Is alexithymia the emotional equivalent of blindsight? Biol Psychiatry 42:834–844. https://doi.org/10.1016/s0006-3223(97)00050-4

Lane RD, Reiman EM, Axelrod B et al (1998) Neural correlates of levels of emotional awareness: evidence of an interaction between emotion and attention in the anterior cingulate cortex. J Cogn Neurosci. https://doi.org/10.1162/089892998562924

Lang PJ (1994) The varieties of emotional experience: a meditation on James-Lange theory. Psychol Rev 101:211–221. https://doi.org/10.1037/0033-295x.101.2.211

Lang PJ, Greenwald MK, Bradley MM, Hamm AO (1993) Looking at pictures: affective, facial, visceral, and behavioral reactions. Psychophysiology 30:261–273. https://doi.org/10.1111/j.1469-8986.1993.tb03352.x

Lange CG (1885) Om Sindsbevaegelser et Psyko-Fysiologisk Studie. Kronar, Copenhagen

Lazarus RS (1966) Psychological stress and the coping process. McGraw-Hill, New York

Lazarus RS (1991) Emotion and adaptation, 1st edn. Oxford University Press, Oxford, England

Lazarus RS, Folkman S (1984) Stress, appraisal, and coping, 1st edn. Springer, New York

Lazarus RS, Smith CA (1988) Knowledge and appraisal in the cognition–emotion relationship. Cognit Emot 2:281–300. https://doi.org/10.1080/02699938808412701
LeDoux JE (1990) Information flow from sensation to emotion plasticity in the neural computation of stimulus value. In: Gabriel M, Moore J (eds) Learning and computational neuroscience: foundations of adaptive networks. Bradford Books/MIT Press, Cambridge, MA, pp 3–52
LeDoux J (1998) The emotional brain: the mysterious underpinnings of emotional life, unknown edition. Simon & Schuster, New York
Levenson RW (2014) The autonomic nervous system and emotion. Emot Rev 6:100–112. https://doi.org/10.1177/1754073913512003
Levenson RW, Ekman P, Friesen WV (1990) Voluntary facial action generates emotion-specific autonomic nervous system activity. Psychophysiology 27:363–384. https://doi.org/10.1111/j.1469-8986.1990.tb02330.x
Leventhal H (1984) A perceptual-motor theory of emotion. In: Berkowitz L (ed) Advances in experimental social psychology. Academic, New York, pp 117–182
Leventhal H, Scherer K (1987) The relationship of emotion to cognition: a functional approach to a semantic controversy. Cognit Emot 1:3–28. https://doi.org/10.1080/02699938708408361
Lewis MD (2005) Bridging emotion theory and neurobiology through dynamic systems modeling. Behav Brain Sci 28:169–194; discussion 194–245. https://doi.org/10.1017/s0140525x0500004x
Longarzo M, D'Olimpio F, Chiavazzo A et al (2015) The relationships between interoception and alexithymic trait. The Self-Awareness Questionnaire in healthy subjects. Front Psychol 6:1149. https://doi.org/10.3389/fpsyg.2015.01149
Ludwick-Rosenthal R, Neufeld RW (1985) Heart beat interoception: a study of individual differences. Int J Psychophysiol 3:57–65
Lumley MA, Smith JA, Longo DJ (2002) The relationship of alexithymia to pain severity and impairment among patients with chronic myofascial pain: comparisons with self-efficacy, catastrophizing, and depression. J Psychosom Res 53:823–830. https://doi.org/10.1016/s0022-3999(02)00337-9
MacLean PD (1949) Psychosomatic disease and the visceral brain; recent developments bearing on the Papez theory of emotion. Psychosom Med 11:338–353. https://doi.org/10.1097/00006842-194911000-00003
Malmo RB, Shagass C, Davis FH (1950) Symptom specificity and bodily reactions during psychiatric interview. Psychosom Med 12:362–376. https://doi.org/10.1097/00006842-195011000-00006
Mandler G, Kremen I (1958) Autonomic feedback: a correlational study. J Pers 26:388–399. https://doi.org/10.1111/j.1467-6494.1958.tb01594.x
Mandler G, Mandler JM, Uviller ET (1958) Autonomic feedback: the perception of autonomic activity. J Abnorm Soc Psychol 56:367–373. https://doi.org/10.1037/h0048083
Mandler G, Mandler JM, Kremen I, Sholiton RD (1961) The response to threat: relations among verbal and physiological indices. Psychol Monogr Gen Appl 75:1–22. https://doi.org/10.1037/h0093803
Manstead ASR, Wagner HL (1981) Arousal, cognition and emotion: an appraisal of two-factor theory. Curr Psychol Rev 1:35–54. https://doi.org/10.1007/BF02979253
Maranon G (1924) Contribution a l'etude de l'action emotive de l'adrenaline. Rev Francaise Endocrinol 2:301–325
Marks IM, Huson J (1973) Physiological aspects of neutral and phobic imagery: further observations. Br J Psychiatry 122:567–572. https://doi.org/10.1192/bjp.122.5.567
Marshall GD, Zimbardo PG (1979) Affective consequences of inadequately explained physiological arousal. J Pers Soc Psychol 37:970–988. https://doi.org/10.1037/0022-3514.37.6.970
Martin I, Levey AB (1978) Evaluative conditioning. Adv Behav Res Ther 1:57–101. https://doi.org/10.1016/0146-6402(78)90013-9
Martin JB, Pihl RO (1985) The stress-alexithymia hypothesis: theorectical and empirical considerations. Psychother Psychosom 43:169–176. https://doi.org/10.1159/000287876
Maslach C (1979) Negative emotional biasing of unexplained arousal. J Pers Soc Psychol 37:953–969. https://doi.org/10.1037/0022-3514.37.6.953

Mason RE (1959) Three studies relating internal sensory experiences and feeling reactions. J Gen Psychol 60:211–228. https://doi.org/10.1080/00221309.1959.9710223

Mata F, Verdejo-Roman J, Soriano-Mas C, Verdejo-Garcia A (2015) Insula tuning towards external eating versus interoceptive input in adolescents with overweight and obesity. Appetite 93:24–30. https://doi.org/10.1016/j.appet.2015.03.024

Metzinger T (2009) The ego tunnel. The science of the mind and the myth of the self. Basic Books, New York

Miller GA, Levin DN, Kozak MJ et al (1987) Individual differences in imagery and the psychophysiology of emotion. Cognit Emot 1:367–390. https://doi.org/10.1080/02699938708408058

Moors A (2009) Theories of emotion causation: a review. Cognit Emot 23:625–662. https://doi.org/10.1080/02699930802645739

Motomura K, Terasawa Y, Natsume A et al (2019) Anterior insular cortex stimulation and its effects on emotion recognition. Brain Struct Funct 224:2167–2181. https://doi.org/10.1007/s00429-019-01895-9

Mul C-L, Stagg SD, Herbelin B, Aspell JE (2018) The feeling of me feeling for you: interoception, alexithymia and empathy in autism. J Autism Dev Disord 48:2953–2967. https://doi.org/10.1007/s10803-018-3564-3

Murphy J, Catmur C, Bird G (2018a) Alexithymia is associated with a multidomain, multidimensional failure of interoception: evidence from novel tests. J Exp Psychol Gen 147:398–408. https://doi.org/10.1037/xge0000366

Murphy J, Geary H, Millgate E et al (2018b) Direct and indirect effects of age on interoceptive accuracy and awareness across the adult lifespan. Psychon Bull Rev 25:1193–1202. https://doi.org/10.3758/s13423-017-1339-z

Mussgay L, Klinkenberg N, Rüddel H (1999) Heart beat perception in patients with depressive, somatoform, and personality disorders. J Psychophysiol 13:27–36. https://doi.org/10.1027//0269-8803.13.1.27

Myers DG (2004) Intuition: its powers and perils, New edition. Yale University Press, New Haven

Nafe JP (1924) An experimental study of the affective qualities. Am J Psychol 35:507–544. https://doi.org/10.2307/1414034

Nakao M, Barsky AJ, Kumano H, Kuboki T (2002) Relationship between somatosensory amplification and alexithymia in a Japanese psychosomatic clinic. Psychosomatics 43:55–60. https://doi.org/10.1176/appi.psy.43.1.55

Nakao M, Barsky AJ, Nishikitani M et al (2007) Somatosensory amplification and its relationship to somatosensory, auditory, and visual evoked and event-related potentials (P300). Neurosci Lett 415:185–189. https://doi.org/10.1016/j.neulet.2007.01.021

Näring GW, van der Staak CP (1995) Perception of heart rate and blood pressure: the role of alexithymia and anxiety. Psychother Psychosom 63:193–200

Nemiah JC, Sifneos PE (1970) Affects and fantasy in patients with psychosomatic disorders. In: Hill O (ed) Modern trends in psychosomatic medicine. Butterworths, London

Nesse RM (1990) Evolutionary explanations of emotions. Hum Nat 1:261–289. https://doi.org/10.1007/BF02733986

Neumann R, Strack F (2000) Approach and avoidance: the influence of proprioceptive and exteroceptive cues on encoding of affective information. J Pers Soc Psychol 79:39–48. https://doi.org/10.1037/0022-3514.79.1.39

Nicholson TM, Williams DM, Grainger C et al (2018) Interoceptive impairments do not lie at the heart of autism or alexithymia. J Abnorm Psychol 127:612–622. https://doi.org/10.1037/abn0000370

Nicholson T, Williams D, Carpenter K, Kallitsounaki A (2019) Interoception is impaired in children, but not adults, with autism spectrum disorder. J Autism Dev Disord 49:3625–3637. https://doi.org/10.1007/s10803-019-04079-w

Noah T, Schul Y, Mayo R (2018) When both the original study and its failed replication are correct: feeling observed eliminates the facial-feedback effect. J Pers Soc Psychol 114:657–664. https://doi.org/10.1037/pspa0000121

Noel J-P, Lytle M, Cascio C, Wallace MT (2018) Disrupted integration of exteroceptive and interoceptive signaling in autism spectrum disorder. Autism Res 11:194–205. https://doi.org/10.1002/aur.1880

Novembre G, Zanon M, Morrison I, Ambron E (2019) Bodily sensations in social scenarios: where in the body? PLoS One 14:e0206270. https://doi.org/10.1371/journal.pone.0206270

Nummenmaa L, Glerean E, Hari R, Hietanen JK (2014) Bodily maps of emotions. Proc Natl Acad Sci U S A 111:646–651. https://doi.org/10.1073/pnas.1321664111

Nummenmaa L, Hari R, Hietanen JK, Glerean E (2018) Maps of subjective feelings. Proc Natl Acad Sci U S A 115:9198–9203. https://doi.org/10.1073/pnas.1807390115

Nyklíček I, Vingerhoets AJJM (2000) Alexithymia is associated with low tolerance to experimental painful stimulation. Pain 85:471–475. https://doi.org/10.1016/S0304-3959(99)00295-X

Oatley K, Johnson-laird PN (1987) Towards a cognitive theory of emotions. Cognit Emot 1:29–50. https://doi.org/10.1080/02699938708408362

Olds J (1967) Emotional centres in the brain. Sci J 3:87–92

Olds J, Olds ME (1961) Interference and learning in paleo-cortical systems. In: Delafresnaye JF (ed) Brain mechanisms and learning. Blackwell, Oxford, England, pp 153–187

Osgood CE, Suci GJ, Tannenbaum PH (1957) The measurement of meaning. University of Illinois Press, Urbana

Pace-Schott EF, Amole MC, Aue T et al (2019) Physiological feelings. Neurosci Biobehav Rev 103:267–304. https://doi.org/10.1016/j.neubiorev.2019.05.002

Palmer CJ, Seth AK, Hohwy J (2015) The felt presence of other minds: predictive processing, counterfactual predictions, and mentalising in autism. Conscious Cogn 36:376–389. https://doi.org/10.1016/j.concog.2015.04.007

Palmer CJ, Lawson RP, Hohwy J (2017) Bayesian approaches to autism: towards volatility, action, and behavior. Psychol Bull 143:521–542. https://doi.org/10.1037/bul0000097

Palser ER, Palmer CE, Galvez-Pol A et al (2018) Alexithymia mediates the relationship between interoceptive sensibility and anxiety. PLoS One 13:e203212. https://doi.org/10.1371/journal.pone.0203212

Palser ER, Fotopoulou A, Pellicano E, Kilner JM (2020) Dissociation in how core autism features relate to interoceptive dimensions: evidence from cardiac awareness in children. J Autism Dev Disord 50:572–582. https://doi.org/10.1007/s10803-019-04279-4

Panksepp J (1982) Toward a general psychobiological theory of emotions. Behav Brain Sci 5:407–467. https://doi.org/10.1017/S0140525X00012759

Panksepp J (1998) Affective neuroscience: the foundations of human and animal emotions, 1st edn. Oxford University Press, New York

Panksepp J, Biven L (2012) The archaeology of mind. Neuroevolutionary origins of human emotions. W. W. Norton & Company, New York

Papez JW (1937) A proposed mechanism of emotion. Arch Neurol Psychiatr 38:725–743. https://doi.org/10.1001/archneurpsyc.1937.02260220069003

Pellicano E, Burr D (2012) When the world becomes "too real": a Bayesian explanation of autistic perception. Trends Cogn Sci 16:504–510. https://doi.org/10.1016/j.tics.2012.08.009

Pennebaker JW (1982) The psychology of physical symptoms. Springer, New York

Pfeifer G, Garfinkel SN, Gould van Praag CD et al (2017) Feedback from the heart: emotional learning and memory is controlled by cardiac cycle, interoceptive accuracy and personality. Biol Psychol 126:19–29. https://doi.org/10.1016/j.biopsycho.2017.04.001

Philippot P, Chapelle G, Blairy S (2002) Respiratory feedback in the generation of emotion. Cognit Emot 16:605–627. https://doi.org/10.1080/02699930143000392

Piech RM, Strelchuk D, Knights J et al (2017) People with higher interoceptive sensitivity are more altruistic, but improving interoception does not increase altruism. Sci Rep 7:15652. https://doi.org/10.1038/s41598-017-14318-8

Plutchik R (1982) A psychoevolutionary theory of emotions. Soc Sci Inf 21:529–553. https://doi.org/10.1177/053901882021004003

Plutchik R, Ax AF (1967) A critique of determinants of emotional state by Schachter and Singer (1962). Psychophysiology 4:79–82. https://doi.org/10.1111/j.1469-8986.1967.tb02740.x
Pollatos O, Ferentzi E (2018) Embodiment of emotion regulation. In: Hauke G, Kritikos A (eds) Embodiment in psychotherapy: a practitioner's guide. Springer International Publishing, Cham, pp 43–55
Pollatos O, Georgiou E (2016) Normal interoceptive accuracy in women with bulimia nervosa. Psychiatry Res 240:328–332. https://doi.org/10.1016/j.psychres.2016.04.072
Pollatos O, Kirsch W, Schandry R (2005) On the relationship between interoceptive awareness, emotional experience, and brain processes. Brain Res Cogn Brain Res 25:948–962. https://doi.org/10.1016/j.cogbrainres.2005.09.019
Pollatos O, Gramann K, Schandry R (2007a) Neural systems connecting interoceptive awareness and feelings. Hum Brain Mapp 28:9–18. https://doi.org/10.1002/hbm.20258
Pollatos O, Herbert BM, Kaufmann C et al (2007b) Interoceptive awareness, anxiety and cardiovascular reactivity to isometric exercise. Int J Psychophysiol 65:167–173. https://doi.org/10.1016/j.ijpsycho.2007.03.005
Pollatos O, Herbert BM, Matthias E, Schandry R (2007c) Heart rate response after emotional picture presentation is modulated by interoceptive awareness. Int J Psychophysiol 63:117–124. https://doi.org/10.1016/j.ijpsycho.2006.09.003
Pollatos O, Schandry R, Auer DP, Kaufmann C (2007d) Brain structures mediating cardiovascular arousal and interoceptive awareness. Brain Res 1141:178–187. https://doi.org/10.1016/j.brainres.2007.01.026
Pollatos O, Traut-Mattausch E, Schroeder H, Schandry R (2007e) Interoceptive awareness mediates the relationship between anxiety and the intensity of unpleasant feelings. J Anxiety Disord 21:931–943. https://doi.org/10.1016/j.janxdis.2006.12.004
Pollatos O, Kurz A-L, Albrecht J et al (2008) Reduced perception of bodily signals in anorexia nervosa. Eat Behav 9:381–388. https://doi.org/10.1016/j.eatbeh.2008.02.001
Pollatos O, Traut-Mattausch E, Schandry R (2009) Differential effects of anxiety and depression on interoceptive accuracy. Depress Anxiety 26:167–173. https://doi.org/10.1002/da.20504
Pollatos O, Dietel A, Gündel H, Duschek S (2015) Alexithymic trait, painful heat stimulation, and everyday pain experience. Front Psychiatry 6:139. https://doi.org/10.3389/fpsyt.2015.00139
Porcelli P, Tulipani C, Maiello E et al (2007) Alexithymia, coping, and illness behavior correlates of pain experience in cancer patients. Psychooncology 16:644–650. https://doi.org/10.1002/pon.1115
Porges SW (1993) Body perception questionnaire. Laboratory of Developmental Assessment, University of Maryland, College Park
Premack D, Woodruff G (1978) Does the chimpanzee have a theory of mind? Behav Brain Sci 1:515–526. https://doi.org/10.1017/S0140525X00076512
Preuschoff K, Quartz SR, Bossaerts P (2008) Human insula activation reflects risk prediction errors as well as risk. J Neurosci 28:2745–2752. https://doi.org/10.1523/JNEUROSCI.4286-07.2008
Prinz J (2004) Gut reactions: a perceptual theory of emotion. Oxford University Press, Oxford, England
Quattrocki E, Friston K (2014) Autism, oxytocin and interoception. Neurosci Biobehav Rev 47:410–430. https://doi.org/10.1016/j.neubiorev.2014.09.012
Quirin M, Kazén M, Kuhl J (2009a) When nonsense sounds happy or helpless: The Implicit Positive and Negative Affect Test (IPANAT). J Pers Soc Psychol 97:500–516. https://doi.org/10.1037/a0016063
Quirin M, Kazén M, Rohrmann S, Kuhl J (2009b) Implicit but not explicit affectivity predicts circadian and reactive cortisol: using the implicit positive and negative affect test. J Pers 77:401–425. https://doi.org/10.1111/j.1467-6494.2008.00552.x
Rachman S, Hodgson R (1974) I. Synchrony and desynchrony in fear and avoidance. Behav Res Ther 12:311–318. https://doi.org/10.1016/0005-7967(74)90005-9

Ranehill E, Dreber A, Johannesson M et al (2015) Assessing the robustness of power posing: no effect on hormones and risk tolerance in a large sample of men and women. Psychol Sci 26:653–656. https://doi.org/10.1177/0956797614553946

Reisenzein R (1983) The Schachter theory of emotion: two decades later. Psychol Bull 94:239–264. https://doi.org/10.1037/0033-2909.94.2.239

Ring C, Brener J (1996) Influence of beliefs about heart rate and actual heart rate on heartbeat counting. Psychophysiology 33:541–546. https://doi.org/10.1111/j.1469-8986.1996.tb02430.x

Ring C, Brener J, Knapp K, Mailloux J (2015) Effects of heartbeat feedback on beliefs about heart rate and heartbeat counting: a cautionary tale about interoceptive awareness. Biol Psychol 104:193–198. https://doi.org/10.1016/j.biopsycho.2014.12.010

Robbins TW, Granon S, Muir JL et al (1998) Neural systems underlying arousal and attention: implications for drug abuse. Ann N Y Acad Sci 846:222–237. https://doi.org/10.1111/j.1749-6632.1998.tb09740.x

Roberts RJ, Weerts TC (1982) Cardiovascular responding during anger and fear imagery. Psychol Rep 50:219–230. https://doi.org/10.2466/pr0.1982.50.1.219

Roessler R, Engel BT (1974) The current status of the concepts of physiological response specificity and activation. Int J Psychiatry Med 5:359–366

Rogers RW, Deckner CW (1975) Effects of fear appeals and physiological arousal upon emotion, attitudes, and cigarette smoking. J Pers Soc Psychol 32:222–230

Rolls ET, Everitt BJ, Roberts A et al (1996) The orbitofrontal cortex. Philos Trans R Soc Lond B Biol Sci 351:1433–1444. https://doi.org/10.1098/rstb.1996.0128

Rosenzweig S (1993) Freud, Jung, and Hall the king-maker: the expedition to America, 2nd edn. Hogrefe & Huber, St. Louis, MO

Russell JA (1980) A circumplex model of affect. J Pers Soc Psychol 39:1161–1178. https://doi.org/10.1037/h0077714

Russell JA (2003) Core affect and the psychological construction of emotion. Psychol Rev 110:145–172. https://doi.org/10.1037/0033-295X.110.1.145

Schachter H (1957) Pain, fear, and anger in hypertensives and normotensives; a psychophysiological study. Psychosom Med 19:17–29. https://doi.org/10.1097/00006842-195701000-00003

Schachter S (1964) The interaction of cognitive and physiological determinants of emotional state. In: Berkowitz L (ed) Advances in experimental social psychology, vol 1. Academic, New York, pp 49–80

Schachter S, Singer JE (1962) Cognitive, social, and physiological determinants of emotional state. Psychol Rev 69:379–399

Schandry R (1981) Heart beat perception and emotional experience. Psychophysiology 18:483–488. https://doi.org/10.1111/j.1469-8986.1981.tb02486.x

Schauder KB, Mash LE, Bryant LK, Cascio CJ (2015) Interoceptive ability and body awareness in autism spectrum disorder. J Exp Child Psychol 131:193–200. https://doi.org/10.1016/j.jecp.2014.11.002

Scherer KR (1999) Appraisal theory. In: Dalgleish T, Power M (eds) Handbook of cognition and emotion. Wiley, Chichester, UK, pp 637–661

Scherer KR (2004) Feelings integrate the central representation of appraisal-driven response organization in emotion. In: Manstead ASR, Frijda N, Fischer A (eds) Feelings and emotions. The Amsterdam Symposium. Cambridge University Press, Cambridge, UK, pp 136–157

Schmidt AF, Eulenbruch T, Langer C, Banger M (2013) Interoceptive awareness, tension reduction expectancies and self-reported drinking behavior. Alcohol Alcohol 48:472–477. https://doi.org/10.1093/alcalc/agt024

Schwarz N, Servay W, Kumpf M (1985) Attribution of arousal as a mediator of the effectiveness of fear-arousing communications. J Appl Soc Psychol 15:178–188. https://doi.org/10.1111/j.1559-1816.1985.tb02343.x

Seth AK (2013) Interoceptive inference, emotion, and the embodied self. Trends Cogn Sci 17:565–573. https://doi.org/10.1016/j.tics.2013.09.007

Seth AK, Critchley HD (2013) Extending predictive processing to the body: emotion as interoceptive inference. Behav Brain Sci 36:227–228. https://doi.org/10.1017/S0140525X12002270
Seth AK, Suzuki K, Critchley HD (2012) An interoceptive predictive coding model of conscious presence. Front Psychol 2:395. https://doi.org/10.3389/fpsyg.2011.00395
Shafir T, Tsachor RP, Welch KB (2015) Emotion regulation through movement: unique sets of movement characteristics are associated with and enhance basic emotions. Front Psychol 6:2030. https://doi.org/10.3389/fpsyg.2015.02030
Shah P, Hall R, Catmur C, Bird G (2016) Alexithymia, not autism, is associated with impaired interoception. Cortex 81:215–220. https://doi.org/10.1016/j.cortex.2016.03.021
Shah P, Catmur C, Bird G (2017) From heart to mind: linking interoception, emotion, and theory of mind. Cortex 93:220–223. https://doi.org/10.1016/j.cortex.2017.02.010
Shamay-Tsoory SG, Abu-Akel A (2016) The social salience hypothesis of oxytocin. Biol Psychiatry 79:194–202. https://doi.org/10.1016/j.biopsych.2015.07.020
Shibata M, Ninomiya T, Jensen MP et al (2014) Alexithymia is associated with greater risk of chronic pain and negative affect and with lower life satisfaction in a general population: the Hisayama study. PLoS One 9:e90984. https://doi.org/10.1371/journal.pone.0090984
Shields SA (1984) Reports of bodily change in anxiety, sadness, and anger. Motiv Emot 8:1–21. https://doi.org/10.1007/BF00992989
Shields SA, Simon A (1991) Is awareness of bodily change in emotion related to awareness of other bodily processes? J Pers Assess 57:96–109. https://doi.org/10.1207/s15327752jpa5701_12
Sifneos PE (1973) The prevalence of 'alexithymic' characteristics in psychosomatic patients. Psychother Psychosom 22:255–262. https://doi.org/10.1159/000286529
Silani G, Bird G, Brindley R et al (2008) Levels of emotional awareness and autism: an fMRI study. Soc Neurosci 3:97–112. https://doi.org/10.1080/17470910701577020
Simonovic B, Stupple E, Gale M, Sheffield D (2019) Sweating the small stuff: a meta-analysis of skin conductance on the Iowa gambling task. Cogn Affect Behav Neurosci 19:1097–1112. https://doi.org/10.3758/s13415-019-00744-w
Singer T, Lamm C (2009) The social neuroscience of empathy. Ann N Y Acad Sci 1156:81–96. https://doi.org/10.1111/j.1749-6632.2009.04418.x
Singer T, Critchley HD, Preuschoff K (2009) A common role of insula in feelings, empathy and uncertainty. Trends Cogn Sci 13:334–340. https://doi.org/10.1016/j.tics.2009.05.001
Sivik T (1993) Alexithymia and hypersensitivity to touch and palpation. Integr Physiol Behav Sci 28:130–136. https://doi.org/10.1007/BF02691215
Smith R, Lane RD (2015) The neural basis of one's own conscious and unconscious emotional states. Neurosci Biobehav Rev 57:1–29. https://doi.org/10.1016/j.neubiorev.2015.08.003
Sönmez MB, Kılıç EK, Çöl IA et al (2017) Decreased interoceptive awareness in patients with substance use disorders. J Subst Use 22:60–65. https://doi.org/10.3109/14659891.2016.1143048
Stephens CL, Christie IC, Friedman BH (2010) Autonomic specificity of basic emotions: evidence from pattern classification and cluster analysis. Biol Psychol 84:463–473. https://doi.org/10.1016/j.biopsycho.2010.03.014
Tajadura-Jiménez A, Tsakiris M (2014) Balancing the "inner" and the "outer" self: interoceptive sensitivity modulates self–other boundaries. J Exp Psychol Gen 143:736–744. https://doi.org/10.1037/a0033171
Taycan O, Ozdemir A, Erdogan-Taycan S, Jurcik T (2015) Associations of somatic symptom attribution in Turkish patients with major depression. Nord J Psychiatry 69:167–173. https://doi.org/10.3109/08039488.2014.950328
Taylor GJ, Parker JD, Bagby RM, Acklin MW (1992) Alexithymia and somatic complaints in psychiatric out-patients. J Psychosom Res 36:417–424. https://doi.org/10.1016/0022-3999(92)90002-j
Terasawa Y, Moriguchi Y, Tochizawa S, Umeda S (2014) Interoceptive sensitivity predicts sensitivity to the emotions of others. Cognit Emot 28:1435–1448. https://doi.org/10.1080/02699931.2014.888988

Terasawa Y, Kurosaki Y, Ibata Y et al (2015) Attenuated sensitivity to the emotions of others by insular lesion. Front Psychol 6:1314. https://doi.org/10.3389/fpsyg.2015.01314
Terhaar J, Viola FC, Bär K-J, Debener S (2012) Heartbeat evoked potentials mirror altered body perception in depressed patients. Clin Neurophysiol 123:1950–1957. https://doi.org/10.1016/j.clinph.2012.02.086
Tichtener EB (1908) Lectures on elementary psychology of feeling and attention. Macmillan, New York
Tichtener EB (1926) A textbook of psychology. Macmillan, New York
Tollenaar MS, Chatzimanoli M, van der Wee NJA, Putman P (2013) Enhanced orienting of attention in response to emotional gaze cues after oxytocin administration in healthy young men. Psychoneuroendocrinology 38:1797–1802. https://doi.org/10.1016/j.psyneuen.2013.02.018
Tominaga T, Choi H, Nagoshi Y et al (2014) Relationship between alexithymia and coping strategies in patients with somatoform disorder. Neuropsychiatr Dis Treat 10:55–62. https://doi.org/10.2147/NDT.S55956
Tomkins SS (1962) Affect, imagery, consciousness, The positive affects, vol 1. Springer, New York
Tomkins SS (1981) The role of facial response in the experience of emotion: a reply to Tourangeau and Ellsworth. J Pers Soc Psychol 40:355–357. https://doi.org/10.1037/0022-3514.40.2.355
Torregrossa LJ, Snodgress MA, Hong SJ et al (2019) Anomalous bodily maps of emotions in schizophrenia. Schizophr Bull 45:1060–1067. https://doi.org/10.1093/schbul/sby179
Tversky A, Kahneman D (1983) Extensional versus intuitive reasoning: the conjunction fallacy in probability judgment. Psychol Rev 90:293–315. https://doi.org/10.1037/0033-295X.90.4.293
Valdés M, Bernardo M, Segarra N et al (2008) La amplificación somatosensorial en la esquizofrenia está relacionada con la preservación del rendimiento neuropsicológico. Revista de Psiquiatría y Salud Mental 1:3–9. https://doi.org/10.1016/S1888-9891(08)72510-2
Valins S (1966) Cognitive effects of false heart-rate feedback. J Pers Soc Psychol 4:400–408. https://doi.org/10.1037/h0023791
Van der Does AJW, Van Dyck R, Spinhoven P (1997) Accurate heartbeat perception in panic disorder: fact and artefact. J Affect Disord 43:121–130
Van der Does AJW, Antony MM, Ehlers A, Barsky AJ (2000) Heartbeat perception in panic disorder: a reanalysis. Behav Res Ther 38:47–62
van der Ploeg MM, Brosschot JF, Thayer JF, Verkuil B (2016) The implicit positive and negative affect test: validity and relationship with cardiovascular stress-responses. Front Psychol 7:425. https://doi.org/10.3389/fpsyg.2016.00425
Weineck F, Messner M, Hauke G, Pollatos O (2019) Improving interoceptive ability through the practice of power posing: a pilot study. PLoS One 14:e0211453. https://doi.org/10.1371/journal.pone.0211453
Weiss S, Sack M, Henningsen P, Pollatos O (2014) On the interaction of self-regulation, interoception and pain perception. Psychopathology 47:377–382. https://doi.org/10.1159/000365107
Wenger MA, Engel BT, Clemens TL (1957) Studies of autonomic response patterns: rationale and methods. Behav Sci 2:216–221. https://doi.org/10.1002/bs.3830020304
Werner NS, Jung K, Duschek S, Schandry R (2009) Enhanced cardiac perception is associated with benefits in decision-making. Psychophysiology 46:1123–1129. https://doi.org/10.1111/j.1469-8986.2009.00855.x
Wiebking C, Northoff G (2015) Neural activity during interoceptive awareness and its associations with alexithymia-an fMRI study in major depressive disorder and non-psychiatric controls. Front Psychol 6:589. https://doi.org/10.3389/fpsyg.2015.00589
Wiebking C, Bauer A, de Greck M et al (2010) Abnormal body perception and neural activity in the insula in depression: an fMRI study of the depressed “material me”. World J Biol Psychiatry 11:538–549. https://doi.org/10.3109/15622970903563794
Wiebking C, de Greck M, Duncan NW et al (2015) Interoception in insula subregions as a possible state marker for depression—an exploratory fMRI study investigating healthy, depressed and remitted participants. Front Behav Neurosci 9:82. https://doi.org/10.3389/fnbeh.2015.00082
Wiens S (2005) Interoception in emotional experience. Curr Opin Neurol 18:442–447

Wiens S, Mezzacappa ES, Katkin ES (2000) Heartbeat detection and the experience of emotions. Cognit Emot 14:417–427. https://doi.org/10.1080/026999300378905
Winkielman P, Berridge KC (2004) Unconscious emotion. Curr Dir Psychol Sci 13:120–123
Wise TN, Mann LS (1994) The relationship between somatosensory amplification, alexithymia, and neuroticism. J Psychosom Res 38:515–521
Wise TN, Mann LS (1995) The attribution of somatic symptoms in psychiatric outpatients. Compr Psychiatry 36:407–410
Wolf S, Wolff HG (1943) Human gastric function. An experimental study of man and his stomach. Oxford University Press, London
Wölk J, Sütterlin S, Koch S et al (2014) Enhanced cardiac perception predicts impaired performance in the Iowa Gambling Task in patients with panic disorder. Brain Behav 4:238–246. https://doi.org/10.1002/brb3.206
Xiang T, Lohrenz T, Montague PR (2013) Computational substrates of norms and their violations during social exchange. J Neurosci 33:1099–1108. https://doi.org/10.1523/JNEUROSCI.1642-12.2013
Yao S, Becker B, Zhao W et al (2018) Oxytocin modulates attention switching between interoceptive signals and external social cues. Neuropsychopharmacology 43:294–301. https://doi.org/10.1038/npp.2017.189
Yokoyama M (1921) The nature of the affective judgment in the method of paired comparisons. Am J Psychol 32:357–369. https://doi.org/10.2307/1414000
Young PT (1927) Studies in affective psychology. Am J Psychol 38:157–193. https://doi.org/10.2307/1415200
Young PT (1932) The relation of bright and dull pressure to affectivity. Am J Psychol 44:780–784. https://doi.org/10.2307/1414542
Young HA, Williams C, Pink AE et al (2017) Getting to the heart of the matter: does aberrant interoceptive processing contribute towards emotional eating? PLoS One 12:e0186312. https://doi.org/10.1371/journal.pone.0186312
Zajonc RB (1980) Feeling and thinking: preferences need no inferences. Am Psychol 35:151–175. https://doi.org/10.1037/0003-066X.35.2.151
Zajonc RB (1984) On the primacy of affect. Am Psychol 39:117–123. https://doi.org/10.1037/0003-066X.39.2.117
Zamariola G, Vlemincx E, Corneille O, Luminet O (2018) Relationship between interoceptive accuracy, interoceptive sensibility, and alexithymia. Pers Individ Differ 125:14–20. https://doi.org/10.1016/j.paid.2017.12.024

Chapter 9
The Dark Side of Body Sensations

9.1 Introduction

The presence of body sensations is usually a meaningful experience. The meaning we attach to somatic percepts is the result of primary and secondary evaluative processes (Chaps. 5 and 7), which impacts our behavior in multiple ways (tenet of relevance). Although these sensations are located in the body and perceived as real (tenet of realness), they are often evoked by top-down information (tenet of malleability) and lack actual peripheral concomitants (tenet of dissociation). In certain cases, these features make up a unique blend: we appraise actually nonexisting but clearly felt body changes as unpleasant and threatening (i.e., as indicators of actual or possible pathological processes) and actively attempt to eliminate or at least reduce them. This is the nocebo reaction; as we will see, this phenomenon is way more widespread and impactful than generally assumed. It can be considered the acute analog of clinical hypochondriasis (see Chap. 6), which is usually evoked by external information. On the other hand, sometimes we do not experience body sensations and take this as evidence of a properly functioning body. This intuition relies on the implicit (and mistaken) assumption that we are able to feel if something is wrong in our body. In the absence of unpleasant sensations, such as pain, we are less motivated to take similarly unpleasant steps (i.e., screening and diagnostic procedures) in order to prevent and diagnose certain diseases. In my view, these two cases represent the most dangerous and harmful aspects of body sensations.

9.2 The Nocebo Phenomenon

Perhaps the most important step in the development of modern scientific medicine was the introduction of a placebo group in clinical trials (Shapiro and Shapiro 1997). Participants of a trial are randomly assigned to one of two groups: one

F. Köteles, *Body Sensations*, https://doi.org/10.1007/978-3-030-63201-4_9

receives the drug under test, whereas the other is administered with seemingly the same treatment; however, their medication does not contain active substance(s). This way, the specific (pharmacological) effects of drugs can be separated from the natural course of the disease and a number of confounding factors and artifacts (Ernst and Reschl 1995; Kienle 1995; Kienle and Kiene 1997). Placebo control became the "gold standard" of the evaluation of the effectiveness of drugs (Lasagna 1955; Shapiro and Shapiro 1997); its systematic use started in the 1950s. To the surprise of the researchers, patients in the placebo groups reported a similarly high number of side effects than those in the drug (aka verum) groups (Greiner et al. 1950; Wolf and Pinsky 1954; Green 1962, 1964). This phenomenon, i.e., drug side effects experienced and reported by patients in the placebo groups of clinical trials, was called nocebo from the Latin verb *nocere* ("I shall harm") (Kennedy 1961; Kissel and Barrucand 1964). Even more surprisingly, certain nocebo symptoms showed dose dependency and were often similar to the side effects of the respective active substance (Pogge 1963; Honigfeld 1964; Green 1964; Rosenzweig et al. 1993). These early findings were partly replicated in more recent studies (Silvestri et al. 2003; Reuter et al. 2003; Amanzio et al. 2009; Cocco 2009; Rief et al. 2009; Mora et al. 2011; Mitsikostas et al. 2011). It also turned out that not only patients but even healthy people can experience nocebo symptoms (Rosenzweig et al. 1993).

The term was defined in a different way in medical anthropology in the 1980s. Here, the nocebo phenomenon refers to symptoms caused by expectations of sickness and the affective states associated with these expectations (Hahn 1985, 1997, 1999). It was proposed that witchcraft and similar ancient practices utilized this mechanism (Edwards et al. 2010). For example, diseases, even death, due to curse of a witch or sorcerer in traditional cultures belong here; such cases were reliably documented by Walter Cannon, the elite physiologist (Cannon 1942, 1957). For such powerful effects, a firm and unquestionable belief system is required; this is why Hahn interpreted nocebo effects as the side effects of culture. Contrary to the concept of many authors (see Chap. 7), expectations in this model can be implicit; a curse or a ritual can automatically and uncontrollably evoke the negative state. As for the mechanism, Cannon assumed sympathetic overactivation in the background; since then, other options were also raised (Sternberg 2002; Samuels 2007; Lester 2009). Automatic physiological and psychological reactions to the symbolic aspects of the environment are called meaning response in medical anthropology (Moerman 2002, 2011); the nocebo (as well as the placebo) phenomenon represents a health-related instance of this broad category. Hahn (1997) makes a distinction between the nocebo reaction and the adverse effects of placebos: whereas the former is mediated by negative (harmful) expectations, the latter develop with positive expectations in his view. Another emblematic figure of early placebo research, Arthur K. Shapiro separated the negative placebo reaction (i.e., symptoms of the disease under treatment become worse) from placebo-induced side effects (Shapiro et al. 1979). These distinctions are often difficult to operationalize in empirical research though.

In the approach of medical anthropology to nocebo, no physical vehicle transmitting the message that evokes negative expectations is needed beyond words and/or rituals. The original medical approach assumed a proxy, typically a pill without active substances (i.e., placebo), which has symbolic features only. In a paper published in 2002, Arthur Barsky and colleagues emphasized the possible role of the nocebo phenomenon in medication side effects (Barsky et al. 2002). They claimed that a proportion of side effects of active drugs may also be nonspecific, i.e., due to the nocebo phenomenon. This idea is widely acknowledged for the placebo phenomenon: the overall effectiveness of a substance is the sum of its specific (i.e., pharmacological) and nonspecific effects (see Chap. 10) (Shapiro and Shapiro 1997), and the share of the nonspecific component can be close to that of the specific for certain conditions, such as pain, anxiety, and depression (Kirsch 2010; Benedetti 2011). In a similar vein, the overall side effect profile of a medication for a patient is the sum of pharmacological and non-pharmacological (nonspecific) reactions, sometimes combined and complicated by interactions between the two (Lyerly et al. 1964; Penick and Fisher 1965; Luparello et al. 1970; Flaten et al. 1999, 2004). Research into nonspecific drug side effects gained momentum in the last 10–15 years as it was realized that it has very serious practical implications (see below) beyond its theoretical importance.

The common point of the various approaches to the nocebo reaction is that negative changes, typically unpleasant body sensations, are evoked by top-down (aka psychological or psychogenic) factors. Such factors can be (explicit or implicit) suggestions, expectancies, or expectations evoked by own experience or observations of others, or simply beliefs. If one accepts such a broad definition, then the perspective will be even wider, and other phenomena, such as various idiopathic environmental intolerances (Chap. 7) and mass psychogenic illness, can also be included (Rubin et al. 2008; Szemerszky et al. 2010). These phenomena are in perfect accordance with the ideas of Hahn. Perhaps the only important difference is that Hahn focused on the belief system of local cultures, whereas the mass media (both printed and electronic) broadcasts such beliefs worldwide today. Media reports on the acute negative effects of monosodium-glutamate, dubbed Chinese restaurant syndrome (Kwok 1968; Tarasoff and Kelly 1993; Geha et al. 2000; Freeman 2006; Williams and Woessner 2009), peanut allergy (Waggoner 2013), concussion (Polich et al. 2020), statins (Matthews et al. 2016), generic medicines (Faasse et al. 2009, 2012a, b, 2015; MacKrill et al. 2019), vaccination (Clements 2003; Dunn et al. 2015; Hornsey et al. 2018), exposure to electromagnetic sources (Elvers et al. 2009; Claassen et al. 2012; Huiberts et al. 2013; Witthöft and Rubin 2013; Witthöft et al. 2018), wind turbines (Chapman et al. 2013), and generally the harmful consequences of modern technologies, dubbed modern health worries (Petrie et al. 2001; Petrie and Wessely 2002; Dömötör et al. 2019), often substantially contribute to the development of symptoms that is often mistakenly ascribed to the aforementioned factors. Also, the perceived importance of (perfect) health and factors that can negatively influence health has increased recently (Barsky 1988; Petrie et al. 2001). Thus, if we call the nocebo the side effect of culture, we can also state that our culture is the culture of side effects (Köteles and Bárdos 2009). Beyond symptoms and sickness, the nocebo phenomenon is present in other areas of everyday life. For example, it impacts physical (Bérdi et al. 2011; Beedie et al. 2018; Hurst et al. 2019; Raglin et al. 2020) and actual and/or perceived cognitive performance (Schwarz and

Büchel 2015; Winkler and Hermann 2019; Mlynski et al. 2020). The next question is how does it work.

9.3 Nocebo Mechanisms

If it comes to mechanisms, nocebo can be regarded an umbrella term for top-down influences on body sensations and functioning. Moreover, the different mechanisms are often difficult to separate from each other because of definition and operationalization-related issues.

The most often mentioned mediating factor between top-down information and perceived changes, expectation, is a vaguely defined term (see Chap. 7). Most importantly, it can be conscious (explicit), as well as nonconscious (implicit), which leads to methodological challenges. Moreover, at least three more or less distinct processes belong here. First, expectations can shape the attribution of already existing symptoms, which results in more symptoms associated with the suspected cause without the increase of the overall number and/or severity of symptoms. This was neatly demonstrated by a natural experiment in New Zealand: worries about harmful effects before a pesticide spraying predicted the number of symptoms attributed to the intervention but not the overall symptom score following the spaying (Petrie et al. 2005). Also, our research group showed that expectations of symptoms caused by sham electromagnetic exposure or a placebo pill given with the suggestion of possible side effects can shape attributions without increasing the total number of symptoms (Szemerszky et al. 2015a). As we experience unpleasant body sensations on a daily basis (Pennebaker 1982), they provide us a symptom pool that can be attributed to medications or other environmental factors if we are primed to do so (Reidenberg and Lowenthal 1968; Khosla et al. 1992; Rosenzweig et al. 1993; Meyer et al. 1996; Röösli et al. 2004). All of us is familiar with the nonspecific symptoms most often reported as drug side effects by patients, such as headache, fatigue, drowsiness, and difficulties with concentration (Rosenzweig et al. 1993; Barsky et al. 2002; Köteles and Bárdos 2009). This mechanism can be particularly relevant when a high number of symptoms are present, for example, in patients taking multiple medications for multiple conditions (Sanderson et al. 2013) or with proneness to somatization or anxiety (Barsky et al. 2002).

Second, expectations initialize a seeking process which may lead to an increase of perceived body sensations (Pennebaker and Skelton 1981). As we discussed in Chaps. 5 and 7, body focused attention can enhance body sensations; if we also have cues about what to feel (for example, because we carefully read the patient information leaflet that comes with our medication, see below), these expectations can easily manifest themselves as unpleasant sensations, in other words, symptoms and nonspecific side effects.

Finally, expectations can also lead to actual changes, for example, in pain perception or airway resistance and dyspnea (Luparello et al. 1968, 1970; Butler and Steptoe 1986; Benedetti 2009; Rodriguez-Raecke et al. 2010; Tracey 2010; Bingel et al. 2011; van Laarhoven et al. 2011). The latter two mechanisms are sometimes hard to separate; still, it is important to emphasize that the nocebo reaction is much more than a simple (re)attribution process. It also highlights the fact that some of the nonspecific symptoms (although this might be a relatively small proportion) may have a physiological background and can even lead to serious physical consequences in certain cases.

Also, expectations can actively trigger the development of new symptoms (or increase the pool of available body sensations) primarily through the associated negative affective states, most importantly, stress and anxiety. As expectations belong to the cognitive domain, this is often considered a distinct mechanism (for example, the aforementioned definition by Hahn includes the affective aspect separately). This distinction is again vague for many cases, most importantly, for pain, which is a body-related emotion (Craig 2003). Interactions between the placebo response and the decrease of anxiety and stress have been described since the beginning of placebo research (Wolff et al. 1946; Beecher 1956, 1960). These early studies primarily focused on pain, the threshold of which can be increased and decreased (analgesia and hyperalgesia, respectively) by top-down factors, and the latter is considered a nocebo response (Benedetti 2009, 2011). Various pain paradigms are often used as a model of the placebo/nocebo response; the fact that pain is the most disturbing and most often reported complaint largely justifies this practice. However, the generalizability of these findings to other, not pain-related symptoms, appears questionable to me because of the special neural mechanisms involved in pain perception. Hyperalgesic (nocebo) suggestions induce anticipatory anxiety which activates the stress response and at the same time lowers the pain threshold (Benedetti 2011; Elsenbruch et al. 2012). However, whereas anxiolytic drugs, such as benzodiazepines, inhibit both processes, proglumide, a colecystokinin antagonist, blocks only hyperalgesia (Benedetti et al. 2007; Flaten et al. 2011). These results clearly show the importance of anxiety in pain processing as well as the speciality of pain processing. How expectation-generated anxiety turns into other complaints unrelated to pain, such as nausea, dizziness, or fatigue, is still to be explored.

Implicit or explicit expectations of side effects can also be evoked by perceptual and other nonspecific characteristics of medication. For example, different colors and size of pills evoked different expectations which led to specific side effects in our studies (Köteles and Bárdos 2011; Köteles et al. 2011). Also, change to a new dosage form with the same active substance led to a marked increase (from 26% to 60%) of adverse events (Uhlenhuth et al. 1998). Generic drugs often evoke more nocebo-related side effects due to the negative expectations associated with them (Weissenfeld et al. 2010; Faasse et al. 2012a, 2015; Faasse and Petrie 2013; Colgan et al. 2015).

Conditioning is also an often-mentioned mechanism of the development of side effects. The pioneering study of Ader and Cohen (1975), which demonstrated the possibility of influencing the immune system by conditioning in animals, actually evoked a nocebo-like response, i.e., the learned suppression of the immune reaction, which was beneficial under the given circumstances (autoimmune condition). Conditioned hyperalgesia was demonstrated in newborn infants (Taddio et al. 2002);

moreover, physiological (i.e., consciously not accessible) effects of drugs, such as respiratory depression, can also be learned (Benedetti et al. 1999). If one does not take into consideration the cognitive aspects of conditioning (Rescorla and Wagner 1972), these findings might represent the purest possible conditioned responses. In adult humans, however, conditioned responses show a substantial overlap with expectations. For example, two frequent acute side effects of chemotherapy, nausea and vomiting, are partly evoked by classical conditioning, i.e., contingency-based learning (Bovbjerg 2006; Stockhorst et al. 2007). As people often realize conditioned effects and develop conscious expectations too, it is not surprising that the same side effects are also predicted by expectations (Colagiuri and Zachariae 2010). In accordance with the laws of classical conditioning, more experience (i.e., more pairing episodes) leads to more robust and long-lasting learning than less trials (Colloca et al. 2010). In a similar vein, a continuous reinforcement schedule evokes stronger effects than partial reinforcement (Colagiuri et al. 2015). However, these findings can also be explained by more firm expectations developed during a longer or more consistent learning process. These issues neatly demonstrate the difficulties posed by the theoretical separation of the two mechanisms in humans (Stewart-Williams and Podd 2004; Stewart-Williams 2004). In placebo studies, the term conditioning often simply refers to expectations developed on the basis of personal experience for consciously accessible changes, such as pain (Voudouris et al. 1985, 1990; Montgomery and Kirsch 1997). Conditioned responses without the induction of explicit expectations were evoked in a recent study (Bräscher et al. 2017); here, the self-reported and behavioral aspect of the conditioned response showed a dissociation, which highlights the importance of the study of pure conditioned responses. In contrast, expectations can be developed via social processes, such as suggestion (verbal or nonverbal communication) and observation. For example, hyperalgesia as well as symptoms attributed to an environmental agent or medication can be evoked via social observational learning (Mazzoni et al. 2010; Swider and Bąbel 2013; Vögtle et al. 2013; Faasse et al. 2018).

Another example for social transmission of symptoms is mass hysteria or mass psychogenic illness (Sirois 1974; Colligan and Murphy 1979; Page et al. 2010). In this case, individuals reach an altered state of consciousness due to high level of uncertainty and fear, and involuntarily adapt the response pattern of others around them. We are social beings thus if everyone shows the signs of, say, a chemical disaster, this makes us bias our perception of reality (Lees-Haley and Brown 1992; Bartholomew and Wessely 2002; Balaratnasingam and Janca 2006). In consequence, we will also perceive the symptoms that the claimed accident evokes. This mechanism is clearly different from conditioning and expectations, even if we use these terms in the broadest possible sense. It is a real-time, automatic response, a social phenomenon. Originally, mass psychogenic illness was described as being provoked by the direct observation of the behavior of other members of the group, for example, in schools and work settings (Colligan and Murphy 1979; Bartholomew and Wessely 2002). Recently, in the era of mass (social) media, both printed and electronic, personal presence is not necessary any more. Media reports on side effects of medication or claimed exposure to chemicals or electromagnetic radiation can easily trigger social contagions for much more people (Faasse et al. 2009, 2012b; MacKrill et al. 2019). Considering the impact of social media, in my view these cases can be explained partly by expectations and partly by social contagion.

Furthermore, the experience and reporting of medication side effects can rely on another social process, i.e., various aspects of the doctor–patients relationship (Colloca and Miller 2011a; Häuser et al. 2012a, b). Just like mass psychogenic illness, this is not a completely expectation-driven phenomenon. Nonverbal communication can include the development of adverse reactions; for example, if the doctor is concerned about side effects, the patient can implicitly or explicitly decode this (Evans and Rogers 2003; Planès et al. 2016). Such unintended messages can lead to the development of implicit or explicit expectations and also to the increase of uncertainty and anxiety. However, side effects can also indicate the general dissatisfaction of the patient with the doctor or the treatment; in other words, they can serve as the expression of therapeutic resistance, primarily based on affective-evaluative processes (Downing and Rickels 1967; Barsky 1983; Barsky et al. 2002; Thompson 2005; Colloca and Barsky 2020). For example, those having no choice with respect to the medication they have to take (this is an indicator of a paternalistic doctor–patient relationship) report more nocebo side effects than those who could choose between two, pharmacologically equivalent, pills (Bartley et al. 2016).

Sex and certain trait-like (temporally stable) individual characteristics also impact the sensitivity to nonspecific side effects. Females and those who are prone to somatosensory amplification, somatic symptom distress, somatic style, and negative affect/neuroticism/anxiety (these are substantially overlapping categories) are characterized by high level of reported side effects (Watson and Pennebaker 1989; Pennebaker 1994; Barsky et al. 1999, 2002; Wendt et al. 2014; Kern et al. 2020). In our own studies, the role of somatosensory amplification proved to be significant in both laboratory and clinical settings. For example, it predicted the number of side effects participants expected (Köteles et al. 2011); this expectation in turn predicted experienced symptoms after they had taken a white or red painkiller with the same active substance (Köteles and Bárdos 2011). In a similar vein, somatosensory amplification tendency predicted side effects reported by patients who started their antihypertensive medication even after 4 weeks (Doering et al. 2016), and it was consistently associated with electromagnetic hypersensitivity and related symptom reports in a number of studies (Köteles et al. 2013, 2019; Szemerszky et al. 2015b; Dömötör et al. 2016). Also, medicine-related beliefs and modern health worries foretold nonspecific symptoms in several experimental studies (Horne et al. 2012; Witthöft and Rubin 2013; Webster et al. 2018). Interestingly it turned out that the proneness to nocebo reactions is partly determined by genetics, i.e., the catechol-*O*-methyltransferase (COMT) gene (Val158Met) (Wendt et al. 2014). Overall, although temporally stable characteristics play a role, the impact of situational factors on the nocebo response appears more pronounced.

Finally, a number of confusing factors should also be taken into consideration. For example, fabrication of symptoms is a viable option if it is associated with possible financial or other benefits (secondary gains) (Lees-Haley and Brown 1992; Köteles and Bárdos 2009). Also, the assessment of symptoms impacts symptom reports: symptom lists act as reminders or suggestions thus they lead to the reporting of more symptoms compared to open questions (Avery et al. 1967; Rosenzweig et al. 1993; Barsky et al. 2002; Rief et al. 2009; Colloca and Miller 2011b). Finally, seemingly nonspecific side effects can also be evoked by additional components of medication and placebo pills; in these cases, specific (pharmacologic) effects are mistakenly regarded as nocebo symptoms (Golomb 1995, 1996, 2002). For example, it was reported as early as in 1933 that a proportion of patients who were administered with lactose-based placebo pills showed nausea, diarrhea, and other gastrointestinal symptoms (Diehl 1933). Today, we would assume lactose intolerance but not psychogenic symptoms in the background.

An obvious question is the applicability of models of symptom perception, presented in Chap. 7, to the nocebo phenomenon. Symptoms of nocebo origin can be considered a subcategory of symptom reports; alternatively, symptom reports can be regarded as an instance of the nocebo phenomenon. Models of symptom perception dominantly focus on complaints that are relatively stable over time, thus they emphasize the role of internal characteristics, such as body focus and negative affect. On the other hand, the nocebo approach primarily assumes external (i.e., social-cognitive) factors in the background of usually transient physical symptoms (Ongaro and Kaptchuk 2019). In doing this, it takes more possible mechanisms into consideration. Idiopathic environmental intolerances might represent a boundary condition as they show nocebo-like characteristics (Rubin et al. 2010; Köteles et al. 2013). Their psychogenic (top-down) origin is much more likely than a biophysical (bottom-up) mechanism (Staudenmayer et al. 2003a, b). Here, the symptoms are attributed to an external factor and depend on the perceived presence of that factor.

It is important to see that people's vulnerability to nocebo messages is increased in certain situations. In medical settings, this is mainly due to the affective changes evoked by the insight of being sick and the accompanied uncertainty, stress, and anxiety (see below). In mass hysteria, including the internet-based version, affective contagion might play the same role (Staudenmayer et al. 2011; Chapman et al. 2013). Thus, classic models of symptom perception should be completed with additional factors and processes to be able to handle the complexity of the nocebo phenomenon in the broader sense. As per usual, the predictive processing approach might be able to explain the development and maintenance of nocebo responses (Büchel et al. 2014; Van den Bergh et al. 2017a, b; Ongaro and Kaptchuk 2019).

9.4 Consequences: Suffering and Nonadherence

The typical nocebo symptoms are unpleasant body sensations with unclear (if any) peripheral background. Also, they are usually transient; they disappear within a couple of hours, at most days. However, certain nocebo reactions can be much more serious (Voelker 1996). For example, a participant of an antidepressant drug trial overdosed the medication and showed such a serious drop in blood pressure that intravenous fluids were necessary to maintain an adequate blood pressure. When it turned out that the person actually belonged to the placebo arm of the trial, the adverse reaction disappeared (Reeves et al. 2007). Wolf and Pinsky (1954) also reported several quite serious reactions, such as dermatitis, urticaria, and diarrhea, for the minority of patients taking placebos in an early clinical trial. Various skin-related and respiratory allergic conditions belong to the most often reported, objectively assessable symptoms (Dekker et al. 1957; Liccardi et al. 2004; Lombardi et al. 2008). I think that the importance of this option (i.e., nocebo symptoms with peripheral changes) is underestimated and would deserve way more research attention. As the origin of the symptoms is clearly top-down, the majority of researchers do not even consider this possibility.

Nocebo sensations and reactions are unpleasant; in other words, they cause suffering. It is important to see that the very same unpleasant body sensations (adverse reactions) can be elicited pharmacologically by an active substance and psychologically by various top-down mechanisms. Nobody, including the individual experiencing these feelings, can make a reliable distinction between the two (this is a consequence of the tenet of realness). In a clinical setting, when a patient reports such a reaction, the clinician cannot be sure about the background either. The assumption that, say, a headache is of nocebo origin is not necessarily valid; it can actually be an idiopathic pharmacological reaction. The probability of nocebo origin can appear much larger for certain symptoms (particularly if the characteristics of the patient are also known to the doctor and support this option), but this is only an intuitive guess. From a statistical point of view, an adverse reaction of an individual to any treatment, such as medication, curse, or scary media report on the adverse effects of 5G, is an elementary event. In fact, we should not even call it reaction (i.e., a response to an event), as the only thing we know is that the treatment preceded the negative condition. This does not necessarily prove causality in an individual case (Gøtzsche 1994). To shed light to a causal relationship, randomization, manipulation of the factor in question, and, most importantly systematic comparison of groups or multiple measurements with the same individual are needed. Thus, the best thing a clinician can do is trying to ameliorate the symptoms, sometimes using another medication with another possible (pharmacological and non-specific) side effects (Barsky et al. 2002). Although such a secondary treatment appears inappropriate, even an overkill, at first sight (to sum the situation up, possibly psychogenic symptoms are treated with pharmacotherapy that in turn can cause further side effect), considering the uncertainty with respect to the background of the symptoms it appears to be the safest solution. Not only in the short run, but also if we consider the long-term consequences of adverse reactions. Placebo treatment would be another option, but it is ethically unacceptable in most countries (see Chap. 10). Simultaneously, information on the nocebo phenomenon can also be given to the patient, although obviously this is not the best time for patient education. Just like in the case of somatization and idiopathic environmental intolerances (Chap. 7), most people might find the idea that their very concrete and localized somatic sensations (remember the tenet of realness again) are of psychogenic origin ("only in their mind") simply embarrassing.

Beyond unpleasantness and suffering, specific and nonspecific medication side effects have a more serious consequence: they weaken the patient's commitment to the therapy (Kardas et al. 2013). This leads to adherence-related issues, such as self-determined reduction of the dosage and quitting the treatment, both in clinical trials and under everyday circumstances (Barsky 1983; Preston et al. 2000; Mitsikostas et al. 2011; Papadopoulos and Mitsikostas 2012; Häuser et al. 2012a; Sanderson et al. 2013; Wendt et al. 2014). Thus, the effectiveness of the treatment for the individual will be reduced at best. Such learned effects can impact future treatments too (Kessner et al. 2013). Moreover, as certain characteristics make individuals more prone to adverse reactions, dropouts in clinical trials do not occur in a random fashion. In technical terms, this leads to a bias in the sample, which reduces the generalizability of the findings of that trial (Colloca and Miller 2011a; Häuser et al. 2012b). This impacts future patients too.

Concerning clinical trials, things are even more complicated. As mentioned, the invention and use of placebo control represented a substantial step toward the measurement of pharmacological therapeutic effects of drugs. It was (or at least appeared to be) a methodologically perfect idea: there are two groups with randomly assigned participants who uniformly think that they receive the active substance or the placebo (Shapiro and Shapiro 1997). As the trial is double-blind, the medical stuff is not aware of the assignment either thus no group-specific demand characteristics arise. At the end point(s) of the clinical trial, the statistician just needs to subtract the mean improvement of the placebo arm from that of the active arm to calculate the specific effectiveness of the drug. Unfortunately, drug side effects, both specific and nonspecific, substantially disturb this clear picture. It turned out that both participants and physicians are able to determine above chance level who belongs to the active arm and who to the placebo arm, often based on the side effect profile of the drug (Rabkin et al. 1986; White et al. 1992; Fischer and Greenberg 1997; Kirsch 2010). Also, patients who do not experience side effects are prone to believe that they take placebo, which can reduce the effectiveness of the drug in multiple ways. How do they know which side effects would likely occur? From the mandatory information given to them before the decision to participate in a trial (Myers et al. 1987; Rief et al. 2009; Mitsikostas 2012). In such a situation, adverse side effects can be desirable for patients as they carry the message that they receive an effective medication (Thomson 1982). This will boost the placebo effect in the placebo group which in turn reduces the difference between the two groups, i.e., the measured effectiveness of the drug (Berthelot et al. 2001). On the other hand, as mentioned, the appearance of side effects is scaring for many patients (both in clinical trials and in the course of normal treatment), thus it can make them quit the trial or the treatment (Myers et al. 1987; Barsky et al. 2002; Rief et al. 2006; Colloca and Barsky 2020). These complex interactions are very difficult to understand and take into consideration.

9.5 Nocebo Responses Cannot Be Eliminated Just Reduced

The nocebo phenomenon is not without serious ethical issues. One category of these issues arises from a conflict between two fundamental principles of medical ethics. On the one hand, any harm should be avoided during medical treatment ("first, do no harm"; this principle was already part of the Hippocratic Oath); on the other hand, the patients should be provided with all information that is necessary for an informed decision about the treatment (Brody 1980; Bok 2002; Fortunato et al. 2017). The possible side effects of the treatment obviously represent important pieces of information that should not be concealed. Thus, mandatory information before participating in a clinical trial as well as patient information leaflets summarizing the most important things about medications already shape expectations with respect to side effects (Flaten et al. 1999; Mondaini et al. 2007; Colloca and Miller 2011a). In other words, just like the placebo, it acts as a self-fulfilling prophecy (Miller and Colloca 2009; Ren and Xu 2018). Of course, this happens not only in clinical trials but other aspects of life. For example, informing the population about a scheduled pesticide spraying that might be harmful is mandatory from an ethical point of view (Fig. 9.1); on the other hand, it can have negative effects too (Petrie et al. 2005). Also, if one wants to experimentally study the nocebo phenomenon in the lab, the manipulation of expectations, including the generation of negative ones, cannot be avoided.

These issues can more or less easily be solved in experiments with healthy participants: they are informed in advance that certain aspects of the experiment cannot be revealed, they can quit at any time, and they can receive information about the concealed details following the experiment. Patients are a different story though. Being sick often means an altered state of consciousness, particularly in life-threatening situations (Häuser et al. 2012b). We are more emotional than usual; more precisely, the relative importance of primary (affective-experiential) processes increases at the expense of rational processes. This increases our receptiveness or vulnerability to suggestions (Bejenke 2011; Häuser et al. 2012b; Varga 2013); actually, the boundary between information and suggestion is very vague in this state.

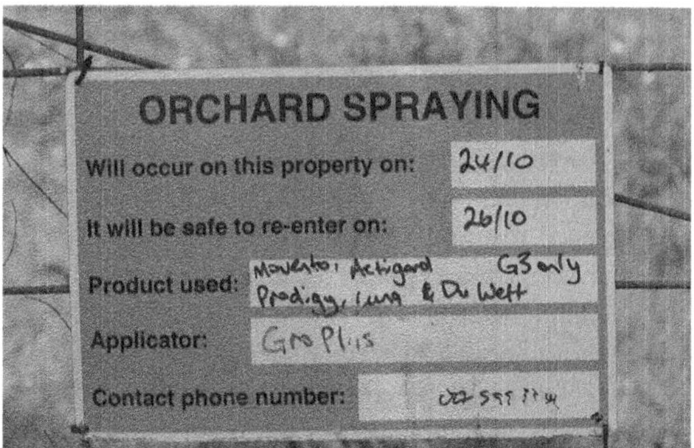

Fig. 9.1 A table presenting safety-related information at the entry of a kiwifruit orchard in Glenbrook, New Zealand

The list of possible side effects represents very concrete suggestions as it mainly consists of body sensations that also belong to primary consciousness. The rational component of the information (e.g., the relative frequency of the symptoms) is processed less precisely as it requires the rational system and effortful thinking. Moreover, symptom-related information evokes negative affective reactions, such as health anxiety, which produce further body sensations that can be ascribed to the medication. On top of this, the topic is much more well-known and popular these days than in the past because of the impact of mass media. Mass media reports on side effects often present individual stories (case studies), based on personal beliefs and attributions, full of emotions—this is the perfect input for experiential processing (Köteles et al. 2016a, b; Dömötör et al. 2019).

What to do? Over time, a number of communication-related techniques were proposed to reduce the amount and severity of nocebo adverse reactions. One option is accepting the experienced unpleasant sensations but reframing them. For example, they can be regarded as indicators of the effectiveness of the drug ("non-nocebo response") (Evans 2003). The presentation of side effects together with the therapeutic effects of the medication can also be helpful (Colloca and Finniss 2012; von Blanckenburg et al. 2013; Heisig et al. 2015). Also, the likelihood of the occurrence of the adverse reactions can be presented in a less threatening manner as the likelihood of not being experienced (O'Connor et al. 1996; Colloca and Finniss 2012; Williams et al. 2013; Enck et al. 2013; Webster et al. 2018; Ren and Xu 2018).

Others recommend a so-called contextualized informed consent, which presents the possible side effects in less detail (Wells and Kaptchuk 2012). An even more drastic solution is dubbed authorized concealment, when patients can voluntarily reject information on certain side effects (Colloca and Miller 2011a). These techniques attempt to find an ethically appropriate balance between information given to the patients and the self-fulfilling nature of nocebo effects: the more information the patient has, the more side effect will likely occur.

Improving both the affective and the informative aspects of doctor–patient communication might also be helpful (Bingel and Placebo Competence Team 2014; Widdershoven et al. 2017; Ren and Xu 2018). The nocebo phenomenon is less known than the placebo (Berthelot et al. 2001; Planès et al. 2016). Therefore, education of patients, healthy people, as well as medical professionals on the nocebo phenomenon is also recommended as it can substantially reduce the number of side effects (Bennett et al. 2011; Häuser et al. 2012b; Colloca and Finniss 2012; Faasse and Petrie 2013; Crichton and Petrie 2015; Pan et al. 2019).

However, as mentioned in Chap. 1, human perception and healthy functioning relies on expectations (I use this term in the broadest possible sense now, including conditioned responses, priors, and everything else that refers to future states). Our brain always tries to prepare for future events, based on previous experiences and the available information (Hohwy 2014; Clark 2015). Beyond the explicit information, the perceptual characteristics (color, size, shape) of medication, the doctor–patient relationship, and many other factors, not necessarily conscious, shape these expectations. We could not live without expectations; even the simple act of drinking a glass of water cannot be completed without a number of expectations, from movement-related predictions to more conceptual beliefs on the association between thirst and water. The brain is a prediction machine, and the benefits of using this way of functioning cannot be compared to the negative aspects. In other words, the nocebo phenomenon can be considered the side effect of the predictive way the brain operates in. Findings of a recent study illustrate this point very well. It was well-known that price of the medication improves the placebo response due to the heightened expectations with respect to its effectiveness (Waber et al. 2008). However, it turned out that the same factor makes the nocebo response (hyperalgesia) worse, too (Tinnermann et al. 2017). Thus, positive and negative expectations often go hand in hand and cannot be separated. This is why I am firmly convinced that nonspecific medication side effects and other nocebo responses cannot completely be eliminated. They can be ameliorated and reduced by certain techniques, but we cannot get rid of them.

9.6 Issues Due to the Lack of Body Sensations

Unpleasant body sensations, most importantly pain, primarily serve as warning signals. They indicate that something is wrong in our physiological machinery, our homeostasis is threatened, or the use of a body part should be minimized because of an injury. These originally useful warning signals can be self-maintaining thus maladaptive or loss their informative value in certain cases. We do not like them: they are disturbing, they cause inconvenience, even suffering, thus our automatic reaction is eliminating them at any cost. In the light of the previous sections of this chapter, it may be surprising that the lack of body sensations can sometimes be even more dangerous.

There are chronic, usually slowly progressing conditions which cannot be detected by the human nervous system in their early phase. Over time, they can evoke symptoms; however, the damage is already there by then and often cannot be reversed any more. Some of these diseases can be lethal, therefore they are also called silent killers. For example, malignant tumors or chronic cardiovascular conditions, such as hypertension and atherosclerosis, belong to this category. There are at least two issues with these so-called asymptomatic diseases that belong to our topic. First, we are often not aware of them at all. Second, even if we learn them (which typically requires a time-consuming and often unpleasant medical diagnostic procedure), the lack of symptoms easily leads to the lack of the insight that we are sick; we might know it but do not feel it. This reduces our motivation to change our behavior, e.g., taking medication as prescribed by the physician or changing the risky components of our lifestyle, from bad dietary habits to the lack of physical activity (Working Party from compliance to concordance 1993; Costello et al. 2008; Kardas et al. 2013). As body sensations belong to the domain of primary consciousness, their actual presence or absence impacts us much deeper than abstract knowledge on possible pathophysiological processes and their long-term consequences (Cioffi 1991).

This phenomenon is perhaps best understood with respect to hypertension. As it was shown in Chap. 4, this condition is not consistently associated with symptoms. In consequence, although it leads to serious cardiovascular degeneration in the long run, many patients are not willing to take their medication. In fact, approximately one-half of the patients do not take the prescribed medication at all, and one-third of the rest take it in a subtherapeutic dose which is insufficient to achieve and maintain the desired blood pressure values (Dunbar-Jacob et al. 1991; Vermeire et al. 2001). The long-term consequence of hypertension, coronary artery disease, is also asymptomatic (Kroenke 2003). Osteoporosis is another example: until the first fracture most patients do not believe that such serious consequences can happen. As medication is not associated with the instant reduction of symptoms, people do not want to take it on a regular basis (Gold et al. 2006). With respect to multiple sclerosis, the disease can be diagnosed before the occurrence of symptoms; this makes the acceptance of the necessity of regular self-injection very difficult for patients (Costello et al. 2008). In summary, the lack of symptoms in the case of serious diseases makes adherence to the treatment very difficult, requiring voluntary effort and the acceptance of unpleasant procedures.

9.7 The Ignorance of Body Sensations

Sometimes the warning signals are there but people ignore them. In fact, we are able to suppress unpleasant body sensations, including symptoms of diseases, or at least distract our attention to some extent. For example, we can completely immerse in our job or various aspects of our activity or the external environment. In these cases, the bothersomeness of our somatic symptoms decreases. Right now, I am sitting with a light pain in my lower back, but do not want to stand up and cease typing because my mind is still active, and the deadline of the submission of this work is approaching. Such decisions can be an adaptive strategy in the short run; however, they can have negative health-related consequences in the long run (Pennebaker 1982). Also, the proneness to minimize somatic sensations (the opposite of somatosensory amplification) can also lead to the ignorance of symptoms (Barsky 1979) that might indicate serious pathology. The malady may already be present—in this case, the ignorance leads to delays in diagnosis and treatment and the possible progression of the disease, which usually means a worse outcome. Sometimes realizing the first signs of a migraine attack can be helpful to make it less severe and long via preventive behavioral steps (Bakal 1999). In other cases, lack of attention and the necessary behavioral steps actively contribute to the progression. For example, sedentary lifestyle and sitting too much in anatomically not optimal positions can lead to chronic low back pain and even herniated disks. If we realize the warning signals, such as ache and stiffness of the lower part of the spinal cord, we can decrease the risk by regularly changing our position while sitting, inserting short breaks, and increasing our overall physical activity (Koes et al. 2006; Shiri et al. 2019).

The Western culture, focusing on office work, rules, schedules, and effectiveness, presses us to the suppression of primary conscious materials, such as body sensations (Leder 2018). Unfortunately, other primary processes, for example, feeling healthy or having the impression that we were cured can further decrease adherence and possible preventive steps (Kardas et al. 2013). These issues will be discussed in Chap. 10.

9.8 Summary

This chapter focuses on two specific negative aspects of body sensations: sometimes they are caused by top-down processes even in the absence of peripheral changes and sometimes they are lacking (or suppressed) when they should be there and indicate disease-related body changes. Both conditions are closely related to the tenet of dissociation; however, the steps they necessitate are seemingly the opposite. Concerning nocebo symptoms, the best thing to do would be understanding that they refer to brain as opposed to body states and doing nothing else for a while unless their severity is not acceptable. As of lacking sensations, counter-intuitive actions are required. Although we do not feel anything threatening in our body, if we have a good reason to suspect pathology, e.g., family background or previous own issues with a disease, we should regularly check it and accept the reality if the diagnostic procedures reveal something. Better safe than sorry.

How can we resolve this apparent conflict and determine the really adaptive steps? The answer is quite simple but difficult to follow: as in other matters in life, we should use our rational system too instead of relying exclusively on the experiential-intuitive system (Epstein 2014). Education (both general information on the nocebo phenomenon and disease and medication-specific knowledge) appears to be a promising path (Colloca and Finniss 2012; von Blanckenburg et al. 2013, 2015; Faasse and Petrie 2013; Crichton and Petrie 2015; Planès et al. 2016; Pan et al. 2019). I think personal experience is of crucial importance beyond providing people with more knowledge on these topics. Personal experience directly impacts primary consciousness, thus it powerfully complements and supports abstract knowledge. Demonstration of some instances of the discordance between actual and perceived body processes, such as heart rate, blood pressure, or electrodermal activity, is quite easy. It can be conducted with inexpensive and widely available devices, takes little time and requires minimal technical knowledge from the instructor. This form of education would be useful for children as well as for patients with a proneness to the nocebo reaction.

Take-Home Messages

For laypeople: Transient symptoms occurring after starting a new medication might be drug side effects or due to the nocebo phenomenon. Never quit the therapy or reduce the dosage without consulting with your physician. Also, bombastic stories broadcast by the mass and social media on the risks of medicines and various environmental factors can cause such nocebo symptoms that usually disappear soon. On the other hand, regularly occurring or long-lasting symptoms need medical supervision even if they do not seem to indicate serious disease.

For clinicians: There is no way to completely separate specific and nonspecific side effects of medical interventions in individual cases. Brief questionnaires, such as the Somatosensory Amplification Scale (Barsky et al. 1990) or the Perceived Sensitivity to Medicines Scale (Horne et al. 2012), can help to assess the proneness of the patient to nonspecific symptoms. Still, severe and disturbing symptoms, regardless of the origin, might need pharmacological treatment and a critical look at the doctor–patient relationship.

For researchers: I see three important but largely neglected issues with respect to the nocebo phenomenon. First, just because a symptom is of top-down origin, it does not mean that it does not have a peripheral background. For example, stress and anxiety can be accompanied by physiological changes. Studies that investigate the possible peripheral effects of nocebo interventions are scarce; existing studies typically remain at the level of the group. The systematic investigation of the prevalence of peripheral effects (e.g., the proportion of individuals impacted, the severity and replicability of the condition) is of huge practical importance. If it turns out that a frequency of such reactions is negligible, this finding will help clinicians in the determination of treatment. On the other hand, the information that certain changes do take place is also significant. For example, even relatively slight increases of blood pressure and heart rate can reach the limits of regulation for patients with chronic cardiovascular disease; for them, a stress-inducing message can be fatal (Lovallo 2016). Perhaps a nocebo can kill people, as reported by Cannon, even in the modern world under certain circumstances. Second, the role of explicit (i.e., assessed via self-report) expectations should be reconsidered in a more systematic way. Expectations sometimes forecast future outcome. For example, the so-called stress mindset, i.e., the individual's beliefs about the harmful effects of stress, predicts the stress response (Crum et al. 2013), and it proved to be a significant predictor of mortality (independently from the actual physical condition) in longitudinal studies (Keller et al. 2012; Nabi et al. 2013). This does not necessarily mean a causal effect though. On the other hand, receiving information about genetic risk factors evoked physiological changes consistent with the expected risk profile in a recent study (Turnwald et al. 2019). Clarifying the role of expectations would help to explain their background mechanisms in more detail. Third, it is important to understand that the nocebo phenomenon is not the exact opposite of the placebo. Even in pain studies, when threshold and/or tolerance are shifted in the opposite directions by top-down information, the activation pattern of the brain shows considerable differences (Benedetti et al. 2006; Kong et al. 2008; Freeman et al. 2015).

References

Ader R, Cohen N (1975) Behaviorally conditioned immunosuppression. Psychosom Med 37:333–340

Amanzio M, Corazzini LL, Vase L, Benedetti F (2009) A systematic review of adverse events in placebo groups of anti-migraine clinical trials. Pain 146:261–269. https://doi.org/10.1016/j.pain.2009.07.010

Avery CW, Ibelle BP, Allison B, Mandell N (1967) Systematic errors in the evaluation of side effects. Am J Psychiatry 123:875–878

Bakal D (1999) Minding the body: clinical uses of somatic awareness. Guilford Press, New York

Balaratnasingam S, Janca A (2006) Mass hysteria revisited. Curr Opin Psychiatry 19:171–174. https://doi.org/10.1097/01.yco.0000214343.59872.7a

Barsky AJ (1979) Patients who amplify bodily sensations. Ann Intern Med 91:63–70. https://doi.org/10.1059/0003-4819-91-1-63

Barsky AJ (1983) Nonpharmacologic aspects of medication. Arch Intern Med 143:1544–1548

Barsky AJ (1988) The paradox of health. N Engl J Med 318:414–418. https://doi.org/10.1056/NEJM198802183180705

Barsky AJ, Wyshak G, Klerman GL (1990) The Somatosensory Amplification Scale and its relationship to hypochondriasis. J Psychiatr Res 24:323–334. https://doi.org/10.1016/0022-3956(90)90004-A

Barsky AJ, Orav EJ, Ahern DK et al (1999) Somatic style and symptom reporting in rheumatoid arthritis. Psychosomatics 40:396–403

Barsky AJ, Saintfort R, Rogers MP, Borus JF (2002) Nonspecific medication side effects and the nocebo phenomenon. JAMA 287:622–627

Bartholomew RE, Wessely S (2002) Protean nature of mass sociogenic illness: from possessed nuns to chemical and biological terrorism fears. Br J Psychiatry 180:300–306. https://doi.org/10.1192/bjp.180.4.300

Bartley H, Faasse K, Horne R, Petrie KJ (2016) You can't always get what you want: the influence of choice on nocebo and placebo responding. Ann Behav Med 50(3):445–451. https://doi.org/10.1007/s12160-016-9772-1

Beecher HK (1956) Evidence for increased effectiveness of placebos with increased stress. Am J Phys 187:163–169

Beecher HK (1960) Increased stress and effectiveness of placebos and "active" drugs. Science 132:91–92

Beedie C, Benedetti F, Barbiani D et al (2018) Consensus statement on placebo effects in sports and exercise: the need for conceptual clarity, methodological rigour, and the elucidation of neurobiological mechanisms. Eur J Sport Sci 18:1383–1389. https://doi.org/10.1080/17461391.2018.1496144

Bejenke CJ (2011) Suggestive communication: its wide applicability in somatic medicine. A clinician's 35 year observations, teaching and practice. In: Varga K (ed) Beyond the words: communication and suggestion in medical practice. Nova Science Publishers, New York, pp 83–96

Benedetti F (2009) Placebo effects: understanding the mechanisms in health and disease. Oxford University Press, New York

Benedetti F (2011) The patient's brain. The neuroscience behind the doctor-patient relationship. Oxford University Press, New York

Benedetti F, Amanzio M, Baldi S et al (1999) Inducing placebo respiratory depressant responses in humans via opioid receptors. Eur J Neurosci 11:625–631

Benedetti F, Amanzio M, Vighetti S, Asteggiano G (2006) The biochemical and neuroendocrine bases of the hyperalgesic nocebo effect. J Neurosci 26:12014–12022. https://doi.org/10.1523/JNEUROSCI.2947-06.2006

Benedetti F, Lanotte M, Lopiano L, Colloca L (2007) When words are painful: unraveling the mechanisms of the nocebo effect. Neuroscience 147:260–271. https://doi.org/10.1016/j.neuroscience.2007.02.020

Bennett MI, Bagnall A-M, Raine G et al (2011) Educational interventions by pharmacists to patients with chronic pain: systematic review and meta-analysis. Clin J Pain 27:623–630. https://doi.org/10.1097/AJP.0b013e31821b6be4

Bérdi M, Köteles F, Szabó A, Bárdos G (2011) Placebo effects in sport and exercise: a meta-analysis. Eur J Mental Health 6:196–212. https://doi.org/10.5708/EJMH.6.2011.2.5

Berthelot JM, Maugars Y, Abgrall M, Prost A (2001) Interindividual variations in beliefs about the placebo effect: a study in 300 rheumatology inpatients and 100 nurses. Joint Bone Spine 68:65–70

Bingel U, Placebo Competence Team (2014) Avoiding nocebo effects to optimize treatment outcome. JAMA 312:693–694. https://doi.org/10.1001/jama.2014.8342

Bingel U, Wanigasekera V, Wiech K et al (2011) The effect of treatment expectation on drug efficacy: imaging the analgesic benefit of the opioid remifentanil. Sci Transl Med 3:70ra14. https://doi.org/10.1126/scitranslmed.3001244

Bok S (2002) Ethical issues in use of placebo in medical practice and clinical trials. In: Guess HA, Kleinman A, Kusek JW, Engel LW (eds) The Science of the placebo. Toward an interdisciplinary research agenda. BMJ Books, London, pp 53–74

Bovbjerg DH (2006) The continuing problem of post chemotherapy nausea and vomiting: contributions of classical conditioning. Auton Neurosci 129:92–98. https://doi.org/10.1016/j.autneu.2006.07.016

Bräscher A-K, Kleinböhl D, Hölzl R, Becker S (2017) Differential classical conditioning of the nocebo effect: increasing heat-pain perception without verbal suggestions. Front Psychol 8:2163. https://doi.org/10.3389/fpsyg.2017.02163

Brody H (1980) Placebos and the philosophy of medicine. Clinical, conceptual, and ethical issues. University of Chicago Press, Chicago

Büchel C, Geuter S, Sprenger C, Eippert F (2014) Placebo analgesia: a predictive coding perspective. Neuron 81:1223–1239. https://doi.org/10.1016/j.neuron.2014.02.042

Butler C, Steptoe A (1986) Placebo responses: an experimental study of psychophysiological processes in asthmatic volunteers. Br J Clin Psychol 25(Pt 3):173–183

Cannon WB (1942) "Voodoo" death. Am Anthropol 44:169–181. https://doi.org/10.1525/aa.1942.44.2.02a00010

Cannon WB (1957) Voodoo death. Psychosom Med 19:182–190

Chapman S, George AS, Waller K, Cakic V (2013) The pattern of complaints about Australian wind farms does not match the establishment and distribution of turbines: support for the psychogenic, 'communicated disease' hypothesis. PLoS One 8:e76584. https://doi.org/10.1371/journal.pone.0076584

Cioffi D (1991) Beyond attentional strategies: cognitive-perceptual model of somatic interpretation. Psychol Bull 109:25–41

Claassen L, Smid T, Woudenberg F, Timmermans DRM (2012) Media coverage on electromagnetic fields and health: content analysis of Dutch newspaper articles and websites. Health Risk Soc 14:681–696. https://doi.org/10.1080/13698575.2012.716820

Clark A (2015) Surfing uncertainty: prediction, action, and the embodied mind, 1st edn. Oxford University Press, Oxford, England

Clements CJ (2003) Mass psychogenic illness after vaccination. Drug Saf 26:599–604

Cocco G (2009) Erectile dysfunction after therapy with Metoprolol: the Hawthorne effect. Cardiology 112:174–177. https://doi.org/10.1159/000147951

Colagiuri B, Zachariae R (2010) Patient expectancy and post-chemotherapy nausea: a meta-analysis. Ann Behav Med 40:3–14. https://doi.org/10.1007/s12160-010-9186-4

Colagiuri B, Quinn VF, Colloca L (2015) Nocebo hyperalgesia, partial reinforcement, and extinction. J Pain 16:995–1004. https://doi.org/10.1016/j.jpain.2015.06.012

Colgan S, Faasse K, Martin LR et al (2015) Perceptions of generic medication in the general population, doctors and pharmacists: a systematic review. BMJ Open 5:e008915. https://doi.org/10.1136/bmjopen-2015-008915

Colligan MJ, Murphy LR (1979) Mass psychogenic illness in organizations: an overview. J Occup Psychol 52:77–90. https://doi.org/10.1111/j.2044-8325.1979.tb00444.x

Colloca L, Barsky AJ (2020) Placebo and nocebo effects. N Engl J Med 382:554–561. https://doi.org/10.1056/NEJMra1907805

Colloca L, Finniss D (2012) Nocebo effects, patient-clinician communication, and therapeutic outcomes. JAMA 307:567–568. https://doi.org/10.1001/jama.2012.115

Colloca L, Miller FG (2011a) The nocebo effect and its relevance for clinical practice. Psychosom Med. https://doi.org/10.1097/PSY.0b013e3182294a50

Colloca L, Miller FG (2011b) Harnessing the placebo effect: the need for translational research. Philos Trans R Soc B Biol Sci 366:1922–1930. https://doi.org/10.1098/rstb.2010.0399

Colloca L, Petrovic P, Wager TD et al (2010) How the number of learning trials affects placebo and nocebo responses. Pain 151:430–439. https://doi.org/10.1016/j.pain.2010.08.007

Costello K, Kennedy P, Scanzillo J (2008) Recognizing nonadherence in patients with multiple sclerosis and maintaining treatment adherence in the long term. Medscape J Med 10:225

Craig AD (2003) A new view of pain as a homeostatic emotion. Trends Neurosci 26:303–307

Crichton F, Petrie KJ (2015) Health complaints and wind turbines: the efficacy of explaining the nocebo response to reduce symptom reporting. Environ Res 140:449–455. https://doi.org/10.1016/j.envres.2015.04.016

Crum AJ, Salovey P, Achor S (2013) Rethinking stress: the role of mindsets in determining the stress response. J Pers Soc Psychol 104:716–733. https://doi.org/10.1037/a0031201

Dekker E, Pelser HE, Groen J (1957) Conditioning as a cause of asthmatic attacks; a laboratory study. J Psychosom Res 2:97–108

Diehl HS (1933) Medicinal treatment of the common cold. JAMA 101:2042–2049. https://doi.org/10.1001/jama.1933.02740510034009

Doering BK, Szécsi J, Bárdos G, Köteles F (2016) Somatosensory amplification is a predictor of self-reported side effects in the treatment of primary hypertension: a pilot study. Int J Behav Med. https://doi.org/10.1007/s12529-016-9536-0

Dömötör Z, Doering BK, Köteles F (2016) Dispositional aspects of body focus and idiopathic environmental intolerance attributed to electromagnetic fields (IEI-EMF). Scand J Psychol 57:136–143. https://doi.org/10.1111/sjop.12271

Dömötör Z, Nordin S, Witthöft M, Köteles F (2019) Modern health worries: a systematic review. J Psychosom Res 124:109781. https://doi.org/10.1016/j.jpsychores.2019.109781

Downing RW, Rickels K (1967) Self-report of hostility and the incidence of side reactions in neurotic outpatients treated with tranquilizing drugs and placebo. J Consult Psychol 31:71–76. https://doi.org/10.1037/h0024213

Dunbar-Jacob J, Dwyer K, Dunning EJ (1991) Compliance with antihypertensive regimen: a review of the research in the 1980s. Ann Behav Med 13:31–39. https://doi.org/10.1093/abm/13.1.31

Dunn AG, Leask J, Zhou X et al (2015) Associations between exposure to and expression of negative opinions about human papillomavirus vaccines on social media: an observational study. J Med Internet Res 17:e144. https://doi.org/10.2196/jmir.4343

Edwards IR, Graedon J, Graedon T (2010) Placebo harm. Drug Saf 33:439–441. https://doi.org/10.2165/11532400-000000000-00000

Elsenbruch S, Kotsis V, Benson S et al (2012) Neural mechanisms mediating the effects of expectation in visceral placebo analgesia: an fMRI study in healthy placebo responders and nonresponders. Pain 153:382–390. https://doi.org/10.1016/j.pain.2011.10.036

Elvers H-D, Jandrig B, Grummich K, Tannert C (2009) Mobile phones and health: media coverage study of German newspapers on possible adverse health effects of mobile phone use. Health Risk Soc 11:165–179. https://doi.org/10.1080/13698570902784273

Enck P, Bingel U, Schedlowski M, Rief W (2013) The placebo response in medicine: minimize, maximize or personalize? Nat Rev Drug Discov 12:191–204. https://doi.org/10.1038/nrd3923

Epstein S (2014) Cognitive-experiential theory. Oxford University Press, Oxford, England

Ernst E, Reschl KL (1995) The concept of the perceived and true placebo effect. Br Med J 311:551–553

Evans RW (2003) The non-nocebo response: can migraine medication efficacy be enhanced? Headache 43:693

Evans RW, Rogers MP (2003) Headaches and the nocebo effect. Headache 43:1113–1115

Faasse K, Petrie KJ (2013) The nocebo effect: patient expectations and medication side effects. Postgrad Med J 89:540–546. https://doi.org/10.1136/postgradmedj-2012-131730

Faasse K, Cundy T, Petrie KJ (2009) Thyroxine: anatomy of a health scare. BMJ 339:20–21. https://doi.org/10.1136/bmj.b5613

Faasse K, Cundy T, Gamble G, Petrie KJ (2012a) The effect of an apparent change to a branded or generic medication on drug effectiveness and side effects. Psychosom Med. https://doi.org/10.1097/PSY.0b013e3182738826

Faasse K, Gamble G, Cundy T, Petrie KJ (2012b) Impact of television coverage on the number and type of symptoms reported during a health scare: a retrospective pre-post observational study. BMJ Open 2:1607. https://doi.org/10.1136/bmjopen-2012-001607

Faasse K, Martin LR, Grey A et al (2015) Impact of brand or generic labeling on medication effectiveness and side effects. Health Psychol. https://doi.org/10.1037/hea0000282

Faasse K, Parkes B, Kearney J, Petrie KJ (2018) The influence of social modeling, gender, and empathy on treatment side effects. Ann Behav Med 52:560–570. https://doi.org/10.1093/abm/kax025

Fischer S, Greenberg RP (1997) The curse of the placebo: fanciful pursuit of a pure biological therapy. In: Fisher S, Greenberg RP (eds) From placebo to pancea - putting psychiatric drugs to the test. Wiley, New York, pp 3–56

Flaten MA, Simonsen T, Olsen H (1999) Drug-related information generates placebo and nocebo responses that modify the drug response. Psychosom Med 61:250–255

Flaten MA, Simonsen T, Zahlsen K et al (2004) Stimulant and relaxant drugs combined with stimulant and relaxant information: a study of active placebo. Psychopharmacology 176:426–434. https://doi.org/10.1007/s00213-004-1886-7

Flaten MA, Aslaksen PM, Lyby PS, Bjørkedal E (2011) The relation of emotions to placebo responses. Philos Trans R Soc B Biol Sci 366:1818–1827. https://doi.org/10.1098/rstb.2010.0407

Fortunato JT, Wasserman JA, Menkes DL (2017) When respecting autonomy is harmful: a clinically useful approach to the nocebo effect. Am J Bioeth 17:36–42. https://doi.org/10.1080/15265161.2017.1314042

Freeman M (2006) Reconsidering the effects of monosodium glutamate: a literature review. J Am Acad Nurse Pract 18:482–486. https://doi.org/10.1111/j.1745-7599.2006.00160.x

Freeman S, Yu R, Egorova N et al (2015) Distinct neural representations of placebo and nocebo effect. NeuroImage. https://doi.org/10.1016/j.neuroimage.2015.03.015

Geha RS, Beiser A, Ren C et al (2000) Review of alleged reaction to monosodium glutamate and outcome of a multicenter double-blind placebo-controlled study. J Nutr 130:1058S–1062S

Gold DT, Alexander IM, Ettinger MP (2006) How can osteoporosis patients benefit more from their therapy? Adherence issues with bisphosphonate therapy. Ann Pharmacother 40:1143–1150. https://doi.org/10.1345/aph.1G534

Golomb BA (1995) Paradox of placebo effect. Nature 375:530. https://doi.org/10.1038/375530a0

Golomb BA (1996) Using placebos. Nature 379:765. https://doi.org/10.1038/379765b0

Golomb BA (2002) When are medication side effects due to the nocebo phenomenon? JAMA 287:2502–2503; author reply 2503–2504

Gøtzsche PC (1994) Is there logic in the placebo? Lancet 344:925–926. https://doi.org/10.1016/S0140-6736(94)92273-X

Green DM (1962) Side effects. Fed Proc 21:179

Green DM (1964) Pre-existing conditions, placebo reactions, and "side effects". Ann Intern Med 60:255–265

Greiner T, Gold H, Cattel M et al (1950) A method for the evaluation of the effects of drugs on cardiac pain in patients with angina of effort; a study of khellin (visammin). Am J Med 9:143–155

Hahn RA (1985) A sociocultural model of illness and healing. In: White L, Thursky B, Schwartz GE (eds) Placebo. Theory, research, and mechanisms. The Guilford Press, New York, pp 167–195
Hahn RA (1997) The nocebo phenomenon: scope and foundations. In: Harrington A (ed) The placebo effect. An interdisciplinary exploration. Harvard University Press, Cambridge, pp 56–76
Hahn RA (1999) Expectations of sickness: concept and evidence of the nocebo phenomenon. In: Kirsch I (ed) How expectancies shape experience. American Psychological Association, Washington, DC, pp 333–356
Häuser W, Bartram C, Bartram-Wunn E, Tölle T (2012a) Adverse events attributable to nocebo in randomized controlled drug trials in fibromyalgia syndrome and painful diabetic peripheral neuropathy: systematic review. Clin J Pain 28:437–451. https://doi.org/10.1097/AJP.0b013e3182321ad8
Häuser W, Hansen E, Enck P (2012b) Nocebo phenomena in medicine. Their relevance in everyday clinical practice. Dtsch Arztebl Int 109:459–465. https://doi.org/10.3238/arztebl.2012.0459
Heisig SR, Shedden-Mora MC, Hidalgo P, Nestoriuc Y (2015) Framing and personalizing informed consent to prevent negative expectations: an experimental pilot study. Health Psychol. https://doi.org/10.1037/hea0000217
Hohwy J (2014) The predictive mind, 1st edn. Oxford University Press, Oxford, England
Honigfeld G (1964) Non-specific factors in treatment. I. Review of placebo reactions and placebo reactors. Dis Nervous Syst 25:145–155
Horne R, Faasse K, Cooper V et al (2012) The perceived sensitivity to medicines (PSM) scale: an evaluation of validity and reliability. Br J Health Psychol. https://doi.org/10.1111/j.2044-8287.2012.02071.x
Hornsey MJ, Harris EA, Fielding KS (2018) The psychological roots of anti-vaccination attitudes: a 24-nation investigation. Health Psychol 37:307–315. https://doi.org/10.1037/hea0000586
Huiberts A, Hjørnevik M, Mykletun A, Skogen JC (2013) Electromagnetic hypersensitivity (EHS) in the media - a qualitative content analysis of Norwegian newspapers. JRSM Short Rep 4:2042533313487332. https://doi.org/10.1177/2042533313487332
Hurst P, Schipof-Godart L, Szabo A et al (2019) The Placebo and Nocebo effect on sports performance: a systematic review. Eur J Sport Sci 20(4):1–26. https://doi.org/10.1080/17461391.2019.1655098
Kardas P, Lewek P, Matyjaszczyk M (2013) Determinants of patient adherence: a review of systematic reviews. Front Pharmacol 4:91. https://doi.org/10.3389/fphar.2013.00091
Keller A, Litzelman K, Wisk LE et al (2012) Does the perception that stress affects health matter? The association with health and mortality. Health Psychol 31:677–684. https://doi.org/10.1037/a0026743
Kennedy WP (1961) The nocebo reaction. Med World 95:203–205
Kern A, Kramm C, Witt CM, Barth J (2020) The influence of personality traits on the placebo/nocebo response: a systematic review. J Psychosom Res 128:109866. https://doi.org/10.1016/j.jpsychores.2019.109866
Kessner S, Wiech K, Forkmann K et al (2013) The effect of treatment history on therapeutic outcome: an experimental approach. JAMA Intern Med 173:1468–1469. https://doi.org/10.1001/jamainternmed.2013.6705
Khosla PP, Bajaj VK, Sharma G, Mishra KC (1992) Background noise in healthy volunteers—a consideration in adverse drug reaction studies. Indian J Physiol Pharmacol 36:259–262
Kienle GS (1995) Der sogenannte Placeboeffekt. Illusion, Fakten, Realitat. Schattauer, Stuttgart
Kienle GS, Kiene H (1997) Placebo effect and placebo concept: a critical methodological and conceptual analysis of reports on the magnitude of the placebo effect. Altern Ther Health Med 2:39–54
Kirsch I (2010) The emperor's new drugs: exploding the antidepressant myth. Basic Books, New York
Kissel P, Barrucand D (1964) Placebos et effet placebo en médecine. Masson, Paris
Koes BW, van Tulder MW, Thomas S (2006) Diagnosis and treatment of low back pain. BMJ 332:1430–1434

Kong J, Gollub RL, Polich G et al (2008) An fMRI study on the neural mechanisms of hyperalgesic nocebo effect. J Neurosci 28:13354–13362. https://doi.org/10.1523/JNEUROSCI.2944-08.2008
Köteles F, Bárdos G (2009) Nil nocere? A nocebo-jelenség. Magyar Pszichológiai Szemle 64:697–727. https://doi.org/10.1556/MPSzle.64.2009.4.5
Köteles F, Bárdos G (2011) What makes us sicker? An experimental study of non-specific adverse drug effects. Clin Exp Med J 5:203–215. https://doi.org/10.1556/CEMED.5.2011.4.4
Köteles F, Tóth A, Bárdos G (2011) Expectations of medicine adverse effects: perceptual characteristics of tablets and personality background. Clin Exp Med J 5:199–202. https://doi.org/10.1556/CEMED.5.2011.4.3
Köteles F, Szemerszky R, Gubányi M et al (2013) Idiopathic environmental intolerance attributed to electromagnetic fields (IEI-EMF) and electrosensibility (ES) – are they connected? Int J Hyg Environ Health 216:362–370. https://doi.org/10.1016/j.ijheh.2012.05.007
Köteles F, Simor P, Czető M et al (2016a) Modern health worries - the dark side of spirituality? Scand J Psychol 57:313–320. https://doi.org/10.1111/sjop.12297
Köteles F, Tarján E, Berkes T (2016b) Artificial concerns. Effects of a commercial advertisement on modern health worries and sympathetic activation. Mentálhigiéné és Pszichoszomatika 17:61–79. https://doi.org/10.1556/0406.17.2016.1.4
Köteles F, Simor P, Szemerszky R (2019) Somatosensory amplification and absorption contribute to electrosensitivity. Ideggyógyászati Szemle 72:165–170. https://doi.org/10.18071/isz.72.0165
Kroenke K (2003) Patients presenting with somatic complaints: epidemiology, psychiatric comorbidity and management. Int J Methods Psychiatr Res 12:34–43. https://doi.org/10.1002/mpr.140
Kwok RHM (1968) Chinese-restaurant syndrome. N Engl J Med 278:796. https://doi.org/10.1056/NEJM196804042781419
Lasagna L (1955) The controlled clinical trial: theory and practice. J Chronic Dis 1:353–367
Leder D (2018) Inside insights: a phenomenology of interoception. In: Tsakiris M, De Preester H (eds) The interoceptive mind. From homeostasis to awareness. Oxford University Press, Oxford, England, pp 307–322
Lees-Haley PR, Brown RS (1992) Biases in perception and reporting following a perceived toxic exposure. Percept Mot Skills 75:531–544
Lester D (2009) Voodoo death. Omega (Westport) 59:1–18
Liccardi G, Senna G, Russo M et al (2004) Evaluation of the nocebo effect during oral challenge in patients with adverse drug reactions. J Investig Allergol Clin Immunol 14:104–107
Lombardi C, Gargioni S, Canonica GW, Passalacqua G (2008) The nocebo effect during oral challenge in subjects with adverse drug reactions. Eur Ann Allergy Clin Immunol 40:138–141
Lovallo WR (2016) Stress and health. Biological and psychological interactions. Sage Publications, Inc, London
Luparello T, Lyons HA, Bleecker ER, McFadden ER Jr (1968) Influences of suggestion on airway reactivity in asthmatic subjects. Psychosom Med 30:819–825
Luparello T, Leist N, Lourie CH, Sweet P (1970) The interaction of psychologic stimuli and pharmacologic agents on airway reactivity in asthmatic subjects. Psychosom Med 32:509–513
Lyerly SB, Ross S, Krugman AD, Clyde DJ (1964) Drugs and placebos: the effects of instructions upon performance and mood under amphetamine sulfate and chloral hydrate. J Abnorm Psychol 68:321–327. https://doi.org/10.1037/h0044351
MacKrill K, Gamble GD, Bean DJ et al (2019) Evidence of a media-induced nocebo response following a nationwide antidepressant drug switch. Clin Psychol Eur 1:1–12. https://doi.org/10.32872/cpe.v1i1.29642
Matthews A, Herrett E, Gasparrini A et al (2016) Impact of statin related media coverage on use of statins: interrupted time series analysis with UK primary care data. BMJ 353:3283. https://doi.org/10.1136/bmj.i3283
Mazzoni G, Foan L, Hyland ME, Kirsch I (2010) The effects of observation and gender on psychogenic symptoms. Health Psychol 29:181–185. https://doi.org/10.1037/a0017860

Meyer FP, Tröger U, Röhl FW (1996) Adverse nondrug reactions: an update. Clin Pharmacol Ther 60:347–352. https://doi.org/10.1016/S0009-9236(96)90062-4
Miller FG, Colloca L (2009) The legitimacy of placebo treatments in clinical practice: evidence and ethics. Am J Bioeth 9:39–47. https://doi.org/10.1080/15265160903316263
Mitsikostas DD (2012) Nocebo in headaches: implications for clinical practice and trial design. Curr Neurol Neurosci Rep 12:132–137. https://doi.org/10.1007/s11910-011-0245-4
Mitsikostas DD, Mantonakis LI, Chalarakis NG (2011) Nocebo is the enemy, not placebo. A meta-analysis of reported side effects after placebo treatment in headaches. Cephalalgia 31:550–561. https://doi.org/10.1177/0333102410391485
Mlynski C, Wright RA, Kelly K (2020) Ability influence on effort and associated cardiovascular responses: Nocebo-Placebo evidence that perception is key. Biol Psychol 152:107867. https://doi.org/10.1016/j.biopsycho.2020.107867
Moerman DE (2002) Meaning, medicine and the "placebo effect". Cambridge University Press, Cambridge
Moerman DE (2011) Examining a powerful healing effect through a cultural lens, and finding meaning. J Mind Body Regul 1:63–72
Mondaini N, Gontero P, Giubilei G et al (2007) Finasteride 5 mg and sexual side effects: how many of these are related to a nocebo phenomenon? J Sex Med 4:1708–1712. https://doi.org/10.1111/j.1743-6109.2007.00563.x
Montgomery GH, Kirsch I (1997) Classical conditioning and the placebo effect. Pain 72:107–113
Mora MS, Nestoriuc Y, Rief W (2011) Lessons learned from placebo groups in antidepressant trials. Philos Trans R Soc B Biol Sci 366:1879–1888. https://doi.org/10.1098/rstb.2010.0394
Myers MG, Cairns JA, Singer J (1987) The consent form as a possible cause of side effects. Clin Pharmacol Ther 42:250–253. https://doi.org/10.1038/clpt.1987.142
Nabi H, Kivimäki M, Batty GD et al (2013) Increased risk of coronary heart disease among individuals reporting adverse impact of stress on their health: the Whitehall II prospective cohort study. Eur Heart J 34:2697–2705. https://doi.org/10.1093/eurheartj/eht216
O'Connor AM, Pennie RA, Dales RE (1996) Framing effects on expectations, decisions, and side effects experienced: the case of influenza immunization. J Clin Epidemiol 49:1271–1276
Ongaro G, Kaptchuk TJ (2019) Symptom perception, placebo effects, and the Bayesian brain. Pain 160:1–4. https://doi.org/10.1097/j.pain.0000000000001367
Page LA, Keshishian C, Leonardi G et al (2010) Frequency and predictors of mass psychogenic illness. Epidemiology 21:744–747. https://doi.org/10.1097/EDE.0b013e3181e9edc4
Pan Y, Kinitz T, Stapic M, Nestoriuc Y (2019) Minimizing drug adverse events by informing about the nocebo effect-an experimental study. Front Psych 10:504. https://doi.org/10.3389/fpsyt.2019.00504
Papadopoulos D, Mitsikostas DD (2012) A meta-analytic approach to estimating nocebo effects in neuropathic pain trials. J Neurol 259:436–447. https://doi.org/10.1007/s00415-011-6197-4
Penick SB, Fisher S (1965) Drug-set interaction: psychological and physiological effects of epinephrine under differential expectations. Psychosom Med 27:177–182
Pennebaker JW (1982) The psychology of physical symptoms. Springer, New York
Pennebaker JW (1994) Psychological bases of symptom reporting: perceptual and emotional aspects of chemical sensitivity. Toxicol Ind Health 10:497–511
Pennebaker JW, Skelton JA (1981) Selective monitoring of physical sensations. J Pers Soc Psychol 41:213–223
Petrie KJ, Wessely S (2002) Modern worries, new technology, and medicine. BMJ 324:690–691
Petrie KJ, Sivertsen B, Hysing M et al (2001) Thoroughly modern worries: the relationship of worries about modernity to reported symptoms, health and medical care utilization. J Psychosom Res 51:395–401. https://doi.org/10.1016/S0022-3999(01)00219-7
Petrie KJ, Broadbent E, Kley N et al (2005) Worries about modernity predict symptom complaints after environmental pesticide spraying. Psychosom Med 67:778–782. https://doi.org/10.1097/01.psy.0000181277.48575.a4

Planès S, Villier C, Mallaret M (2016) The nocebo effect of drugs. Pharmacol Res Perspect 4:e00208. https://doi.org/10.1002/prp2.208
Pogge RC (1963) The toxic placebo. I. Side and toxic effects reported during the administration of placebo medicine. Med Times 91:773–778
Polich G, Iaccarino MA, Kaptchuk TJ et al (2020) Nocebo effects in concussion: is all that is told beneficial? Am J Phys Med Rehabil 99:71–80. https://doi.org/10.1097/PHM.0000000000001290
Preston RA, Materson BJ, Reda DJ, Williams DW (2000) Placebo-associated blood pressure response and adverse effects in the treatment of hypertension: observations from a Department of Veterans Affairs Cooperative Study. Arch Intern Med 160:1449–1454
Rabkin JG, Markowitz JS, Stewart J et al (1986) How blind is blind? Assessment of patient and doctor medication guesses in a placebo-controlled trial of imipramine and phenelzine. Psychiatry Res 19:75–86
Raglin J, Szabo A, Lindheimer JB, Beedie C (2020) Understanding placebo and nocebo effects in the context of sport: a psychological perspective. Eur J Sport Sci 20(3):293–301. https://doi.org/10.1080/17461391.2020.1727021
Reeves RR, Ladner ME, Hart RH, Burke RS (2007) Nocebo effects with antidepressant clinical drug trial placebos. Gen Hosp Psychiatry 29:275–277. https://doi.org/10.1016/j.genhosppsych.2007.01.010
Reidenberg MM, Lowenthal DT (1968) Adverse nondrug reactions. N Engl J Med 279:678–679. https://doi.org/10.1056/NEJM196809262791304
Ren Y, Xu F (2018) How patients should be counseled on adverse drug reactions: avoiding the nocebo effect. Res Soc Adm Pharm 14:705. https://doi.org/10.1016/j.sapharm.2018.04.007
Rescorla RA, Wagner AR (1972) A theory of Pavlovian conditioning: variations in effectiveness of reinforcement and nonreinforcement. In: Black AH, Prokasy WF (eds) Conditioning II: current research and theory. Appleton-Century-Crofts, New York, pp 64–99
Reuter U, Sanchez del Rio M, Carpay JA et al (2003) Placebo adverse events in headache trials: headache as an adverse event of placebo. Cephalalgia 23:496–503
Rief W, Avorn J, Barsky AJ (2006) Medication-attributed adverse effects in placebo groups: implications for assessment of adverse effects. Arch Intern Med 166:155–160. https://doi.org/10.1001/archinte.166.2.155
Rief W, Nestoriuc Y, von Lilienfeld-Toal A et al (2009) Differences in adverse effect reporting in placebo groups in SSRI and tricyclic antidepressant trials. Drug Saf 32:1041–1056. https://doi.org/10.2165/11316580-000000000-00000
Rodriguez-Raecke R, Doganci B, Breimhorst M et al (2010) Insular cortex activity is associated with effects of negative expectation on nociceptive long-term habituation. J Neurosci 30:11363–11368. https://doi.org/10.1523/JNEUROSCI.2197-10.2010
Röösli M, Moser M, Baldinini Y et al (2004) Symptoms of ill health ascribed to electromagnetic field exposure – a questionnaire survey. Int J Hyg Environ Health 207:141–150. https://doi.org/10.1078/1438-4639-00269
Rosenzweig P, Brohier S, Zipfel A (1993) The placebo effect in healthy volunteers: influence of experimental conditions on the adverse events profile during phase I studies. Clin Pharmacol Ther 54:578–583. https://doi.org/10.1038/clpt.1993.190
Rubin GJ, Cleare AJ, Wessely S (2008) Psychological factors associated with self-reported sensitivity to mobile phones. J Psychosom Res 64:1–12. https://doi.org/10.1016/j.jpsychores.2007.05.006
Rubin GJ, Nieto-Hernandez R, Wessely S (2010) Idiopathic environmental intolerance attributed to electromagnetic fields (formerly 'electromagnetic hypersensitivity'): an updated systematic review of provocation studies. Bioelectromagnetics 31:1–11. https://doi.org/10.1002/bem.20536
Samuels MA (2007) "Voodoo" death revisited: the modern lessons of neurocardiology. Cleve Clin J Med 74(Suppl 1):S8–S16

Sanderson C, Hardy J, Spruyt O, Currow DC (2013) Placebo and nocebo effects in randomized controlled trials: the implications for research and practice. J Pain Symptom Manag 46:722–730. https://doi.org/10.1016/j.jpainsymman.2012.12.005
Schwarz KA, Büchel C (2015) Cognition and the placebo effect - dissociating subjective perception and actual performance. PLoS One 10:e0130492. https://doi.org/10.1371/journal.pone.0130492
Shapiro AK, Shapiro E (1997) The powerful placebo: from ancient priest to modern physician. Johns Hopkins University Press, Baltimore
Shapiro AK, Struening EL, Shapiro E (1979) The reliability and validity of a placebo test. J Psychiatr Res 15:253–290. https://doi.org/10.1016/0022-3956(79)90016-5
Shiri R, Falah-Hassani K, Heliövaara M et al (2019) Risk factors for low back pain: a population-based longitudinal study. Arthritis Care Res (Hoboken) 71:290–299. https://doi.org/10.1002/acr.23710
Silvestri A, Galetta P, Cerquetani E et al (2003) Report of erectile dysfunction after therapy with beta-blockers is related to patient knowledge of side effects and is reversed by placebo. Eur Heart J 24:1928–1932
Sirois F (1974) Epidemic hysteria. Acta Psychiatr Scand Suppl 252:1–46
Staudenmayer H, Binkley KE, Leznoff A, Phillips S (2003a) Idiopathic environmental intolerance: Part 1: A causation analysis applying Bradford Hill's criteria to the toxicogenic theory. Toxicol Rev 22:235–246
Staudenmayer H, Binkley KE, Leznoff A, Phillips S (2003b) Idiopathic environmental intolerance: Part 2: A causation analysis applying Bradford Hill's criteria to the psychogenic theory. Toxicol Rev 22:247–261
Staudenmayer H, Christopher KL, Repsher L, Hill RH (2011) Mass psychogenic illness: psychological predisposition and iatrogenic pseudo-vocal cord dysfunction and pseudo-reactive airways disease syndrome. J Med Toxicol 7:109–117. https://doi.org/10.1007/s13181-011-0136-8
Sternberg EM (2002) Walter B. Cannon and "'Voodoo' Death": a perspective from 60 years on. Am J Public Health 92:1564–1566
Stewart-Williams S (2004) The placebo puzzle: putting together the pieces. Health Psychol 23:198–206. https://doi.org/10.1037/0278-6133.23.2.198
Stewart-Williams S, Podd J (2004) The placebo effect: dissolving the expectancy versus conditioning debate. Psychol Bull 130:324–340. https://doi.org/10.1037/0033-2909.130.2.324
Stockhorst U, Enck P, Klosterhalfen S (2007) Role of classical conditioning in learning gastrointestinal symptoms. World J Gastroenterol 13:3430–3437
Swider K, Bąbel P (2013) The effect of the sex of a model on nocebo hyperalgesia induced by social observational learning. Pain 154:1312–1317. https://doi.org/10.1016/j.pain.2013.04.001
Szemerszky R, Köteles F, Lihi R, Bárdos G (2010) Polluted places or polluted minds? An experimental sham-exposure study on background psychological factors of symptom formation in "Idiophatic Environmental Intolerance attributed to electromagnetic fields". Int J Hyg Environ Health 213:387–394. https://doi.org/10.1016/j.ijheh.2010.05.001
Szemerszky R, Dömötör Z, Berkes T, Köteles F (2015a) Attribution-based nocebo effects. Perceived effects of a placebo pill and a sham magnetic field on cognitive performance and somatic symptoms. Int J Behav Med 23:204–213. https://doi.org/10.1007/s12529-015-9511-1
Szemerszky R, Gubányi M, Árvai D et al (2015b) Is there a connection between electrosensitivity and electrosensibility? A replication study. Int J Behav Med 22:755–763. https://doi.org/10.1007/s12529-015-9477-z
Taddio A, Shah V, Gilbert-MacLeod C, Katz J (2002) Conditioning and hyperalgesia in newborns exposed to repeated heel lances. JAMA 288:857–861. https://doi.org/10.1001/jama.288.7.857
Tarasoff L, Kelly MF (1993) Monosodium L-glutamate: a double-blind study and review. Food Chem Toxicol 31:1019–1035. https://doi.org/10.1016/0278-6915(93)90012-N
Thompson WG (2005) Placebo effect and health: combining science with compassionate care. Prometheus Books, Amherst, MA
Thomson R (1982) Side effects and placebo amplification. Br J Psychiatry 140:64–68

Tinnermann A, Geuter S, Sprenger C et al (2017) Interactions between brain and spinal cord mediate value effects in nocebo hyperalgesia. Science 358:105–108. https://doi.org/10.1126/science.aan1221

Tracey I (2010) Getting the pain you expect: mechanisms of placebo, nocebo and reappraisal effects in humans. Nat Med 16:1277–1283. https://doi.org/10.1038/nm.2229

Turnwald BP, Goyer JP, Boles DZ et al (2019) Learning one's genetic risk changes physiology independent of actual genetic risk. Nat Hum Behav 3:48–56. https://doi.org/10.1038/s41562-018-0483-4

Uhlenhuth EH, Alexander PE, Dempsey GM et al (1998) Medication side effects in anxious patients: negative placebo responses? J Affect Disord 47:183–190. https://doi.org/10.1016/S0165-0327(97)00125-0

Van den Bergh O, Brown RJ, Petersen S, Witthöft M (2017a) Idiopathic environmental intolerance: a comprehensive model. Clin Psychol Sci 5:551–567. https://doi.org/10.1177/2167702617693327

Van den Bergh O, Witthöft M, Petersen S, Brown RJ (2017b) Symptoms and the body: taking the inferential leap. Neurosci Biobehav Rev 74(Pt A):185–203. https://doi.org/10.1016/j.neubiorev.2017.01.015

van Laarhoven AIM, Vogelaar ML, Wilder-Smith OH et al (2011) Induction of nocebo and placebo effects on itch and pain by verbal suggestions. Pain 152:1486–1494. https://doi.org/10.1016/j.pain.2011.01.043

Varga K (2013) Suggestive techniques connected to medical interventions. Interv Med Appl Sci 5:95–100. https://doi.org/10.1556/IMAS.5.2013.3.1

Vermeire E, Hearnshaw H, Royen PV, Denekens J (2001) Patient adherence to treatment: three decades of research. A comprehensive review. J Clin Pharm Ther 26:331–342. https://doi.org/10.1046/j.1365-2710.2001.00363.x

Voelker R (1996) Nocebos contribute to host of ills. JAMA 275:345–347. https://doi.org/10.1001/jama.1996.03530290015007

Vögtle E, Barke A, Kröner-Herwig B (2013) Nocebo hyperalgesia induced by social observational learning. Pain 154:1427–1433. https://doi.org/10.1016/j.pain.2013.04.041

von Blanckenburg P, Schuricht F, Albert U-S et al (2013) Optimizing expectations to prevent side effects and enhance quality of life in breast cancer patients undergoing endocrine therapy: study protocol of a randomized controlled trial. BMC Cancer 13:426. https://doi.org/10.1186/1471-2407-13-426

von Blanckenburg P, Schuricht F, Heisig SR et al (2015) Psychologische Optimierung von Erwartungen zur Prävention von Nocebo-Nebenwirkungen bei Brustkrebs - 2 Fallberichte. Verhaltenstherapie. https://doi.org/10.1159/000377711

Voudouris NJ, Peck CL, Coleman G (1985) Conditioned placebo responses. J Pers Soc Psychol 48:47–53

Voudouris NJ, Peck CL, Coleman G (1990) The role of conditioning and verbal expectancy in the placebo response. Pain 43:121–128. https://doi.org/10.1016/0304-3959(90)90057-K

Waber RL, Shiv B, Carmon Z, Ariely D (2008) Commercial features of placebo and therapeutic efficacy. JAMA 299:1016–1017. https://doi.org/10.1001/jama.299.9.1016

Waggoner MR (2013) Parsing the peanut panic: the social life of a contested food allergy epidemic. Soc Sci Med 90:49–55. https://doi.org/10.1016/j.socscimed.2013.04.031

Watson D, Pennebaker JW (1989) Health complaints, stress, and distress: exploring the central role of negative affectivity. Psychol Rev 96:234–254

Webster RK, Weinman J, Rubin GJ (2018) Medicine-related beliefs predict attribution of symptoms to a sham medicine: a prospective study. Br J Health Psychol 23:436–454. https://doi.org/10.1111/bjhp.12298

Weissenfeld J, Stock S, Lüngen M, Gerber A (2010) The nocebo effect: a reason for patients' non-adherence to generic substitution? Pharmazie 65:451–456

Wells RE, Kaptchuk TJ (2012) To tell the truth, the whole truth, may do patients harm: the problem of the nocebo effect for informed consent. Am J Bioeth 12:22–29. https://doi.org/10.1080/15265161.2011.652798

Wendt L, Albring A, Benson S et al (2014) Catechol-O-methyltransferase Val158Met polymorphism is associated with somatosensory amplification and nocebo responses. PLoS One 9:e107665. https://doi.org/10.1371/journal.pone.0107665

White K, Kando J, Park T et al (1992) Side effects and the “blindability” of clinical drug trials. Am J Psychiatry 149:1730–1731

Widdershoven G, Meynen G, Metselaar S (2017) Dealing with the nocebo effect: taking physician–patient interaction seriously. Am J Bioeth 17:48–50. https://doi.org/10.1080/15265161.2017.1314046

Williams AN, Woessner KM (2009) Monosodium glutamate “allergy”: menace or myth? Clin Exp Allergy 39:640–646. https://doi.org/10.1111/j.1365-2222.2009.03221.x

Williams JB, Sade RM, D’Amico TA (2013) Framing for success: nocebo effects in thoracic surgery. Ann Thorac Surg 95:9–11. https://doi.org/10.1016/j.athoracsur.2012.10.048

Winkler A, Hermann C (2019) Placebo- and nocebo-effects in cognitive neuroenhancement: when expectation shapes perception. Front Psych 10:498. https://doi.org/10.3389/fpsyt.2019.00498

Witthöft M, Rubin GJ (2013) Are media warnings about the adverse health effects of modern life self-fulfilling? An experimental study on idiopathic environmental intolerance attributed to electromagnetic fields (IEI-EMF). J Psychosom Res 74:206–212. https://doi.org/10.1016/j.jpsychores.2012.12.002

Witthöft M, Freitag I, Nußbaum C et al (2018) On the origin of worries about modern health hazards: experimental evidence for a conjoint influence of media reports and personality traits. Psychol Health 33:361–380. https://doi.org/10.1080/08870446.2017.1357814

Wolf S, Pinsky RH (1954) Effects of placebo administration and occurrence of toxic reactions. J Am Med Assoc 155:339–341

Wolff HG, DuBois EF, Cattell M et al (1946) The use of placebos in therapy. N Y State J Med 46:1718–1727

Working Party from compliance to concordance (1993) Patient compliance. Ann Pharmacother 27:S5–S19. https://doi.org/10.1177/10600280930270S902

Wang J, Ahrens A, Hudson S et al (2014) Catechol O-methyltransferase Val158Met polymorphism is associated with [illegible] amplification and nocebo responses. PLoS One 9 [illegible] https://doi.org/10.1371/journal.pone.0107965

[illegible]

Chapter 10
The Bright Side of Body Sensations

10.1 Introduction

In this chapter, we will consider the pleasant aspects of body sensations from the viewpoint of health and illness. As we will see, pleasant does not necessarily mean adaptive though. There are sensations that we interpret as signs of good health and functioning (both biological and psychological), of recovering from a disease, or, using the everyday meaning of the word, of energy. As mentioned with respect to symptoms and nocebo reactions, these sensations belong to the domain of primary consciousness, thus their impact on our perceived state is direct and often overwhelming (tenet of realness).

Which sensations can belong here? Surprisingly, almost all of them, sometimes even including uncomfortable and displeasing ones. As we know very well, some body sensations, such as sexual satisfaction, affective touch, or mild to moderate physical activity, are inherently pleasant (Chap. 4). The tingling sensation appears to be neutral, thus it can be interpreted in a positive way if the cues of the situation suggest this. Finally, unpleasant sensations, most importantly pain, can be used as indicators of a positive change if they are *getting reduced*. Also, having food or water after a period of hunger or thirst makes us satisfied and improves our overall evaluation about the momentary state of affairs. What changes? If we look for one single term, I think subjective well-being suits best our needs. As its name indicates, subjective well-being is strictly based on personal experience and evaluation; the high level of subjective well-being is a mix of high level of positive affect, low level of negative affect, and an overall positive global assessment of one' life, also called happiness (Diener 1984; Lucas et al. 1996; Arthaud-Day et al. 2005).

Body sensations and changes can directly lead to improvements in subjective well-being, for example, by contributing to positive affect and/or by the reduction of negative affect. The gentle stroke of our romantic partner and the beneficial effect of a painkiller are able to effectively improve our well-being state (I am not saying, of course, that these two things are equal or mean the same; they just push us to the

F. Köteles, *Body Sensations*, https://doi.org/10.1007/978-3-030-63201-4_10

same direction). Awareness of body changes and sensations showed a positive association with subjective well-being in number of studies (Köteles 2014; Tihanyi et al. 2017; Ferentzi et al. 2019). Even spirituality, one of the most complex and sophisticated human experiences, can be affected by the state of the body. In our empirical study, beginner-level yoga classes were delivered to two groups: the physical exercises were identical, but participants in one group received spiritual instructions, including description of chakras, prana, and energy, whereas the other group was instructed in a pure physical manner (Csala et al. 2020). Spirituality increased equally in both groups compared to a control group, which suggests that verbal instructions are way less important in yoga practice than the actual pattern of physical activity.

Beyond these primary and automatic processes, secondary evaluation, i.e., the meaning attached to body sensations and generally the body, also plays a role here. Dissatisfaction with one's own body is perhaps the most thoroughly investigated aspect of this appraisal; it typically refers to the external appearance of the body. As the standards set by the media are practically unreachable for everyday people, both women and men, this comparison usually generates negative affective states, and even can contribute to mental disorders (Cash and Pruzinsky 1990, 2002; Cash and Deagle 1997; Emanuelsen et al. 2015). Also, overemphasizing the importance of perfect health, or simply the suggestion that there is such a thing as perfect health and life is inferior without it, often lead to a dissatisfaction with the body with respect to its functioning and health status (Barsky 1988; Petrie et al. 2001). The term interoceptive appraisal refers to both primary and secondary evaluation of our interoceptive state, with an emphasis on the perceived aspect of interoception (Farb and Logie 2018). It substantially contributes to well-being and the overall feeling of being in good health (Paulus and Stein 2010; Farb and Logie 2018).

Body sensations evoking positive affective states are able to improve negative states. Affect and mood usually comes and goes spontaneously, it cannot be directly controlled. Still, the fact that the motor system is voluntarily controllable also means that we have several options to influence our negative affective states. Physical activity, including sports with a play component, can serve as a useful tool to improve mood state (Ekkekakis 2003; Ekkekakis et al. 2011; Panksepp and Biven 2012; Köteles et al. 2020); it can be useful even in pathological conditions, such as depression and anxiety (Conn 2010; Blake 2012; Brown et al. 2013; Josefsson et al. 2014; Stonerock et al. 2015). Breathing exercises are able to reduce sympathetic activation and negative affect as well as increase parasympathetic activation and positive affect (Calabrese et al. 2000; Sovik 2000; Brown and Gerbarg 2005; Zautra et al. 2010; Zeidan et al. 2015; Schmalzl et al. 2015; Strigo and Craig 2016; Gibson 2019). Even a simple sigh can relieve anxiety and physiological tension in the short run (Vlemincx et al. 2010, 2016). Also, although its neuroendocrine effects are questioned, power posing positively impacts our subjective experience, including the affective state (Carney et al. 2010; Ranehill et al. 2015; Weineck et al. 2019). Finally, affective touch delivered by significant ones, such as romantic partners, parents, or healers, also has beneficial subjective effects (Lloyd et al. 2013; Calsius et al. 2016; von Mohr and Fotopoulou 2018; von Mohr et al. 2018).

Beyond these direct effects on affective state, pleasant body sensations play a decisive role in the evaluation of healing interventions in the broader sense. With respect to physical diseases, however, this is a double-edged sword; we will discuss the pros and cons of this phenomenon in the light of the placebo response and complementary and alternative medicine (CAM). In the second part of this chapter, we will try to approach certain aspects of mind–body practices from the viewpoint of body sensations.

10.2 The Placebo Response

The placebo response, just like the nocebo response, is basically a self-fulfilling prophecy (Benedetti 2011; Brody et al. 2012; Petrie and Rief 2019). Expectations of positive changes, using the term expectation in the broadest possible sense, including conditioning, beliefs, conscious and nonconscious predictions, can lead to the expected changes. This phenomenon is often presented by laypeople as an example for the dominance of the mind over the body or the entire material world. It is claimed that one's body can get better in an instance if the mind changes. The reality is less mysterious though, and this is a lucky thing as such a direct causal link between the physical and psychological state would be very dangerous. The scope of the placebo response is limited, and it is particularly so with respect to bodily (i.e., peripheral) changes. I think the general rule is self-evident: the placebo response cannot impact body functions and changes that cannot be reversed and for which the organism has no regulatory and/or reparatory mechanism (Köteles 2013). Moreover, constraints of the existing repair mechanisms necessarily limit the extent and swiftness of the response. Degenerative diseases, such as atherosclerosis, the loss of the muscles of the heart due to infarction, or arthrosis, cannot be directly impacted by placebo interventions. Independent of our expectations and positive thoughts, the wall of our arteries will not be more flexible and free from plaque and cholesterol, cardiac muscle will not regenerate, and the cartilage in our joints will not become thicker. However, the subjective aspects can be improved for sure; placebo interventions can make us feel healthier and are able to reduce our pain independent of our actual body state. Sometimes the behavioral change enabled by the reduction of pain, for example, the increased mobility and use of the joints, can further improve their state. Still, placebo interventions cannot do miracles.

Findings of biofeedback research clearly indicated the limitations of the impact of top-down processes on physiology. For example, participants were able to learn to control their blood pressure and heart rate to some extent. The problem is that this extent was too small, i.e., clinically insignificant (Seer 1979; Nakao et al. 2003). In my view, these findings are also relevant and applicable for changes potentially elicitable by placebo interventions (Köteles 2013).

In fact, the conditions that are most sensitive to placebo interventions are typically situated in our brain not in our body. The best examples are pain, depression, and anxiety. For these conditions, the effectiveness of placebo interventions is often comparable with that of medications (Benedetti 2009, 2011). Actually, the magnitude of the placebo effect in these cases often makes the demonstration of the specific effect of a drug (i.e., the significant difference between the improvement measured in drug group and the placebo group) difficult in clinical trials (Thompson 2005; Kirsch 2008, 2010). For example, using the words of Schweizer and Rickels (1997), the high placebo-response rate sabotages the development of new treatments for generalized anxiety disorder. The placebo phenomenon, similar to the nocebo, impacts many (perhaps all) aspects of everyday functioning beyond medical conditions, such as physical (Bérdi et al. 2011, 2015; Vercelli et al. 2013; Szabo and Müller 2015; Beedie et al. 2018; Hurst et al. 2019; Raglin et al. 2020) and cognitive performance (Schwarz and Büchel 2015; Turi et al. 2018; Winkler and Hermann 2019; Mlynski et al. 2020) and mood states (Szabo et al. 2018).

However, just because these improvements are located in the brain, it does not mean that they are imaginary, fabricated, or simply biased. Reported intensity of pain correlates well with the activation of centers of the so-called pain network (thalamus, anterior cingulum, etc.) (Petrovic et al. 2002), and expectations influence the early steps of pain processing that are not accessible for conscious awareness (Lorenz et al. 2005; Wager et al. 2006; Aslaksen et al. 2011; Bjørkedal and Flaten 2011). The major neural mechanisms of placebo analgesia are comparatively well described (Price et al. 2008; Benedetti 2009, 2011; Colloca 2018; Colloca and Barsky 2020); we know that not only the brain (including the brainstem) but also the spinal cord is impacted (Eippert et al. 2009). Moreover, placebo responses were elicited in animals (Guo et al. 2010) that are not really prone to fake symptoms or distort their behavior in favor of the research hypothesis. In summary, these responses are real and demonstrate the substantial involvement of top-down factors in the respective conditions. Also, being located in the brain does not mean less significance than peripheral pathologies. Actually, the aforementioned three conditions, i.e., pain, depression, and anxiety, are among the major causes (if not THE major causes) of human suffering.

What makes these conditions suspicious to many is that they are primarily accessible via self-report, and self-reported symptoms and states are often considered less reliable by physicians than objective (i.e., externally verifiable) conditions. At the end of the 1970's, Jon Levine and colleagues reported a very interesting (and at that time very surprising) finding: the placebo response could be blocked by a chemical substance, naloxone (Levine et al. 1978). From the viewpoint of neuropharmacology, this simply indicates that placebo analgesia is mediated by the endogenous opioid system (naloxone is an opioid antagonist). However, the fact that such a suspicious process as placebo analgesia, generally assumed to belong to the domain of the mind, can be influenced by objective means was much more interesting from a philosophical point of view. Placebo reactors have been described as having low IQ, being susceptible to external influences and prone to yea-saying (acquiescence) and similar less desirable features over decades (Claridge 1970; McNair et al. 1979; Buckalew et al. 1981; Fischer and Greenberg 1997; Shapiro and Shapiro 1997; Harrington 2008). In accordance with this view, the placebo response itself was also often regarded as a kind of self-deception or the deception of the medical staff; for example, a positive response to a placebo treatment was considered a proof of fabrication of symptoms in certain clinics (Brown 2013). However, if it can be influenced by a drug, as shown by the findings of Levine and colleagues, it must be real. Since this significant first step and its replications (Gracely et al. 1983; Grevert et al. 1983; Levine and Gordon 1984; Benedetti 1996), many details were revealed about the neuropharmacology of these processes (Fields and Price 1997; Benedetti 2009, 2011).

In 2000, a very intriguing professional debate on the nature of the placebo response emerged. A number of renowned scientists who were strongly convinced of the importance and therapeutic significance of the placebo response, including Seymour Fisher, one of the giants of the research of body sensations and Howard Spiro, the author of a great book on this topic (Spiro 1998), claimed that it belongs to the subjective domain (Brody 2000; Fisher 2000; Price 2000; Spiro 2000). In other words, it cannot cause objective changes (the term objective refers to peripheral here). Others (Ader 2000; Bierman 2000; Cardeña and Kirsch 2000) found this statement overly restrictive; still, they accepted that the majority of placebo responses are subjective, i.e., do not include peripheral changes. They dominantly happen in the brain; thus, they impact the representation of the state of the body rather than the actual body.

The therapeutic use of placebos without informing the patients about the fact that they receive a placebo has been considered unethical, thus it has been prohibited for decades (Bok 1974, 2002; Brody 1980; Gold and Lichtenberg 2013; Roeser 2015). Still, data from a considerable number of surveys show that physicians (general practitioners, internists, rheumatologists, etc.) keep delivering placebo interventions to their patients (Nitzan and Lichtenberg 2004; Tilburt et al. 2008; Fässler et al. 2009; Meissner et al. 2012; Howick et al. 2013). In our own study with Hungarian GPs (Ferentzi et al. 2010, 2011), almost 80% of the doctors had used a placebo in their practice in the preceding year; on average, 6% of all cases belonged to this category. One-fourth of them intentionally carried something with them (e.g., a vitamin product) to use it as a placebo if necessary. Most importantly, 46.9% of our respondents believed that placebos are effective only for subjective complaints (e.g., anxiety, fatigue, insomnia, functional problems, pain), and further 44.4% thought that they can be successfully used for both subjective and objective symptoms. The vast majority believed that this practice is ethical under certain circumstances; first of all, if it is beneficial for the patient. The bottom line is that physicians apply placebos because they work in certain cases; typically, they are often able to relieve subjective symptoms that cause huge stress and suffering.

If we combine this insight with the marked dissociation between actual and perceived bodily changes, the conclusion is quite straightforward: placebos are more effective in the alleviation of suffering than in healing the body. They have more to do with care than with cure (Spiro 1998). Once again, this is a great thing, and I completely agree with those who think that this unique phenomenon should be used in clinical practice after solving the difficult ethical and practical issues (Enck et al. 2013; Savvas et al. 2014; Roeser 2015; Beedie et al. 2015; Kaptchuk 2018).

Beyond the well-known ethical considerations (Bok 1974, 2002; Brody 1982), one of the biggest risks of placebo interventions relies exactly on their ability to make patients feel better. Feeling better often means feeling cured for laypeople (Oehl et al. 2000; Munro et al. 2007); this mistaken conviction can then lead to non-adherence or to quitting potentially effective but unpleasant treatments (Kardas et al. 2013). Fortunately, the placebo response is usually limited in time; from a conditioning point of view, it usually requires reinforcement on a regular basis (Wickramasekera 1985; Ader 1997; Ádám 1998; Ader et al. 2010). Expectations can be self-reinforcing thus evoke long-term changes, at least in theory (Kirsch 1985). In fact, our knowledge on the long-term pattern of the placebo response is poor. Experimental studies typically focus on acute changes. To my knowledge, only a handful of studies reported long-term placebo effects (Klinger et al. 2007; Colloca et al. 2010). Overall, I believe that beneficial effects of placebo treatment, e.g., the alleviation of pain and anxiety, are rarely long-lasting; this feature might be disappointing, but it definitely has positive consequences (see below).

The placebo response is a prototypical top-down interoceptive process, neatly explainable by the predictive processing approach (Büchel et al. 2014; Grahl et al. 2018). In a nutshell, expectations (priors) have a decisive impact on the processing of sensory input; moreover, it is hard to separate afferent (ascending) and efferent (descending) processes as they are in constant interaction. Well before the application of the predictive processing approach to body sensations and changes, a number of authors proposed that regeneration from a pathological condition can be regarded as the impact of the template of healthy functioning on the actual state of the body. Such a template can also be considered a prior. For example, Benson and Friedman (1996) called the placebo response remembered wellness; my older mentor György Ádám (1998, 2011) was convinced that the healthy template is primarily developed by learning during the first period of life. The spontaneous recovery process, controlled by the template as a prior at medium levels of the processing hierarchy, can be stimulated by higher-level priors, i.e., external factors, such as feeling safe and cared for (Brody 1997; Brody and Brody 2000; Humphrey 2002).

10.3 Complementary and Alternative Medicine (CAM)

The placebo response can be evoked by many contextual or nonspecific factors, such as the perceptual characteristics of the medication and components of the healing environment (de Craen et al. 1996, 2000; Di Blasi et al. 2001; Moerman 2002). The most important placebogenic factor is, however, the person delivering the treatment. Howard Brody called the doctor "walking placebo" (Brody 1997), which neatly illustrates the decisive impact of the significant person on the overall outcome of the treatment. The doctor–patient relationship and communication influence the patient's adherence to the treatment, his or her reaction to side effects (see Chap. 9), and much more (Barsky 1983; Barsky et al. 2002; Shea 2006); further, it can directly contribute to the improvement due to the enhancement of the placebo response. Brody describes the placebo response as a psychobiological response elicited by the symbolic features of the treatment, including the healer (Brody and Brody 2000). In more detail, the meaning of the illness changes in a positive direction because (1) the patient feels listened to and given an acceptable explanation for the illness in question; (2) the patient feels cared for by the healer; and (3) the patient's perceived control over the condition increases. The process requires time and personal attention from the doctor; the approach of conventional medicine with its step-by-step protocols, specialists, and limited time is not optimal for this (Spiro 1998; Thompson 2005). There is another branch of medicine, called complementary and alternative medicine, that appears to utilize the placebo response in a more effective way.

CAM is impossible to define in a consistent way as it is an umbrella term for treatments and procedures that are not accepted and practiced officially by the medical profession of a given country or culture (Ernst et al. 1995; Institute of Medicine (US) Committee on the Use of Complementary and Alternative Medicine by the American Public 2005; NCCIH 2020). Thus, the status of a certain method can change over time and over countries, depending on local medical tradition and many other factors. Also, methods belonging here are very heterogeneous with respect to their philosophical background, claimed mechanism, relation to scientific medicine, and effectiveness. Thus, I believe that putting them in the same category is a major source of error and misunderstanding. Perhaps one end of the spectrum is represented by phytotherapy, the medical use of active substances of plant origin. A proportion of these molecules, such as opioids of the poppy plant or glicosides of foxglove, have been used as components of medication for centuries. The constitution, biochemical interactions, effectiveness, and dosage of molecules synthesized by plants can be studied and described using the usual scientific approach. At the other extremity, one can find the so-called spiritual and/or bioenergetic methods, relying on claimed energies and dimensions beyond the physical world and highly subjective diagnostic and treatment procedures. They are much closer to witchcraft and religious practices than to science. What can have these methods (and the rest between these extremities) in common? Perhaps the most important shared characteristic of these methods is that they heavily rely on the nonspecific or placebo component of treatment (Buckman and Sabbagh 1993; Walach and Jonas 2004; Thompson 2005; Brown 2013). Whereas their specific therapeutic effectiveness, as demonstrated by double-blind, placebo-controlled studies, shows huge differences from poor, even non-existing, to considerable (Ernst et al. 2006), they put a special emphasis on the nonspecific aspects. For starters, they claim that they rely on natural substances and procedures, which is a synonym for secure and effective for many people (Dömötör et al. 2017). Their worldview and diagnostic procedure is holistic, i.e., takes psychological, social, sometimes even spiritual factors into consideration (NCCIH 2020), which leads to longer and more detailed conversations and a deeper relationship between the patient and the healer (Hawkins 2001; Reilly 2001; Richardson 2001). As a consequence, the patient has the feeling that his or her illness is thoroughly understood by the healer and its meaning was explained. Touch is also often involved in such therapeutic procedures (Latey 2001; Wright and

Sayre-Adams 2001), which instantly improves the affective state of the patient. To put it together, these factors dominantly target our primary consciousness (Saher and Lindeman 2005; Köteles et al. 2016) and result in a rapid increase of subjective well-being. If everything goes well, we feel much better after these treatments.

This is a good thing without doubt. Sick people deserve compassionate care and improvement of mood. The problem is, however, that we are prone to confuse the perceived state of our body with the actual state (tenet of realness) and the acute state with the chronic state. Primary consciousness is a real-time subjective experience, heavily influenced by biasing factors, such as pleasant experiences and positive messages. As mentioned above, feeling better means feeling cured for many of us, particularly for those with less knowledge of health and disease. Unfortunately, things are much more difficult, and particularly so for asymptomatic diseases (see also Chap. 9). In my opinion, this represents the most risky aspect of CAM methods. CAM often explicitly claims to deliver treatment and cure, not just temporary relief or subjective improvement, for a wide variety of diseases, including conditions for which there is no conventional treatment. In such a therapeutic context, an improvement in subjective well-being is automatically interpreted as an indicator of physical improvement by the patient. The healer, often not being aware of the tenet of dissociation, typically supports and reinforces this interpretation.

Unfortunately, the associations between self-reported and objective indicators of health status are moderate only. For example, subjective well-being explained approximately 12.6% of the total variance of perceived health status and 10.7% of the variance of actual health status in a recent meta-analysis (Ngamaba et al. 2017). This is a considerable overlap but far from complete. Also, subjective well-being does contribute to health status and longevity, but this does not mean that it also impacts serious diseases such as cancer (Diener and Chan 2011). Issues with causality are often mentioned; the possibility that good physiological functioning improves well-being is as likely (to say the least) as a direct impact of well-being on health, not to speak of interactions between these two and the possible existence of latent third variables (Diener and Chan 2011). Also, subjective health status, as assessed with a single question, is associated with indicators of objective health (Kaplan et al. 1996; Wu et al. 2013) and even mortality (Idler and Angel 1990; Appels et al. 1996; Benjamins et al. 2004). However, the strength of the association is in the weak to moderate domain, and self-assessment of health status is partly based on knowledge of existing diseases, such as diabetes or cardiovascular diseases (Wu et al. 2013). Negative affectivity, a component of subjective well-being, is also associated with perceived but not with objective heath status (Costa and McCrae 1985, 1987; Watson and Pennebaker 1989). In conclusion, acute positive changes in mood and subjective well-being are not valid and reliable indicators of objective health status; they can be regarded as biasing factors rather than signs of good health. Consciously accessible aspects of the actual status of the body are designed to indicate the factors that acutely threat homeostasis and need behavioral steps (hunger, thirst, breathlessness, pain, and so forth) (Ádám 1998). They cannot provide us with an objective and thorough diagnosis of the body state, particularly not with respect to chronic degenerative diseases. When we instantly feel better after a CAM intervention, it is a mistake to take this as proof of long-term therapeutic effectiveness. Of course, long-term therapeutic effects can occur, depending on the condition and the intervention, but this is not necessarily felt subjectively. Our body sensations, impacted by non-interoceptive factors (tenet of malleability), can easily fool us.

What can we do to avoid this pitfall? Once again (see Chap. 9), we simply need to use our rational thinking, i.e., higher consciousness. First, we should not take perceived improvement as a decisive sign of therapeutic effect. Second, we need to use these practices appropriately. As the name states, these methods can be used in two different ways: as an alternative of a conventional medical treatment or as an auxiliary treatment in addition to the conventional procedure. The complementary option might be risky in many cases as it often leads to the rejection of a potentially effective conventional therapy. If and when timing matters, for example, in rapidly progressing malignant tumors, such a rejection, or even a several months long delay before starting a potentially effective therapy, can be irreversible and fatal. As CAM is typically characterized by a non-scientific worldview, even with anti-scientific attitudes, practitioners sometimes advise their patients to alter or even abandon conventional therapy (Moody et al. 1998; Macpherson et al. 2004; Ernst 2006). In the case of slowly progressing diseases, such as arthrosis accompanied by intense pain in the impacted joints or hypertension, a pause in the medication does not lead to such tough consequences. The alternative approach, however, appears more rational to me. Interactions between conventional medical and CAM treatment can occur; for example, active substances of plants can decrease or increase the effectiveness of certain drugs (Brinker 2010). Still, the vast majority of CAM treatments appear to be safely usable simultaneously with their conventional counterparts. Also, we have empirical evidence on the specific effects of a number of CAM interventions beyond their nonspecific impact (Ernst et al. 2006). Still, the use of CAM techniques "just" to alleviate suffering and improve well-being is a valid and reasonable goal.

10.4 Mind–Body Techniques

The typical placebo response takes place in a non-healthy body and represents a shift toward the healthy state. If one conceptualizes the placebo phenomenon more generally, including cognitive and physical performance (Ross and Olson 1981), it still refers to an improvement toward a desired state. CAM is primarily involved in pathological conditions; however, it strongly emphasizes the prevention of diseases and the maintenance of healthy functioning (Stevinson and Ernst 2006). Mind–body practices, i.e., various movement and meditation systems that typically originate in Eastern cultures, such as yoga, tai chi chuan, chi kung, or breathing exercises, attempt to improve physical, mental, and spiritual aspects of well-being (Telles et al. 2015). Thus, practicing such techniques on a regular basis is regarded as an option for healthy people to keep themselves healthy.

Practicing mind–body techniques is a pleasurable activity, the pleasantness of which clearly originates in one's experience with his or her body. This positive feeling can partly be explained by the physical activity itself, as mild to moderate activity generates positive affect and reduces negative affect (Ekkekakis 2003; Ekkekakis et al. 2011, 2020). However, this explanation is not satisfying for static exercises, such as controlled breathing, and I am convinced that it is also insufficient for techniques that involve more physical activity. The quality and extent of the pleasure generated by jogging, a physical activity characterized by similar activation, cannot be compared to that of doing yoga. What makes these activities so special? I think partly the harmonic postures and movements; one cannot forget that it took thousands of experts hundreds of years to elaborate those seemingly simple techniques (Iyengar 1991; Ross and Thomas 2010; Govindaraj et al. 2016). Also, appropriately controlled breathing shifts the sympathetic–parasympathetic balance to the latter direction; meditation can even decrease the physiological aspects of stress (Sanada et al. 2016; Pascoe et al. 2017; Koncz et al. 2020), and parasympathetic activation is accompanied by positive feelings (Schmalzl et al. 2015; Strigo and Craig 2016).

From a psychological point of view, the positivity of these activities can be explained in multiple ways. For example, the state of being absorbed in an activity, called subjective self-awareness, is assumed to eliminate negative affect generated by objective self-awareness (Duval and Wicklund 1972). In more detail, the feeling of agency and being in the present changes the way we see ourselves; we apply a first-person perspective instead of the usual third-person perspective to ourselves. As the third-person perspective (i.e., objective self-awareness) is necessarily characterized by social comparison and we are very rarely good enough to completely reach social norms, it generates negative affect (Chap. 5). If we abandon this perspective, we simply eliminate a major source of negative affect, which in turn improves our mood state. Flow theory says something similar: our attentional capacity is finite, thus if we fill it with the here and now by an appropriately challenging activity, there will remain no room for higher self-awareness (self-reflection) with its worries, uncontrollable thoughts, and so forth (Jackson and Csikszentmihalyi 1999). Basically, these models operate with the reduction of negative affect.

However, we do not worry all the time and sometimes we are satisfied with ourselves—there are moments when the actual level of negative affect is already low. Practicing yoga or tai chi chuan can further improve our mood state even in these calm periods of life. How? I think being in the present oriented existence of primary consciousness is a positive experience *per se* for the majority of people. In addition, the experience of self-efficacy, i.e., being able to control our body in a sophisticated and relaxed way, is also rewarding (White 1959; Bandura 1977, 1997; Lox et al. 2010). Last but not least, many of these practices work with the concept of vital energy (Hui et al. 2007; Farb et al. 2015), something that belongs to primary consciousness (Lindeman and Saher 2007). Vital energy is not an abstract concept in many mind–body practices: it is clearly sensible as tingling and warmth in parts of the body one focuses on (Raingruber and Robinson 2007; Tihanyi and Köteles 2017). It can be evoked by healing practices claimed to work with spiritual energies, such as reiki; it is able to reduce pain and anxiety (Thrane and Cohen 2014; Billot et al. 2019), sometimes more efficiently than a simple placebo (McManus 2017). Energy-like sensations can also be evoked after intense and tenacious practice in Eastern mind–body systems (Farb et al. 2015). I have had an almost 10-year-long period in my life when I practiced tai chi chuan and chi kung on a daily basis. I was fascinated by the plethora of energy-like body feelings one can evoke and govern, directing them from one part of the body to another. One can clearly feel the flow of energy cycling in the centerline of the body or from the shoulders to the palms, following one's imagination and the rhythm of breath. This is a direct and distinct body sensation, obviously in the very absence of any external stimulation. As it belongs to primary consciousness, it is so strong and convincing (recall the tenet of realness) that is serves as perhaps the most important proof of the existence of spiritual energy for many people. As a scientist, I think that it can sufficiently be explained by the tingling sensation introduced in Chap. 5; attention focused on certain regions of the body, enhanced by expectation and imagination of the feelings, can trigger such sensations. This "materialistic" explanation may sound disappointing; however, I do not think that this insight reduces the pleasantness and peculiarity of the experience in any way, or devalue the altered state of consciousness it is accompanied by. Although psychotic states can also lead to odd body experiences (Fisher and Cleveland 1968; Fisher 1986), I do not regard these sensations as a pathological experience, at least as far as one can integrate them into the self and utilize it in an adaptive way (Raingruber and Robinson 2007). From a scientific point of view, the so-called deqi experience, a composite of sensations interpreted as the flow of vital energy (qi) in traditional Chinese medicine (Hui et al. 2007), received special attention. Deqi is considered essential for the clinical efficacy of acupuncture by experts of this method, thus a number of studies investigated its background (Raingruber and Robinson 2007; Litscher 2013; Zhou and Benharash 2014). Deqi is a sensation often propagated along lines on the surface of the body; these lines are similar to meridians described by traditional Chinese medicine (Beissner and Marzolff 2012; Li et al. 2013). The deqi sensation can be evoked by both real and sham acupuncture treatment, thus it was concluded that it is a nonspecific effect of the overall treatment procedure and closely related to conscious awareness (Salih et al. 2010; Irnich et al. 2011). Visual cues claimed to indicate stimulation in certain acupuncture points led to more intense deqi sensation than a cue outside the body even in the absence of actual stimulation (Jung et al. 2018), and the spatial pattern of the sensation can also be changed via the manipulation of beliefs with respect to the background physiological changes (Song et al. 2019). Sensations initialized by real acupuncture were more intense than those evoked by tactile stimulation (Jung et al. 2016). Moreover, stimulation delivered on real acupuncture points led to more intense deqi sensations than that on sham points (Li et al. 2013; Spaeth et al. 2013; Razavy et al. 2017). To sum up, the deqi sensation can be explained as an attention-related body sensation (Tihanyi et al. 2018), which can be particularly easily evoked by acupuncture-like interventions (Beissner et al. 2015). Bottom-up input caused by actual acupuncture can interact with the top-down component and further intensifies it (Chae et al. 2015). In consequence, acupuncture interventions are able to evoke a greater than usual placebo response (Beissner et al. 2015) thus can be considered a mega-placebo (Musial 2019), i.e., a nonspecific intervention that evokes an enhanced placebo response (Kaptchuk et al. 2000). In my opinion, the body sensations triggered by these practices represent a component that substantially boost their effectiveness in certain pathological conditions, most importantly pain, as well as in improving subjective well-being.

10.5 Summary

The human brain has multiple functional layers that are shaped to fulfill different jobs at different levels. These functional levels cannot be completely separated; if everything goes well, they interact with each other and complement each other in an adaptive manner. The most ancient layer takes care of automatic physiological regulation, thus it is relatively separated from higher levels of functioning. Experiential thinking, dominantly belonging to primary consciousness and a relatively older functional layer, is anchored in the present (Epstein et al. 1996). Its main aim is to maximize actual pleasure and minimize suffering; its decisions are made relying on previous experiences with repetitive co-occurrence of events (Epstein 2003). I think the topics discussed in this chapter (the placebo response, complementary and alternative medicine, mind–body practices) neatly illustrate the benefits and also the pitfalls of experiential thinking. Typically, we have no detailed conscious information on the actual (patho)physiological state of our body. We just estimate it, relying on affectively colored body sensations, which are often misleading after an intervention with a marked placebo component. This might be dangerous, if we accept it as proof of physical improvement. Thus, despite the positive state, we need to accept difficulties and unpleasantness, such as preventive care interventions or sophisticated diagnostic procedures. These decisions and behaviors should be made and controlled by rational thinking, our most recent functional layer (Chap. 9). It is future-oriented and typically less strong and convincing than direct experience, thus its use requires conscious effort (Epstein 2003; Kahneman 2011).

I think the bottom line of this chapter (in concert with Chap. 9) is quite simple: never trust your body in health issues. The lack of unpleasant body sensations, even the presence of positive feelings, may refer to a healthy state as well as a variety of asymptomatic diseases or transient effects of placebo interventions. As the tenet of dissociation states, we cannot decide whether the representation of the body described the actual state of the body or not. This conclusion sounds tough, I know. Still, it is the best advice one can follow in health-related issues. Further, it does not mean that the placebo response or alternative healing practices are worthless or simple quackery. We can use them wisely to alleviate suffering and further improve our well-being in healthy states. A huge benefit of CAM and mind–body practices is that they emphasize the importance of prevention and try to change our lifestyle dominantly through pleasurable activities.

Take-Home Messages

For laypeople: After receiving a treatment or intervention with a marked placebo component, such as complementary and alternative or mind–body techniques, never quit conventional therapy just because you feel much better at the moment. The improvement can be temporary and may not impact your body.

For clinicians: Placebo interventions and other interventions with a nonspecific component have a strong impact on pain, anxiety, and depression; however, their peripheral effects are much weaker. Thus, they can be used to alleviate suffering and subjective symptoms rather than cure organic diseases.

For researchers: The three points mentioned with respect to the nocebo effect (i.e., research into the peripheral effects, the role of expectations, and understanding the asymmetries between nocebo and placebo mechanisms) are valid here too.

References

Ádám G (1998) Visceral perception: understanding internal cognition. Plenum Press, New York

Ádám G (2011) A placebojelenség kognitív éntérkép-elmélete. Magy Tud 285–289

Ader R (1997) The role of conditioning in pharmacotherapy. In: Harrington A (ed) The placebo effect. An interdisciplinary exploration. Harvard University Press, Cambridge, pp 138–165

Ader R (2000) The placebo effect: if it's all in your head, does that mean you only think you feel better? Adv Mind Body Med 16:7–11. https://doi.org/10.1054/ambm.2000.0181

Ader R, Mercurio MG, Walton J et al (2010) Conditioned pharmacotherapeutic effects: a preliminary study. Psychosom Med 72:192–197. https://doi.org/10.1097/PSY.0b013e3181cbd38b

Appels A, Bosma H, Grabauskas V et al (1996) Self-rated health and mortality in a Lithuanian and a Dutch population. Soc Sci Med 42:681–689. https://doi.org/10.1016/0277-9536(95)00195-6

Arthaud-Day ML, Rode JC, Mooney CH, Near JP (2005) The subjective well-being construct: a test of its convergent, discriminant, and factorial validity. Soc Indic Res 74:445–476. https://doi.org/10.1007/s11205-004-8209-6

Aslaksen PM, Bystad M, Vambheim SM, Flaten MA (2011) Gender differences in placebo analgesia: event-related potentials and emotional modulation. Psychosom Med 73:193–199. https://doi.org/10.1097/PSY.0b013e3182080d73

Bandura A (1977) Self-efficacy: toward a unifying theory of behavioral change. Psychol Rev 84:191–215

Bandura A (1997) Self-efficacy: the exercise of control. W.H. Freeman & Co Ltd, New York

Barsky AJ (1983) Nonpharmacologic aspects of medication. Arch Intern Med 143:1544–1548

Barsky AJ (1988) The paradox of health. N Engl J Med 318:414–418. https://doi.org/10.1056/NEJM198802183180705

Barsky AJ, Saintfort R, Rogers MP, Borus JF (2002) Nonspecific medication side effects and the nocebo phenomenon. JAMA 287:622–627

Beedie C, Foad A, Hurst P (2015) Capitalizing on the placebo component of treatments. Curr Sports Med Rep 14:284–287. https://doi.org/10.1249/JSR.0000000000000172

Beedie C, Benedetti F, Barbiani D et al (2018) Consensus statement on placebo effects in sports and exercise: the need for conceptual clarity, methodological rigour, and the elucidation of neurobiological mechanisms. Eur J Sport Sci 18:1383–1389. https://doi.org/10.1080/17461391.2018.1496144

Beissner F, Marzolff I (2012) Investigation of acupuncture sensation patterns under sensory deprivation using a geographic information system. Evid Based Complement Altern Med 2012:e591304. https://doi.org/10.1155/2012/591304

Beissner F, Brünner F, Fink M et al (2015) Placebo-induced somatic sensations: a multi-modal study of three different placebo interventions. PLoS One 10:e0124808. https://doi.org/10.1371/journal.pone.0124808

Benedetti F (1996) The opposite effects of the opiate antagonist naloxone and the cholecystokinin antagonist proglumide on placebo analgesia. Pain 64:535–543

Benedetti F (2009) Placebo effects: understanding the mechanisms in health and disease. Oxford University Press, New York

Benedetti F (2011) The patient's brain. The neuroscience behind the doctor-patient relationship. New York, Oxford University Press

Benjamins MR, Hummer RA, Eberstein IW, Nam CB (2004) Self-reported health and adult mortality risk: an analysis of cause-specific mortality. Soc Sci Med 59:1297–1306. https://doi.org/10.1016/j.socscimed.2003.01.001

Benson H, Friedman R (1996) Harnessing the power of the placebo effect and renaming it "remembered wellness". Annu Rev Med 47:193–199. https://doi.org/10.1146/annurev.med.47.1.193

Bérdi M, Köteles F, Szabó A, Bárdos G (2011) Placebo effects in sport and exercise: a meta-analysis. Eur J Ment Health 6:196–212. https://doi.org/10.5708/EJMH.6.2011.2.5

Bérdi M, Köteles F, Hevesi K et al (2015) Elite athletes' attitudes towards the use of placebo-induced performance enhancement in sports. Eur J Sport Sci 15:315–321. https://doi.org/10.1080/17461391.2014.955126

Bierman SF (2000) Of course mental events affect physical events. Adv Mind Body Med 16:11–13. https://doi.org/10.1054/ambm.2000.0124

Billot M, Daycard M, Wood C, Tchalla A (2019) Reiki therapy for pain, anxiety and quality of life. BMJ Support Palliat Care 9:434–438. https://doi.org/10.1136/bmjspcare-2019-001775

Bjørkedal E, Flaten MA (2011) Interaction between expectancies and drug effects: an experimental investigation of placebo analgesia with caffeine as an active placebo. Psychopharmacology 215:537–548. https://doi.org/10.1007/s00213-011-2233-4

Blake H (2012) Physical activity and exercise in the treatment of depression. Front Psych 3:106. https://doi.org/10.3389/fpsyt.2012.00106

Bok S (1974) The ethics of giving placebos. Sci Am 231:17–23

Bok S (2002) Ethical issues in use of placebo in medical practice and clinical trials. In: Guess HA, Kleinman A, Kusek JW, Engel LW (eds) The science of the placebo. Toward an interdisciplinary research agenda. BMJ Books, London, pp 53–74

Brinker F (2010) Herbal contraindications and drug interactions: plus herbal adjuncts with medicines, 4th edn. Eclectic Medical Publications, Sandy, OR

Brody H (1980) Placebos and the philosophy of medicine. Clinical, conceptual, and ethical issues. University of Chicago Press, Chicago

Brody H (1982) The lie that heals: the ethics of giving placebos. Ann Intern Med 97:112–118

Brody H (1997) The doctor as a therapeutic agent: a placebo effect research agenda. In: Harrington A (ed) The placebo effect. An interdisciplinary exploration. Harvard University Press, Cambridge, pp 77–92

Brody H (2000) The vast majority of drug studies offer little if any evidence to support the physiologic effects of the placebo. Adv Mind Body Med 16:13–16. https://doi.org/10.1054/ambm.2000.0125

Brody H, Brody D (2000) The placebo response. Harper Collins, New York

Brody H, Colloca L, Miller FG (2012) The placebo phenomenon: implications for the ethics of shared decision-making. J Gen Intern Med 27:739–742. https://doi.org/10.1007/s11606-011-1977-1

Brown WA (2013) The placebo effect in clinical practice. Oxford University Press, New York

Brown RP, Gerbarg PL (2005) Sudarshan Kriya yogic breathing in the treatment of stress, anxiety, and depression: Part I-Neurophysiologic model. J Altern Complement Med 11:189–201. https://doi.org/10.1089/acm.2005.11.189

Brown HE, Pearson N, Braithwaite RE et al (2013) Physical activity interventions and depression in children and adolescents: a systematic review and meta-analysis. Sports Med 43:195–206. https://doi.org/10.1007/s40279-012-0015-8

Büchel C, Geuter S, Sprenger C, Eippert F (2014) Placebo analgesia: a predictive coding perspective. Neuron 81:1223–1239. https://doi.org/10.1016/j.neuron.2014.02.042

Buckalew LW, Ross S, Starr BJ (1981) Nonspecific factors in drug effects: placebo personality. Psychol Rep 48:3–8

Buckman R, Sabbagh K (1993) Magic or medicine? An investigation into healing. Maxmillan, London

Calabrese P, Perrault H, Dinh TP et al (2000) Cardiorespiratory interactions during resistive load breathing. Am J Physiol Regul Integr Comp Physiol 279:R2208–R2213. https://doi.org/10.1152/ajpregu.2000.279.6.R2208

Calsius J, De Bie J, Hertogen R, Meesen R (2016) Touching the lived body in patients with medically unexplained symptoms. How an integration of hands-on bodywork and body awareness in psychotherapy may help people with alexithymia. Front Psychol 7:253. https://doi.org/10.3389/fpsyg.2016.00253

Cardeña E, Kirsch I (2000) What is so special about the placebo effect? Adv Mind Body Med 16:16–18. https://doi.org/10.1054/ambm.2000.0126

Carney DR, Cuddy AJC, Yap AJ (2010) Power posing: brief nonverbal displays affect neuroendocrine levels and risk tolerance. Psychol Sci 21:1363–1368. https://doi.org/10.1177/0956797610383437

Cash TF, Deagle EA (1997) The nature and extent of body-image disturbances in anorexia nervosa and bulimia nervosa: a meta-analysis. Int J Eat Disord 22:107–125

Cash TF, Pruzinsky T (eds) (1990) Body images: development, deviance, and change. Guilford Press, New York

Cash TF, Pruzinsky T (eds) (2002) Body image: a handbook of theory, research and clinical practice. Guilford Press, New York

Chae Y, Lee I-S, Jung W-M et al (2015) Psychophysical and neurophysiological responses to acupuncture stimulation to incorporated rubber hand. Neurosci Lett 591:48–52. https://doi.org/10.1016/j.neulet.2015.02.025

Claridge G (1970) Drugs and human behaviour. Praeger Publishers, New York

Colloca L (2018) The fascinating mechanisms and implications of the placebo effect. Int Rev Neurobiol 138:xv–xx. https://doi.org/10.1016/S0074-7742(18)30027-8

Colloca L, Barsky AJ (2020) Placebo and nocebo effects. N Engl J Med 382:554–561. https://doi.org/10.1056/NEJMra1907805

Colloca L, Petrovic P, Wager TD et al (2010) How the number of learning trials affects placebo and nocebo responses. Pain 151:430–439. https://doi.org/10.1016/j.pain.2010.08.007

Conn VS (2010) Anxiety outcomes after physical activity interventions: meta-analysis findings. Nurs Res 59:224–231. https://doi.org/10.1097/NNR.0b013e3181dbb2f8

Costa PT, McCrae RR (1985) Hypochondriasis, neuroticism, and aging. When are somatic complaints unfounded? Am Psychol 40:19–28

Costa PT, McCrae RR (1987) Neuroticism, somatic complaints, and disease: is the bark worse than the bite? J Pers 55:299–316. https://doi.org/10.1111/j.1467-6494.1987.tb00438.x

Csala B, Ferentzi E, Tihanyi BT et al (2020) Verbal cuing is not the path to enlightenment. Psychological effects of a 10-session hatha yoga practice. Front Psychol 11:1375. https://doi.org/10.3389/fpsyg.2020.01375

de Craen AJ, Roos PJ, de Vries AL, Kleijnen J (1996) Effect of colour of drugs: systematic review of perceived effect of drugs and of their effectiveness. Br Med J 313:1624–1626

de Craen AJ, Tijssen JG, de Gans J, Kleijnen J (2000) Placebo effect in the acute treatment of migraine: subcutaneous placebos are better than oral placebos. J Neurol 247:183–188

Di Blasi Z, Harkness E, Ernst E et al (2001) Influence of context effects on health outcomes: a systematic review. Lancet 357:757–762

Diener E (1984) Subjective well-being. Psychol Bull 95:542–575

Diener E, Chan MY (2011) Happy people live longer: subjective well-being contributes to health and longevity. Appl Psychol Health Well Being 3:1–43. https://doi.org/10.1111/j.1758-0854.2010.01045.x

Dömötör Z, Szemerszky R, Köteles F (2017) Nature relatedness is connected with modern health worries and electromagnetic hypersensitivity. J Health Psychol 24:1359105317699681. https://doi.org/10.1177/1359105317699681

Duval S, Wicklund RA (1972) A theory of objective self-awareness. Academic, New York

Eippert F, Finsterbusch J, Bingel U, Büchel C (2009) Direct evidence for spinal cord involvement in placebo analgesia. Science 326:404. https://doi.org/10.1126/science.1180142

Ekkekakis P (2003) Pleasure and displeasure from the body: perspectives from exercise. Cogn Emot 17:213–239. https://doi.org/10.1080/02699930302292

Ekkekakis P, Parfitt G, Petruzzello SJ (2011) The pleasure and displeasure people feel when they exercise at different intensities: decennial update and progress towards a tripartite rationale for exercise intensity prescription. Sports Med 41:641–671. https://doi.org/10.2165/11590680-000000000-00000

Ekkekakis P, Hartman ME, Ladwig MA (2020) Affective responses to exercise. In: Tenenbaum G, Eklund RC (eds) Handbook of sport psychology, 4th edn. Wiley, Hoboken, NJ, pp 233–253

Emanuelsen L, Drew R, Köteles F (2015) Interoceptive sensitivity, body image dissatisfaction, and body awareness in healthy individuals. Scand J Psychol 56:167–174. https://doi.org/10.1111/sjop.12183

Enck P, Bingel U, Schedlowski M, Rief W (2013) The placebo response in medicine: minimize, maximize or personalize? Nat Rev Drug Discov 12:191–204. https://doi.org/10.1038/nrd3923

Epstein S (2003) Cognitive-experiential self-theory of personality. In: Millon T, Lerner MJ (eds) Comprehensive handbook of psychology, Personality and social psychology, vol 5. Wiley, Hoboken, NJ, pp 159–184

Epstein S, Pacini R, Denes-Raj V, Heier H (1996) Individual differences in intuitive-experiential and analytical-rational thinking styles. J Pers Soc Psychol 71:390–405

Ernst E (2006) Safety issues in complementary and alternative medicine. In: Ernst E, Pittler MH, Wider B (eds) The desktop guide to complementary and alternative medicine. An evidence-based approach. Mosby Elsevier, Maryland Heights, pp 520–524

Ernst E, Resch KL, Mills S et al (1995) Complementary medicine—a definition. Br J Gen Pract 45:506

Ernst E, Pittler MH, Wider B (eds) (2006) A desktop guide to complementary and alternative medicine. An evidence-based approach. Mosby Elsevier, Maryland Heights

Farb NA, Logie K (2018) Interoceptive appraisal and mental health. In: Tsakiris M, De Preester H (eds) The interoceptive mind. From homeostasis to awareness. Oxford University Press, Oxford, England, pp 227–241

Farb NA, Daubenmier JJ, Price CJ et al (2015) Interoception, contemplative practice, and health. Front Psychol 6:1–26. https://doi.org/10.3389/fpsyg.2015.00763

Fässler M, Gnädinger M, Rosemann T, Biller-Andorno N (2009) Use of placebo interventions among Swiss primary care providers. BMC Health Serv Res 9:144. https://doi.org/10.1186/1472-6963-9-144

Ferentzi E, Köteles F, Bárdos G (2010) The therapeutic use of placebos among Hungarian GPs - a preliminary research report. Clin Exp Med J 5:1–5. https://doi.org/10.1556/CEMED.4.2010.2.5

Ferentzi E, Köteles F, Bárdos G (2011) The use of placebos in medical practice. A questionnaire survey among GPs of Hungary. Clin Exp Med J 5:73–84. https://doi.org/10.1556/CEMED.5.2011.2.2

Ferentzi E, Horváth Á, Köteles F (2019) Do body-related sensations make feel us better? Subjective well-being is associated only with the subjective aspect of interoception. Psychophysiology 56:e13319. https://doi.org/10.1111/psyp.13319

Fields HL, Price DD (1997) Toward a neurobiology of placebo analgesia. In: Harrington A (ed) The placebo effect: an interdisciplinary exploration. Harvard University Press, Cambridge, MA, pp 93–116

Fischer S, Greenberg RP (1997) The curse of the placebo: fanciful pursuit of a pure biological therapy. In: Fisher S, Greenberg RP (eds) From placebo to pancea - putting psychiatric drugs to the test. Wiley, New York, pp 3–56

Fisher S (1986) Development and structure of the body image, vol 1. L. Erlbaum Associates, New York

Fisher S (2000) "Is there really a placebo effect, professor"? Adv Mind Body Med 16:19–21. https://doi.org/10.1054/ambm.2000.0127

Fisher S, Cleveland SE (1968) Body image and personality, 2nd rev edn. Dover Publications, New York

Gibson J (2019) Mindfulness, interoception, and the body: a contemporary perspective. Front Psychol 10:2012. https://doi.org/10.3389/fpsyg.2019.02012

Gold A, Lichtenberg P (2013) The moral case for the clinical placebo. J Med Ethics 40(4):219–224. https://doi.org/10.1136/medethics-2012-101314

Govindaraj R, Karmani S, Varambally S, Gangadhar BN (2016) Yoga and physical exercise - a review and comparison. Int Rev Psychiatry 28:242–253. https://doi.org/10.3109/09540261.2016.1160878

Gracely RH, Dubner R, Wolskee PJ, Deeter WR (1983) Placebo and naloxone can alter post-surgical pain by separate mechanisms. Nature 306:264–265

Grahl A, Onat S, Büchel C (2018) The periaqueductal gray and Bayesian integration in placebo analgesia. Elife 7:e32930. https://doi.org/10.7554/eLife.32930

Grevert P, Albert LH, Goldstein A (1983) Partial antagonism of placebo analgesia by naloxone. Pain 16:129–143

Guo J-Y, Wang J-Y, Luo F (2010) Dissection of placebo analgesia in mice: the conditions for activation of opioid and non-opioid systems. J Psychopharmacol (Oxford) 24:1561–1567. https://doi.org/10.1177/0269881109104848

Harrington A (2008) The cure within. A history of mind-body medicine. W. W. Norton, New York

Hawkins J (2001) How can we optimize non-specific effects. In: Peters D (ed) Understanding the placebo effect in complementary medicine. Churchill Livingstone, Edinburgh, pp 69–88

Howick J, Bishop FL, Heneghan C et al (2013) Placebo use in the United Kingdom: results from a National Survey of Primary Care Practitioners. PLoS One 8:e58247. https://doi.org/10.1371/journal.pone.0058247

Hui KK, Nixon EE, Vangel MG et al (2007) Characterization of the "deqi" response in acupuncture. BMC Complement Altern Med 7:33. https://doi.org/10.1186/1472-6882-7-33

Humphrey N (2002) Great expectations: the evolutionary psychology of faith healing and the placebo effect. In: Humphrey N (ed) The mind made flesh: essays from the frontier of evolution and psychology. Oxford University Press, Oxford, England, pp 255–285

Hurst P, Schipof-Godart L, Szabo A et al (2019) The placebo and nocebo effect on sports performance: a systematic review. Eur J Sport Sci 20:1–14. https://doi.org/10.1080/17461391.2019.1655098

Idler EL, Angel RJ (1990) Self-rated health and mortality in the NHANES-I Epidemiologic Follow-up Study. Am J Public Health 80:446–452

Institute of Medicine (US) Committee on the Use of Complementary and Alternative Medicine by the American Public (2005) Complementary and alternative medicine in the United States. National Academies Press (US), Washington, DC

Irnich D, Salih N, Offenbächer M, Fleckenstein J (2011) Is sham laser a valid control for acupuncture trials? Evid Based Complement Alternat Med 2011:1–8. https://doi.org/10.1093/ecam/neq009

Iyengar BKS (1991) Light on yoga. The Aquarian Press, London

Jackson SA, Csikszentmihalyi M (1999) Flow in sports. Human Kinetics, Champaign, IL

Josefsson T, Lindwall M, Archer T (2014) Physical exercise intervention in depressive disorders: meta-analysis and systematic review. Scand J Med Sci Sports 24:259–272. https://doi.org/10.1111/sms.12050

Jung W-M, Shim W, Lee T et al (2016) More than DeQi: spatial patterns of acupuncture-induced bodily sensations. Front Neurosci 10:462. https://doi.org/10.3389/fnins.2016.00462

Jung W-M, Ryu Y, Park H-J et al (2018) Brain activation during the expectations of sensory experience for cutaneous electrical stimulation. Neuroimage Clin 19:982–989. https://doi.org/10.1016/j.nicl.2018.06.022

Kahneman D (2011) Thinking, fast and slow. Farrar, Straus and Giroux, New York

Kaplan GA, Goldberg DE, Everson SA et al (1996) Perceived health status and morbidity and mortality: evidence from the Kuopio ischaemic heart disease risk factor study. Int J Epidemiol 25:259–265. https://doi.org/10.1093/ije/25.2.259

Kaptchuk TJ (2018) Open-label placebo: reflections on a research agenda. Perspect Biol Med 61:311–334. https://doi.org/10.1353/pbm.2018.0045

Kaptchuk TJ, Goldman P, Stone DA, Stason WB (2000) Do medical devices have enhanced placebo effects? J Clin Epidemiol 53:786–792. https://doi.org/10.1016/S0895-4356(00)00206-7

Kardas P, Lewek P, Matyjaszczyk M (2013) Determinants of patient adherence: a review of systematic reviews. Front Pharmacol 4:91. https://doi.org/10.3389/fphar.2013.00091

Kirsch I (1985) Response expectancy as a determinant of experience and behavior. Am Psychol 40:1189–1202

Kirsch I (2008) Challenging received wisdom: antidepressants and the placebo effect. Mcgill J Med 11:219–222

Kirsch I (2010) The Emperor's new drugs: exploding the antidepressant myth. Basic Books, New York

Klinger R, Soost S, Flor H, Worm M (2007) Classical conditioning and expectancy in placebo hypoalgesia: a randomized controlled study in patients with atopic dermatitis and persons with healthy skin. Pain 128:31–39. https://doi.org/10.1016/j.pain.2006.08.025

Koncz A, Demetrovics Z, Takacs ZK (2020) Meditation interventions efficiently reduce cortisol levels of at-risk samples: a meta-analysis. Health Psychol Rev 1–29. https://doi.org/10.1080/17437199.2020.1760727

Köteles F (2013) A placebo-válasz. Medicina Könyvkiadó Zrt, Budapest

Köteles F (2014) A Testi Tudatosság Kérdőív magyar verziójának (BAQ-H) vizsgálata jógázó és fiatal felnőtt kontroll mintán. Mentálhigiéné és Pszichoszomatika 15:373–391. https://doi.org/10.1556/Mental.15.2014.4.4

Köteles F, Simor P, Czető M et al (2016) Modern health worries - the dark side of spirituality? Scand J Psychol 57:313–320. https://doi.org/10.1111/sjop.12297

Köteles F, Teufel B, Körmendi J et al (2020) Cardioceptive accuracy is associated with arousal but not with valence and perceived exertion under physical load. Psychophysiology 57:e13620. https://doi.org/10.1111/psyp.13620

Latey P (2001) Placebo responses in bodywork. In: Peters D (ed) Understanding the placebo effect in complementary medicine. Churchill Livingstone, Edinburgh, pp 147–163

Levine JD, Gordon NC (1984) Influence of the method of drug administration on analgesic response. Nature 312:755–756

Levine JD, Gordon NC, Fields HL (1978) The mechanism of placebo analgesia. Lancet 2:654–657

Li C-R, Lin Y, Guan H-Y et al (2013) Different surface electromyography of propagated sensation along meridians produced by acupuncturing Quchi acupoint (LI11) or control points. Evid Based Complement Alternat Med 2013:198451. https://doi.org/10.1155/2013/198451

Lindeman M, Saher M (2007) Vitalism, purpose and superstition. Br J Psychol 98:33–44

Litscher G (2013) Yes, there is Deqi sensation in laser acupuncture. Evid Based Complement Alternat Med 2013:198254. https://doi.org/10.1155/2013/198254

Lloyd DM, Gillis V, Lewis E et al (2013) Pleasant touch moderates the subjective but not objective aspects of body perception. Front Behav Neurosci 7:207. https://doi.org/10.3389/fnbeh.2013.00207

Lorenz J, Hauck M, Paur RC et al (2005) Cortical correlates of false expectations during pain intensity judgments—a possible manifestation of placebo/nocebo cognitions. Brain Behav Immun 19:283–295. https://doi.org/10.1016/j.bbi.2005.03.010

Lox CL, Martin Ginis KA, Petruzzello SJ (2010) The psychology of exercise. Integrating theory and practice, 3rd edn. Holcomb Hathaway, Scottsdale, AZ

Lucas RE, Diener E, Suh E (1996) Discriminant validity of well-being measures. J Pers Soc Psychol 71:616–628. https://doi.org/10.1037//0022-3514.71.3.616

Macpherson H, Scullion A, Thomas KJ, Walters S (2004) Patient reports of adverse events associated with acupuncture treatment: a prospective national survey. Qual Saf Health Care 13:349–355. https://doi.org/10.1136/qhc.13.5.349

McManus DE (2017) Reiki is better than placebo and has broad potential as a complementary health therapy. J Evid Based Complement Altern Med 22:1051–1057. https://doi.org/10.1177/2156587217728644

McNair DM, Gardos G, Haskell DS, Fisher S (1979) Placebo response, placebo effect, and two attributes. Psychopharmacology 63:245–250

Meissner K, Höfner L, Fässler M, Linde K (2012) Widespread use of pure and impure placebo interventions by GPs in Germany. Fam Pract 29:79–85. https://doi.org/10.1093/fampra/cmr045

Mlynski C, Wright RA, Kelly K (2020) Ability influence on effort and associated cardiovascular responses: nocebo-placebo evidence that perception is key. Biol Psychol 152:107867. https://doi.org/10.1016/j.biopsycho.2020.107867

Moerman DE (2002) Meaning, medicine and the "placebo effect". Cambridge University Press, Cambridge

Moody G, Eaden J, Bhakta P et al (1998) The role of complementary medicine in European and Asian patients with inflammatory bowel disease. Public Health 112:269–271. https://doi.org/10.1016/S0033-3506(98)00245-5

Munro SA, Lewin SA, Smith HJ et al (2007) Patient adherence to tuberculosis treatment: a systematic review of qualitative research. PLoS Med 4:e238. https://doi.org/10.1371/journal.pmed.0040238

Musial F (2019) Acupuncture for the treatment of pain – a mega-placebo? Front Neurosci 13:1110. https://doi.org/10.3389/fnins.2019.01110

Nakao M, Yano E, Nomura S, Kuboki T (2003) Blood pressure-lowering effects of biofeedback treatment in hypertension: a meta-analysis of randomized controlled trials. Hypertens Res 26:37–46. https://doi.org/10.1291/hypres.26.37

NCCIH (2020) Complementary, alternative, or integrative health: what's in a name?. NCCIH. https://www.nccih.nih.gov/health/complementary-alternative-or-integrative-health-whats-in-a-name. Accessed 30 Jun 2020

Ngamaba KH, Panagioti M, Armitage CJ (2017) How strongly related are health status and subjective well-being? Systematic review and meta-analysis. Eur J Pub Health 27:879–885. https://doi.org/10.1093/eurpub/ckx081

Nitzan U, Lichtenberg P (2004) Questionnaire survey on use of placebo. BMJ 329:944–946. https://doi.org/10.1136/bmj.38236.646678.55

Oehl M, Hummer M, Fleischhacker WW (2000) Compliance with antipsychotic treatment. Acta Psychiatr Scand Suppl 102:83–86. https://doi.org/10.1034/j.1600-0447.2000.00016.x

Panksepp J, Biven L (2012) The archaeology of mind. Neuroevolutionary origins of human emotions. W. W. Norton & Company, New York

Pascoe MC, Thompson DR, Jenkins ZM, Ski CF (2017) Mindfulness mediates the physiological markers of stress: systematic review and meta-analysis. J Psychiatr Res 95:156–178. https://doi.org/10.1016/j.jpsychires.2017.08.004

Paulus MP, Stein MB (2010) Interoception in anxiety and depression. Brain Struct Funct 214:451–463. https://doi.org/10.1007/s00429-010-0258-9

Petrie KJ, Rief W (2019) Psychobiological mechanisms of placebo and nocebo effects: pathways to improve treatments and reduce side effects. Annu Rev Psychol 70:599–625. https://doi.org/10.1146/annurev-psych-010418-102907

Petrie KJ, Sivertsen B, Hysing M et al (2001) Thoroughly modern worries: the relationship of worries about modernity to reported symptoms, health and medical care utilization. J Psychosom Res 51:395–401. https://doi.org/10.1016/S0022-3999(01)00219-7

Petrovic P, Kalso E, Petersson KM, Ingvar M (2002) Placebo and opioid analgesia—imaging a shared neuronal network. Science 295:1737–1740. https://doi.org/10.1126/science.1067176
Price DD (2000) Do placebo effects in analgesic drug studies demonstrate powerful mind-body interactions? Adv Mind Body Med 16:21–24. https://doi.org/10.1054/ambm.2000.0128
Price DD, Finniss D, Benedetti F (2008) A comprehensive review of the placebo effect: recent advances and current thought. Annu Rev Psychol 59:565–590. https://doi.org/10.1146/annurev.psych.59.113006.095941
Raglin J, Szabo A, Lindheimer JB, Beedie C (2020) Understanding placebo and nocebo effects in the context of sport: a psychological perspective. Eur J Sport Sci 20:1–9. https://doi.org/10.1080/17461391.2020.1727021
Raingruber B, Robinson C (2007) The effectiveness of tai chi, yoga, meditation, and reiki healing sessions in promoting health and enhancing problem solving abilities of registered nurses. Issues Ment Health Nurs 28:1141–1155. https://doi.org/10.1080/01612840701581255
Ranehill E, Dreber A, Johannesson M et al (2015) Assessing the robustness of power posing: no effect on hormones and risk tolerance in a large sample of men and women. Psychol Sci 26:653–656. https://doi.org/10.1177/0956797614553946
Razavy S, Gadau M, Zhang SP et al (2017) Investigation of the phenomenon of propagated sensation along the channels in the upper limb following administration of acupuncture and mock laser. J Acupunct Meridian Stud 10:307–316. https://doi.org/10.1016/j.jams.2017.06.007
Reilly D (2001) Some reflections on creating therapeutic consultations. In: Peters D (ed) Understanding the placebo effect in complementary medicine. Churchill Livingstone, Edinburgh, pp 89–110
Richardson J (2001) Intersubjectivity and the therapeutic relationship. In: Peters D (ed) Understanding the placebo effect in complementary medicine. Churchill Livingstone, Edinburgh, pp 131–146
Roeser S (2015) Placebo, nocebo, informed consent, and moral technologies. Am J Bioeth 15:15–17. https://doi.org/10.1080/15265161.2015.1074317
Ross M, Olson JM (1981) An expectancy-attribution model of the effects of placebos. Psychol Rev 88:408–437
Ross A, Thomas S (2010) The health benefits of yoga and exercise: a review of comparison studies. J Altern Complement Med 16:3–12. https://doi.org/10.1089/acm.2009.0044
Saher M, Lindeman M (2005) Alternative medicine: a psychological perspective. Personal Individ Differ 39:1169–1178. https://doi.org/10.1016/j.paid.2005.04.008
Salih N, Bäumler PI, Simang M, Irnich D (2010) Deqi sensations without cutaneous sensory input: results of an RCT. BMC Complement Altern Med 10:81. https://doi.org/10.1186/1472-6882-10-81
Sanada K, Montero-Marin J, Alda Díez M et al (2016) Effects of mindfulness-based interventions on salivary cortisol in healthy adults: a meta-analytical review. Front Physiol 7:471. https://doi.org/10.3389/fphys.2016.00471
Savvas SM, Zelencich LM, Gibson SJ (2014) Should placebo be used routinely for chronic pain in older people? Maturitas 79:389–400. https://doi.org/10.1016/j.maturitas.2014.09.006
Schmalzl L, Powers C, Henje Blom E (2015) Neurophysiological and neurocognitive mechanisms underlying the effects of yoga-based practices: towards a comprehensive theoretical framework. Front Hum Neurosci 9:235. https://doi.org/10.3389/fnhum.2015.00235
Schwarz KA, Büchel C (2015) Cognition and the placebo effect - dissociating subjective perception and actual performance. PLoS One 10:e0130492. https://doi.org/10.1371/journal.pone.0130492
Schweizer E, Rickels K (1997) Placebo response in generalized anxiety: its effect on the outcome of clinical trials. J Clin Psychiatry 58(Suppl 11):30–38
Seer P (1979) Psychological control of essential hypertension: review of the literature and methodological critique. Psychol Bull 86:1015–1043. https://doi.org/10.1037/0033-2909.86.5.1015
Shapiro AK, Shapiro E (1997) The powerful placebo: from ancient priest to modern physician. Johns Hopkins University Press, Baltimore

Shea SC (2006) Improving medication adherence. Lippincott Williams & Wilkins, Philadelphia
Song H-S, Jung W-M, Lee Y-S et al (2019) Expectations of the physiological responses can change the somatosensory experience for acupuncture stimulation. Front Neurosci 13:74. https://doi.org/10.3389/fnins.2019.00074
Sovik R (2000) The science of breathing—the yogic view. Prog Brain Res 122:491–505. https://doi.org/10.1016/s0079-6123(08)62159-7
Spaeth RB, Camhi S, Hashmi JA et al (2013) A longitudinal study of the reliability of acupuncture deqi sensations in knee osteoarthritis. Evid Based Complement Altern Med 2013:204259. https://doi.org/10.1155/2013/204259
Spiro HM (1998) The power of hope: a doctor's perspective. Yale University Press, New Haven
Spiro HM (2000) A contribution to the debate. Adv Mind Body Med 16:26–27. https://doi.org/10.1054/ambm.2000.0130
Stevinson C, Ernst E (2006) Why patients use complementary and alternative medicine. In: Ernst E, Pittler MH, Wider B (eds) The desktop guide to complementary and alternative medicine. An evidence-based approach, vol 2. Mosby Elsevier, Maryland Heights, pp 501–510
Stonerock GL, Hoffman BM, Smith PJ, Blumenthal JA (2015) Exercise as treatment for anxiety: systematic review and analysis. Ann Behav Med 49:542–556. https://doi.org/10.1007/s12160-014-9685-9
Strigo IA, Craig AD (2016) Interoception, homeostatic emotions and sympathovagal balance. Philos Trans R Soc B Biol Sci 371:20160010. https://doi.org/10.1098/rstb.2016.0010
Szabo A, Müller A (2015) Coaches' attitudes towards placebo interventions in sport. Eur J Sport Sci 16:1–8. https://doi.org/10.1080/17461391.2015.1019572
Szabo A, Jobbágy L, Köteles F (2018) Super pill is less effective than an ordinary mint in altering subjective psychological feeling states within a few minutes. J Gen Psychol 145:208–222. https://doi.org/10.1080/00221309.2018.1459454
Telles S, Gerbarg P, Kozasa EH (2015) Physiological effects of mind and body practices. Biomed Res Int 2015:1. https://doi.org/10.1155/2015/983086
Thompson WG (2005) Placebo effect and health: combining science with compassionate care. Prometheus Books, Amherst, MA
Thrane S, Cohen SM (2014) Effect of reiki therapy on pain and anxiety in adults: an in-depth literature review of randomized trials with effect size calculations. Pain Manag Nurs 15:897–908. https://doi.org/10.1016/j.pmn.2013.07.008
Tihanyi BT, Köteles F (2017) Physiological and psychological correlates of attention-related body sensations (tinglind and warmth). Physiol Int 104:235–246. https://doi.org/10.1556/2060.104.2017.3.4
Tihanyi BT, Ferentzi E, Daubenmier JJ et al (2017) Body responsiveness questionnaire: validation on a European sample, mediation between body awareness and affect, connection with mindfulness, body image, and physical activity. Int Body Psychotherapy J 16:56–73
Tihanyi BT, Ferentzi E, Beissner F, Köteles F (2018) The neuropsychophysiology of tingling. Conscious Cogn 58:97–110. https://doi.org/10.1016/j.concog.2017.10.015
Tilburt JC, Emanuel EJ, Kaptchuk TJ et al (2008) Prescribing "placebo treatments": results of national survey of US internists and rheumatologists. BMJ 337:a1938
Turi Z, Bjørkedal E, Gunkel L et al (2018) Evidence for cognitive placebo and nocebo effects in healthy individuals. Sci Rep 8:17443. https://doi.org/10.1038/s41598-018-35124-w
Vercelli S, Ferriero G, Bravini E, Sartorio F (2013) How much is Kinesio taping a psychological crutch? Man Ther 18:e11. https://doi.org/10.1016/j.math.2012.10.008
Vlemincx E, Taelman J, Van Diest I, Van den Bergh O (2010) Take a deep breath: the relief effect of spontaneous and instructed sighs. Physiol Behav 101:67–73. https://doi.org/10.1016/j.physbeh.2010.04.015
Vlemincx E, Van Diest I, Van den Bergh O (2016) A sigh of relief or a sigh to relieve: the psychological and physiological relief effect of deep breaths. Physiol Behav 165:127–135. https://doi.org/10.1016/j.physbeh.2016.07.004

von Mohr M, Fotopoulou A (2018) The cutaneous borders of interoception: active and social inference of pain and pleasure on the skin. In: Tsakiris M, De Preester H (eds) The interoceptive mind. From homeostasis to awareness. Oxford University Press, Oxford, England, pp 102–120
von Mohr M, Krahé C, Beck B, Fotopoulou A (2018) The social buffering of pain by affective touch: a laser-evoked potential study in romantic couples. Soc Cogn Affect Neurosci 13:1121–1130. https://doi.org/10.1093/scan/nsy085
Wager TD, Matre D, Casey KL (2006) Placebo effects in laser-evoked pain potentials. Brain Behav Immun 20:219–230. https://doi.org/10.1016/j.bbi.2006.01.007
Walach H, Jonas WB (2004) Placebo research: the evidence base for harnessing self-healing capacities. J Altern Complement Med 10(Suppl 1):S103–S112
Watson D, Pennebaker JW (1989) Health complaints, stress, and distress: exploring the central role of negative affectivity. Psychol Rev 96:234–254
Weineck F, Messner M, Hauke G, Pollatos O (2019) Improving interoceptive ability through the practice of power posing: a pilot study. PLoS One 14:e0211453. https://doi.org/10.1371/journal.pone.0211453
White RW (1959) Motivation reconsidered: the concept of competence. Psychol Rev 66:297–333. https://doi.org/10.1037/h0040934
Wickramasekera I (1985) A conditioned response model of the placebo effect: predictions from the model. In: White L, Thursky B, Schwartz GE (eds) Placebo. Theory, research, and mechanisms. The Guilford Press, New York, pp 255–287
Winkler A, Hermann C (2019) Placebo- and nocebo-effects in cognitive neuroenhancement: when expectation shapes perception. Front Psych 10:498. https://doi.org/10.3389/fpsyt.2019.00498
Wright SG, Sayre-Adams J (2001) Healing and therapeutic touch: is it all in the mind? In: Peters D (ed) Understanding the placebo effect in complementary medicine. Churchill Livingstone, Edinburgh, pp 165–175
Wu S, Wang R, Zhao Y et al (2013) The relationship between self-rated health and objective health status: a population-based study. BMC Public Health 13:320. https://doi.org/10.1186/1471-2458-13-320
Zautra AJ, Fasman R, Davis MC, Craig ADB (2010) The effects of slow breathing on affective responses to pain stimuli: an experimental study. Pain 149:12–18. https://doi.org/10.1016/j.pain.2009.10.001
Zeidan F, Emerson NM, Farris SR et al (2015) Mindfulness meditation-based pain relief employs different neural mechanisms than placebo and sham mindfulness meditation-induced analgesia. J Neurosci 35:15307–15325. https://doi.org/10.1523/JNEUROSCI.2542-15.2015
Zhou W, Benharash P (2014) Significance of "Deqi" response in acupuncture treatment: myth or reality. J Acupunct Meridian Stud 7:186–189. https://doi.org/10.1016/j.jams.2014.02.008

Chapter 11
Epilog

At the beginning of this monograph, I summarized the most important features of body sensations in six tenets. According to these tenets, body sensations are (1) realistic and convincing, (2) difficult to identify and describe, (3) subject to evaluation, (4) often dissociated from the actual state of the body, (5) largely shaped by top-down factors, and (6) impact our behavior in multiple ways. Throughout this work, the biological and psychological aspects of body sensations and the most important consequences of the interaction of the basic six tenets were systematically described and discussed.

If one needs to choose the most striking tenet of the six, dissociation appears first in my mind. After facing so many examples of the discordance between subjective experience and actual physiological processes, the first question to answer necessarily concerns the reliability of somatic percepts. Body sensations, like other types of feelings, are complex representations of internal states (Damasio 2004); this inevitably leads to a gap between the biological and the felt body. Can we then trust these sensations or not? I think the best answer is, as per usual, that it depends. On the one hand, natural selection had tens of millions of years to fine-tune our conscious experience. It might not be the best possible solution; still, considering that we are still here and even read books, it appears a reasonably good one. Under everyday circumstances and healthy functioning, materials that reach conscious awareness represent important messages about our condition and also push us toward the necessary behavioral steps (Ádám 1998). Sometimes these sensations do have a marked peripheral background, other times top-down information considerably reshapes this background or no body changes take place at all. What should be kept in mind that these sensations refer to those aspects of the actual state of the organism which require conscious management and their attractiveness always reflect their actual salience (in other words, they are weighted with their relative actual importance; if you wish, they are biased). We can clearly sense hunger or the urge to urinate under resting conditions, whereas the same condition can completely be forgotten in a stressful situation when bigger things are at stake. On the other hand, such in the case of nocebo and placebo effects, body sensations can deceive us. This may have

F. Köteles, *Body Sensations*, https://doi.org/10.1007/978-3-030-63201-4_11

its own evolutionary background (Evans 2002; Humphrey 2002; Kirmayer 2011; Trimmer et al. 2013); alternatively, these reactions are not adaptive on their own just side effects or by-products of other adaptive features (Adler and Hammett 1973; Evans 2004; Köteles and Bárdos 2007). Anyway, sensations evoked by external interventions and factors, both pleasant and unpleasant, should always be taken with a grain of salt in the context of health and disease.

How can we avoid such traps? I think the key is education. Features of body sensations and the risk posed by the presence or absence of certain sensations should be taught and demonstrated in the elementary school. Demonstration is particularly important because personal experience belongs to primary consciousness, thus its impact on other primary materials is more direct and powerful than that of knowledge. Of course, knowledge can also play a role because the major issues with body sensations are tightly connected with behavior, and behavior can be influenced by rational thinking and voluntary decision-making. Moreover, the boundary between primary and secondary consciousness is not sharp and impenetrable; culture and secondary consciousness (rational thinking) are able to shape primary percepts in the long run (Mason 1961; Metzinger 2009). Unfortunately, the world sometimes appears to move in the opposite direction and favors primary experience and intuition (not to speak of conspiracy theories) over rationality and scientific knowledge with respect to health and disease-related decisions. Issues with vaccination or the recent Covid-19 pandemic clearly demonstrate the risks of such thinking. Also, the striking abundance of novel alternative "treatments" with claimed miraculous effects on various conditions, such as obesity, sexual dysfunction, or even cancer, should be taken as a warning message that highlights the need of education. Such frauds often trigger and utilize a transient placebo response to convince people about their effectiveness; then the positive reaction disappears within days. Too many people pay a high price, in terms of both money and health, for the lack of their awareness of the existence and features of the placebo response. In a similar vein, somatic symptoms can be evoked by bombastic messages on the risks of various environmental factors, and products with claimed protective features can be sold for the masses who know nothing about the nocebo phenomenon (Köteles et al. 2016). Education and knowledge are key to avoid these pitfalls.

Concerning the recently accepted tripartite approach to interoception (Garfinkel and Critchley 2013; Garfinkel et al. 2015), I think empirical findings only weakly support the importance of interoceptive accuracy and awareness, whereas the impact of interoceptive sensibility (i.e., the awareness of and attentiveness to body sensations) on human functioning is profound. Thus, this model needs revision from this point of view too. A sensory detection threshold (the "accuracy" of interoception) assessed under sterile laboratory conditions is meaningless without taking into consideration the actual state of the individual and the environment. It is not even clear in certain cases, such as heartbeat, whether it represents information meant to conscious awareness or simply a by-product of sensory processing. As natural selection was able to mark the salience of a number of body sensations, it is not reasonable to assume that it could not do so for others. Also, the concordance between accuracy and confidence (i.e., interoceptive awareness) appears less important to me than the dissociation between actual peripheral events and their perceived counterparts, particularly in pathological conditions. Once again, it is important to recall that conscious experience refers to a special area of human functioning and homeostatic regulation (including social aspects of homeostasis), and body sensations should always be studied through this lens. To this end, the better conceptualization of interoceptive sensibility with respect to (1) sensory versus evaluative, (2) acute versus chronic (trait-like), and (3) modality-specific versus generalizable aspects is perhaps the most important task for the future. Once these relationships and their associations with the physiological background are clarified, the systematic investigation of the impact of interoception on human functioning will gain a huge momentum. Despite its importance, the evaluative aspect of body sensations is largely neglected in laboratory measurements. This is why the tenet of evaluation and the affective aspects of body percepts were mentioned again and again throughout this work.

Scientists involved in research in a particular area often have a special experience. More expertise is typically accompanied by more uncertainty and the feeling of knowing less and less instead of more and more on the topic at hand. This may hold particularly true for human behavior, a phenomenon impacted by so many factors. Robert M. Sapolsky, the renowned physiologist, draws the following final conclusion at the end of an almost 700 pages long book on human behavior: "it's complicated" (2017, p. 674). I like this conclusion as it reminds us of the complexity of human functioning. Still, I am convinced that the deeper understanding of body sensations and their impact on human behavior might lead to a better understanding of this complicated phenomenon, as well as to the alleviation of human suffering.

References

Ádám G (1998) Visceral perception: understanding internal cognition. Plenum Press, New York

Adler HM, Hammett VB (1973) The doctor-patient relationship revisited. An analysis of the placebo effect. Ann Intern Med 78:595–598

Damasio A (2004) Emotions and feelings: a neurobiological perspective. In: Manstead ASR, Frijda N, Fischer A (eds) Feelings and emotions. The Amsterdam Symposium. Cambridge University Press, Cambridge, UK, pp 49–57

Evans D (2002) Pain, evolution, and the placebo response. Behav Brain Sci 25:459–560

Evans D (2004) Placebo. Mind over matter in modern medicine. Oxford University Press, Oxford, England

Garfinkel SN, Critchley HD (2013) Interoception, emotion and brain: new insights link internal physiology to social behaviour. Commentary on: "Anterior insular cortex mediates bodily sensibility and social anxiety" by Terasawa et al. (2012). Soc Cogn Affect Neurosci 8:231–234. https://doi.org/10.1093/scan/nss140

Garfinkel SN, Seth AK, Barrett AB et al (2015) Knowing your own heart: distinguishing interoceptive accuracy from interoceptive awareness. Biol Psychol 104:65–74. https://doi.org/10.1016/j.biopsycho.2014.11.004

Humphrey N (2002) Great expectations: the evolutionary psychology of faith healing and the placebo effect. In: Humphrey N (ed) The mind made flesh: essays from the frontier of evolution and psychology. Oxford University Press, Oxford, England, pp 255–285

Kirmayer LJ (2011) Unpacking the placebo response: insights from ethnographic studies of healing. J Mind Body Regul 1:112–124
Köteles F, Bárdos G (2007) A placebo - evolúciós szemmel. Magyar Pszichológiai Szemle 62:239–252. https://doi.org/10.1556/MPSzle.62.2007.2.6
Köteles F, Tarján E, Berkes T (2016) Artificial concerns. Effects of a commercial advertisement on modern health worries and sympathetic activation. Mentálhigiéné és Pszichoszomatika 17:61–79. https://doi.org/10.1556/0406.17.2016.1.4
Mason RE (1961) Internal perception and bodily functioning. International Universities Press, New York
Metzinger T (2009) The ego tunnel. The science of the mind and the myth of the self. Basic Books, New York
Sapolsky RM (2017) Behave: the biology of humans at our best and worst. Penguin, New York
Trimmer PC, Marshall JAR, Fromhage L et al (2013) Understanding the placebo effect from an evolutionary perspective. Evol Hum Behav 34:8–15. https://doi.org/10.1016/j.evolhumbehav.2012.07.004

Index

F. Köteles, *Body Sensations*, https://doi.org/10.1007/978-3-030-63201-4

Washburn DA, Smith JD, Taglialatela LA (2005) Individual differences in metacognitive responsiveness: cognitive and personality correlates. J Gen Psychol 132:446–461. https://doi.org/10.3200/GENP.132.4.446-461

The manufacturer's authorised representative in the EU is Springer Nature Customer Service Centre GmbH, Europaplatz 3, 69115 Heidelberg, Germany. If you have any concerns regarding our products, please contact ProductSafety@springernature.com

Printed and bound by CPI Group (UK) Ltd, Croydon, CR0 4YY
22/09/2026
02220985-0004